Linear Methods: A General Education Course

Textbooks in Mathematics
Series editors:
Al Boggess and Ken Rosen

Financial Mathematics:
A Comprehensive Treatment
Giuseppe Campolieti and Roman N. Makarov

Advanced Linear Algebra
Nicholas Loehr

Differential Equations with MATLAB: Exploration, Applications, and Theory
Mark McKibben and Micah D. Webster

Counter Examples:
From Elementary Calculus to the Beginnings of Analysis
Andrei Bourchtein and Ludmila Bourchtein

Convex Analysis
Steven G. Krantz

Differential Equations:
Theory, Technique and Practice, Second Edition
Steven G. Krantz

Exploring Linear Algebra:
Labs and Projects with Mathematica®
Crista Arangala

Analysis with Ultra Small Numbers
Karel Hrbacek, Olivier Lessmann, and Richard O'Donovan

The Mathematics of Games:
An Introduction to Probability
David G. Taylor

Transformational Plane Geometry
Ronald N. Umble and Zhigang Han

Linear Algebra, Geometry and Transformation
Bruce Solomon

Mathematical Modeling with Case Studies:
Using Maple and MATLAB®, Third Edition
B. Barnes and G.R. Fulford

Applied Differential Equations: The Primary Course
Vladimir A. Dobrushkin

Introduction to Mathematical Proofs, Second Edition
Charles Roberts

Measure Theory and Fine Properties of Functions, Revised Edition
Lawrence Craig Evans, and Ronald F. Gariepy

Advanced Linear Algebra, Second Edition
Bruce Cooperstein

A Course in Abstract Harmonic Analysis, Second Edition
Gerald B. Folland

Numerical Analysis for Engineers:
Methods and Applications, Second Edition
Bilal Ayyub and Richard H. McCuen

Differential Equations:
Theory, Technique and Practice with Boundary Value Problems
Steven G. Krantz

Introduction to Abstract Algebra, Second Edition
Jonathan D. H. Smith

Mathematics in Games, Sports, and Gambling:
The Games People Play, Second Edition
Ronald J. Gould

Computational Mathematics:
Models, Methods, and Analysis with MATLAB® and MPI, Second Edition
Robert E. White

Applied Abstract Algebra with Maple™ and MATLAB®, Third Edition
Richard Klima, Neil Sigmon, and Ernest Stitzinger

Introduction to Number Theory, 2nd Edition
Anthony Vazzana and David Garth

Ordinary Differential Equations:
An Introduction to the Fundamentals
Kenneth B. Howell

Advanced Linear Algebra
Hugo Woerdeman

Graphs & Digraphs, Sixth Edition
Gary Chartrand, Linda Lesniak, and Ping Zhang

Abstract Algebra: An Interactive Approach, Second Edition
William Paulsen

A MatLab® Companion to Complex Variables
A. David Wunsch

Real Analysis and Foundations, Third Edition
Steven G. Krantz

Exploring Calculus:
Labs and Projects with Mathematica
Crista Arangala and Karen A. Yokley

Differential Equations with Applications and Historical Notes, Third Edition
George F. Simmons

A Bridge to Higher Mathematics
Valentin Deaconu, Donald C. Pfaff

Exploring Geometry, Second Edition
Michael Hvidsten

Discovering Group Theory:
A Transition to Advanced Mathematics
Tony Barnard, Hugh Neill

Real Analysis and Foundations, Fourth Edition
Steven G. Krantz

Principles of Fourier Analysis, Second Edition
Kenneth B. Howell

Exploring the Infinite: An Introduction to Proof and Analysis
Jennifer Brooks

Abstract Algebra: A Gentle Introduction
Gary L. Mullen, James A. Sellers

A Course in Differential Equations with Boundary Value Problems, Second Edition
Stephen A. Wirkus, Randall J. Swift, and Ryan Szypowski

Invitation to Linear Algebra
David C. Mello

Discrete Mathematics and Applications, Second Edition
Kevin Ferland

Essentials of Topology with Applications
Steven G. Krantz

Introduction to Analysis
Corey M. Dunn

Applied Differential Equations with Boundary Value Problems
Vladimir Dobrushkin

The Elements of Advanced Mathematics, Fourth Edition
Steven G. Krantz

A Tour Through Graph Theory
Karin R Saoub

Transition to Analysis with Proof
Steven Krantz

Essentials of Mathematical Thinking
Steven G. Krantz

Elementary Differential Equations
Kenneth Kuttler

A Concrete Introduction to Real Analysis, Second Edition
Robert Carlson

Mathematical Modeling for Business Analytics
William P. Fox

Elementary Linear Algebra
James R. Kirkwood and Bessie H. Kirkwood

Applied Functional Analysis, Third Edition
J. Tinsley Oden and Leszek Demkowicz

An Introduction to Number Theory with Cryptography, Second Edition
James Kraft and Lawrence Washington

Mathematical Modeling:
Branching Beyond Calculus
Crista Arangala, Nicolas S. Luke, and Karen A. Yokley

Linear Methods: A General Education Course
David Hecker and Stephen Andrilli

https://www.crcpress.com/Textbooks-in-Mathematics/book-series/CANDHTEXBOOMTH

Linear Methods: A General Education Course

David Hecker
Department of Mathematics
Saint Joseph's University
Philadelphia, PA

Stephen Andrilli
Department of Mathematics and
Computer Science
La Salle University
Philadelphia, PA

CRC Press
Taylor & Francis Group
Boca Raton London New York

CRC Press is an imprint of the
Taylor & Francis Group, an **informa** business

Cover photo is courtesy of Christine Hecker

CRC Press
Taylor & Francis Group
6000 Broken Sound Parkway NW, Suite 300
Boca Raton, FL 33487-2742

Printed on acid-free paper
Version Date: 20180601

International Standard Book Number-13: 978-1-138-06292-4 (Hardback)
International Standard Book Number-13: 978-1-138-04921-5 (Paperback)

**Visit the Taylor & Francis Web site at
http://www.taylorandfrancis.com**

**and the CRC Press Web site at
http://www.crcpress.com**

Dedication

To our parents
Robert and Genevieve Hecker,
and
Francis and Leatrice Andrilli,
who made many sacrifices on our behalf.

Contents

Preface for the Instructor

Rationale and Intended Audience for the Text

In this text, we present a course in elementary linear algebra to satisfy a general core mathematics requirement. Typically, a core curriculum requirement in mathematics for students not taking calculus consists of a course with a wide breadth of topics, and consequently, it is a course with little depth in any one branch of the subject. Instead, in order to provide students with a deeper, richer experience of mathematics, this text provides a more detailed study of a single topic, but on a level that is still accessible to non-math majors. To illustrate the many uses of linear algebra in today's world, we include applications in such relevant areas as business and economics, elementary graph theory, Markov chains, elementary physics, least-squares polynomials, and geometric transformations of the plane. There is no formal prerequisite for this course, aside from some familiarity with simple concepts from high school algebra and geometry. Students are expected to use a graphing calculator that can perform matrix operations — and in particular, one that can compute the reduced row echelon form of a matrix for matrix sizes up to 7×7.

Proofs

In this text, students are also re-introduced (since their study of high-school geometry) to the nature of mathematical proof. Proofs are included in the text for some elementary results. In the exercises, students are asked to provide similar and/or related proofs. This involves a careful understanding of the definitions of important terms on the students' part – a skill to which they may not be accustomed at first. More difficult results are presented in the text without proof.

Instructors who want to supplement this textbook with material on logic and reasoning can download a supplementary chapter on the basics of elementary logic. Topics covered include DeMorgan's Laws, truth tables, conditional statements, quantifiers, and logical arguments – along with many exercises for students. The supplement is freely available at the following URL: sju.edu/LinearMethods

Exercises, True/False Questions, and Appendix A

Every section of the text is accompanied by a set of exercises. In addition, at
the conclusion of each chapter, there is a Chapter Test with problems in the
same spirit as those the students have already encountered. The problems in
each exercise set throughout the text have been designed in pairs (except for
the Chapter Tests), so that in almost every case, an odd-numbered exercise
has a corresponding similar even-numbered exercise.

Each Chapter Test ends with a series of True/False questions to help the
students to review their knowledge of the material. Answers are provided
in Appendix A for all of the odd-numbered exercises throughout the text —
except for the exercises in the Chapter Tests, for which only the answers to
the True/False questions are given.

Altogether, aside from the True/False exercises, there are 437 other num-
bered exercises in the text, many of which are multi-part exercises. If the
additional portions of these exercises are counted separately, there actually
are 979 such exercises in the text. Finally, there are also 227 True/False
exercises, for an overall total of 1206 exercises.

Assignments

The book also contains several "Assignments" within the body of the text.
These are strong suggestions that students learn some particular skill or carry
out some calculations. For example, Assignment 6 in Chapter 1 directs the
students to "Learn how to perform matrix multiplication on your calculator.
Try several different examples, including an attempt to multiply two matrices
that do not have compatible sizes. Then create a 3×4 matrix **A** and a 4×3
matrix **B** and multiply them in both orders, **AB** and **BA**, observing the
difference between the two results."

Appendix B

Appendix B contains a summary of important theorems and methods from
each chapter for convenient reference.

Instructor's Manual

An Instructor's Manual is available online for instructors who adopt the textbook. The Instructor's Manual contains solutions for all of the exercises in
the text.

Glossaries and Index

At the conclusion of each section in the text, there is a Glossary of newly
introduced terms along with their definitions, in order to assist students in
mastering new vocabulary. In the Index, an entry in **boldface** type indicates

the page on which the Glossary entry for that item in the Index can be found, making it easier for students to find definitions of terms.

Optional Sections

Sections 3.6, 4.6 (somewhat dependent on 3.6), 5.3 (dependent on 3.6), and 6.3 (dependent on 4.6) contain more challenging material that can be omitted depending on the time available and the interests of the students. If Section 7.3 is covered, the material related to composition of various matrix transformations with translations in Section 7.2 can be skipped in favor of the more elegant approach in Section 7.3.

Website for the Text

Further information about the text can be found at the following website:

https://www.crcpress.com/Linear-Methods-A-General-Education-Course/ Hecker-Andrilli/p/book/9781138049215

Acknowledgments

The authors gratefully acknowledge our Acquisitions Editor, Robert Ross, our Editorial Project Manager, Jose Soto, our Project Manager, Robin Lloyd-Starkes, and our copy editor Elissa Rudolph. We also express our sincere gratitude to Saint Joseph's University which has allowed us to pilot preliminary versions of this textbook over the last decade, as well as to all of the instructors and students who have suggested improvements and additions to the text. We also express our thanks to both La Salle University and Saint Joseph's University for granting sabbaticals and course reductions to assist in the completion of this manuscript.

Preface for the Student

What is Linear Algebra?

Linear algebra is the study of matrices and vectors and "linear" operations performed upon them. Matrices are essentially arrays of numbers arranged in rows and columns. Vectors are matrices having only a single column. We will see that vectors have many geometric interpretations. However, the "algebra" portion of linear algebra means that we will be analyzing these geometric properties by performing algebraic computations. Once we have established the necessary algebraic framework, many useful and powerful applications of linear algebra using vectors and matrices will be at our disposal, including applications involving business and economics, connections in physical and social networks (elementary graph theory), probabilities of ending in a particular state in a dynamic situation (Markov chains), forces and velocities in elementary physics, formulas that approximate empirical data (least-squares polynomials), and the use of matrices to compute the effects of various geometric transformations in the plane.

Terms and Definitions

In this book we need to define a large number of other terms and concepts that will be new to you. It is important that you absorb new vocabulary as it is introduced, because once a new term has been explained, we often use it throughout the rest of the text. If you do not understand new vocabulary terms when they are first defined, you will find succeeding pages more and more difficult to understand. In fact, one of the biggest obstacles students have in comprehending linear algebra is properly understanding the definitions of crucial terms.

To assist you in this process, each section of the text ends with a Glossary of the important new terms introduced in that section. Also, the page number in the Index corresponding to that Glossary entry is set in **boldface** type to make it especially easy to find. Therefore, if you are reading and encounter an unfamiliar linear algebra term, look the term up in the Index to find the page containing its Glossary entry, and then re-read the definition of that term. In general, the Glossaries and the Index are helpful resources that will improve your efficiency when studying the material.

Exercises and Chapter Tests

This textbook contains a set of Exercises at the end of each section. We strongly urge you to do as many of these exercises as you reasonably can. You will only truly become skillful in using the methods of linear algebra by practicing them. It is a poor test-taking strategy to assume that you will succeed in doing a certain type of problem on a test if you have never previously succeeded in doing such a problem for homework.

Appendix A contains answers for the odd-numbered exercises. When doing these exercises, if you do not get a correct answer, go back and try to figure out *why* your answer is different. Each even-numbered exercise is generally similar to the odd-numbered exercise that directly precedes it. In this way, you have an opportunity to practice on a problem whose answer is known before you attempt a related problem whose answer is not provided.

Also, at the end of each chapter there is a Chapter Test to help you review the main concepts and skills learned in that chapter. Each Chapter Test concludes with an extensive collection of True/False questions to test your knowledge of the material. Although the answers for most Chapter Test problems are not provided, the answers to the True/False questions do appear in Appendix A. If you find your answer to a True/False question is not correct, go back and try to figure out *why*. Sometimes you may get a True/False question wrong because there is a particular situation that you are not considering. In general, if you are not confident about your solution to a Chapter Test question, you should go back to the corresponding section in the chapter and review the material, doing similar exercises in that section for which the answer appears in Appendix A.

Use of Calculators

In order to do the exercises and implement the various methods presented in this text, you must have a calculator that can perform matrix operations. Most modern graphing calculators have this ability. Please check to make sure that your calculator can handle matrices up to size 7×7, and that it has a "rref" command for row reduction (see Chapter 4) built in.

Assignments

In addition to the exercise sets, there are other tasks throughout the text labeled as "Assignments." These are instructions to complete specific procedures, such as performing a particular function on your calculator. It is important for you to do each of these Assignments when you encounter them, since you will need these skills to do the exercises, and will undoubtedly need them on tests as well.

Theorems and Proofs

This text will also introduce you to some theoretical aspects of mathematics by illustrating why some of the methods of linear algebra work and why some of the properties of vectors and matrices are valid. This involves the concept of mathematical proof. We present proofs for some elementary theorems, and we ask you to prove a few statements whose proofs are relatively direct and/or similar to those appearing in the text. However, we will not present proofs for the more abstract or technical statements and methods that are covered. Appendix B contains a handy summary of the major properties, theorems, and methods presented throughout the text.

Reading a Mathematics Textbook

Finally, a word of advice on how to read a mathematics textbook. You should not read mathematics the way you read a novel. When reading mathematics, you should reflect on each sentence, making sure you truly understand its meaning. If you do not understand a comment, try to figure it out. If you cannot, make a note of it, and then ask your instructor for an explanation at the earliest opportunity.

Keep paper, pencil and a calculator at your side while reading. You should use them to perform each of the computations you encounter in the text to make sure that you can do them on your own. Do not just skim over them. If you do not obtain the same results as the text, look back over the text material and then try again. If you still cannot duplicate a result in the text, ask your instructor about it. If you do not resolve your problem, your difficulties will only get worse, because each new concept is built upon those that come before. Eventually this will result in less and less understanding of the material as you continue onward. Therefore, take your time and read each sentence carefully. Speed reading a mathematics text is never a good strategy!

Chapter 1

Matrices

1.1   Introduction to Matrices

Definition of a Matrix

The matrix is one of the fundamental objects used in the study of linear algebra. A **matrix** is a rectangular array of numbers, enclosed in brackets, such as

$$\begin{bmatrix} 1 & 2 & 0 & -1 \\ 4 & 3 & 9 & 7 \\ -6 & 8 & 0 & 4 \end{bmatrix}.$$

The plural of "matrix" is "matrices." Matrices can have any number of rows and columns. The **size** of a matrix is designated by its

$$\text{(number of rows)} \times \text{(number of columns)}.$$

For example, the matrix given above is a 3×4 matrix, because it has 3 rows and 4 columns. Here are some of examples of matrices and their corresponding sizes:

$$\begin{array}{cc} 2 \times 2 \\ \text{matrix} \end{array} : \begin{bmatrix} 1 & 2 \\ 5 & -3 \end{bmatrix} \qquad \begin{array}{cc} 1 \times 4 \\ \text{matrix} \end{array} : \begin{bmatrix} 4 & 0 & 1 & 2 \end{bmatrix}$$

$$\begin{array}{cc} 2 \times 3 \\ \text{matrix} \end{array} : \begin{bmatrix} 3 & 1 & 1 \\ 7 & -9 & 0 \end{bmatrix} \quad \begin{array}{cc} 3 \times 2 \\ \text{matrix} \end{array} : \begin{bmatrix} 2 & 4 \\ 6 & 8 \\ 10 & 12 \end{bmatrix} \quad \begin{array}{cc} 3 \times 1 \\ \text{matrix} \end{array} : \begin{bmatrix} -6 \\ 2 \\ 1 \end{bmatrix}.$$

Two matrices are considered to be **equal** to each other if they have the same size and the same numbers as entries in corresponding rows and columns.

Representing Matrices

For easier reference, we always use uppercase bold letters to represent matrices. For example, we could refer to the first matrix above as follows:

$$\mathbf{A} = \begin{bmatrix} 1 & 2 & 0 & -1 \\ 4 & 3 & 9 & 7 \\ -6 & 8 & 0 & 4 \end{bmatrix}.$$

We could refer to other unknown matrices as follows:

"Let $\mathbf{B}$ be a 4×3 matrix." or "Suppose $\mathbf{C}$ is an $m \times n$ matrix."

Then $\mathbf{B}$ represents the same unknown 4×3 matrix until indicated otherwise. Similarly, $\mathbf{C}$ represents an unknown matrix of unknown size, where the variable m represents the number of rows in $\mathbf{C}$ and the variable n represents the number of columns in $\mathbf{C}$.

The various entries in the matrix $\mathbf{A}$ can be referred to by their row and column numbers. For example, we say that the number 9 in $\mathbf{A}$ is the $(2,3)$ entry of $\mathbf{A}$, because it appears in the 2nd row, 3rd column of $\mathbf{A}$. Also, since $\mathbf{A}$ represents the entire matrix, we also use the corresponding subscripted lower case variable a_{23} to represent the $(2,3)$ entry of $\mathbf{A}$; that is, $a_{23} = 9$. Similarly, b_{12} would represent the $(1,2)$ entry of a matrix $\mathbf{B}$, and c_{ij} would represent the number in the ith row and jth column of a matrix $\mathbf{C}$.

A matrix having every entry equal to zero is called a **zero matrix**. There are zero matrices of every size. The $m \times n$ zero matrix is written symbolically as $\mathbf{O}_{mn}$. So, for example,

$$\mathbf{O}_{24} = \begin{bmatrix} 0 & 0 & 0 & 0 \\ 0 & 0 & 0 & 0 \end{bmatrix}.$$

Assignment 1: Learn how to enter matrices into your calculator, how to give names to them, and how to recall them from memory as needed.

Square Matrices

A matrix that has the same number of rows as columns is called a **square matrix**. For example,

$$[1], \quad \begin{bmatrix} 2 & 3 \\ 3 & 5 \end{bmatrix}, \quad \begin{bmatrix} 6 & 7 & 8 \\ 9 & 0 & 1 \\ 2 & 3 & 4 \end{bmatrix}, \quad \text{and} \quad \begin{bmatrix} 3 & 2 & 1 & 0 \\ 9 & 8 & 7 & 6 \\ 5 & 6 & 7 & 8 \\ 9 & 0 & 1 & 2 \end{bmatrix}$$

are all square matrices, of sizes 1×1, 2×2, 3×3, and 4×4, respectively. In general, an $n \times n$ matrix is a square matrix having n rows and n columns.

The Main Diagonal of a Matrix

A **main diagonal entry** of a matrix is an entry having the same row and column numbers. All of these entries together form the **main diagonal** of the matrix. So, the main diagonal entries of the 3×3 matrix above are 6, 0, and 4. Notice how the main diagonal entries of the matrix occur as we move down diagonally from the top left-hand corner of the matrix. For the matrix **A** given above, the three main diagonal entries are $a_{11} = 1$, $a_{22} = 3$, and $a_{33} = 0$.

If **D** represents a (square) $n \times n$ matrix, then the main diagonal entries of **D** are $d_{11}, d_{22}, \ldots, d_{nn}$. Notice that we have used "ellipsis" notation (the three dots) to represent the missing entries. Ellipsis notation is frequently used to denote items that are omitted from a list that follows an obvious pattern. In this particular case, there are actually n main diagonal entries in **D**, not just the three specifically mentioned above.

Some Word Problem Examples

Matrices often provide a convenient way to store numerical data, or trends in that data, as in the following examples.

Example 1. Judy runs the 5K race on her cross-country team. The following 3×10 matrix **J** gives Judy's times in her first ten races. The first row of **J** is the race number. The second and third rows of **J** give the number of minutes and seconds, respectively, for her time in the race:

$$\mathbf{J} = \begin{array}{l} \text{Race Number} \\ \text{Minutes} \\ \text{Seconds} \end{array} \begin{bmatrix} 1 & 2 & 3 & 4 & 5 & 6 & 7 & 8 & 9 & 10 \\ 23 & 23 & 22 & 22 & 22 & 22 & 21 & 21 & 21 & 21 \\ 33 & 16 & 59 & 43 & 25 & 8 & 50 & 48 & 38 & 25 \end{bmatrix}. \blacksquare$$

Example 2. Suppose U.S. citizens move in or out of the Northeast region each decade in accordance with the following matrix:

$$\mathbf{N} = \begin{array}{l} \\ \text{(to) NE} \\ \text{(to) Rest of U.S.} \end{array} \begin{array}{cc} \text{(from)} & \text{(from)} \\ \text{NE} & \text{Rest} \\ & \text{of U.S.} \end{array} \begin{bmatrix} 0.95 & 0.02 \\ 0.05 & 0.98 \end{bmatrix}.$$

The entries of the first column of **N** indicate that 95% of the population of the Northeast will stay in the Northeast in a ten-year period, and 5% will move into some other region. For those living in the rest of the U.S., the entries of the second column of **N** indicate that 2% will move into the Northeast, while 98% will remain outside the Northeast. Notice that the columns of **N** add up to 100% = 1. $\blacksquare$

As we will see in Section 1.5, the matrix **N** in Example 2 is an example of a special type of matrix called a stochastic matrix. Matrices of this type are used in an application of matrices called Markov chains, which we will study in Chapter 3.

Glossary

- Ellipsis notation: A method of expressing the missing elements in a list of objects that follow a similar pattern. The ellipsis, "...", represents the objects that are not specifically listed. For example, the numbers from 1 to 10 can be listed in ellipsis notation as $1, 2, \ldots, 10$.

- Entry in a matrix: A number in a matrix. The entry in row i and column j of a matrix is referred to as the (i, j) entry.

- Equal matrices: Two matrices are equal if and only if they have the same size and every pair of corresponding entries are equal.

- Main diagonal of a matrix: The collection of the main diagonal entries of a matrix: the $(1, 1)$ entry, the $(2, 2)$ entry, etc.

- Main diagonal entry of a matrix: An entry of a matrix whose row number equals its column number.

- Matrix: A rectangular array of numbers. An $m \times n$ matrix has m rows and n columns of numbers.

- Square matrix: A matrix that has the same number of rows as it has columns. Examples of square matrices are matrices of size 1×1, 2×2, 3×3, ..., $n \times n$.

- Zero matrix: A matrix for which every entry equals zero. The $m \times n$ zero matrix is denoted as $\mathbf{O}_{mn}$.

Exercises for Section 1.1

1. If $\mathbf{A} = \begin{bmatrix} 3 & 5 & 0 & -1 \\ 2 & 6 & 7 & 4 \end{bmatrix}$, what is the size of $\mathbf{A}$, and what are the values of a_{12}, a_{21}, and a_{24}? What is the sum of the entries on the main diagonal of $\mathbf{A}$?

2. If $\mathbf{B} = \begin{bmatrix} 6 & 1 & 4 & 0 \\ -1 & 5 & 9 & -3 \\ -4 & 7 & 8 & 3 \end{bmatrix}$, what is the size of $\mathbf{B}$, and what are the values of b_{32}, b_{23}, b_{33} and b_{24}? What is the sum of the entries on the main diagonal of $\mathbf{B}$?

3. In each of the following matrices, how many total entries are there, and how many of these entries are *not* on the main diagonal?

   (a) $\mathbf{C}$, a 5×7 matrix

   (b) $\mathbf{D}$, a 8×6 matrix

4. In each of the following matrices, how many total entries are there, and how many of these entries are *not* on the main diagonal?

   (a) **E**, a 6×5 matrix

   (b) **F**, a 7×9 matrix

5. For a general 75×73 matrix **G**, list the main diagonal elements symbolically using ellipsis notation.

6. For a general 80×83 matrix **H**, list the main diagonal elements symbolically using ellipsis notation.

7. At Saint Joshua's University, students generally take a two-course sequence in mathematics. Of those students that receive an A or a B in the first course, 78% also receive an A or a B in the second course, while 19% receive a C or a D in the second course, and the remaining 3% either fail the second course or do not take it. Of those that receive a C or a D in the first course, only 36% receive an A or a B in the second course, while 51% receive a C or D in the second course, and 13% fail the second course or do not take it. None of those that fail the first course move on to take the second course. Place this data into a matrix **M** so that the columns of **M** add up to 1, as in Example 2, and label your rows and columns appropriately. (The students should be separated into three categories: those receiving A or B, those receiving C or D, and those either receiving F or not taking the course.)

8. At the Baby Tree Nursery, a certain chemical treatment for a tree disease is generally applied every other year. So, if a tree has not been treated the previous year, it is scheduled for treatment in the current year. However, about $\frac{1}{3}$ of the trees treated in the previous year continue to exhibit poor symptoms, requiring these trees to be treated anyway. Place this information into a matrix **M** so that the columns of **M** add up to 1, as in Example 2, and label your rows and columns appropriately. (The trees should be separated into two categories: those treated, and those not treated.)

1.2   Matrix Addition and Scalar Multiplication

Definition of Addition

Two matrices that are the same size can be added together simply by adding corresponding entries. For example, if

$$\mathbf{A} = \begin{bmatrix} 3 & 4 & -2 \\ 0 & 1 & 3 \end{bmatrix} \text{ and } \mathbf{B} = \begin{bmatrix} 4 & -5 & 0 \\ 2 & 3 & 8 \end{bmatrix},$$

then

$$\mathbf{A} + \mathbf{B} = \begin{bmatrix} 3 & 4 & -2 \\ 0 & 1 & 3 \end{bmatrix} + \begin{bmatrix} 4 & -5 & 0 \\ 2 & 3 & 8 \end{bmatrix} = \begin{bmatrix} 7 & -1 & -2 \\ 2 & 4 & 11 \end{bmatrix}.$$

Notice that the result is another matrix of the same size (in this case, 2×3). However, matrices having different sizes, either in the number of rows or in the number of columns, or both, cannot be added together. Hence, the matrices

$$\mathbf{C} = \begin{bmatrix} 3 & 2 \\ -7 & 8 \end{bmatrix} \quad \text{and} \quad \mathbf{D} = \begin{bmatrix} 4 & 1 \\ 0 & -6 \\ 5 & 2 \end{bmatrix}$$

may not be added. That is, if $\mathbf{G}$ is an $m \times n$ matrix and $\mathbf{H}$ is a $p \times q$ matrix, then $\mathbf{G} + \mathbf{H}$ makes sense if and only if $m = p$ and $n = q$. When $\mathbf{G}$ and $\mathbf{H}$ do have the same size, then the **addition** (sum) of $\mathbf{G} + \mathbf{H}$ is another matrix of the same size whose (i, j) entry equals $g_{ij} + h_{ij}$, for every value of i and j.

A Word Problem Example

Example 1. DVDTV is planning a sale on DVD sets of seasons of the popular TV series *Vampire Romance*. After doing an inventory at each of its three stores, the store owners created the following matrix to record the data:

$$\mathbf{V} = \begin{array}{c} \\ \text{Store 1} \\ \text{Store 2} \\ \text{Store 3} \end{array} \begin{array}{cccc} \text{Season 1} & \text{Season 2} & \text{Season 3} & \text{Season 4} \\ \begin{bmatrix} 5 & 3 & 2 & 2 \\ 6 & 3 & 4 & 1 \\ 4 & 3 & 0 & 3 \end{bmatrix} \end{array}.$$

Since $v_{31} = 4$, there are currently 4 copies of Season 1 of *Vampire Romance* in stock in Store 3.

For the sale, the owner of DVDTV decides to order more copies of each season for each store. Using the same row and column labels as above, the following matrix summarizes the information regarding this order:

$$\mathbf{R} = \begin{bmatrix} 3 & 4 & 5 & 5 \\ 4 & 5 & 4 & 6 \\ 3 & 3 & 6 & 2 \end{bmatrix}.$$

Hence, the matrix $\mathbf{V} + \mathbf{R}$ gives the inventories at the three stores once the order has arrived. In particular,

$$\mathbf{T} = \mathbf{V} + \mathbf{R} = \begin{bmatrix} 5 & 3 & 2 & 2 \\ 6 & 3 & 4 & 1 \\ 4 & 3 & 0 & 3 \end{bmatrix} + \begin{bmatrix} 3 & 4 & 5 & 5 \\ 4 & 5 & 4 & 6 \\ 3 & 3 & 6 & 2 \end{bmatrix}$$

$$= \begin{array}{c} \\ \text{Store 1} \\ \text{Store 2} \\ \text{Store 3} \end{array} \begin{array}{cccc} \text{Season 1} & \text{Season 2} & \text{Season 3} & \text{Season 4} \\ \begin{bmatrix} 8 & 7 & 7 & 7 \\ 10 & 8 & 8 & 7 \\ 7 & 6 & 6 & 5 \end{bmatrix} \end{array}.$$

So, for example, after the order arrives, the number of copies of Season 3 at Store 2 equals $t_{23} = 8$. Note that $t_{32} = 6$ represents the number of copies of Season 2 in Store 3. ∎

Assignment 2: Learn how to add two matrices using your calculator. Practice by adding the matrices **V** and **R** from Example 1, and save the result for future use. Then try to add two matrices of different sizes to see how your calculator responds.

Definition of Scalar Multiplication

Single numbers, of course, are different from matrices, each of which contains an entire array of numbers. Numbers by themselves are called **scalars**.[1] The next operation we define combines a scalar with a matrix. This operation is called **scalar multiplication**, and is performed by multiplying every entry in a matrix by the same scalar. Scalar multiplication is denoted by writing the scalar and the matrix directly next to each other: either $c\mathbf{A}$ or $\mathbf{A}c$ denotes the scalar multiplication of c times $\mathbf{A}$. For example, if $c = 3$ and

$$\mathbf{A} = \begin{bmatrix} 2 & 4 & 8 \\ 3 & 6 & 9 \end{bmatrix}, \text{ then } c\mathbf{A} = \mathbf{A}c = \begin{bmatrix} 6 & 12 & 24 \\ 9 & 18 & 27 \end{bmatrix}.$$

In general, if c is a scalar, and $\mathbf{A}$ is an $m \times n$ matrix, then the (i, j) entry of $c\mathbf{A} = \mathbf{A}c$ is $c(a_{ij})$.

We say that an operation on two objects has the **commutative** property if it can be performed in either order with the same result. For instance, addition of numbers is commutative ($3+4 = 4+3$), as is multiplication of numbers ($3 \times 4 = 4 \times 3$). However, subtraction of numbers is *not* commutative. For example, $8 - 5 = 3$, but $5 - 8 = -3$, and the answers are different. In the next section, we will prove that addition of matrices is commutative. However, notice that scalar multiplication is commutative by its very definition, meaning that for any scalar c and any matrix $\mathbf{A}$, we have $c\mathbf{A} = \mathbf{A}c$.

A Word Problem Example

Example 2. DVDTV is planning another sale on DVD sets of seasons of the popular TV series *Vampire Romance*. The current prices charged, in dollars, in-store and on-line for the various seasons of the show are given in the following matrix:

	Season 1	Season 2	Season 3	Season 4	
$\mathbf{P} = $	25	25	22	20	in-store
	23	23	19	17	on-line

During the sale, prices will be marked down by 20%; that is, the sale price will be 80% of the current price. Hence, the scalar multiplication $.80\mathbf{P}$ is a matrix giving the sale price for each season from each venue:

	Season 1	Season 2	Season 3	Season 4	
Sale price $= .80\mathbf{P} = $	20.00	20.00	17.60	16.00	in-store
	18.40	18.40	15.20	13.60	on-line

∎

---

[1]Sometimes, when convenient, a 1×1 matrix is also treated as if it is a scalar, since it has only one entry.

Assignment 3: Learn how to perform scalar multiplication on matrices using your calculator. Practice by performing the computation in Example 2, and save the result for future use.

Subtraction of Matrices

Two matrices having the same size can be subtracted merely by subtracting corresponding entries. For example,

$$\begin{bmatrix} 8 & 2 & -1 \\ 3 & -4 & 5 \end{bmatrix} - \begin{bmatrix} 5 & -3 & -2 \\ 6 & 1 & -1 \end{bmatrix} = \begin{bmatrix} 3 & 5 & 1 \\ -3 & -5 & 6 \end{bmatrix}.$$

Subtraction is another operation on matrices, which takes two matrices and produces another matrix of the same size. Rather than have to deal with the properties of yet another operation, mathematicians tend to think of the difference of two matrices simply as adding the first matrix to (-1) times the second matrix. Symbolically,

$$\mathbf{A} - \mathbf{B} = \mathbf{A} + ((-1)\mathbf{B}).$$

So, for example,

$$\begin{aligned}
\begin{bmatrix} 8 & 2 & -1 \\ 3 & -4 & 5 \end{bmatrix} - \begin{bmatrix} 5 & -3 & -2 \\ 6 & 1 & -1 \end{bmatrix} &= \begin{bmatrix} 8 & 2 & -1 \\ 3 & -4 & 5 \end{bmatrix} + \left((-1) \begin{bmatrix} 5 & -3 & -2 \\ 6 & 1 & -1 \end{bmatrix} \right) \\
&= \begin{bmatrix} 8 & 2 & -1 \\ 3 & -4 & 5 \end{bmatrix} + \begin{bmatrix} -5 & 3 & 2 \\ -6 & -1 & 1 \end{bmatrix} \\
&= \begin{bmatrix} 3 & 5 & 1 \\ -3 & -5 & 6 \end{bmatrix}.
\end{aligned}$$

Hence, the subtraction operation is really just a combination of the operations of addition and scalar multiplication that we introduced earlier. Also notice that subtraction is not commutative, since $\mathbf{B} - \mathbf{A} \neq \mathbf{A} - \mathbf{B}$ in most cases. (Check this yourself for the matrices $\mathbf{A}$ and $\mathbf{B}$ above.)

Order of Operations

Whenever it is necessary to perform both an addition (or subtraction) and a scalar multiplication, as in the expression $c\mathbf{A} + \mathbf{B}$, the scalar multiplication is always performed first. This rule holds unless parentheses indicate otherwise, as with $c(\mathbf{A} + \mathbf{B})$, where the addition would be performed first. That is, in the order of matrix operations, scalar multiplication of matrices has a *higher* order of precedence than addition (or subtraction) of matrices.

Glossary

- Addition of matrices: An operation on matrices, in which two matrices $\mathbf{A}$ and $\mathbf{B}$ may be added if and only if they have the same size. Under this condition, the sum, $\mathbf{A} + \mathbf{B}$, is a matrix having the same size as $\mathbf{A}$ (and $\mathbf{B}$), whose (i, j) entry is $(a_{ij} + b_{ij})$.

- Commutative property: An operation on two objects has the commutative property if the same result is obtained regardless of the order the objects are listed in. Addition and multiplication of numbers is commutative, but subtraction is not. Addition of matrices (as we will see shortly) and scalar multiplication of matrices are both commutative, but subtraction of matrices is not.

- Operation: An action on one or more objects that creates another object. For example, the addition operation "+" combines two matrices to generate another matrix.

- Order of matrix operations: Scalar multiplication has a higher precedence than addition (or subtraction), unless otherwise indicated. That is, if both a scalar multiplication and an addition (or subtraction) are to be performed, the scalar multiplication is done first.

- Scalar: A single real number. Sometimes, when convenient, a 1×1 matrix is also treated as if it is a scalar, since it has only one entry.

- Scalar multiplication of matrices: An operation that combines a scalar and a matrix to produce another matrix. If c is a scalar and $\mathbf{A}$ is an $m \times n$ matrix, then $c\mathbf{A} = \mathbf{A}c$, the scalar multiplication of c and $\mathbf{A}$, is an $m \times n$ matrix whose (i, j) entry equals ca_{ij}. Scalar multiplication is commutative *by definition*. Also, scalar multiplication takes precedence over addition (and subtraction) unless indicated otherwise.

- Subtraction of matrices: If matrices $\mathbf{A}$ and $\mathbf{B}$ are the same size, then $\mathbf{A} - \mathbf{B}$ is the matrix whose (i, j) entry is $a_{ij} - b_{ij}$. The matrix $\mathbf{A} - \mathbf{B}$ is equal to $\mathbf{A} + (-1)\mathbf{B}$.

Exercises for Section 1.2

1. Let $\mathbf{A} = \begin{bmatrix} 2 & 0 \\ -3 & 7 \\ 6 & -2 \end{bmatrix}$, and $\mathbf{B} = \begin{bmatrix} -5 & -8 \\ 3 & 2 \\ 0 & -6 \end{bmatrix}$.

   (a) Calculate $\mathbf{A} + \mathbf{B}$, $\mathbf{A} - \mathbf{B}$, and $\mathbf{B} - \mathbf{A}$.

   (b) What is the sum of the entries on the main diagonal of $\mathbf{A} + \mathbf{B}$? of $\mathbf{A} - \mathbf{B}$?

2. Let $\mathbf{A} = \begin{bmatrix} 4 & -2 & -3 & 0 \\ -5 & -4 & -1 & 6 \end{bmatrix}$, and $\mathbf{B} = \begin{bmatrix} -3 & 5 & -2 & -7 \\ -5 & 0 & -9 & 1 \end{bmatrix}$.

(a) Calculate $\mathbf{A} + \mathbf{B}$, $\mathbf{A} - \mathbf{B}$, and $\mathbf{B} - \mathbf{A}$.

(b) What is the sum of the entries on the main diagonal of $\mathbf{A} + \mathbf{B}$? of $\mathbf{A} - \mathbf{B}$?

3. The current inventory of various CDs at branches of the retail stores BetterBuy and Floormart in a certain region is given in the following matrices, respectively.

	CD 1	CD 2	CD 3
North Branch	18	14	23
West Branch	17	19	16
South Branch	15	11	21
East Branch	19	13	22

$\mathbf{B} = $ (above)

	CD 1	CD 2	CD 3
North Branch	22	16	20
West Branch	19	14	13
South Branch	17	13	24
East Branch	14	16	21

$\mathbf{F} = $ (above)

(a) If these two retail stores merge their corresponding branches into a larger Megastore, state a matrix that represents the inventory of each type of CD at each branch of Megastore.

(b) State a matrix that represents the difference in the inventories of each type of CD at the corresponding branches before the merger.

4. The current inventory of certain bestselling books at rival branches of the retail stores Narnes & Boble and Boundaries in a certain city is given in the following matrices, respectively.

	Book 1	Book 2	Book 3	Book 4
Downtown Branch	9	12	15	19
Suburban Branch	8	13	18	15
University Branch	11	8	13	16

$\mathbf{J} = $ (above)

	Book 1	Book 2	Book 3	Book 4
Downtown Branch	11	14	13	15
Suburban Branch	6	12	17	13
University Branch	13	10	16	18

$\mathbf{K} = $ (above)

(a) If these two retail stores merge their corresponding branches into a larger BooksLtd store, state a matrix that represents the inventory of each bestseller at each branch of BooksLtd.

(b) State a matrix that represents the difference in the inventories of each bestseller at the corresponding branches before the merger.

5. The current prices (in dollars) of various items at three area supermarkets are given in the following matrix:

$$L = \begin{array}{c} \\ \text{Market 1} \\ \text{Market 2} \\ \text{Market 3} \end{array} \begin{array}{ccc} \text{Item 1} & \text{Item 2} & \text{Item 3} \\ \left[\begin{array}{ccc} 5.60 & 7.40 & 2.80 \\ 6.00 & 8.00 & 3.00 \\ 6.40 & 8.20 & 3.20 \end{array}\right] \end{array}.$$

(a) State a matrix that represents the new prices for each item at each market after a 25% price increase.

(b) State a matrix that represents the new prices for each item at each market after a 15% price decrease.

6. The current prices (in dollars) of various items at two local hardware stores are given in the following matrix:

$$H = \begin{array}{c} \\ \text{Store 1} \\ \text{Store 2} \end{array} \begin{array}{cccc} \text{Item 1} & \text{Item 2} & \text{Item 3} & \text{Item 4} \\ \left[\begin{array}{cccc} 13.00 & 20.00 & 15.50 & 18.50 \\ 14.00 & 19.50 & 16.00 & 18.00 \end{array}\right] \end{array}.$$

(a) State a matrix that represents the new prices for each item at each market after an 18% price increase.

(b) State a matrix that represents the new prices for each item at each market after a 6% price decrease.

1.3 Fundamental Properties of Addition and Scalar Multiplication

Eight Special Properties of Matrices

The following properties show that many familiar rules from high school algebra do carry over to addition and scalar multiplication of matrices. Because these properties are so fundamental to the entire subject of linear algebra, we will refer to them as the "Eight Special Properties" or "ESP" for short.

If **A**, **B**, and **C** represent matrices of the same size, **O** is the matrix the same size as **A** having all zero entries, and x and y are scalars, then these properties are:

ESP1: A + B = B + A (**Commutative Property of Addition**)

ESP2: (A+B)+C = A+(B+C) (**Associative Property of Addition**)

ESP3: A + O = O + A = A (**Additive Identity Property of Zero Matrix**)

ESP4: $\mathbf{A} + ((-1)\mathbf{A}) = ((-1)\mathbf{A}) + \mathbf{A} = \mathbf{O}$ (**Additive Inverse Property**)

ESP5: $x(\mathbf{A} \pm \mathbf{B}) = x\mathbf{A} \pm x\mathbf{B}$ (**First Distributive Law for Scalar Multiplication over Addition**)

ESP6: $(x \pm y)\mathbf{A} = x\mathbf{A} \pm y\mathbf{A}$ (**Second Distributive Law for Scalar Multiplication over Addition**)

ESP7: $x(y\mathbf{A}) = (xy)\mathbf{A}$ (**Associative Property of Scalar Multiplication**)

ESP8: $1\mathbf{A} = \mathbf{A}$ (**Scalar Multiplicative Identity Property of** "1")

In both ESP5 and ESP6, the symbol "$\pm$" means that the formula is valid as long as a "+" sign is used on both sides, or as long as a "−" sign is used on both sides.

For now, the main significance of the Eight Special Properties is that they allow us to perform algebraic operations involving addition and scalar multiplication with matrices using many of the same rules with which we are already accustomed. For instance, we can rearrange the orders of sums (from ESP1), and we can distribute a scalar multiplication over a sum of matrices (from ESP5).

ESP3 says that adding the zero matrix to any matrix does not affect that matrix, and ESP8 tells us that a scalar multiplication of 1 times any matrix does not affect that matrix. In effect, the zero matrix "**O**" acts as an **identity element for matrix addition**, and the scalar "1" acts as an **identity element for scalar multiplication**.

ESP4 says that every matrix has an "additive inverse" which, when added to the original matrix, gives the zero matrix as a result. In fact, the additive inverse (that is, the **additive inverse element**) of any matrix is not hard to find – it is simply the scalar multiple of "−1" times the original matrix. For example,

$$\left(\begin{bmatrix} 7 & -2 \\ -3 & 5 \end{bmatrix} + (-1) \begin{bmatrix} 7 & -2 \\ -3 & 5 \end{bmatrix} \right) = \begin{bmatrix} 0 & 0 \\ 0 & 0 \end{bmatrix}$$
$$= \left((-1) \begin{bmatrix} 7 & -2 \\ -3 & 5 \end{bmatrix} + \begin{bmatrix} 7 & -2 \\ -3 & 5 \end{bmatrix} \right)$$

Why is ESP2 needed? Well, addition is an operation for adding two matrices at a time. But what if we have three matrices to add, such as "**A**+**B**+**C**"? Which two matrices do we add first? ESP2 tells us that it doesn't matter – we get the same answer no matter which way we pair up the matrices. For example,

$$\left(\begin{bmatrix} 4 & 3 \\ 1 & 2 \end{bmatrix} + \begin{bmatrix} 6 & -1 \\ 3 & 0 \end{bmatrix} \right) + \begin{bmatrix} 5 & 4 \\ 8 & -7 \end{bmatrix} = \begin{bmatrix} 10 & 2 \\ 4 & 2 \end{bmatrix} + \begin{bmatrix} 5 & 4 \\ 8 & -7 \end{bmatrix} = \begin{bmatrix} 15 & 6 \\ 12 & -5 \end{bmatrix},$$

while

$$\begin{bmatrix} 4 & 3 \\ 1 & 2 \end{bmatrix} + \left(\begin{bmatrix} 6 & -1 \\ 3 & 0 \end{bmatrix} + \begin{bmatrix} 5 & 4 \\ 8 & -7 \end{bmatrix} \right) = \begin{bmatrix} 4 & 3 \\ 1 & 2 \end{bmatrix} + \begin{bmatrix} 11 & 3 \\ 11 & -7 \end{bmatrix} = \begin{bmatrix} 15 & 6 \\ 12 & -5 \end{bmatrix}$$

as well. In effect, ESP2 tells us that the *parentheses are not necessary* when writing a sum of three (or more!) matrices.

ESP6 allows us to distribute a matrix over a scalar addition. This property tells us that we can either add the scalars first and then multiply by the matrix, or we can multiply each scalar by the matrix and then add the results. For example,

$$(4 + 3) \begin{bmatrix} -1 & 6 \\ 2 & 5 \end{bmatrix} = 4 \begin{bmatrix} -1 & 6 \\ 2 & 5 \end{bmatrix} + 3 \begin{bmatrix} -1 & 6 \\ 2 & 5 \end{bmatrix},$$

because if we add the scalars first (as on the left hand side), we have

$$(4 + 3) \begin{bmatrix} -1 & 6 \\ 2 & 5 \end{bmatrix} = 7 \begin{bmatrix} -1 & 6 \\ 2 & 5 \end{bmatrix} = \begin{bmatrix} -7 & 42 \\ 14 & 35 \end{bmatrix},$$

while instead, if we perform both scalar multiplications first (as on the right hand side), we obtain the same result:

$$4 \begin{bmatrix} -1 & 6 \\ 2 & 5 \end{bmatrix} + 3 \begin{bmatrix} -1 & 6 \\ 2 & 5 \end{bmatrix} = \begin{bmatrix} -4 & 24 \\ 8 & 20 \end{bmatrix} + \begin{bmatrix} -3 & 18 \\ 6 & 15 \end{bmatrix} = \begin{bmatrix} -7 & 42 \\ 14 & 35 \end{bmatrix}.$$

ESP6 also tells us that when we have "like terms" involving matrices, such as "$x\mathbf{A}$" and "$y\mathbf{A}$," we can collect them together as a single term "$(x + y)\mathbf{A}$," just as we are accustomed to collecting like terms together in algebraic expressions. The other six Special Properties help us regroup or simplify algebraic expressions in similar ways.

Assignment 4: Write out an additional example of each of the Eight Special Properties for matrices above, checking that the rule is valid for the particular scalars and matrices you have chosen.

Proof of the Addition Case of ESP5

One of our goals in this book is to introduce the theoretical side of mathematics as well as its applications, and so from time to time we will prove some of the properties we introduce. There will also be occasional exercises asking you to provide simple proofs of statements. As an example, we provide a proof of one case of ESP5, the first distributive law. The proof given here is only for the case when the operation on both sides is addition, but the proof for the subtraction case is done in a similar manner.

Addition Case of ESP5: If $\mathbf{A}$ and $\mathbf{B}$ are matrices of the same size and x is a scalar, then $x(\mathbf{A} + \mathbf{B}) = x\mathbf{A} + x\mathbf{B}$.

Helpful Remarks: It is important to realize that we cannot prove a general statement such as this by merely providing one particular example.

We must give a *general* proof that the Addition Case of ESP5 is true for *all* possible situations.

Our goal is to show that the matrix $x(\mathbf{A} + \mathbf{B})$ is equal to the matrix $x\mathbf{A} + x\mathbf{B}$, for all possible values of x, $\mathbf{A}$ and $\mathbf{B}$. As we saw earlier, two matrices are equal if and only if they have the same size, and all of their corresponding entries have equal values. Hence, we must prove two things:

1. Matrix $x(\mathbf{A} + \mathbf{B})$ has the *same size* as matrix $x\mathbf{A} + x\mathbf{B}$; and

2. Corresponding entries of $x(\mathbf{A} + \mathbf{B})$ and $x\mathbf{A} + x\mathbf{B}$ have the *same values*.

Now, in the first part of this proof, we cannot simply work with one specific size for the matrices $\mathbf{A}$ and $\mathbf{B}$. We must show that the Addition Case of ESP5 is true for every possible (common) size that $\mathbf{A}$ and $\mathbf{B}$ can have. Therefore, in our proof, we must be careful to give $\mathbf{A}$ and $\mathbf{B}$ a *generic* size, such as $m \times n$, so that our proof will hold for every possible matrix size.

Also, in the second part of the proof, we have to show that if we choose any entry from $x(\mathbf{A} + \mathbf{B})$, then its corresponding entry in $x\mathbf{A} + x\mathbf{B}$ has the same value. To make our work easier, we will choose a *generic* entry, the (i, j) entry, from $x(\mathbf{A} + \mathbf{B})$, and show that it is equal to the corresponding entry in $x\mathbf{A} + x\mathbf{B}$. (In other words, since i and j can represent *any* row and column, we are, in effect, actually proving that *all* corresponding entries of both matrices are equal.)

Proof of the Addition Case of ESP5:

Step (1): (Proving both $x(\mathbf{A} + \mathbf{B})$ and $x\mathbf{A} + x\mathbf{B}$ have the same size)
Suppose $\mathbf{A}$ and $\mathbf{B}$ are both $m \times n$ matrices.

Size of $x(\mathbf{A} + \mathbf{B})$: By the definition of matrix addition, $\mathbf{A} + \mathbf{B}$ is also an $m \times n$ matrix. Then, since scalar multiplication does not change the size of a matrix, $x(\mathbf{A} + \mathbf{B})$ is also an $m \times n$ matrix.

Size of $x\mathbf{A} + x\mathbf{B}$: Since scalar multiplication does not affect the matrix size, $x\mathbf{A}$ and $x\mathbf{B}$ are also $m \times n$ matrices, and so their sum $x\mathbf{A} + x\mathbf{B}$ is another $m \times n$ matrix.

Conclusion: Both $x(\mathbf{A} + \mathbf{B})$ and $x\mathbf{A} + x\mathbf{B}$ are $m \times n$ matrices, so they both have the same size.

Step (2): (Proving the (i, j) entry of $x(\mathbf{A} + \mathbf{B})$ equals the (i, j) entry of $x\mathbf{A} + x\mathbf{B}$)
Recall that the (i, j) entries of $\mathbf{A}$ and $\mathbf{B}$, respectively, are a_{ij} and b_{ij}.

(i, j) entry of $x(\mathbf{A} + \mathbf{B})$: By the definition of matrix addition, the (i, j) entry of $(\mathbf{A} + \mathbf{B})$ is $a_{ij} + b_{ij}$. Then, by the definition of scalar multiplication, the (i, j) entry of $x(\mathbf{A} + \mathbf{B})$ is $x(a_{ij} + b_{ij})$. This last expression involves only numbers, and so we can apply the distributive law of multiplication over addition for *numbers* to express the (i, j) entry of $x(\mathbf{A} + \mathbf{B})$ as $xa_{ij} + xb_{ij}$.

(i, j) entry of $x\mathbf{A} + x\mathbf{B}$: By the definition of scalar multiplication, the (i, j) entry of $x\mathbf{A}$ is xa_{ij} and the (i, j) entry of $x\mathbf{B}$ is xb_{ij}. Therefore, by the definition of matrix addition, the (i,j) entry of $x\mathbf{A} + x\mathbf{B}$ equals the sum of these, namely, $xa_{ij} + xb_{ij}$.

Conclusion: Both the (i,j) entries of $x(\mathbf{A} + \mathbf{B})$ and $x\mathbf{A} + x\mathbf{B}$ equal $xa_{ij} + xb_{ij}$, so *all* corresponding entries of $x(\mathbf{A}+\mathbf{B})$ and $x\mathbf{A}+x\mathbf{B}$ are equal.

Since we have completed both steps (1) and (2), the proof of the Addition Case of ESP5 is finished.◆

We have been very careful in the proof of the Addition Case of ESP5 to provide all of the necessary details. As we progress through the textbook and provide more examples of proofs, we will occasionally skip some of the more obvious steps and reasons to save time.

The proofs of the other Special Properties for matrices are done in a similar manner, using the fact that in each case a similar law is true for numbers.

Linear Combinations of Matrices and Sigma Notation

We frequently have more than two matrices on which we are performing operations. A common method of combining matrices is to multiply each matrix by a (possibly different) scalar, and then add up all of these scalar multiples. Such an expression is called a **linear combination** of the matrices. For example, the following matrix $\mathbf{A}$ represents a linear combination of three matrices:

$$\mathbf{A} = 3 \begin{bmatrix} 4 & -2 \\ 1 & 0 \end{bmatrix} + 2 \begin{bmatrix} -6 & 3 \\ 2 & -7 \end{bmatrix} - 5 \begin{bmatrix} -3 & 2 \\ 4 & -1 \end{bmatrix}.$$

To evaluate such a linear combination, first multiply the individual matrices by the scalars, and then add. ESP2 assures us the answer will be the same regardless of which pair of matrices we add first. Thus, the matrix $\mathbf{A}$, above, equals

$$\begin{bmatrix} 12 & -6 \\ 3 & 0 \end{bmatrix} + \begin{bmatrix} -12 & 6 \\ 4 & -14 \end{bmatrix} + \begin{bmatrix} 15 & -10 \\ -20 & 5 \end{bmatrix} = \begin{bmatrix} 15 & -10 \\ -13 & -9 \end{bmatrix}.$$

In general, if we have a list of matrices $\mathbf{A}_1, \ldots, \mathbf{A}_n$ and corresponding scalars $c_1, \ldots, c_n$, we can form the linear combination $c_1 \mathbf{A}_1 + \cdots + c_n \mathbf{A}_n$. Notice that we have used the ellipsis notation here.

Instead of using ellipsis notation, we can also express the same linear combination $c_1 \mathbf{A}_1 + \cdots + c_n \mathbf{A}_n$ using **sigma notation** as follows:

$$\sum_{i=1}^{n} c_i \mathbf{A}_i.$$

The upper case Greek letter $\sum$ (sigma) means that we are performing a sum of terms. Each term is of the form $c_i \mathbf{A}_i$ for some value of i. The notation below and above the $\sum$ sign gives the range of values for i. In this case, the first term in the sum has $i = 1$, which is the term $c_1 \mathbf{A}_1$. Then, the next term has $i = 2$: $c_2 \mathbf{A}_2$. This continues up to $i = n$, since n is the number above the $\sum$ sign. This last term is $c_n\mathbf{A}_n$. So, the expression $\sum_{i=1}^{n} c_i\mathbf{A}_i$ means $c_1\mathbf{A}_1 + \cdots + c_n\mathbf{A}_n$. We will often use both sigma and ellipsis notation for sums throughout the text.

Finally, the single scalar multiple $c_1\mathbf{A}_1$ is also considered to be a linear combination (having just a single term), even though there is really no addition involved. In sigma notation, this equals $\sum_{i=1}^{1} c_i\mathbf{A}_i$, since the range of values for i both begins and ends with "1."

Glossary

- **Additive identity element:** The matrix "$\mathbf{O}$" is an identity element for addition of matrices, because if $\mathbf{O}$ is added to any matrix having the same size, it does not change that matrix.

- **Additive identity property:** $\mathbf{A} + \mathbf{O} = \mathbf{O} + \mathbf{A} = \mathbf{A}$. That is, adding a zero matrix to any matrix having the same size does not change that matrix.

- **Additive inverse element:** $(-1)\mathbf{A} = -\mathbf{A}$ is the additive inverse of $\mathbf{A}$. That is, if $(-1)\mathbf{A} = -\mathbf{A}$ is added to $\mathbf{A}$, the result is $\mathbf{O}$.

- **Additive inverse property:** $\mathbf{A} + ((-1)\mathbf{A}) = ((-1)\mathbf{A}) + \mathbf{A} = \mathbf{O}$. That is, adding the additive inverse of a matrix to that matrix gives the zero matrix.

- **Associative property for addition:** $(\mathbf{A} + \mathbf{B}) + \mathbf{C} = \mathbf{A} + (\mathbf{B} + \mathbf{C})$. That is, when adding three matrices *without changing their order*, any pair can be added first, and the result is the same. (Subtraction of matrices is *not* associative.)

- **Associative property for scalar multiplication:** For matrices: $x(y\mathbf{A}) = (xy)\mathbf{A}$. That is, when performing two scalar multiplications on a matrix, either the scalars can be multiplied first, or the second scalar multiplication can be performed first.

- **Commutative property for addition:** $\mathbf{A} + \mathbf{B} = \mathbf{B} + \mathbf{A}$. That is, addition of two matrices can be performed in either order.

- **Distributive properties of scalar multiplication over addition:** $x(\mathbf{A}+\mathbf{B}) = x\mathbf{A}+x\mathbf{B}$, and $(x+y)\mathbf{A} = x\mathbf{A}+y\mathbf{A}$. That is, when performing scalar multiplication with a sum, either the sum could be added first, or the scalar multiples could be calculated first.

- **Generic expression:** An expression (for example, a variable) that represents the general case. Generic expressions typically appear in proofs when we must prove that a result is true in all possible cases.

- **Identity properties:** Two such properties have been introduced in this section: the additive identity property for matrices, and the scalar multiplicative identity property for matrices.

- Linear combination of matrices: A sum of scalar multiples of matrices. A linear combination of the matrices $\mathbf{A}_1, \ldots, \mathbf{A}_n$ with corresponding scalars $c_1, \ldots, c_n$, would have the form $c_1 \mathbf{A}_1 + \cdots + c_n \mathbf{A}_n$. (It is possible to have a linear combination consisting of just a single scalar multiplication, such as $c_1 \mathbf{A}_1$.)

- Scalar multiplicative identity: The scalar "1" is an identity element for scalar multiplication of matrices, since $1\mathbf{A} = \mathbf{A}$.

- Scalar multiplicative identity property: $1\mathbf{A} = \mathbf{A}$. That is, multiplying the scalar "1" by any matrix does not change that matrix

- Sigma notation: A way to express the sum of many similar terms having a common variable. We use the upper case Greek letter sigma, $\sum$, to indicate the sum. The values below and above the $\sum$ sign indicate the range of values for the variable. This is an alternative to ellipsis notation. For example, $\sum_{i=1}^{5} \mathbf{A}_i = \mathbf{A}_1 + \cdots + \mathbf{A}_5$.

Exercises for Section 1.3

In Exercises 1 through 16, suppose

$$\mathbf{A}_1 = \begin{bmatrix} 2 & 3 \\ 5 & 7 \end{bmatrix}, \quad \mathbf{A}_2 = \begin{bmatrix} -1 & 0 \\ 4 & 6 \end{bmatrix}, \quad \mathbf{A}_3 = \begin{bmatrix} 9 & 2 \\ 1 & -7 \end{bmatrix},$$

$$\mathbf{A}_4 = \begin{bmatrix} 5 & 8 \\ -2 & 0 \end{bmatrix}, \quad \mathbf{A}_5 = \begin{bmatrix} 8 & -4 \\ 11 & 3 \end{bmatrix}, \quad \mathbf{A}_6 = \begin{bmatrix} 0 & 0 \\ 0 & 0 \end{bmatrix},$$

$$\mathbf{A}_7 = \begin{bmatrix} 6 & 1 & 2 \\ 4 & 5 & 9 \end{bmatrix}, \quad \text{and} \quad \mathbf{A}_8 = \begin{bmatrix} 3 & 0 & 5 \\ -1 & 7 & -4 \end{bmatrix}.$$

1. Compute $\mathbf{A}_1 + \mathbf{A}_2 + \mathbf{A}_3$ and $5\mathbf{A}_1 - 3\mathbf{A}_2$.

2. Compute $\mathbf{A}_3 + \mathbf{A}_4 + \mathbf{A}_5$ and $-2\mathbf{A}_5 + 4\mathbf{A}_3$.

3. Compute $3(\mathbf{A}_2 + \mathbf{A}_5)$ two different ways by using ESP5.

4. Compute $-4(\mathbf{A}_1 + \mathbf{A}_4)$ two different ways by using ESP5.

5. Compute $(4 + 7)\mathbf{A}_8$ two different ways by using ESP6.

6. Compute $(-6 + 2)\mathbf{A}_7$ two different ways by using ESP6.

7. Compute $5(2\mathbf{A}_5)$ two different ways by using ESP7.

8. Compute $7(-3\mathbf{A}_4)$ two different ways by using ESP7.

9. Compute $\mathbf{A}_1 + \cdots + \mathbf{A}_4$, and write the expression $\mathbf{A}_1 + \cdots + \mathbf{A}_4$ in sigma notation.

10. Compute $\mathbf{A}_3 + \cdots + \mathbf{A}_6$, and write the expression $\mathbf{A}_3 + \cdots + \mathbf{A}_6$ in sigma notation.

11. Compute $\sum_{i=2}^{5} \mathbf{A}_i$ and $\sum_{j=7}^{7} \mathbf{A}_j$. Then write the first of these expressions in ellipsis notation.

12. Compute $\sum_{i=2}^{4} \mathbf{A}_i$ and $\sum_{j=8}^{8} \mathbf{A}_j$. Then write the first of these expressions in ellipsis notation.

13. If $b_7 = 4$ and $b_8 = -3$, compute $\sum_{i=7}^{8} b_i \mathbf{A}_i$.

14. If $c_1 = 3$, $c_2 = 5$, $c_3 = -2$, $c_4 = 0$, $c_5 = -6$, and $c_6 = 1$, compute $\sum_{i=1}^{6} c_i \mathbf{A}_i$.

15. Explain why it is not possible to compute $\sum_{i=1}^{8} \mathbf{A}_i$.

16. Explain why it is not possible to compute the linear combination $4\mathbf{A}_5 - 3\mathbf{A}_6 + 5\mathbf{A}_7$.

17. Consider ESP6: $(x + y)\mathbf{A} = x\mathbf{A} + y\mathbf{A}$. What is different about the two addition operations in this equation?

18. Consider ESP7: $x(y\mathbf{A}) = (xy)\mathbf{A}$. There are four multiplication operations in this equation. One of them is different than the others. Why?

19. State and prove ESP2.

20. State and prove ESP6.

1.4   Transpose of a Matrix

We begin this section by introducing a new operation on matrices called the transpose.

Definition of the Transpose

The transpose operation acts on a matrix by converting all of the rows of the original matrix into columns, and all the columns of the original matrix into rows. We use the notation $\mathbf{A}^T$ to denote the transpose of a matrix $\mathbf{A}$. Note that the "T" is not an exponent, but denotes the transpose operation.

For example,

$$\text{if } \mathbf{A} = \begin{bmatrix} 2 & 1 & 4 \\ 5 & 3 & 6 \end{bmatrix}, \text{ then } \mathbf{A}^T = \begin{bmatrix} 2 & 5 \\ 1 & 3 \\ 4 & 6 \end{bmatrix}.$$

Notice how the two rows of $\mathbf{A}$ become the two columns of $\mathbf{A}^T$. Similarly, the three columns of $\mathbf{A}$ become the three rows of $\mathbf{A}^T$. Here, the 2×3 matrix $\mathbf{A}$ has a 3×2 matrix for its transpose $\mathbf{A}^T$. The transpose $\mathbf{A}^T$ does *not* necessarily have the same size as the original matrix $\mathbf{A}$! In general, if $\mathbf{A}$ is an $m \times n$ matrix, then its transpose $\mathbf{A}^T$ is an $n \times m$ matrix.

Also notice that the (i, j) entry of the transpose is the (j, i) entry of the original matrix, and vice versa. Just as we use a_{ij} to signify the (i, j) entry of $\mathbf{A}$, we will use a_{ij}^t to denote the (i, j) entry of $\mathbf{A}^T$. Since the (i, j) entry of $\mathbf{A}^T$ is equal to the (j, i) entry of $\mathbf{A}$, we always have

$$a_{ij}^t = a_{ji}.$$

You can see this for the matrix $\mathbf{A}$ above: $a_{12}^t = 5 = a_{21}$, $a_{21}^t = 1 = a_{12}$, $a_{31}^t = 4 = a_{13}$, and $a_{32}^t = 6 = a_{23}$. Only the entries along the main diagonal remain unchanged, because there, the row and column subscripts are the same; that is, $a_{11}^t = 2 = a_{11}$ and $a_{22}^t = 3 = a_{22}$.

As we have seen, a matrix and its transpose generally have different sizes. However, if a matrix is square, then its transpose has the same size as the original matrix. For example,

$$\text{if } \mathbf{B} = \begin{bmatrix} 1 & 2 & 3 \\ 4 & 5 & 6 \\ 7 & 8 & 9 \end{bmatrix}, \text{ then } \mathbf{B}^T = \begin{bmatrix} 1 & 4 & 7 \\ 2 & 5 & 8 \\ 3 & 6 & 9 \end{bmatrix};$$

both are 3×3 matrices. Observe that the main diagonal remains unchanged, while the other entries are "reflected" across the main diagonal to new positions on the opposite side.

Assignment 5: Learn how to perform the transpose operation on your calculator. Practice using matrices having various sizes. Also, try entering a square matrix $\mathbf{A}$, and then computing $\mathbf{A}^T$, $\mathbf{A} + \mathbf{A}^T$, and $\mathbf{A} - \mathbf{A}^T$.

A Word Problem Example

Example 1. In Example 1 in Section 1.2, we discussed the inventory matrix

		Season 1	Season 2	Season 3	Season 4
	Store 1	5	3	2	2
$\mathbf{V} =$	Store 2	6	3	4	1
	Store 3	4	3	0	3

that told us the number of DVD sets of seasons of the popular TV series *Vampire Romance* that were available at each of the three stores in the chain DVDTV.

Suppose, however, that in the chain's inventory database of other DVD sets, the seasons are represented by the rows, while the stores are represented by the columns. In order to make the data for *Vampire Romance* compatible with the other data in the chain's database, the bookkeeper can simply use the transpose operation to switch rows with columns:

		Store 1	Store 2	Store 3
	Season 1	5	6	4
$\mathbf{V}^T =$	Season 2	3	3	3
	Season 3	2	4	0
	Season 4	2	1	3

Basic Properties of the Transpose

We next introduce three basic properties of the transpose operation, which show how the transpose interacts with all of the matrix operations we have studied so far: addition, scalar multiplication, and the transpose itself. Regarding the order of operations, the transpose operation is performed before either addition and scalar multiplication, unless parentheses force a different order.

Suppose that $\mathbf{A}$ and $\mathbf{B}$ are both matrices of the same size and that c is a scalar. Then,

TR1: $\left(\mathbf{A}^T\right)^T = \mathbf{A}$

TR2: $(c\mathbf{A})^T = c\mathbf{A}^T$

TR3: $(\mathbf{A} \pm \mathbf{B})^T = \mathbf{A}^T \pm \mathbf{B}^T$.

TR1 states that taking the transpose of a transpose returns us to the original matrix we started with. This makes sense, since each time we take the transpose of a matrix we are switching rows with columns. Taking the transpose a second time simply returns the rows and columns to where they originally were. For example,

$$\left(\begin{bmatrix} 1 & 2 \\ 3 & 4 \end{bmatrix}^T\right)^T = \left(\begin{bmatrix} 1 & 3 \\ 2 & 4 \end{bmatrix}\right)^T = \begin{bmatrix} 1 & 2 \\ 3 & 4 \end{bmatrix}.$$

TR2 shows the relationship between transpose and scalar multiplication. For an expression such as $c\mathbf{A}^T$, even though we are expected to perform the transpose operation before scalar multiplication, TR2 tells us that we could actually do either operation first and get the same result. For example,

$$\left(3\begin{bmatrix} 1 & 2 \\ 3 & 4 \end{bmatrix}\right)^T = \begin{bmatrix} 3 & 6 \\ 9 & 12 \end{bmatrix}^T = \begin{bmatrix} 3 & 9 \\ 6 & 12 \end{bmatrix},$$

while

$$3\left(\begin{bmatrix} 1 & 2 \\ 3 & 4 \end{bmatrix}^T\right) = 3\begin{bmatrix} 1 & 3 \\ 2 & 4 \end{bmatrix} = \begin{bmatrix} 3 & 9 \\ 6 & 12 \end{bmatrix}.$$

Finally, TR3 shows that the transpose operation distributes over matrix addition and matrix subtraction. For an example involving addition,

$$\text{if } \mathbf{A} = \begin{bmatrix} 4 & 2 & -6 \\ 5 & -3 & 8 \end{bmatrix} \text{ and } \mathbf{B} = \begin{bmatrix} 0 & 9 & 7 \\ -4 & 1 & -5 \end{bmatrix},$$

$$\text{then } (\mathbf{A}+\mathbf{B})^T = \left(\begin{bmatrix} 4 & 2 & -6 \\ 5 & -3 & 8 \end{bmatrix} + \begin{bmatrix} 0 & 9 & 7 \\ -4 & 1 & -5 \end{bmatrix}\right)^T$$

$$= \begin{bmatrix} 4 & 11 & 1 \\ 1 & -2 & 3 \end{bmatrix}^T = \begin{bmatrix} 4 & 1 \\ 11 & -2 \\ 1 & 3 \end{bmatrix},$$

$$\text{while } \mathbf{A}^T + \mathbf{B}^T = \begin{bmatrix} 4 & 2 & -6 \\ 5 & -3 & 8 \end{bmatrix}^T + \begin{bmatrix} 0 & 9 & 7 \\ -4 & 1 & -5 \end{bmatrix}^T$$

$$= \begin{bmatrix} 4 & 5 \\ 2 & -3 \\ -6 & 8 \end{bmatrix} + \begin{bmatrix} 0 & -4 \\ 9 & 1 \\ 7 & -5 \end{bmatrix} = \begin{bmatrix} 4 & 1 \\ 11 & -2 \\ 1 & 3 \end{bmatrix},$$

so we have verified in this case that $(\mathbf{A} + \mathbf{B})^T = \mathbf{A}^T + \mathbf{B}^T$.

Proof of the Addition Case of TR3

We now give a formal proof of TR3 for addition.

Addition Case of TR3: Suppose that $\mathbf{A}$ and $\mathbf{B}$ are both matrices of the same size. Then, $(\mathbf{A} + \mathbf{B})^T = \mathbf{A}^T + \mathbf{B}^T$.

TR3 is a statement claiming that two expressions involving matrices are equal to each other. As we saw earlier in Section 1.3 in the proof of ESP5, such a proof has two parts: (1) we must prove that the two matrices $(\mathbf{A} + \mathbf{B})^T$ and $\mathbf{A}^T + \mathbf{B}^T$ have the same size, and (2) we must prove that their corresponding entries are equal. As before, we use a generic size for the given matrices $\mathbf{A}$ and $\mathbf{B}$, and in Step (2) we compare a generic entry (the (i, j) entry) of both matrices.

Proof of the Addition Case of TR3:

Step (1): (Proving $(\mathbf{A} + \mathbf{B})^T$ and $\mathbf{A}^T + \mathbf{B}^T$ have the same size)
Suppose $\mathbf{A}$ and $\mathbf{B}$ are both $m \times n$ matrices.
$(\mathbf{A} + \mathbf{B})^T$: By the definition of matrix addition, $\mathbf{A} + \mathbf{B}$ is also an $m \times n$ matrix. Then, by the definition of the transpose operation, $(\mathbf{A} + \mathbf{B})^T$ is an $n \times m$ matrix.
$\mathbf{A}^T + \mathbf{B}^T$: By the definition of the transpose operation, $\mathbf{A}^T$ and $\mathbf{B}^T$ are both $n \times m$ matrices. Hence, by the definition of matrix addition, $\mathbf{A}^T + \mathbf{B}^T$ is also an $n \times m$ matrix.
Conclusion: Therefore, $(\mathbf{A} + \mathbf{B})^T$ and $\mathbf{A}^T + \mathbf{B}^T$ have the same size; namely, $n \times m$.

Step (2): (Proving that corresponding entries of $(\mathbf{A} + \mathbf{B})^T$ and $\mathbf{A}^T + \mathbf{B}^T$ are equal)
We will show that the (i, j) entries of both matrices are equal.
(i, j) entry of $(\mathbf{A} + \mathbf{B})^T$: The (i, j) entry of $(\mathbf{A} + \mathbf{B})^T$ is equal to the (j, i) entry of $\mathbf{A} + \mathbf{B}$. But, by the definition of matrix addition, the (j, i) entry of $\mathbf{A} + \mathbf{B}$ equals $a_{ji} + b_{ji}$.
(i, j) entry of $\mathbf{A}^T + \mathbf{B}^T$: The (i, j) entry of $\mathbf{A}^T$ equals a_{ij}^t, and the (i, j) entry of $\mathbf{B}^T$ equals b_{ij}^t. Therefore, by the definition of matrix addition, the (i, j) entry of $\mathbf{A}^T + \mathbf{B}^T$ equals $a_{ij}^t + b_{ij}^t$. But, by the definition of the transpose operation, $a_{ij}^t + b_{ij}^t = a_{ji} + b_{ji}$.
Conclusion: Hence, the (i, j) entries of $(\mathbf{A} + \mathbf{B})^T$ and $\mathbf{A}^T + \mathbf{B}^T$ both equal $a_{ji} + b_{ji}$.

Since we have completed both steps (1) and (2), the proof of the Addition Case of TR3 is finished.◆

Glossary

- Transpose: An operation on a matrix that converts its rows into columns, and vice versa. In particular, the transpose of an $m \times n$ matrix $\mathbf{A}$ is an $n \times m$ matrix, denoted $\mathbf{A}^T$, whose (i, j) entry is the (j, i) entry of $\mathbf{A}$. Regarding order of operations, the transpose operation is performed before either addition of matrices or scalar multiplication.

Exercises for Section 1.4

1. Compute the transpose of each of the following matrices:

    (a) $[-4]$

    (c) $\begin{bmatrix} 4 & 3 & -8 \\ 0 & 6 & 5 \\ 0 & 0 & 9 \end{bmatrix}$

    (e) $\begin{bmatrix} -2 & 6 & -3 \\ 4 & -5 & 7 \\ -1 & 0 & -4 \\ -6 & 8 & 9 \end{bmatrix}$

    (b) $\begin{bmatrix} 1 & 5 \\ -2 & 3 \end{bmatrix}$

    (d) $\begin{bmatrix} 8 & -2 & 7 \\ 10 & -3 & -6 \end{bmatrix}$

2. Compute the transpose of each of the following matrices:

    (a) $\begin{bmatrix} 2 & -1 \\ 3 & 6 \\ 9 & 4 \end{bmatrix}$

    (d) $\begin{bmatrix} 6 & 0 & 0 \\ -8 & 5 & 0 \\ 2 & -9 & 3 \end{bmatrix}$

    (b) $\begin{bmatrix} 4 & 0 \\ 0 & 3 \end{bmatrix}$

    (e) $\begin{bmatrix} 5 & -9 & 0 & -2 \\ -7 & 1 & -1 & 4 \\ 2 & -6 & 3 & -5 \end{bmatrix}$

    (c) $\begin{bmatrix} 0 & -3 & 4 \\ 3 & 0 & 5 \\ -4 & -5 & 0 \end{bmatrix}$

3. Let $\mathbf{A} = \begin{bmatrix} -5 & 4 & 0 & -6 & -3 \\ 1 & -8 & -2 & 3 & -7 \\ 9 & 6 & -1 & 2 & 5 \end{bmatrix}$. State the following entries: $a_{12}, a_{32},$

    $a_{25}, a_{12}^t, a_{32}^t, a_{53}^t$.

4. Let $\mathbf{C} = \begin{bmatrix} -2 & -9 & 3 & -1 \\ 4 & 0 & 7 & -5 \\ 6 & -6 & -3 & -4 \\ 8 & 1 & -9 & 5 \end{bmatrix}$. State the following entries: $c_{24}, c_{43}, c_{31},$

    $c_{24}^t, c_{43}^t, c_{31}^t$.

5. Verify that each of the rules TR1, TR2, and TR3 are valid for the

particular matrices $\mathbf{A} = \begin{bmatrix} 3 & -2 & -1 & 4 \\ -6 & -5 & 0 & 2 \\ -4 & -3 & 7 & 1 \end{bmatrix}$, $\mathbf{B} = \begin{bmatrix} -6 & 8 & 3 & -2 \\ 4 & -4 & -1 & -3 \\ 0 & 9 & 1 & -5 \end{bmatrix}$,

with $c = -3$.

6. Verify that each of the rules TR1, TR2, and TR3 are valid for the

particular matrices $\mathbf{A} = \begin{bmatrix} 7 & -3 \\ -2 & -1 \\ 9 & -6 \\ 0 & 4 \end{bmatrix}$, $\mathbf{B} = \begin{bmatrix} -4 & 8 \\ 5 & 3 \\ 0 & 2 \\ -2 & 6 \end{bmatrix}$, with $c = 4$.

7. Suppose $\mathbf{A}$ and $\mathbf{B}$ are matrices of the same size, and c and d are scalars. Is the rule $(c\mathbf{A} + d\mathbf{B})^T = c\mathbf{A}^T + d\mathbf{B}^T$ valid? Why or why not?

8. Suppose $\mathbf{A}$ and $\mathbf{B}$ are both 3×4 matrices, and c and d are scalars. Is the rule $(c\mathbf{A} - d\mathbf{B})^T = c\mathbf{A} - (d\mathbf{B})^T$ valid? Why or why not?

9. Prove TR1.

10. Prove TR2.

1.5 Special Matrices

Upper Triangular, Lower Triangular, and Diagonal Matrices

There are many special types of matrices that are useful in linear algebra. We have already introduced square matrices and zero matrices. We now describe several additional kinds of square matrices and examine their properties.

A square matrix $\mathbf{A}$ is **upper triangular** if all of the entries of $\mathbf{A}$ *below* the main diagonal are equal to zero. Another way of expressing this property is to say that $a_{ij} = 0$ whenever $i > j$, since the entries below the main diagonal all have their row numbers greater than their column numbers. That is, they are farther "down" the matrix than they are "across" the matrix. The following matrices are all upper triangular:

$$\begin{bmatrix} 2 & 3 \\ 0 & 4 \end{bmatrix}, \quad \begin{bmatrix} 2 & 5 & -1 \\ 0 & 0 & 6 \\ 0 & 0 & 7 \end{bmatrix}, \quad \text{and} \quad \begin{bmatrix} 9 & 0 & 2 & -4 \\ 0 & -1 & 7 & 0 \\ 0 & 0 & 4 & 2 \\ 0 & 0 & 0 & -5 \end{bmatrix}.$$

Notice that the entries in an upper triangular matrix *on or above* the main diagonal *may* equal zero, but all the entries *below* the main diagonal *must* equal zero. To prove that a square matrix $\mathbf{A}$ is upper triangular, we must show that $a_{ij} = 0$ whenever $i > j$.

A square matrix $\mathbf{B}$ is **lower triangular** if all of the entries of $\mathbf{B}$ *above* the main diagonal are equal to zero. This is equivalent to saying that $b_{ij} = 0$

whenever $i < j$. The following matrices are all lower triangular:

$$\begin{bmatrix} 4 & 0 \\ -7 & 6 \end{bmatrix}, \quad \begin{bmatrix} 3 & 0 & 0 \\ 0 & 1 & 0 \\ 5 & 2 & -9 \end{bmatrix}, \quad \text{and} \quad \begin{bmatrix} 6 & 0 & 0 & 0 \\ 18 & 0 & 0 & 0 \\ 9 & -4 & 16 & 0 \\ 3 & 5 & -7 & 6 \end{bmatrix}.$$

To prove that a square matrix $\mathbf{B}$ is lower triangular, we must show that $b_{ij} = 0$ whenever $i < j$.

One important property of upper and lower triangular matrices is the following: The transpose of an upper triangular matrix is lower triangular, and vice versa. This is because the transpose operation simply switches all elements below the main diagonal with their counterparts above the main diagonal. As an example, consider

$$\mathbf{A} = \begin{bmatrix} 4 & 3 & 2 \\ 0 & 1 & -1 \\ 0 & 0 & -2 \end{bmatrix}, \quad \text{whose transpose is } \mathbf{A}^T = \begin{bmatrix} 4 & 0 & 0 \\ 3 & 1 & 0 \\ 2 & -1 & 2 \end{bmatrix}.$$

Notice that $\mathbf{A}$ is upper triangular while $\mathbf{A}^T$ is lower triangular.

Another important type of square matrix is a **diagonal matrix**. A diagonal matrix is a square matrix for which all entries above or below the main diagonal equal zero. (That is, nonzero entries can only appear on the main diagonal.) Symbolically, a square matrix $\mathbf{D}$ is diagonal if $d_{ij} = 0$ whenever $i \neq j$. For example, the following matrices are diagonal matrices:

$$\begin{bmatrix} 3 & 0 \\ 0 & 4 \end{bmatrix} \quad \text{and} \quad \begin{bmatrix} 6 & 0 & 0 \\ 0 & -2 & 0 \\ 0 & 0 & 5 \end{bmatrix} \quad \text{and} \quad \begin{bmatrix} 2 & 0 & 0 & 0 \\ 0 & 0 & 0 & 0 \\ 0 & 0 & -4 & 0 \\ 0 & 0 & 0 & 3 \end{bmatrix}.$$

Every diagonal matrix is both upper and lower triangular. Note that in a diagonal matrix, some or all of the entries on the main diagonal could equal zero. In particular, the square zero matrix $\mathbf{O}_{nn}$ is a diagonal matrix for every value of n.

Symmetric and Skew-Symmetric Matrices

A square matrix $\mathbf{A}$ is **symmetric** if it is equal to its transpose; that is, if $\mathbf{A}^T = \mathbf{A}$. This is equivalent to the condition that $a_{ij} = a_{ji}$ for every i and j. The following matrices are symmetric:

$$[3], \quad \begin{bmatrix} 2 & 5 \\ 5 & -3 \end{bmatrix}, \quad \begin{bmatrix} 6 & 2 & 4 \\ 2 & 9 & -1 \\ 4 & -1 & 0 \end{bmatrix}, \quad \text{and} \quad \begin{bmatrix} 1 & 2 & 3 & 4 \\ 2 & 5 & 6 & 7 \\ 3 & 6 & 8 & 9 \\ 4 & 7 & 9 & 10 \end{bmatrix}.$$

Notice in the given 4×4 matrix that the $(1,2)$ entry equals the $(2,1)$ entry, the $(2,4)$ entry equals the $(4,2)$ entry, etc. Thus, an entry above or below the main diagonal equals its "reflection" across the main diagonal. However, an entry on the main diagonal can take on any value, since it does not pair up with some other entry in the matrix. Also, because the transpose of a symmetric matrix does not change the matrix, each row of a symmetric matrix must have the same entries as its corresponding column. For example, in the given 3×3 symmetric matrix above, the 2nd row has entries 2, 9, and -1, which are the same numbers appearing, in that order, in the 2nd column.

A square matrix $\mathbf{A}$ is **skew-symmetric** if it is the negative of its transpose; that is, if $\mathbf{A}^T = -\mathbf{A}$. This is equivalent to the condition that $a_{ij} = -a_{ji}$ for every i and j. The following matrices are skew-symmetric:

$$[0], \quad \begin{bmatrix} 0 & 5 \\ -5 & 0 \end{bmatrix}, \quad \begin{bmatrix} 0 & -2 & 4 \\ 2 & 0 & 1 \\ -4 & -1 & 0 \end{bmatrix}, \quad \text{and} \quad \begin{bmatrix} 0 & 1 & 2 & 3 \\ -1 & 0 & 4 & 5 \\ -2 & -4 & 0 & 6 \\ -3 & -5 & -6 & 0 \end{bmatrix}.$$

Notice in the given 4×4 matrix that the $(1, 2)$ entry equals the negative of the $(2, 1)$ entry, the $(2, 4)$ entry equals the negative of the $(4, 2)$ entry, etc. Thus, any entry above or below the main diagonal equals the negative of its "reflection" across the main diagonal. Notice that an entry on the main diagonal must equal 0. This is because a main diagonal entry has identical row and column subscripts: a_{11}, a_{22}, a_{33}, etc., and since $a_{ij} = -a_{ji}$ for every i and j, each main diagonal entry must equal the negative of itself. For example, $a_{11} = -a_{11}$. But since the only number equal to its own negative is zero, all entries on the main diagonal must be zero.

Also, because taking the transpose of a skew-symmetric matrix produces the negative (i.e., additive inverse) of the matrix, each row of a skew-symmetric matrix must be the negative of the entries in its corresponding column. For example, in the given 3×3 symmetric matrix above, the 2nd row has entries 2, 0, and 1, which are the negatives of the numbers -2, 0, and -1, appearing, in that order, in the 2nd column.

Symmetric and skew-symmetric matrices have many important properties and appear frequently in the study of linear algebra, as we will see later.

Proving That a Matrix is Symmetric or Skew-Symmetric

At several places in this text we will need to prove that a particular matrix (or type of matrix) is symmetric or skew-symmetric.

Obviously, if you are given the entries of a particular matrix and you want to show that it is symmetric, all you need to do is check that the matrix is square and that the entries are equal to their "reflections" across the main diagonal. However, sometimes we are given just the general form of a matrix and need to show that it is symmetric. For example, it turns out that if $\mathbf{A}$ is any square matrix, then the matrix $(\mathbf{A} + \mathbf{A}^T)$ is symmetric. We now illustrate how the proof of such a statement is done.

Statement: If $\mathbf{A}$ is any square matrix, then the matrix $(\mathbf{A} + \mathbf{A}^T)$ is symmetric.

Proof: Recall that a matrix is symmetric if it is a square matrix and if its transpose equals the original matrix. Notice that $(\mathbf{A} + \mathbf{A}^T)$ is square, since $\mathbf{A}$ and $\mathbf{A}^T$ are both square matrices of the same size.

Hence, we only need to show that the transpose of $(\mathbf{A} + \mathbf{A}^T)$ is equal to $(\mathbf{A} + \mathbf{A}^T)$ itself. While we could do this by verifying that the (i, j) entries of both matrices are equal, the proof is actually simpler if we take advantage of the transpose properties from Section 1.4. Our strategy here is to take the transpose of the matrix $(\mathbf{A} + \mathbf{A}^T)$, and simplify the result using the rules we established earlier, in order to show that it equals the original expression $(\mathbf{A} + \mathbf{A}^T)$ again. Now,

$$
\begin{aligned}
(\mathbf{A} + \mathbf{A}^T)^T &= \mathbf{A}^T + \left(\mathbf{A}^T\right)^T \quad &\text{(by TR3)} \\
&= \mathbf{A}^T + \mathbf{A} \quad &\text{(by TR1)} \\
&= \mathbf{A} + \mathbf{A}^T \quad &\text{(by ESP1)},
\end{aligned}
$$

and so the transpose of $(\mathbf{A} + \mathbf{A}^T)$ is indeed equal to $\mathbf{A} + \mathbf{A}^T$.♦

We can also prove that if $\mathbf{A}$ is any square matrix, then the matrix $(\mathbf{A} - \mathbf{A}^T)$ is skew-symmetric. The proof is similar, except that, after taking the transpose, the result simplifies to the *negative* (*additive inverse*) of the original expression; that is, $(\mathbf{A} - \mathbf{A}^T)^T$ simplifies to $-(\mathbf{A} - \mathbf{A}^T)$.

A similar strategy can also be used to prove that if c is a scalar, and if $\mathbf{A}$ is either symmetric or skew-symmetric, then $c\mathbf{A}$ is a matrix of the same type.

A Decomposition for Square Matrices

It is often helpful to split a matrix into a combination of other matrices that have special properties. For example, notice that we can express the square matrix $\mathbf{A}$ below as the sum of two matrices, where the first matrix in the sum is symmetric and the second is skew-symmetric:

$$
\mathbf{A} = \begin{bmatrix} 1 & 2 & 3 \\ 4 & 5 & 6 \\ 7 & 8 & 9 \end{bmatrix} = \begin{bmatrix} 1 & 3 & 5 \\ 3 & 5 & 7 \\ 5 & 7 & 9 \end{bmatrix} + \begin{bmatrix} 0 & -1 & -2 \\ 1 & 0 & -1 \\ 2 & 1 & 0 \end{bmatrix}.
$$

In fact, as we will see, there is no other possible sum of a symmetric and a skew-symmetric matrix that adds up to $\mathbf{A}$; that is, there is a *unique* way to express $\mathbf{A}$ in this manner!

It turns out that a similar result is true for *any* square matrix. This is such an important property of square matrices, that we now state it formally as a theorem. Throughout the text, we label a result as a **theorem** if it is one of the most significant that we will encounter in this text.

Theorem 1. *(**A Decomposition for Square Matrices**) Every square matrix* **A** *can be expressed as the sum of a symmetric matrix* **B** *and a skew-symmetric matrix* **C** *in precisely one way:* **B** *must equal* $\frac{1}{2}\left(\mathbf{A} + \mathbf{A}^T\right)$ *and* **C** *must equal* $\frac{1}{2}\left(\mathbf{A} - \mathbf{A}^T\right)$.

Mathematicians use the term *decomposition* for a separation of some object into two or more objects with an operation between them. Here, the original square matrix is split into two other matrices, one symmetric and the other skew-symmetric, and in this case, the two matrices in the decomposition are *added* together. The theorem says that this type of decomposition is *unique*; that is, there is actually only one way to separate **A** into a sum of a symmetric and a skew-symmetric matrix.

To prove Theorem 1 for any given square matrix **A**, we must show: (1) there is a symmetric matrix **B** and there is a skew-symmetric matrix **C** so that $\mathbf{A} = \mathbf{B} + \mathbf{C}$, and, (2) the matrices **B** and **C** are the only possible symmetric and skew-symmetric choices whose sum is **A**. We will prove only the first part here. (Later in the text we will discuss how to prove statements that claim uniqueness.)

Proof of Step (1): (Proving there is a symmetric matrix **B** and there is a skew-symmetric matrix **C** so that $\mathbf{A} = \mathbf{B} + \mathbf{C}$)

Let **A** be a given square matrix. Consider the matrices $\mathbf{B} = \frac{1}{2}\left(\mathbf{A} + \mathbf{A}^T\right)$ and $\mathbf{C} = \frac{1}{2}\left(\mathbf{A} - \mathbf{A}^T\right)$. Now, earlier in this section we proved that for any square matrix **A**, the matrix $\left(\mathbf{A} + \mathbf{A}^T\right)$ is symmetric, and after that proof we noted that any scalar multiple of a symmetric matrix is symmetric, so **B** $= \frac{1}{2}\left(\mathbf{A} + \mathbf{A}^T\right)$ is *symmetric*. Similarly, earlier in this section we also pointed out that for any square matrix **A**, the matrix $\left(\mathbf{A} - \mathbf{A}^T\right)$ is skew-symmetric, and since any scalar multiple of a skew-symmetric matrix is skew-symmetric, we know that $\mathbf{C} = \frac{1}{2}\left(\mathbf{A} - \mathbf{A}^T\right)$ is *skew-symmetric*. Finally, notice that

$$
\begin{aligned}
\mathbf{B} + \mathbf{C} &= \frac{1}{2}\left(\mathbf{A} + \mathbf{A}^T\right) + \frac{1}{2}\left(\mathbf{A} - \mathbf{A}^T\right) \\
&= \left(\frac{1}{2}\mathbf{A} + \frac{1}{2}\mathbf{A}^T\right) + \left(\frac{1}{2}\mathbf{A} - \frac{1}{2}\mathbf{A}^T\right) \quad \text{(by ESP5)} \\
&= \frac{1}{2}\mathbf{A} + \frac{1}{2}\mathbf{A} + \frac{1}{2}\mathbf{A}^T - \frac{1}{2}\mathbf{A}^T \quad\quad \text{(by ESP1 and ESP2)} \\
&= 1\mathbf{A} + 0\mathbf{A}^T = \mathbf{A} \quad\quad\quad\quad\quad \text{(by ESP6)}.
\end{aligned}
$$

Hence, **B** + **C** is a valid decomposition for **A**. ◆

Example 1. Suppose

$$
\mathbf{A} = \begin{bmatrix} 4 & -2 & 3 \\ 6 & 1 & 5 \\ -7 & 9 & 8 \end{bmatrix}.
$$

We decompose **A** as the sum of a symmetric matrix **B** and a skew-symmetric matrix **C** as follows:

$$\mathbf{B} = \frac{1}{2}\left(\mathbf{A} + \mathbf{A}^T\right) = \frac{1}{2}\left(\begin{bmatrix} 4 & -2 & 3 \\ 6 & 1 & 5 \\ -7 & 9 & 8 \end{bmatrix} + \begin{bmatrix} 4 & 6 & -7 \\ -2 & 1 & 9 \\ 3 & 5 & 8 \end{bmatrix}\right)$$

$$= \frac{1}{2}\begin{bmatrix} 8 & 4 & -4 \\ 4 & 2 & 14 \\ -4 & 14 & 16 \end{bmatrix} = \begin{bmatrix} 4 & 2 & -2 \\ 2 & 1 & 7 \\ -2 & 7 & 8 \end{bmatrix},$$

and

$$\mathbf{C} = \frac{1}{2}\left(\mathbf{A} - \mathbf{A}^T\right) = \frac{1}{2}\left(\begin{bmatrix} 4 & -2 & 3 \\ 6 & 1 & 5 \\ -7 & 9 & 8 \end{bmatrix} - \begin{bmatrix} 4 & 6 & -7 \\ -2 & 1 & 9 \\ 3 & 5 & 8 \end{bmatrix}\right)$$

$$= \frac{1}{2}\begin{bmatrix} 0 & -8 & 10 \\ 8 & 0 & -4 \\ -10 & 4 & 0 \end{bmatrix} = \begin{bmatrix} 0 & -4 & 5 \\ 4 & 0 & -2 \\ -5 & 2 & 0 \end{bmatrix}.$$

Notice that **B** is symmetric, **C** is skew-symmetric, and that **B** + **C** = **A**. ∎

Stochastic Matrices

A **stochastic matrix A** is a square matrix in which all entries are positive or zero, and the entries in each column add up to 1. For example, the following matrices are all stochastic:

$$\begin{bmatrix} \frac{1}{2} & \frac{1}{3} \\ \frac{1}{2} & \frac{2}{3} \end{bmatrix}, \qquad \begin{bmatrix} 1 & \frac{1}{4} & 0 \\ 0 & \frac{1}{2} & \frac{1}{5} \\ 0 & \frac{1}{4} & \frac{4}{5} \end{bmatrix}, \qquad \begin{bmatrix} \frac{2}{3} & 0 & \frac{1}{5} & 0 \\ \frac{1}{6} & 1 & 0 & \frac{1}{10} \\ 0 & 0 & \frac{4}{5} & \frac{2}{5} \\ \frac{1}{6} & 0 & 0 & \frac{1}{2} \end{bmatrix}.$$

Notice that all the nonzero entries of all three matrices are positive, and the sum of the entries in every column equals 1. One of the most important applications of stochastic matrices is in Markov chains, which will be covered in Chapter 3.

Glossary

- Decomposition: The separation of some mathematical object into two or more objects with an operation between them. For example, in this section we proved that a square matrix can always be expressed (uniquely) as the sum of a symmetric matrix and a skew-symmetric matrix.

- Diagonal matrix: A square matrix **D** is diagonal if $d_{ij} = 0$ whenever $i \neq j$. That is, all entries above and below the main diagonal are zero.

- Lower triangular matrix: A square matrix **B** is lower triangular if $b_{ij} = 0$ whenever $i < j$. That is, all entries above the main diagonal are zero.

- Skew-symmetric matrix: A square matrix $\mathbf{A}$ is skew-symmetric if $\mathbf{A}^T = -\mathbf{A}$. Equivalently, $\mathbf{A}$ is skew-symmetric if $a_{ij} = -a_{ji}$ for every i and j.

- Stochastic matrix: A square matrix $\mathbf{A}$ is stochastic if every nonzero entry is positive, and the sum of the entries of each column is 1.

- Symmetric matrix: A square matrix $\mathbf{A}$ is symmetric if $\mathbf{A}^T = \mathbf{A}$. Equivalently, $\mathbf{A}$ is symmetric if $a_{ij} = a_{ji}$ for every i and j.

- Theorem: A very significant result.

- Upper triangular matrix: A square matrix $\mathbf{A}$ is upper triangular if $a_{ij} = 0$ whenever $i > j$. That is, all entries below the main diagonal are zero.

Exercises for Section 1.5

1. For each of the following matrices, indicate which of the following categories the matrix is in: zero, square, upper triangular, lower triangular, diagonal, symmetric, skew-symmetric, stochastic. (A matrix could be in more than one category. List all that apply.)

(a) $\begin{bmatrix} 4 & 3 \\ -3 & 1 \end{bmatrix}$

(b) $\begin{bmatrix} 2 & 5 & 0 \\ 5 & 1 & 0 \end{bmatrix}$

(c) $\begin{bmatrix} 0 & 0 \\ 0 & 0 \end{bmatrix}$

(d) $\begin{bmatrix} \frac{1}{2} & \frac{1}{3} \\ 0 & \frac{1}{6} \\ \frac{1}{2} & \frac{1}{2} \end{bmatrix}$

(e) $\begin{bmatrix} 0 & 2 & 3 \\ 0 & 0 & 4 \\ 0 & 0 & 0 \end{bmatrix}$

(f) $\begin{bmatrix} 2 & -3 & 4 \\ -3 & 6 & 7 \\ 4 & 7 & 8 \end{bmatrix}$

(g) $\begin{bmatrix} \frac{1}{4} & \frac{1}{2} & \frac{1}{4} \\ \frac{1}{2} & 0 & \frac{1}{2} \\ \frac{1}{4} & \frac{1}{2} & \frac{1}{4} \end{bmatrix}$

(h) $\begin{bmatrix} 5 & 0 & 0 \\ 0 & 4 & 0 \\ 0 & 0 & 0 \end{bmatrix}$

(i) $\begin{bmatrix} 1 & 2 \\ 0 & 8 \end{bmatrix}^T$

2. For each of the following matrices, indicate which of the following categories the matrix is in: zero, square, upper triangular, lower triangular, diagonal, symmetric, skew-symmetric, stochastic. (A matrix could be in more than one category. List all that apply.)

(a) $\begin{bmatrix} 0 & 3 & -2 \\ -3 & 0 & 4 \\ 2 & -4 & 0 \end{bmatrix}$

(b) $\begin{bmatrix} 2 & 0 \\ 0 & -1 \end{bmatrix}$

(c) $\begin{bmatrix} 0 & 0 & 0 & 0 \\ 0 & 0 & 0 & 0 \end{bmatrix}$

(d) $\begin{bmatrix} \frac{1}{5} & 0 & \frac{5}{6} \\ \frac{2}{5} & 0 & \frac{1}{12} \\ \frac{2}{5} & 1 & \frac{1}{12} \end{bmatrix}$

(e) $\begin{bmatrix} 4 & 0 & 0 \\ 8 & 0 & 0 \\ -6 & 4 & 1 \end{bmatrix}$

(f) $\begin{bmatrix} 0 & 0 & 0 & 0 \\ \frac{5}{3} & \frac{1}{5} & 0 & 0 \\ -\frac{4}{3} & \frac{2}{5} & \frac{3}{4} & 0 \\ \frac{2}{3} & \frac{2}{5} & \frac{1}{4} & 1 \end{bmatrix}$

(g) $\begin{bmatrix} 3 & 0 \\ 2 & 1 \end{bmatrix}^T$

(h) $\begin{bmatrix} 4 & 7 \\ 7 & -8 \end{bmatrix}^T$

3. Find the decomposition of each of the following matrices into the sum of a symmetric matrix **B** and a skew-symmetric matrix **C**:

(a) $\mathbf{A} = \begin{bmatrix} 4 & 3 & 9 \\ 5 & 6 & -2 \\ -7 & 8 & 1 \end{bmatrix}$

(c) $\mathbf{A} = \begin{bmatrix} 6 & 10 & 10 & -5 \\ -4 & -4 & 2 & 8 \\ 6 & 0 & 5 & -6 \\ 1 & 8 & 4 & -5 \end{bmatrix}$

(b) $\mathbf{A} = \begin{bmatrix} -3 & -6 & 7 \\ 2 & 1 & 3 \\ 3 & -11 & 6 \end{bmatrix}$

4. Find the decomposition of each of the following matrices into the sum of a symmetric matrix **B** and a skew-symmetric matrix **C**:

(a) $\mathbf{A} = \begin{bmatrix} 4 & 0 & 3 \\ -6 & 10 & 5 \\ 7 & -9 & -2 \end{bmatrix}$

(c) $\mathbf{A} = \begin{bmatrix} -6 & -2 & 5 \\ -2 & 4 & -3 \\ 5 & -3 & 7 \end{bmatrix}$

(b) $\mathbf{A} = \begin{bmatrix} 2 & 5 & 9 \\ -13 & -2 & -11 \\ 1 & 5 & 6 \end{bmatrix}$

(d) $\mathbf{A} = \begin{bmatrix} 8 & 14 & 6 & 17 \\ 8 & -7 & 2 & -13 \\ -14 & 10 & -5 & 11 \\ 7 & -13 & -5 & 9 \end{bmatrix}$

5. Prove that, for any square matrix **A**, the matrix $(\mathbf{A} - \mathbf{A}^T)$ is skew-symmetric.

6. Prove that if c is a scalar, and **A** is a symmetric matrix, then $c\mathbf{A}$ is also symmetric.

7. If **A** and **B** are stochastic matrices of the same size, explain why neither $\mathbf{A} + \mathbf{B}$ nor $\mathbf{A} - \mathbf{B}$ is stochastic. (Hint: What will be the sum of the entries on a column of $\mathbf{A} + \mathbf{B}$? of $\mathbf{A} - \mathbf{B}$?) Also, illustrate this with a particular example involving 3×3 matrices **A** and **B**.

8. If **A** is a stochastic matrix and c is a scalar with $c \neq 1$, then prove that $c\mathbf{A}$ is not stochastic. Also, illustrate this with a particular example involving a 3×3 matrix **A** and a scalar $c \neq 1$.

1.6   Matrix Multiplication

Definition of Matrix Multiplication

Matrix multiplication is an operation that combines two matrices to create a third matrix. In order to perform matrix multiplication, the two matrices involved must have **compatible sizes**. By "compatible sizes," we mean that the number of *columns* in the *first* matrix must equal the number of *rows* in the *second* matrix. For example, if **A** is a 3×4 matrix and **B** is a 4×2 matrix, then the sizes of **A** and **B** are compatible, because the first matrix **A** has 4 columns while the second matrix $\mathbf{B}$ has 4 rows, and so we are allowed to perform the matrix product $\mathbf{AB}$.

If a matrix product $\mathbf{AB}$ is defined, then the size of $\mathbf{AB}$ is determined from the number of *rows* of the *first* matrix, and the number of *columns* of the *second* matrix. In this case, $\mathbf{AB}$ is a 3×2 matrix, since $\mathbf{A}$ has 3 rows and $\mathbf{B}$ has 2 columns.

Notice, however, that the sizes of $\mathbf{B}$ and $\mathbf{A}$ are *not* compatible for performing the matrix product $\mathbf{BA}$. This is because the first matrix $\mathbf{B}$ has 2 columns, but the second matrix $\mathbf{A}$ has 3 rows, and these numbers are not equal.

We now formally define matrix multiplication as follows:

If $\mathbf{A}$ is an $m \times n$ matrix and $\mathbf{B}$ is an $n \times p$ matrix, then $\mathbf{C} = \mathbf{AB}$ is an $m \times p$ matrix having

$$c_{ij} = a_{i1}b_{1j} + a_{i2}b_{2j} + \cdots + a_{in}b_{nj}.$$

Notice in this definition that the number of columns, n, of $\mathbf{A}$ is equal to the number of rows, n, of $\mathbf{B}$. Also notice that the number of rows of the product $\mathbf{AB}$ equals the number of rows, m, of $\mathbf{A}$, and the number of columns of $\mathbf{AB}$ equals the number of columns, p, of $\mathbf{B}$.

For example, if $\mathbf{A}$ and $\mathbf{B}$ are given by

$$\mathbf{A} = \underbrace{\begin{bmatrix} 2 & 8 & -7 & 1 \\ -3 & 2 & 5 & 0 \\ 6 & -1 & 3 & 4 \end{bmatrix}}_{3\times 4 \text{ matrix}} \text{ and } \mathbf{B} = \underbrace{\begin{bmatrix} 6 & 2 \\ 4 & 9 \\ 3 & 7 \\ 5 & -4 \end{bmatrix}}_{4\times 2 \text{ matrix}},$$

then the product $\mathbf{C} = \mathbf{AB}$ is a 3×2 matrix. Hence, we must compute 6 entries for $\mathbf{C}$. We do this as follows:

$$
\begin{aligned}
c_{11} &= a_{11}b_{11} + a_{12}b_{21} + a_{13}b_{31} + a_{14}b_{41} \\
&= 2(6) + 8(4) + (-7)(3) + 1(5) = 28, \\
c_{12} &= a_{11}b_{12} + a_{12}b_{22} + a_{13}b_{32} + a_{14}b_{42} \\
&= 2(2) + 8(9) + (-7)(7) + 1(-4) = 23, \\
c_{21} &= a_{21}b_{11} + a_{22}b_{21} + a_{23}b_{31} + a_{24}b_{41} \\
&= (-3)(6) + 2(4) + 5(3) + 0(5) = 5, \\
c_{22} &= a_{21}b_{12} + a_{22}b_{22} + a_{23}b_{32} + a_{24}b_{42} \\
&= (-3)(2) + 2(9) + 5(7) + 0(-4) = 47, \\
c_{31} &= a_{31}b_{11} + a_{32}b_{21} + a_{33}b_{31} + a_{34}b_{41} \\
&= 6(6) + (-1)(4) + 3(3) + 4(5) = 61, \\
c_{32} &= a_{31}b_{12} + a_{32}b_{22} + a_{33}b_{32} + a_{34}b_{42} \\
&= 6(2) + (-1)(9) + 3(7) + 4(-4) = 8.
\end{aligned}
$$

Therefore,

$$C = AB = \begin{bmatrix} 2 & 8 & -7 & 1 \\ -3 & 2 & 5 & 0 \\ 6 & -1 & 3 & 4 \end{bmatrix} \begin{bmatrix} 6 & 2 \\ 4 & 9 \\ 3 & 7 \\ 5 & -4 \end{bmatrix} = \begin{bmatrix} 28 & 23 \\ 5 & 47 \\ 61 & 8 \end{bmatrix}.$$

Notice how the formula for c_{ij} involves the ith row of the first matrix, A, and the jth column of the second matrix, B. So, for example, c_{21} is computed using the 2nd row of A and the 1st column of B:

$$\begin{bmatrix} 2 & 8 & -7 & 1 \\ \boxed{-3 \quad 2 \quad 5 \quad 0} \\ 6 & -1 & 3 & 4 \end{bmatrix} \begin{bmatrix} \boxed{\begin{matrix} 6 \\ 4 \\ 3 \\ 5 \end{matrix}} & \begin{matrix} 2 \\ 9 \\ 7 \\ -4 \end{matrix} \end{bmatrix} = \begin{bmatrix} 28 & 23 \\ \boxed{5} & 47 \\ 61 & 8 \end{bmatrix}.$$

The formula we gave above for c_{ij} used ellipsis notation. We can write the same expression using sigma notation as follows:

$$c_{ij} = \sum_{k=1}^{n} a_{ik} b_{kj}.$$

In particular, for the matrices A, B, and C in our example, above,

$$c_{21} = \sum_{k=1}^{4} a_{2k} b_{k1}, \text{ which equals } a_{21}b_{11} + a_{22}b_{21} + a_{23}b_{31} + a_{24}b_{41}.$$

Assignment 6: Learn how to perform matrix multiplication on your calculator. Try several different examples, including an attempt to multiply two matrices that do not have compatible sizes. Then create a 3×4 matrix A and a 4×3 matrix B and multiply them in both orders, AB and BA, observing the difference between the two results.

Matrix Multiplication is Not Commutative

As we saw in the previous assignment, matrix multiplication does *not* follow the commutative property; that is, it is generally *not* the case that $AB = BA$. There are several reasons for this.

The first is that the sizes of the two matrices may be compatible one way but not the other. We have already seen, for example, that if A is a 3×4 matrix and B is a 4×2 matrix, then the product AB can be computed (the result is a 3×2 matrix), but the product BA cannot, because the number of columns in B does not match the number of rows in A. Hence, the equation $AB = BA$ does not even make any sense in this case, because the right-hand side of the equation does not exist!

Even when A and B have sizes that allow both AB and BA to be computed, another possible problem is that the sizes of AB and BA could be different. For example, if **A** is a 3×5 matrix and **B** is a 5×3 matrix, then both **AB** and **BA** can be computed. However, **AB** is a 3×3 matrix while **BA** is a 5×5 matrix. Therefore in this case, $\mathbf{AB} \neq \mathbf{BA}$ because the sizes of the two matrices are different.

Finally, even in the case where both **A** and **B** are square matrices of the same size (so that both **AB** and **BA** also have that same size), in most cases the corresponding entries of the products will not agree. For example, suppose

$$\mathbf{A} = \begin{bmatrix} 1 & 2 \\ 3 & 4 \end{bmatrix} \text{ and } \mathbf{B} = \begin{bmatrix} 5 & 6 \\ 7 & 8 \end{bmatrix}.$$

Then,

$$\mathbf{AB} = \begin{bmatrix} 19 & 22 \\ 43 & 50 \end{bmatrix}, \text{ while } \mathbf{BA} = \begin{bmatrix} 23 & 34 \\ 31 & 46 \end{bmatrix}.$$

Here, both **AB** and **BA** are 2×2 matrices but their corresponding entries are not identical. In this case, $\mathbf{AB} \neq \mathbf{BA}$.

In rare cases, it is *possible* for two square matrices **A** and **B** of the same size to commute, even though that typically does not happen. For example, if

$$\mathbf{A} = \begin{bmatrix} 2 & 4 \\ 3 & 1 \end{bmatrix} \text{ and } \mathbf{B} = \begin{bmatrix} 7 & 4 \\ 3 & 6 \end{bmatrix},$$

then
$$\mathbf{AB} = \mathbf{BA} = \begin{bmatrix} 26 & 32 \\ 24 & 18 \end{bmatrix}.$$

We will see other special cases of commuting matrices throughout the text. However, we must remain vigilant because most pairs of square matrices do *not* commute with each other. Also, just because a matrix **A** commutes with a second matrix **B** does not imply that **A** commutes with a third matrix **C**.

Assignment 7: Choose two different 4×4 matrices **A** and **B** with random nonzero entries. Use your calculator to multiply both **AB** and **BA**. Did you expect the answers to agree? Why?

Some Word Problem Examples

Although matrix multiplication may appear to be a rather unusual operation, it has quite a large number of practical applications. We will consider some simple ones here, and explore more complex applications throughout the text.

Example 1. Recall Example 1 in Section 1.2, in which we discussed the inventory matrix

$$\mathbf{V} = \begin{array}{c} \\ \text{Store 1} \\ \text{Store 2} \\ \text{Store 3} \end{array} \begin{bmatrix} \overset{\text{Season 1}}{5} & \overset{\text{Season 2}}{3} & \overset{\text{Season 3}}{2} & \overset{\text{Season 4}}{2} \\ 6 & 3 & 4 & 1 \\ 4 & 3 & 0 & 3 \end{bmatrix}.$$

giving the number of DVD sets of seasons of the popular TV series *Vampire Romance* that were available at each of the three stores in the chain DVDTV.

Suppose that the matrix **P**, below, gives the wholesale and retail prices (in dollars) for each of the DVD sets.

$$\mathbf{P} = \begin{array}{c} \\ \text{Season 1} \\ \text{Season 2} \\ \text{Season 3} \\ \text{Season 4} \end{array} \begin{array}{c} \text{Wholesale} \quad \text{Retail} \\ \begin{bmatrix} 6.50 & 25.00 \\ 6.25 & 24.00 \\ 6.00 & 20.00 \\ 5.50 & 18.00 \end{bmatrix} \end{array}.$$

Then, the 3×2 matrix **VP** gives the total of all the wholesale values and the total of all the retail values of these DVD sets for each store:

$$\mathbf{VP} = \begin{array}{c} \\ \text{Store 1} \\ \text{Store 2} \\ \text{Store 3} \end{array} \begin{array}{c} \text{Wholesale} \quad \text{Retail} \\ \begin{bmatrix} 74.25 & 273.00 \\ 87.25 & 320.00 \\ 61.25 & 226.00 \end{bmatrix} \end{array}.$$

For example, to compute the combined wholesale value of all the DVD sets in Store 2, we calculate the value of all the sets for each season and add them up. That is, we compute

(# of sets of Season 1 in Store 2) × (wholesale value of a Season 1 set)

+ (# of sets of Season 2 in Store 2) × (wholesale value of a Season 2 set)

+ (# of sets of Season 3 in Store 2) × (wholesale value of a Season 3 set)

+ (# of sets of Season 4 in Store 2) × (wholesale value of a Season 4 set).

Equivalently, we can write this computation in terms of the entries of the matrices as

$$v_{21}p_{11} + v_{22}p_{21} + v_{23}p_{31} + v_{24}p_{41}$$
$$= 6 \times 6.50 + 3 \times 6.25 + 4 \times 6.00 + 1 \times 5.50$$
$$= 39.00 + 18.75 + 24.00 + 5.50$$
$$= 87.25.$$

But, this is precisely the way in which we calculate the $(2, 1)$ entry of **VP**. The other entries of **VP** are analyzed and calculated in a similar manner. ∎

In Example 1, we were allowed to perform the matrix product **VP** because the number of columns of **V** is equal to the number of rows of **P**. In fact, looking back, we see that the columns of **V** have the same labels as the rows of **P**. They are the four different seasons of *Vampire Romance*. This is no coincidence! In general, word problems involving matrix multiplication follow this pattern; that is, the labels on the columns of the first matrix should match the labels on the rows of the second matrix. Also, the labels on the rows of the product should match the labels on the rows of the first matrix, and the labels on the columns of the product should match the labels on the columns of the second matrix. Check this for yourself by comparing the labels on all the rows and columns for the matrices in Example 1.

Next, we consider an example where the transpose of one of the matrices is needed in order for a matrix multiplication to make sense.

Example 2. A nutritionist is designing meals at a nursing home. Matrix **A** tells us the vitamin content of various foods in appropriate units[2] per serving. Matrix **B** gives the number of servings of each food given to each of three patients on a given day.

$$
\mathbf{A} = \begin{array}{c} \text{Food 1} \\ \text{Food 2} \\ \text{Food 3} \end{array}
\begin{array}{ccc} \text{Vitamin A} & \text{Vitamin C} & \text{Vitamin D} \\ \left[\begin{array}{ccc} 20 & 35 & 0 \\ 10 & 50 & 40 \\ 25 & 5 & 12 \end{array}\right], \end{array}
$$

$$
\mathbf{B} = \begin{array}{c} \text{Food 1} \\ \text{Food 2} \\ \text{Food 3} \end{array}
\begin{array}{ccc} \text{Fred} & \text{Martha} & \text{Veronica} \\ \left[\begin{array}{ccc} 2 & 1 & 0 \\ 1 & 2 & 1 \\ 1 & 0 & 2 \end{array}\right]. \end{array}
$$

We want to use matrix multiplication to compute the total number of units of each vitamin consumed by each patient.

Now, even though the sizes of **A** and **B** are compatible, allowing us to calculate **AB**, such a computation would be meaningless. Consider what we would do to determine the $(2, 1)$ entry of **AB**. This is $a_{21}b_{11} + a_{22}b_{12} + a_{23}b_{13}$. The first term, $a_{21}b_{11}$, of this represents

(the number of units of Vitamin A in each serving of Food 2)

multiplied by

(the number of servings of Food 1 eaten by Fred).

These two numbers are not related to each other in any way! Multiplying them makes no sense! The only reason that the number of columns in **A** equals the number of rows in **B** is pure coincidence; we are considering the same number of types of vitamins as we are types of food.

Instead, we need to find a way to make the column labels on the first matrix match the row labels on the second matrix so that the match-up of the sizes is not just coincidence. Taking inspiration from Example 1 in Section 1.4, we take the transpose of one of the matrices so that the row and column headings get switched. In particular, we take the transpose of **A**:

$$
\mathbf{A}^T = \begin{array}{c} \text{Vitamin A} \\ \text{Vitamin C} \\ \text{Vitamin D} \end{array}
\begin{array}{ccc} \text{Food 1} & \text{Food 2} & \text{Food 3} \\ \left[\begin{array}{ccc} 20 & 10 & 25 \\ 35 & 50 & 5 \\ 0 & 40 & 12 \end{array}\right]. \end{array}
$$

---

[2]Different units are used in measuring each of these vitamins.

Now the labels on the columns of $\mathbf{A}^T$ match the labels on the rows of $\mathbf{B}$. Computing $\mathbf{A}^T\mathbf{B}$ actually makes sense, and its entries represent the number of units of each vitamin that each person consumes on that day:

$$
\mathbf{A}^T\mathbf{B} = \begin{array}{c} \\ \text{Vitamin A} \\ \text{Vitamin C} \\ \text{Vitamin D} \end{array}
\begin{array}{ccc} \text{Fred} & \text{Martha} & \text{Veronica} \\ \left[\begin{array}{ccc} 75 & 40 & 60 \\ 125 & 135 & 60 \\ 52 & 80 & 64 \end{array}\right] \end{array}.
$$

Notice that the row labels on $\mathbf{A}^T\mathbf{B}$ match the row labels on $\mathbf{A}^T$, and the column headings on $\mathbf{A}^T\mathbf{B}$ match the column headings on $\mathbf{B}$. ∎

Matrix Multiplication and the Transpose

We saw in Example 2 how matrix multiplication and the transpose operation could both appear in the same algebraic expression. Thus, we need to understand how these two operations interact with each other. In the order of operations, the transpose is performed before matrix multiplication, unless indicated otherwise by parentheses, such as: $(\mathbf{AB})^T$. In such a situation, we have the following useful result.

TR4: If $\mathbf{A}$ is an $m \times n$ matrix and $\mathbf{B}$ is an $n \times p$ matrix, then

$$(\mathbf{AB})^T = \mathbf{B}^T\mathbf{A}^T.$$

TR4 says that to distribute the transpose over a matrix product, we must take the transpose of each *and switch the order* before we multiply those transposes!

For example, if

$$
\mathbf{A} = \begin{bmatrix} 1 & 3 & 5 \\ 6 & 4 & 2 \end{bmatrix} \quad \text{and} \quad \mathbf{B} = \begin{bmatrix} 2 & 4 & 6 & 8 \\ 3 & 6 & 9 & 12 \\ 18 & 14 & 10 & 5 \end{bmatrix},
$$

then

$$
(\mathbf{AB})^T = \left(\begin{bmatrix} 1 & 3 & 5 \\ 6 & 4 & 2 \end{bmatrix} \begin{bmatrix} 2 & 4 & 6 & 8 \\ 3 & 6 & 9 & 12 \\ 18 & 14 & 10 & 5 \end{bmatrix} \right)^T
$$

$$
= \begin{bmatrix} 101 & 92 & 83 & 69 \\ 60 & 76 & 92 & 106 \end{bmatrix}^T = \begin{bmatrix} 101 & 60 \\ 92 & 76 \\ 83 & 92 \\ 69 & 106 \end{bmatrix}.
$$

Also,

$$
\mathbf{B}^T\mathbf{A}^T = \begin{bmatrix} 2 & 4 & 6 & 8 \\ 3 & 6 & 9 & 12 \\ 18 & 14 & 10 & 5 \end{bmatrix}^T \begin{bmatrix} 1 & 3 & 5 \\ 6 & 4 & 2 \end{bmatrix}^T
$$

$$= \begin{bmatrix} 2 & 3 & 18 \\ 4 & 6 & 14 \\ 6 & 9 & 10 \\ 8 & 12 & 5 \end{bmatrix} \begin{bmatrix} 1 & 6 \\ 3 & 4 \\ 5 & 2 \end{bmatrix} = \begin{bmatrix} 101 & 60 \\ 92 & 76 \\ 83 & 92 \\ 69 & 106 \end{bmatrix},$$

and so in this case, $(\mathbf{AB})^T$ does equal $\mathbf{B}^T \mathbf{A}^T$. (You should verify these computations either by hand or using your calculator.)

However, if we had attempted to compute $\mathbf{A}^T \mathbf{B}^T$ instead of $\mathbf{B}^T \mathbf{A}^T$, we would have quickly found that the sizes of the matrices would not have been compatible! $\mathbf{A}^T$ is a 3×2 matrix and $\mathbf{B}^T$ is a 4×3 matrix, and so they cannot be multiplied in that fashion. Even in cases where it *is* possible to perform the product $\mathbf{A}^T \mathbf{B}^T$, the resulting matrix entries would generally not match the entries of $(\mathbf{AB})^T$.

Proof of TR4: Let $\mathbf{A}$ be an $m \times n$ matrix and $\mathbf{B}$ be an $n \times p$ matrix. To prove that $(\mathbf{AB})^T = \mathbf{B}^T \mathbf{A}^T$, we must show: (1) the sizes of $(\mathbf{AB})^T$ and $\mathbf{B}^T \mathbf{A}^T$ agree, and (2) the (i, j) entry of $(\mathbf{AB})^T$ equals the (i, j) entry of $\mathbf{B}^T \mathbf{A}^T$, for every possible i and j.

Step 1: (Showing the sizes of $(\mathbf{AB})^T$ and $\mathbf{B}^T \mathbf{A}^T$ agree)

Size of $(\mathbf{AB})^T$: The matrices $\mathbf{A}$ and $\mathbf{B}$ are compatible. By the definition of matrix multiplication, $\mathbf{AB}$ has size $m \times p$, so $(\mathbf{AB})^T$ has size $p \times m$.

Size of $\mathbf{B}^T \mathbf{A}^T$: The size of $\mathbf{B}^T$ is $p \times n$, and the size of $\mathbf{A}^T$ is $n \times m$. These matrices are compatible, so by the definition of matrix multiplication, $\mathbf{B}^T \mathbf{A}^T$ has size $p \times m$. This agrees with the size of $(\mathbf{AB})^T$.

Step 2: (Showing (i, j) entry of $(\mathbf{AB})^T = (i, j)$ entry of $\mathbf{B}^T \mathbf{A}^T$)

(i, j) entry of $(\mathbf{AB})^T$: The (i, j) entry of $(\mathbf{AB})^T$ equals the (j, i) entry of $\mathbf{AB}$, which is $a_{j1}b_{1i} + a_{j2}b_{2i} + a_{j3}b_{3i} + \cdots + a_{jn}b_{ni}$.

(i, j) entry of $\mathbf{B}^T \mathbf{A}^T$: The (i, j) entry of $\mathbf{B}^T \mathbf{A}^T$ equals

$$b_{i1}^t a_{1j}^t + b_{i2}^t a_{2j}^t + b_{i3}^t a_{3j}^t + \cdots + b_{in}^t a_{nj}^t$$

$$= a_{1j}^t b_{i1}^t + a_{2j}^t b_{i2}^t + a_{3j}^t b_{i3}^t + \cdots + a_{nj}^t b_{in}^t$$

$$= a_{j1}b_{1i} + a_{j2}b_{2i} + a_{j3}b_{3i} + \cdots + a_{jn}b_{ni}.$$

This agrees with the (i, j) entry of $(\mathbf{AB})^T$, so *all* corresponding entries of both matrices agree.

Conclusion: Since both Steps (1) and (2) have been verified, the proof is complete. ◆

Finally, we consider another interesting result involving matrix multiplication and the transpose.

Theorem 2. *If $\mathbf{A}$ is an $m \times n$ matrix, then $\mathbf{A}^T \mathbf{A}$ is a symmetric $n \times n$ matrix.*

For example, if $\mathbf{A} = \begin{bmatrix} 2 & 3 & 5 \\ 7 & 4 & 1 \end{bmatrix}$, then

$$\mathbf{A}^T\mathbf{A} = \begin{bmatrix} 2 & 7 \\ 3 & 4 \\ 5 & 1 \end{bmatrix} \begin{bmatrix} 2 & 3 & 5 \\ 7 & 4 & 1 \end{bmatrix} = \begin{bmatrix} 53 & 34 & 17 \\ 34 & 25 & 19 \\ 17 & 19 & 26 \end{bmatrix},$$

which is symmetric.

Proof of Theorem 2: Let $\mathbf{A}$ be an $m \times n$ matrix. We must show: (1) $\mathbf{A}^T\mathbf{A}$ is an $n \times n$ matrix, and (2) the transpose of $\mathbf{A}^T\mathbf{A}$ is equal to $\mathbf{A}^T\mathbf{A}$ again.

Step 1: This is easy. Since $\mathbf{A}^T$ is an $n \times m$ matrix, the matrices $\mathbf{A}^T$ and $\mathbf{A}$ are compatible, and then by the definition of matrix multiplication, $\mathbf{A}^T\mathbf{A}$ is an $n \times n$ matrix.

Step 2: We will prove this in a manner similar to the proof of the Statement in Section 1.5; that is, by taking the transpose of $\mathbf{A}^T\mathbf{A}$, and simplifying the result to show that it equals $\mathbf{A}^T\mathbf{A}$ itself:

$$\begin{aligned} \left(\mathbf{A}^T\mathbf{A}\right)^T &= \left(\mathbf{A}^T\right)\left(\mathbf{A}^T\right)^T && \text{by TR4} \\ &= \mathbf{A}^T\mathbf{A} && \text{by TR1.} \end{aligned}$$

Conclusion: Since both Steps (1) and (2) have been verified, the proof is complete. ♦

Glossary

- **Commuting matrices:** Matrices $\mathbf{A}$ and $\mathbf{B}$ commute with each other if $\mathbf{AB} = \mathbf{BA}$. In order for a pair of matrices to commute, both must be square and have the same size. However, most pairs of square matrices do *not* commute with each other.

- **Conjecture:** A general statement that is believed to be true, but for which we do not yet have conclusive proof. In other words, an educated guess regarding a general principle. (This term is introduced in Exercise 12 below.)

- **Matrix multiplication:** If $\mathbf{A}$ is an $m \times n$ matrix and $\mathbf{B}$ is an $n \times p$ matrix, then $\mathbf{C} = \mathbf{AB}$ is an $m \times p$ matrix having $c_{ij} = a_{i1}b_{1j} + a_{i2}b_{2j} + \cdots + a_{in}b_{nj}$.

- **Matrix product:** A synonym for matrix multiplication.

Exercises for Section 1.6

In Exercises 1 to 12, use the following matrix definitions:

$$\mathbf{A} = \begin{bmatrix} 4 & 1 & 3 \\ 9 & 0 & 5 \end{bmatrix} \quad \mathbf{B} = \begin{bmatrix} -9 & 6 \\ 3 & -2 \\ 12 & -8 \end{bmatrix} \quad \mathbf{C} = \begin{bmatrix} 2 & 4 & 6 & 8 \\ 3 & 2 & 5 & 7 \\ 0 & -9 & 1 & -3 \end{bmatrix}$$

$$D = \begin{bmatrix} 7 & -2 \\ 3 & 8 \end{bmatrix} \quad E = \begin{bmatrix} 4 & 6 \\ 6 & 9 \end{bmatrix} \quad F = \begin{bmatrix} 5 & -8 \\ 12 & 9 \end{bmatrix}$$

$$G = \begin{bmatrix} 6 & 2 & -5 \\ 4 & 1 & 0 \\ 9 & -8 & 3 \end{bmatrix} \quad H = \begin{bmatrix} 2 & 0 & 0 \\ 0 & 3 & 0 \\ 0 & 0 & -1 \end{bmatrix} \quad I = \begin{bmatrix} 1 & 0 & 0 \\ 0 & 1 & 0 \\ 0 & 0 & 1 \end{bmatrix}$$

$$J = \begin{bmatrix} 5 & -1 \\ -3 & 4 \\ 2 & -6 \\ 8 & 1 \end{bmatrix}$$

1. Using the given matrices, determine which pairs can be multiplied, and in which order. (Do not actually multiply the matrices. There are 100 different pairs to consider, including the possibility of multiplying each matrix by itself.)

2. Give the size of the resulting matrix for each possible product in Exercise 1.

3. Compute each of the following matrix products without using a calculator: **AB**, **DA**, **GC**, and **JA**. Then use a calculator to check your work.

4. Compute each of the following matrix products without using a calculator: **BD**, **DF**, **AG**, and **JE**. Then use a calculator to check your work.

5. Consider each of the following pairs of matrices. Do they commute? Why or why not? **A** and **B**, **A** and **C**, **D** and **E**, **G** and **I**.

6. Consider each of the following pairs of matrices. Do they commute? Why or why not? **B** and **D**, **D** and **F**, **E** and **F**, **G** and **H**.

7. Compute $J^T C^T$ and verify that it is equal to $(CJ)^T$ (as predicted by TR4).

8. Compute $B^T G^T$ and verify that it is equal to $(GB)^T$ (as predicted by TR4).

9. Compute each of **IB**, **IC**, **IG**, **AI**, and **GI**. What pattern did you notice?

10. Suppose $K = \begin{bmatrix} 1 & 0 \\ 0 & 1 \end{bmatrix}$. Compute each of **KA**, **KD**, **BK**, **FK**, and **JK**. What pattern did you notice?

11. Compute each of **HB**, **HC**, **HG**, **AH**, and **GH**. What pattern did you notice?

12. Suppose $\mathbf{N} = \begin{bmatrix} 3 & 0 \\ 0 & 4 \end{bmatrix}$. Compute each of **NA**, **ND**, **NF**, **BN**, **FN**, and **JN**. What pattern did you notice? Make a general conjecture[3] about the result obtained when we multiply by a diagonal matrix.

13. Solar Express, Inc. assembles three models of solar panels in its factories in three different cities from parts purchased from another manufacturer. Each solar panel needs to have its parts assembled. Next, the panel must be tested, and then, finally, packed for shipment. Matrix **A** indicates the number of hours needed to prepare each model of solar panel at each stage. Matrix **B** represents the wages paid to workers (in \$/hour) for each of the three types of labor in each city.

		Assembly	Testing	Packaging
A =	Model X	0.75	0.25	0.25
	Model Y	1.00	0.20	0.30
	Model Z	1.25	0.30	0.25

		Philadelphia	San Diego	Detroit
B =	Assembly	22	28	17
	Testing	20	21	15
	Packaging	11	13	10

(a) State a matrix product (with or without using the transpose operation) that gives the total labor costs for producing each model of solar panel in each city.

(b) Compute the $(2,3)$ entry of the product in part (a), and interpret the meaning of that number.

(c) Compute the $(3,1)$ entry of the product in part (a), and interpret the meaning of that number.

(d) State a different matrix product than the answer to part (a) that gives the total labor costs for producing each model of solar panel in each city.

(e) Compute the $(3,2)$ entry of the product in part (d), and interpret the meaning of that number.

(f) Compute the $(2,3)$ entry of the product in part (d), and interpret the meaning of that number.

14. Green Lawns, Inc. packages three different mixtures of grass seed in its processing plants in three different cities. Matrix **A** indicates the number of pounds of bluegrass, rye, and fescue seed used in a single package of each mixture. Matrix **B** represents the cost (in dollars) per

---

[3]See the Glossary.

pound for each type of grass seed in each city.

$$
\mathbf{A} = \begin{array}{c} \text{Bluegrass} \\ \text{Rye} \\ \text{Fescue} \end{array}
\begin{array}{ccc}
\text{Standard Mix} & \text{Shade Mix} & \text{Dry Area Mix} \\
\left[\begin{array}{ccc}
15 & 3 & 3 \\
7 & 11 & 0 \\
3 & 11 & 22
\end{array} \right]
\end{array}
$$

$$
\mathbf{B} = \begin{array}{c} \text{Kansas City} \\ \text{St. Louis} \\ \text{Des Moines} \end{array}
\begin{array}{ccc}
\text{Bluegrass} & \text{Rye} & \text{Fescue} \\
\left[\begin{array}{ccc}
0.15 & 0.12 & 0.14 \\
0.20 & 0.17 & 0.18 \\
0.18 & 0.15 & 0.16
\end{array} \right]
\end{array}
$$

(a) State a matrix product (with or without using the transpose operation) that gives the cost of a single package of each mixture for each city.

(b) Compute the $(1,2)$ entry of the product in part (a), and interpret the meaning of that number.

(c) Compute the $(3,3)$ entry of the product in part (a), and interpret the meaning of that number.

(d) State a different matrix product than the answer to part (a) that gives the cost of a single package of each mixture for each city.

(e) Compute the $(2,1)$ entry of the product in part (d), and interpret the meaning of that number.

(f) Compute the $(1,2)$ entry of the product in part (d), and interpret the meaning of that number.

15. Comfy Chairs, Inc. makes three styles of kitchen chair in its factories in three different cities. Each chair needs to have its parts fabricated, then the chair must be assembled, and then, finally, packed for shipment. Matrix $\mathbf{A}$ represents the number of hours needed to make each style of chair at each stage of manufacture. Matrix $\mathbf{B}$ indicates the wages paid to workers (in \$/hour) for each of the three types of labor in each city.

$$
\mathbf{A} = \begin{array}{c} \text{Style 1} \\ \text{Style 2} \\ \text{Style 3} \end{array}
\begin{array}{ccc}
\text{Fabricating} & \text{Assembly} & \text{Packaging} \\
\left[\begin{array}{ccc}
2.00 & 0.75 & 0.25 \\
2.50 & 1.25 & 0.30 \\
3.00 & 1.50 & 0.25
\end{array} \right]
\end{array}
$$

$$
\mathbf{B} = \begin{array}{c} \text{Philadelphia} \\ \text{San Francisco} \\ \text{Houston} \end{array}
\begin{array}{ccc}
\text{Fabricating} & \text{Assembly} & \text{Packaging} \\
\left[\begin{array}{ccc}
12 & 11 & 9 \\
18 & 14 & 11 \\
11 & 10 & 8
\end{array} \right]
\end{array}
$$

(a) State a matrix product that gives the total labor costs for producing each style of chair in each city.

(b) Compute the $(2, 3)$ entry of the product in part (a), and interpret the meaning of that number.

(c) Compute the $(1, 1)$ entry of the product in part (a), and interpret the meaning of that number.

(d) State a different matrix product than the answer to part (a) that gives the total labor costs for producing each style of chair in each city.

(e) Compute the $(3, 2)$ entry of the product in part (d), and interpret the meaning of that number.

16. Grill Scouts, Inc. packages three different grades of frozen hamburger patties in its factories in three different cities. Matrix **A** represents the number of units of lean beef, regular beef, and soy powder used in the recipe for a single batch of each grade of burger. Matrix **B** indicates the cost in dollars per unit for each ingredient in each city.

		Grade A	Grade B	Grade C
	Lean Beef	15	12	5
A =	Regular Beef	5	6	11
	Soy Powder	0	2	4

		New York	Chicago	Los Angeles
	Lean Beef	200	187	212
B =	Regular Beef	155	124	177
	Soy Powder	25	18	32

(a) State a matrix product that gives the total cost of ingredients for a single batch of each grade of hamburger in each city.

(b) Compute the $(3, 2)$ entry of the product in part (a), and interpret the meaning of that number.

(c) Compute the $(1, 3)$ entry of the product in part (a), and interpret the meaning of that number.

(d) State a different matrix product than the answer to part (a) that gives the total cost of ingredients for a single batch of each grade of hamburger in each city.

(e) Compute the $(2, 3)$ entry of the product in part (d), and interpret the meaning of that number.

17. Oil Czars, Inc. mixes four different blends of gasoline in four different cities. Matrix **A** represents the number of units of each type of fuel used in one tanker truck-load of each blend. Matrix **B** indicates the cost in dollars per unit for each ingredient in each city.

		Blend A	Blend B	Blend C	Blend D
	Gasoline	110	100	85	90
A =	Octane	10	14	15	9
	Ethanol	0	6	20	11

$$
\mathbf{B} = \begin{array}{c} \text{Gasoline} \\ \text{Octane} \\ \text{Ethanol} \end{array}
\begin{array}{cccc} \text{Atlanta} & \text{Portland} & \text{Cincinnati} & \text{Dallas} \\ \left[\begin{array}{cccc} 33 & 38 & 35 & 29 \\ 45 & 42 & 43 & 37 \\ 22 & 21 & 15 & 12 \end{array} \right] \end{array}
$$

(a) State a matrix product that gives the total cost in dollars of ingredients for one tanker truck-load of each blend in each city.

(b) Compute the $(2,1)$ entry of the product in part (a), and interpret the meaning of that number.

(c) Compute the $(1,3)$ entry of the product in part (a), and interpret the meaning of that number.

(d) State a different matrix product than the answer to part (a) that gives the total cost in dollars of ingredients for one tanker truck-load of each blend in each city.

(e) Compute the $(3,1)$ entry of the product in part (d), and interpret the meaning of that number.

18. Major Mills, Inc. makes three different kinds of breakfast cereals in its processing plants in three different countries. The cereal boxes are packed into cases. Matrix **A** represents the number of pounds of each type of grain used in one case of each kind of cereal. Matrix **B** indicates the price per pound in dollars for each type of grain in each country.

$$
\mathbf{A} = \begin{array}{c} \text{Crazy Puffs} \\ \text{Sweet Krisps} \\ \text{Flaky Flakes} \end{array}
\begin{array}{cccc} \text{Corn} & \text{Wheat} & \text{Rice} & \text{Barley} \\ \left[\begin{array}{cccc} 1.5 & 2.5 & 1.8 & 2.2 \\ 2.2 & 1.6 & 2.0 & 1.8 \\ 4.0 & 2.4 & 3.2 & 3.6 \end{array} \right] \end{array}
$$

$$
\mathbf{B} = \begin{array}{c} \text{France} \\ \text{India} \\ \text{Estonia} \end{array}
\begin{array}{cccc} \text{Corn} & \text{Wheat} & \text{Rice} & \text{Barley} \\ \left[\begin{array}{cccc} 0.30 & 0.80 & 0.50 & 0.60 \\ 0.50 & 1.20 & 0.80 & 0.90 \\ 0.40 & 0.90 & 0.70 & 0.80 \end{array} \right] \end{array}
$$

(a) State a matrix product that gives the total cost of the grains in one case of each kind of cereal in each country.

(b) Compute the $(3,2)$ entry of the product in part (a), and interpret the meaning of that number.

(c) Compute the $(2,1)$ entry of the product in part (a), and interpret the meaning of that number.

(d) State a different matrix product than the answer to part (a) that gives the total cost of the grains in one case of each kind of cereal in each country.

(e) Compute the $(2,3)$ entry of the product in part (d), and interpret the meaning of that number.

19. Disprove the following statement by giving a specific counterexample: If **A** and **B** are two square matrices having the same size, then **AB** = **BA**.

20. Disprove the following statement by giving a specific counterexample: If **A** and **B** are two square matrices having the same size, then $(\mathbf{AB})^T = \mathbf{A}^T\mathbf{B}^T$.

21. Use a proof similar to that used for Theorem 2 to prove that if **A** is an $m \times n$ matrix, then $\mathbf{AA}^T$ is a $m \times m$ symmetric matrix.

1.7   Algebraic Rules for Matrix Multiplication

The Associative and Distributive Laws

The operation of matrix multiplication follows four basic rules that we will use frequently. Two of these are associative laws, and two are distributive laws:

If s is a scalar, **A** is an $m \times n$ matrix, **B** and **C** are $n \times p$ matrices, and **D** is a $p \times q$ matrix, then:

MM1: $s(\mathbf{AB}) = (s\mathbf{A})\mathbf{B} = \mathbf{A}(s\mathbf{B})$
   (**Associative Law for Scalar and Matrix Multiplication**)

MM2: $\mathbf{A}(\mathbf{BD}) = (\mathbf{AB})\mathbf{D}$
   (**Associative Law for Matrix Multiplication**)

MM3: $\mathbf{A}(\mathbf{B} \pm \mathbf{C}) = \mathbf{AB} \pm \mathbf{AC}$
   (**First Distributive Law for Matrix Multiplication over Matrix Addition and Subtraction**)

MM4: $(\mathbf{B} \pm \mathbf{C})\mathbf{D} = \mathbf{BD} \pm \mathbf{CD}$
   (**Second Distributive Law for Matrix Multiplication over Matrix Addition and Subtraction**)

We will not supply the proofs of these laws here, although we ask for a proof of MM1 in the exercises. The proofs for MM2, MM3, and MM4 are more complicated.

These four rules allow us to perform simple algebraic simplifications and manipulations on expressions involving matrix multiplication. However, beware! As we have seen, matrix multiplication is *not* commutative, and so we cannot perform all of the manipulations that we might like.

Here are some examples of these properties. Suppose $s = 3$,

$$\mathbf{A} = \begin{bmatrix} 1 & 2 & 3 \\ 4 & 5 & 6 \end{bmatrix}, \ \mathbf{B} = \begin{bmatrix} 4 & 2 \\ 1 & 5 \\ 6 & 3 \end{bmatrix}, \ \mathbf{C} = \begin{bmatrix} 7 & 0 \\ 2 & 4 \\ 5 & 1 \end{bmatrix} \text{ and } \mathbf{D} = \begin{bmatrix} 8 & 1 \\ 3 & 5 \end{bmatrix}.$$

MM1 says that when we are multiplying a scalar and two matrices together, we can multiply the scalar by either of the matrices first, or we can multiply the matrices together first. We obtain the same result either way. (Notice we *cannot* change the *order* of the matrices.) In this particular case:

$$s(\mathbf{AB}) = 3\left(\begin{bmatrix} 1 & 2 & 3 \\ 4 & 5 & 6 \end{bmatrix}\begin{bmatrix} 4 & 2 \\ 1 & 5 \\ 6 & 3 \end{bmatrix}\right) = 3\begin{bmatrix} 24 & 21 \\ 57 & 51 \end{bmatrix} = \begin{bmatrix} 72 & 63 \\ 171 & 153 \end{bmatrix};$$

$$(s\mathbf{A})\mathbf{B} = \left(3\begin{bmatrix} 1 & 2 & 3 \\ 4 & 5 & 6 \end{bmatrix}\right)\begin{bmatrix} 4 & 2 \\ 1 & 5 \\ 6 & 3 \end{bmatrix} = \begin{bmatrix} 3 & 6 & 9 \\ 12 & 15 & 18 \end{bmatrix}\begin{bmatrix} 4 & 2 \\ 1 & 5 \\ 6 & 3 \end{bmatrix} = \begin{bmatrix} 72 & 63 \\ 171 & 153 \end{bmatrix};$$

$$\mathbf{A}(s\mathbf{B}) = \begin{bmatrix} 1 & 2 & 3 \\ 4 & 5 & 6 \end{bmatrix}\left(3\begin{bmatrix} 4 & 2 \\ 1 & 5 \\ 6 & 3 \end{bmatrix}\right) = \begin{bmatrix} 1 & 2 & 3 \\ 4 & 5 & 6 \end{bmatrix}\begin{bmatrix} 12 & 6 \\ 3 & 15 \\ 18 & 9 \end{bmatrix} = \begin{bmatrix} 72 & 63 \\ 171 & 153 \end{bmatrix},$$

and all three of these results are equal.

MM2 says that when multiplying three matrices together without changing their order, we can start either by multiplying the first pair together, or the last pair together. Each time we obtain the same result. In this particular case:

$$\mathbf{A}(\mathbf{BD}) = \begin{bmatrix} 1 & 2 & 3 \\ 4 & 5 & 6 \end{bmatrix}\left(\begin{bmatrix} 4 & 2 \\ 1 & 5 \\ 6 & 3 \end{bmatrix}\begin{bmatrix} 8 & 1 \\ 3 & 5 \end{bmatrix}\right) = \begin{bmatrix} 1 & 2 & 3 \\ 4 & 5 & 6 \end{bmatrix}\begin{bmatrix} 38 & 14 \\ 23 & 26 \\ 57 & 21 \end{bmatrix}$$

$$= \begin{bmatrix} 255 & 129 \\ 609 & 312 \end{bmatrix};$$

$$(\mathbf{AB})\mathbf{D} = \left(\begin{bmatrix} 1 & 2 & 3 \\ 4 & 5 & 6 \end{bmatrix}\begin{bmatrix} 4 & 2 \\ 1 & 5 \\ 6 & 3 \end{bmatrix}\right)\begin{bmatrix} 8 & 1 \\ 3 & 5 \end{bmatrix} = \begin{bmatrix} 24 & 21 \\ 57 & 51 \end{bmatrix}\begin{bmatrix} 8 & 1 \\ 3 & 5 \end{bmatrix}$$

$$= \begin{bmatrix} 255 & 129 \\ 609 & 312 \end{bmatrix}.$$

Again, we see that the results are equal, as expected.

MM3 and MM4 tell us that when a matrix is multiplied by the sum or difference of two matrices (or vice versa), we can either add (or subtract) first and then multiply, or we can first multiply by each matrix in the sum, and then add (or subtract). Either way, we will obtain the same result. Checking MM3 in this particular case, we have:

$$\mathbf{A}(\mathbf{B}+\mathbf{C}) = \begin{bmatrix} 1 & 2 & 3 \\ 4 & 5 & 6 \end{bmatrix}\left(\begin{bmatrix} 4 & 2 \\ 1 & 5 \\ 6 & 3 \end{bmatrix} + \begin{bmatrix} 7 & 0 \\ 2 & 4 \\ 5 & 1 \end{bmatrix}\right)$$

$$= \begin{bmatrix} 1 & 2 & 3 \\ 4 & 5 & 6 \end{bmatrix}\begin{bmatrix} 11 & 2 \\ 3 & 9 \\ 11 & 4 \end{bmatrix} = \begin{bmatrix} 50 & 32 \\ 125 & 77 \end{bmatrix};$$

$$\mathbf{AB} + \mathbf{AC} = \begin{bmatrix} 1 & 2 & 3 \\ 4 & 5 & 6 \end{bmatrix} \begin{bmatrix} 4 & 2 \\ 1 & 5 \\ 6 & 3 \end{bmatrix} + \begin{bmatrix} 1 & 2 & 3 \\ 4 & 5 & 6 \end{bmatrix} \begin{bmatrix} 7 & 0 \\ 2 & 4 \\ 5 & 1 \end{bmatrix}$$

$$= \begin{bmatrix} 24 & 21 \\ 57 & 51 \end{bmatrix} + \begin{bmatrix} 26 & 11 \\ 68 & 26 \end{bmatrix} = \begin{bmatrix} 50 & 32 \\ 125 & 77 \end{bmatrix}.$$

You should verify that MM4 holds as well for the given matrices $\mathbf{B}, \mathbf{C}$, and $\mathbf{D}$.

Simplification of Matrix Expressions

The "MM" and the "TR" Properties are useful for simplifying various algebraic expressions involving matrix addition, transpose, and multiplication. For example, suppose $\mathbf{A}$ and $\mathbf{B}$ are matrices that are the same size, and we wish to simplify the expression $\left((\mathbf{A} + 2\mathbf{B})^T (4\mathbf{A} - 5\mathbf{B}) \right)^T$ to the point so that no parentheses appear in the expression. To do this, we could use the algebraic properties of matrix operations, as follows:

$$\left((\mathbf{A} + 2\mathbf{B})^T (4\mathbf{A} - 5\mathbf{B}) \right)^T$$

$$= (4\mathbf{A} - 5\mathbf{B})^T \left((\mathbf{A} + 2\mathbf{B})^T \right)^T \quad \text{by TR4}$$

$$= (4\mathbf{A} - 5\mathbf{B})^T (\mathbf{A} + 2\mathbf{B}) \quad \text{by TR1}$$

$$= \left((4\mathbf{A})^T - (5\mathbf{B})^T \right) (\mathbf{A} + 2\mathbf{B}) \quad \text{by TR3}$$

$$= \left(4\mathbf{A}^T - 5\mathbf{B}^T \right) (\mathbf{A} + 2\mathbf{B}) \quad \text{by TR2}$$

$$= \left(4\mathbf{A}^T - 5\mathbf{B}^T \right) (\mathbf{A}) + \left(4\mathbf{A}^T - 5\mathbf{B}^T \right) (2\mathbf{B}) \quad \text{by MM3}$$

$$= \left(4\mathbf{A}^T \right) \mathbf{A} - \left(5\mathbf{B}^T \right) \mathbf{A} + \left(4\mathbf{A}^T \right) (2\mathbf{B}) - \left(5\mathbf{B}^T \right) (2\mathbf{B}) \quad \text{by MM4}$$

$$= 4\mathbf{A}^T \mathbf{A} - 5\mathbf{B}^T \mathbf{A} + 8\mathbf{A}^T \mathbf{B} - 10\mathbf{B}^T \mathbf{B} \quad \text{by MM1}.$$

Unfortunately, none of the four terms in the final expression can be combined.

Concerns about Matrix Multiplication

When working with equations containing matrices, we have to take special care since there is no commutative law for matrix multiplication. For example, suppose we have an equation such as $\mathbf{C} = \mathbf{B} + \mathbf{D}$, and we want to multiply both sides of this equation by a matrix $\mathbf{A}$ (assuming the matrices are compatible). Because matrix multiplication is not commutative, if we multiply by $\mathbf{A}$ on the *left* side of *one half* of the equation, we have to multiply by $\mathbf{A}$ on the *left* side of *the other half* of the equation. (A similar rule is true when multiplying on the *right* side.) Thus, from $\mathbf{C} = \mathbf{B} + \mathbf{D}$,

$$\text{either} \quad \mathbf{AC} = \mathbf{A}(\mathbf{B} + \mathbf{D})$$
$$\text{or} \quad \mathbf{CA} = (\mathbf{B} + \mathbf{D})\mathbf{A} \quad \text{is valid.}$$

In the first, we multiplied *both* sides of the equation by $\mathbf{A}$ on the *left*, and in the second, we multiplied *both* sides of the equation by $\mathbf{A}$ on the *right*. Either of these is fine as long as the products make sense. However,

$$\text{neither} \quad \mathbf{AC} = (\mathbf{B} + \mathbf{D})\mathbf{A}$$
$$\text{nor} \quad \mathbf{CA} = \mathbf{A}(\mathbf{B} + \mathbf{D}) \quad \text{is valid.}$$

In the first, we multiplied the first half of the equation by $\mathbf{A}$ on the *left*, but multiplied the other half by $\mathbf{A}$ on the *right*. The second equation has a similar error. Both are invalid conclusions because the *order* in which matrix multiplication is performed makes a difference in the answer. While we started with two equal matrices (namely, $\mathbf{C}$ and $(\mathbf{B} + \mathbf{D})$), we did not end up with equal results because we did not do exactly the same operation on both sides.

It is worth mentioning that we needed to state both distributive laws (MM3 and MM4) above, because matrix multiplication is not commutative. (Either one of these laws would follow from the other if matrix multiplication were commutative.)

No Cancellation Law

In high school algebra, we frequently use a cancellation law that allows us to remove a common nonzero factor from both sides of an equation. For example, if we have $3x = 3y$, we can cancel the 3's to get $x = y$. This does *not* usually work with matrix equations. Even if we know that $\mathbf{A}$ is not a zero matrix and $\mathbf{AB} = \mathbf{AC}$, we can not necessarily conclude that $\mathbf{B} = \mathbf{C}$. That is, there is no general cancellation law for matrix multiplication. For example, you can easily verify that

$$\begin{bmatrix} 4 & 2 \\ 6 & 3 \end{bmatrix} \begin{bmatrix} 3 & 8 \\ 2 & 3 \end{bmatrix} = \begin{bmatrix} 16 & 38 \\ 24 & 57 \end{bmatrix}$$

and

$$\begin{bmatrix} 4 & 2 \\ 6 & 3 \end{bmatrix} \begin{bmatrix} 1 & 5 \\ 6 & 9 \end{bmatrix} = \begin{bmatrix} 16 & 38 \\ 24 & 57 \end{bmatrix},$$

and hence

$$\begin{bmatrix} 4 & 2 \\ 6 & 3 \end{bmatrix} \begin{bmatrix} 3 & 8 \\ 2 & 3 \end{bmatrix} = \begin{bmatrix} 4 & 2 \\ 6 & 3 \end{bmatrix} \begin{bmatrix} 1 & 5 \\ 6 & 9 \end{bmatrix}.$$

However, if we try to cancel out the common matrix factor $\begin{bmatrix} 4 & 2 \\ 6 & 3 \end{bmatrix}$ from both sides, we get

$$\begin{bmatrix} 3 & 8 \\ 2 & 3 \end{bmatrix} = \begin{bmatrix} 1 & 5 \\ 6 & 9 \end{bmatrix},$$

which is clearly not true!

A related result is that matrix multiplication allows for zero divisors. A **zero divisor** is a nonzero matrix that can be multiplied by another nonzero matrix to produce a zero matrix. For example,

$$\begin{bmatrix} 4 & 2 \\ 6 & 3 \end{bmatrix} \begin{bmatrix} 2 & 3 \\ -4 & -6 \end{bmatrix} = \begin{bmatrix} 0 & 0 \\ 0 & 0 \end{bmatrix},$$

and so, both $\begin{bmatrix} 4 & 2 \\ 6 & 3 \end{bmatrix}$ and $\begin{bmatrix} 2 & 3 \\ -4 & -6 \end{bmatrix}$ are zero divisors. For matrices, it is possible to have $\mathbf{AB} = \mathbf{O}$ with neither $\mathbf{A}$ nor $\mathbf{B}$ equal to a zero matrix. Because many matrices are zero divisors, the familiar zero product rule ("If $xy = 0$, then either $x = 0$ or $y = 0$") from high school algebra does *not* apply in general to matrices.

Positive Integer Powers of a Square Matrix

Since square matrices have the same number of rows as columns, they can be multiplied by themselves. For example,

$$\text{if } \mathbf{A} = \begin{bmatrix} 1 & 2 \\ 3 & 4 \end{bmatrix}, \text{ then } \mathbf{AA} = \begin{bmatrix} 1 & 2 \\ 3 & 4 \end{bmatrix}\begin{bmatrix} 1 & 2 \\ 3 & 4 \end{bmatrix} = \begin{bmatrix} 7 & 10 \\ 15 & 22 \end{bmatrix}.$$

We generally express a product of the form $\mathbf{AA}$ as $\mathbf{A}^2$. In general, for any square matrix $\mathbf{A}$, we define the kth power of $\mathbf{A}$ as

$$\mathbf{A}^k = \underbrace{\mathbf{AA}\cdots\mathbf{A}}_{k \text{ times}},$$

where k is a positive integer power of $\mathbf{A}$. (The positive integers are the counting numbers: $1, 2, 3, ...$) For example, $\mathbf{A}^3 = \mathbf{AAA} = \mathbf{A}^2\mathbf{A} = (\mathbf{AA})\mathbf{A} = \mathbf{A}(\mathbf{AA}) = \mathbf{AA}^2$, $\mathbf{A}^4 = \mathbf{AAAA} = \mathbf{A}^3\mathbf{A}$, and, as a special case, $\mathbf{A}^1 = \mathbf{A}$.

Next, we present four basic principles involving powers of matrices. We will refer to them as the "Rules for Exponents" or "RFE" for short. The first two (RFE1 and RFE2) tell us that powers of matrices follow the usual rules for exponents from high school algebra. RFE3 shows how exponents interact with the transpose operator. Finally, RFE4 shows how exponents interact with scalar multiplication. In particular:

If $\mathbf{A}$ is a square matrix and k and l are positive integers, and c is a scalar, then

RFE1: $\mathbf{A}^{k+l} = \mathbf{A}^k\mathbf{A}^l$

RFE2: $\left(\mathbf{A}^k\right)^l = \mathbf{A}^{kl}$

RFE3: $\left(\mathbf{A}^T\right)^k = \left(\mathbf{A}^k\right)^T$

RFE4: $(c\mathbf{A})^k = c^k\mathbf{A}^k$.

So, for example, if $k = 2$, and $l = 3$, then RFE1 shows that we can break up the multiplication for $\mathbf{A}^5$ as follows:

$$\mathbf{A}^5 = \underbrace{\mathbf{A}\mathbf{A}}_{\mathbf{A}^2}\underbrace{\mathbf{A}\mathbf{A}\mathbf{A}}_{\mathbf{A}^3}.$$

Similarly, RFE2 shows that the product $\mathbf{A}^6$ can be split up as follows:

$$\mathbf{A}^6 = \underbrace{\underbrace{\mathbf{A}\mathbf{A}}_{\mathbf{A}^2}\underbrace{\mathbf{A}\mathbf{A}}_{\mathbf{A}^2}\underbrace{\mathbf{A}\mathbf{A}}_{\mathbf{A}^2}}_{(\mathbf{A}^2)^3}.$$

RFE3 is valid because it simply amounts to a repeated use of TR4 and the Associative Law (MM2). For example, for $k = 3$, we have

$$
\begin{aligned}
\left(\mathbf{A}^T\right)^3 &= \mathbf{A}^T\mathbf{A}^T\mathbf{A}^T \\
&= \left(\mathbf{A}^T\mathbf{A}^T\right)\mathbf{A}^T &&\text{(by MM2)} \\
&= (\mathbf{A}\mathbf{A})^T\mathbf{A}^T &&\text{(by TR4)} \\
&= (\mathbf{A}(\mathbf{A}\mathbf{A}))^T &&\text{(by TR4)} \\
&= \left(\mathbf{A}^3\right)^T &&\text{(by MM2).}
\end{aligned}
$$

RFE4 holds because it merely involves a repeated use of MM1. For example, if $c = 5$, then

$$
\begin{aligned}
(5\mathbf{A})^3 &= (5\mathbf{A})(5\mathbf{A})(5\mathbf{A}) \\
&= (5)(5)(5)\mathbf{A}\mathbf{A}\mathbf{A} &&\text{(by MM1)} \\
&= 5^3\mathbf{A}^3.
\end{aligned}
$$

One consequence of RFE1 is that any square matrix commutes with all positive integer powers of itself. For example, $\mathbf{A} = \mathbf{A}^1$ commutes with $\mathbf{A}^2$ because $\mathbf{A}^1\mathbf{A}^2$ and $\mathbf{A}^2\mathbf{A}^1$ both equal $\mathbf{A}^3$ by RFE1. In particular, we saw earlier that

$$\text{if } \mathbf{A} = \begin{bmatrix} 1 & 2 \\ 3 & 4 \end{bmatrix}, \text{ then } \mathbf{A}^2 = \begin{bmatrix} 7 & 10 \\ 15 & 22 \end{bmatrix},$$

and it is easy to check that these two matrices commute:

$$
\mathbf{A}\mathbf{A}^2 = \begin{bmatrix} 1 & 2 \\ 3 & 4 \end{bmatrix}\begin{bmatrix} 7 & 10 \\ 15 & 22 \end{bmatrix} = \begin{bmatrix} 37 & 54 \\ 81 & 118 \end{bmatrix}
$$

$$
\text{and } \mathbf{A}^2\mathbf{A} = \begin{bmatrix} 7 & 10 \\ 15 & 22 \end{bmatrix}\begin{bmatrix} 1 & 2 \\ 3 & 4 \end{bmatrix} = \begin{bmatrix} 37 & 54 \\ 81 & 118 \end{bmatrix},
$$

both of which, of course, equal $\mathbf{A}^3$. Similarly, all positive integer powers of $\mathbf{A}$ commute with all other positive integer powers of $\mathbf{A}$; that is, $\mathbf{A}^k$ commutes with $\mathbf{A}^l$, because both $\mathbf{A}^k\mathbf{A}^l$ and $\mathbf{A}^l\mathbf{A}^k$ equal $\mathbf{A}^{k+l}$.

Assignment 8: Learn how to raise a square matrix to a positive integer power on your calculator. Then, create a 3×3 matrix $\mathbf{A}$ with random nonzero entries. Compute (and save) $\mathbf{A}^2$, $\mathbf{A}^4$, and $\mathbf{A}^6$. Multiply $(\mathbf{A}^2)(\mathbf{A}^4)$ to verify that you obtain $\mathbf{A}^6$. Also, multiply $(\mathbf{A}^4)(\mathbf{A}^2)$, and observe that you also get $\mathbf{A}^6$. (That verifies RFE1 in this particular case as well as the fact that $\mathbf{A}^2$ and $\mathbf{A}^4$ commute.) Finally, raise $\mathbf{A}^2$ to the third power and notice that the answer is also $\mathbf{A}^6$. (That verifies RFE2 in this particular case.)

Squaring the Sum of Two Matrices

If two matrices $\mathbf{A}$ and $\mathbf{B}$ are square matrices of the same size, then when we add them, we get another square matrix $\mathbf{A} + \mathbf{B}$ of the same size. In such a case, it is possible to take powers of $\mathbf{A} + \mathbf{B}$. For example,

$$
\begin{aligned}
(\mathbf{A} + \mathbf{B})^2 &= (\mathbf{A} + \mathbf{B})(\mathbf{A} + \mathbf{B}) \\
&= (\mathbf{A} + \mathbf{B})\mathbf{A} + (\mathbf{A} + \mathbf{B})\mathbf{B} \\
&\qquad \text{(distributing the first } (\mathbf{A} + \mathbf{B}) \text{ over the two matrices} \\
&\qquad\qquad \text{in the second sum } (\mathbf{A} + \mathbf{B}) \text{, using MM3)} \\
&= \mathbf{A}^2 + \mathbf{B}\mathbf{A} + \mathbf{A}\mathbf{B} + \mathbf{B}^2 \\
&\qquad \text{(distributing the two matrices in each sum } (\mathbf{A} + \mathbf{B}) \\
&\qquad\qquad \text{over the remaining matrices, using MM4)}
\end{aligned}
$$

Notice in particular that $(\mathbf{A} + \mathbf{B})^2 \neq \mathbf{A}^2 + \mathbf{B}^2$. (It is a common error to substitute one of these expressions for the other.) Also notice that it is incorrect to express the final answer above as $\mathbf{A}^2 + \mathbf{A}\mathbf{B} + \mathbf{A}\mathbf{B} + \mathbf{B}^2$ $(= \mathbf{A}^2 + 2\mathbf{A}\mathbf{B} + \mathbf{B}^2)$, because we cannot replace $\mathbf{B}\mathbf{A}$ by $\mathbf{A}\mathbf{B}$ in general. In other words, the rule "$(a + b)^2 = a^2 + 2ab + b^2$" from high school algebra does *not* apply in general to matrices. Exercises 1(j) and 2(j) ask you to verify that $(\mathbf{A} + \mathbf{B})^2 \neq \mathbf{A}^2 + 2\mathbf{A}\mathbf{B} + \mathbf{B}^2$ for particular matrices $\mathbf{A}$ and $\mathbf{B}$.

Product of Stochastic Matrices

The next result states a useful property of stochastic matrices. (Recall that stochastic matrices are square matrices in which every nonzero entry is positive, and for which the sum of the entries in each column equals 1.)

Theorem 3. *If* $\mathbf{A}$ *and* $\mathbf{B}$ *are both* $n \times n$ *stochastic matrices, then* $\mathbf{AB}$ *is also an* $n \times n$ *stochastic matrix.*

For example, consider the 3×3 stochastic matrices

$$
\mathbf{A} = \begin{bmatrix} \frac{1}{4} & \frac{1}{2} & \frac{1}{4} \\ \frac{1}{2} & 0 & \frac{1}{2} \\ \frac{1}{4} & \frac{1}{2} & \frac{1}{4} \end{bmatrix} \quad \text{and} \quad \mathbf{B} = \begin{bmatrix} \frac{1}{3} & \frac{3}{8} & \frac{1}{2} \\ \frac{2}{3} & \frac{1}{8} & \frac{1}{4} \\ 0 & \frac{1}{2} & \frac{1}{4} \end{bmatrix} .
$$

You are asked to verify in Exercise 9 that $\mathbf{AB} = \begin{bmatrix} \frac{5}{12} & \frac{9}{32} & \frac{5}{16} \\ \frac{1}{6} & \frac{7}{16} & \frac{3}{8} \\ \frac{5}{12} & \frac{9}{32} & \frac{5}{16} \end{bmatrix}$, which is also

a stochastic matrix, showing that Theorem 3 holds in this particular case. (Notice that each column of $\mathbf{AB}$ adds up to 1.) We will not give a general proof of Theorem 3, but in Exercise 25 you are asked to show that this theorem holds for 3×3 matrices.

From Theorem 3, it follows that a product of any number of stochastic matrices of the same size must be stochastic. In particular, we get the following result:

STOC: If $\mathbf{A}$ is a stochastic matrix, then for any positive integer n, $\mathbf{A}^n$ is also a stochastic matrix.

That is, any positive integer power of a stochastic matrix is again stochastic. For example, consider again the matrix $\mathbf{A} = \begin{bmatrix} \frac{1}{4} & \frac{1}{2} & \frac{1}{4} \\ \frac{1}{2} & 0 & \frac{1}{2} \\ \frac{1}{4} & \frac{1}{2} & \frac{1}{4} \end{bmatrix}$. You are asked

to show in Exercise 10 that $\mathbf{A}^2 = \begin{bmatrix} \frac{3}{8} & \frac{1}{4} & \frac{3}{8} \\ \frac{1}{4} & \frac{1}{2} & \frac{1}{4} \\ \frac{3}{8} & \frac{1}{4} & \frac{3}{8} \end{bmatrix}$ and $\mathbf{A}^3 = \begin{bmatrix} \frac{5}{16} & \frac{3}{8} & \frac{5}{16} \\ \frac{3}{8} & \frac{1}{4} & \frac{3}{8} \\ \frac{5}{16} & \frac{3}{8} & \frac{5}{16} \end{bmatrix}$. Notice

that both $\mathbf{A}^2$ and $\mathbf{A}^3$ are also stochastic matrices since all of their columns sum to 1, and so STOC holds for these two powers of $\mathbf{A}$. We will not give a general proof of STOC, but in Exercise 26 you are asked to show that the first few positive powers of a stochastic matrix are also stochastic.

Glossary

- Associative properties for multiplication: MM1 and MM2, which state that the associative law holds for products involving scalar multiplication and/or matrix multiplication.

- Cancellation law: A rule from high school algebra indicating that we can cancel a common nonzero factor from both sides of an equation. This rule is *not* valid for matrices: we can *not* cancel a common nonzero matrix from both sides of an equation.

- Distributive properties for matrix multiplication: MM3 and MM4, which state that the distributive laws for matrix multiplication over addition hold.

- Integer: The integers are the numbers in the set $\{..., -3, -2, -1, 0, 1, 2, 3, ...\}$. The positive integers are the counting numbers: $1, 2, 3, ...$

- Nilpotent matrix: A square matrix $\mathbf{A}$ for which some positive integer power of $\mathbf{A}$ equals a zero matrix. (This term is introduced in the Exercises for Section 1.7)

- Powers of a square matrix: If $\mathbf{A}$ is a square matrix, we define $\mathbf{A}^k$ when k is a positive integer, as follows: $\mathbf{A}^1 = \mathbf{A}$, and, for $k > 1$, $\mathbf{A}^k = \underbrace{\mathbf{AA}\cdots\mathbf{A}}_{k \text{ times}}$.

- Zero divisor: A nonzero matrix that can be multiplied by another nonzero matrix to produce a zero matrix.

Exercises for Section 1.7

1. Let $\mathbf{A} = \begin{bmatrix} 1 & 3 \\ -2 & 4 \end{bmatrix}$, $\mathbf{B} = \begin{bmatrix} 6 & -1 \\ -3 & 2 \end{bmatrix}$, $\mathbf{C} = \begin{bmatrix} 2 & -5 \\ -4 & 1 \end{bmatrix}$,

   $\mathbf{D} = \begin{bmatrix} 5 & -3 & 4 \\ -2 & 1 & 3 \end{bmatrix}$, $\mathbf{E} = \begin{bmatrix} -2 & 4 & -5 \\ -1 & 0 & -2 \end{bmatrix}$, $\mathbf{F} = \begin{bmatrix} 2 & -1 & 0 \\ -4 & -2 & -3 \\ 5 & 1 & 6 \end{bmatrix}$.

   (a) Calculate $\mathbf{A}^2$, $\mathbf{A}^3$, $\mathbf{F}^2$, and $\mathbf{F}^3$.

   (b) Use your answers to part (a) and RFE1 to calculate $\mathbf{A}^5$ and $\mathbf{F}^5$. (Do not use the matrices $\mathbf{A}$ and $\mathbf{F}$ themselves, but only your answers to part (a).)

   (c) Use your answers to part (a) and RFE2 to calculate $\mathbf{A}^6$ and $\mathbf{F}^6$. (Do not use the matrices $\mathbf{A}$ and $\mathbf{F}$ themselves, but only your answers to part (a).)

   (d) Use your answers to part (c) and RFE3 to calculate $(\mathbf{A}^T)^6$ and $(\mathbf{F}^T)^6$. (Do not use the matrices $\mathbf{A}$ and $\mathbf{F}$ themselves, but only your answers to part (c).)

   (e) For $k = 3$, verify that $(2\mathbf{A})^k = 2^k\mathbf{A}^k$. What property from this section are you checking?

   (f) For $s = -4$, verify that $s(\mathbf{AB}) = (s\mathbf{A})\mathbf{B}$. What property from this section are you checking?

   (g) Verify that $\mathbf{A}(\mathbf{BC}) = (\mathbf{AB})\mathbf{C}$. What property from this section are you checking?

   (h) Verify that $(\mathbf{D}+\mathbf{E})\mathbf{F} = \mathbf{DF}+\mathbf{EF}$. What property from this section are you checking?

   (i) Show that $\mathbf{A}$ and $\mathbf{B}$ do not commute.

   (j) Calculate $(\mathbf{A}+\mathbf{B})^2$, $\mathbf{A}^2+\mathbf{BA}+\mathbf{AB}+\mathbf{B}^2$, and $\mathbf{A}^2+2\mathbf{AB}+\mathbf{B}^2$, and compare your answers. Explain why the answer to part (i) affects the results you obtained.

2. Let $\mathbf{A} = \begin{bmatrix} 5 & -2 \\ 3 & 1 \end{bmatrix}$, $\mathbf{B} = \begin{bmatrix} -1 & 2 \\ 1 & -1 \end{bmatrix}$, $\mathbf{C} = \begin{bmatrix} -1 & 0 \\ 9 & 4 \\ -3 & -2 \end{bmatrix}$,

$\mathbf{D} = \begin{bmatrix} 7 & -1 \\ -8 & -2 \\ 0 & 4 \end{bmatrix}$, $\mathbf{E} = \begin{bmatrix} 2 & -1 & 0 \\ -1 & 1 & 2 \\ -2 & 0 & 1 \end{bmatrix}$, $\mathbf{F} = \begin{bmatrix} -3 & 1 & -2 \\ 5 & -2 & 6 \\ -1 & 2 & 1 \end{bmatrix}$,

and $\mathbf{G} = \begin{bmatrix} 2 & -5 & -3 \\ 1 & -4 & 2 \\ -6 & -1 & 3 \end{bmatrix}$.

(a) Calculate $\mathbf{B}^3$, $\mathbf{B}^4$, $\mathbf{E}^3$, and $\mathbf{E}^4$.

(b) Use your answers to part (a) and RFE1 to calculate $\mathbf{B}^7$ and $\mathbf{E}^7$. (Do not use the matrices $\mathbf{B}$ and $\mathbf{E}$ themselves, but only your answers to part (a).)

(c) Use your answers to part (a) and RFE2 to calculate $\mathbf{B}^{12}$ and $\mathbf{E}^{12}$. (Do not use the matrices $\mathbf{B}$ and $\mathbf{E}$ themselves, but only your answers to part (a).)

(d) Use your answers to part (b) and RFE3 to calculate $(\mathbf{B}^T)^7$ and $(\mathbf{E}^T)^7$. (Do not use the matrices $\mathbf{B}$ and $\mathbf{E}$ themselves, but only your answers to part (b).)

(e) For $k = 3$, verify that $(4\mathbf{A})^k = 4^k \mathbf{A}^k$. What property from this section are you checking?

(f) For $t = 5$, verify that $t(\mathbf{EF}) = (t\mathbf{E})\mathbf{F}$. What property from this section are you checking?

(g) Verify that $\mathbf{E}(\mathbf{FG}) = (\mathbf{EF})\mathbf{G}$. What property from this section are you checking?

(h) Verify that $\mathbf{F}(\mathbf{C}+\mathbf{D}) = \mathbf{FC}+\mathbf{FD}$. What property from this section are you checking?

(i) Show that $\mathbf{E}$ and $\mathbf{F}$ do not commute.

(j) Calculate $(\mathbf{E}+\mathbf{F})^2$, $\mathbf{E}^2 + \mathbf{EF} + \mathbf{FE} + \mathbf{F}^2$, and $\mathbf{E}^2 + 2\mathbf{EF} + \mathbf{F}^2$, and compare your answers. Explain why the answer to part (i) affects the results you obtained.

3. Suppose $\mathbf{A}$ is a matrix such that $\mathbf{A}^3 = \begin{bmatrix} -52 & 15 \\ -20 & -47 \end{bmatrix}$ and

$\mathbf{A}^5 = \begin{bmatrix} 596 & 303 \\ -404 & 697 \end{bmatrix}$. In each part, calculate the given expression.

(a) $\mathbf{A}^{12}$ (b) $(\mathbf{A}^T)^8$ (c) $2\mathbf{A}^{11}$ (d) $(-\mathbf{A}^T)^9$

4. Suppose $\mathbf{A}$ is a matrix such that $\mathbf{A}^4 = \begin{bmatrix} -11 & -84 \\ 84 & 45 \end{bmatrix}$ and

$\mathbf{A}^5 = \begin{bmatrix} 230 & -33 \\ 33 & 252 \end{bmatrix}$. In each part, calculate the given expression.

(a) $\mathbf{A}^{12}$ (b) $\left(\mathbf{A}^T\right)^{10}$ (c) $3\mathbf{A}^{13}$ (d) $\left(-2\mathbf{A}^T\right)^8$

5. Consider the matrices $\mathbf{R} = \begin{bmatrix} 3 & 2 & -6 \\ 2 & 6 & 3 \end{bmatrix}$ and $\mathbf{S} = \begin{bmatrix} 6 & -18 \\ -3 & 9 \\ 2 & -6 \end{bmatrix}$. Show that $\mathbf{RS} = \mathbf{O}_{22}$, making $\mathbf{R}$ and $\mathbf{S}$ zero divisors. Then compute $\mathbf{SR}$. What conclusion can you deduce from this?

6. Consider the matrices $\mathbf{U} = \begin{bmatrix} 1 & 2 & -1 \\ 3 & -1 & 4 \end{bmatrix}$, $\mathbf{V} = \begin{bmatrix} 5 & 2 \\ -3 & 4 \\ 1 & 0 \end{bmatrix}$, and $\mathbf{W} = \begin{bmatrix} 8 & -1 \\ -6 & 7 \\ -2 & 3 \end{bmatrix}$. Compute $\mathbf{UV}$ and $\mathbf{UW}$. These results illustrate that a certain "rule" does not hold for matrix multiplication. Which "rule" is that?

7. Calculate each of the following matrix products:

$$\begin{bmatrix} 3 & 1 \\ 0 & 2 \end{bmatrix}\begin{bmatrix} 4 & 5 \\ 0 & -2 \end{bmatrix}, \quad \begin{bmatrix} 1 & -3 & -1 \\ 0 & 5 & 3 \\ 0 & 0 & -7 \end{bmatrix}\begin{bmatrix} 2 & 8 & 1 \\ 0 & 4 & 5 \\ 0 & 0 & 3 \end{bmatrix},$$

and $\begin{bmatrix} 2 & 1 & -1 & 3 \\ 0 & -4 & 3 & 7 \\ 0 & 0 & 5 & 4 \\ 0 & 0 & 0 & 9 \end{bmatrix}\begin{bmatrix} 8 & -1 & 2 & 4 \\ 0 & 1 & 3 & 5 \\ 0 & 0 & -6 & 7 \\ 0 & 0 & 0 & 5 \end{bmatrix}.$

What pattern do you notice? Make a general conjecture about the result expected in similar cases.

8. Calculate each of the following matrix products:

$$\begin{bmatrix} 4 & 0 \\ 8 & 3 \end{bmatrix}\begin{bmatrix} 1 & 0 \\ -5 & 7 \end{bmatrix}, \quad \begin{bmatrix} 3 & 0 & 0 \\ 2 & -6 & 0 \\ 5 & 1 & 7 \end{bmatrix}\begin{bmatrix} 8 & 0 & 0 \\ -3 & 1 & 0 \\ 7 & 4 & 6 \end{bmatrix},$$

and $\begin{bmatrix} 7 & 0 & 0 & 0 \\ 2 & 3 & 0 & 0 \\ -1 & -5 & 8 & 0 \\ 6 & 2 & 9 & 1 \end{bmatrix}\begin{bmatrix} 3 & 0 & 0 & 0 \\ 4 & 5 & 0 & 0 \\ 2 & 9 & -1 & 0 \\ 8 & 7 & 6 & -2 \end{bmatrix}.$

What pattern do you notice? Make a general conjecture about the result expected in similar cases.

9. Verify the computation given after Theorem 3. That is, if

$$\mathbf{A} = \begin{bmatrix} \frac{1}{4} & \frac{1}{2} & \frac{1}{4} \\ \frac{1}{2} & 0 & \frac{1}{2} \\ \frac{1}{4} & \frac{1}{2} & \frac{1}{4} \end{bmatrix} \text{ and } \mathbf{B} = \begin{bmatrix} \frac{1}{3} & \frac{3}{8} & \frac{1}{2} \\ \frac{2}{3} & \frac{1}{8} & \frac{1}{4} \\ 0 & \frac{1}{2} & \frac{1}{4} \end{bmatrix}, \text{ show } \mathbf{AB} = \begin{bmatrix} \frac{5}{12} & \frac{9}{32} & \frac{5}{16} \\ \frac{1}{6} & \frac{7}{16} & \frac{3}{8} \\ \frac{5}{12} & \frac{9}{32} & \frac{5}{16} \end{bmatrix}.$$

10. Verify the result given just after the STOC principle is introduced. That is, if

$$\mathbf{A} = \begin{bmatrix} \frac{1}{4} & \frac{1}{2} & \frac{1}{4} \\ \frac{1}{2} & 0 & \frac{1}{2} \\ \frac{1}{4} & \frac{1}{2} & \frac{1}{4} \end{bmatrix}, \text{ show } \mathbf{A}^2 = \begin{bmatrix} \frac{3}{8} & \frac{1}{4} & \frac{3}{8} \\ \frac{1}{4} & \frac{1}{2} & \frac{1}{4} \\ \frac{3}{8} & \frac{1}{4} & \frac{3}{8} \end{bmatrix}, \text{ and } \mathbf{A}^3 = \begin{bmatrix} \frac{5}{16} & \frac{3}{8} & \frac{5}{16} \\ \frac{3}{8} & \frac{1}{4} & \frac{3}{8} \\ \frac{5}{16} & \frac{3}{8} & \frac{5}{16} \end{bmatrix}.$$

11. For the stochastic matrix $\mathbf{B} = \begin{bmatrix} \frac{1}{3} & \frac{1}{4} & \frac{1}{2} \\ 0 & \frac{1}{2} & \frac{1}{2} \\ \frac{2}{3} & \frac{1}{4} & 0 \end{bmatrix}$, calculate $\mathbf{B}^2$ and $\mathbf{B}^3$ and verify that both are stochastic.

12. For the stochastic matrix $\mathbf{C} = \begin{bmatrix} \frac{1}{2} & 0 & \frac{1}{3} \\ \frac{1}{3} & 0 & \frac{1}{6} \\ \frac{1}{6} & 1 & \frac{1}{2} \end{bmatrix}$, calculate $\mathbf{C}^2$ and $\mathbf{C}^3$ and verify that both are stochastic.

13. Calculate each of the following matrix products:

$$\begin{bmatrix} 5 & -2 \\ -2 & 6 \end{bmatrix} \begin{bmatrix} 4 & 1 \\ 1 & 3 \end{bmatrix} \text{ and } \begin{bmatrix} 7 & 2 & 3 \\ 2 & -1 & 4 \\ 3 & 4 & 0 \end{bmatrix} \begin{bmatrix} 5 & -1 & 2 \\ -1 & 8 & 7 \\ 2 & 7 & 6 \end{bmatrix}.$$

These results indicate that a certain "property" does *not* hold for matrix multiplication. What is that "property"? (Hint: What special form does each given matrix have?)

14. Calculate each of the following products of skew-symmetric matrices:

$$\begin{bmatrix} 0 & 8 \\ -8 & 0 \end{bmatrix} \begin{bmatrix} 0 & -5 \\ 5 & 0 \end{bmatrix} \text{ and } \begin{bmatrix} 0 & -3 & 4 \\ 3 & 0 & 8 \\ -4 & -8 & 0 \end{bmatrix} \begin{bmatrix} 0 & 5 & 2 \\ -5 & 0 & -3 \\ -2 & 3 & 0 \end{bmatrix}.$$

These results indicate that a certain "property" does *not* hold for matrix multiplication. What is that "property"?

15. Let $\mathbf{A}$ and $\mathbf{B}$ be matrices of the same size.

(a) Expand the expression $((3\mathbf{A} - 5\mathbf{B})\mathbf{A}^T)^T$ so that the final answer contains no parentheses. Explain at each step which property you used.

(b) Simplify your answer to part (a) assuming that $\mathbf{A}$ and $\mathbf{B}$ are symmetric.

16. Let $\mathbf{A}$ and $\mathbf{B}$ be matrices of the same size.

(a) Expand the expression $(\mathbf{B}^T(4\mathbf{A} - 3\mathbf{B}))^T$ so that the final answer contains no parentheses. Explain at each step which property you used.

(b) Simplify your answer to part (a) assuming that $\mathbf{A}$ and $\mathbf{B}$ are skew-symmetric.

17. A **nilpotent** matrix is a square matrix for which some positive integer power is a zero matrix. Show that $\mathbf{A} = \begin{bmatrix} 0 & 1 & 0 & 0 \\ 0 & 0 & 1 & 0 \\ 0 & 0 & 0 & 1 \\ 0 & 0 & 0 & 0 \end{bmatrix}$ is nilpotent, and find the smallest positive integer power of $\mathbf{A}$ that equals $\mathbf{O}_{44}$.

18. Use RFE3 to prove that the transpose of a nilpotent matrix is also nilpotent.

19. Disprove the following statement by giving a specific counterexample: If $\mathbf{A}$ and $\mathbf{B}$ are two symmetric matrices having the same size, then $\mathbf{AB}$ is symmetric.

20. Disprove the following statement by giving a specific counterexample: If $\mathbf{A}$ is a skew-symmetric matrix, then $\mathbf{A}^2$ is skew-symmetric.

21. Prove that if $\mathbf{A}$ is a symmetric matrix and k is a nonnegative integer, then $\mathbf{A}^k$ is also symmetric.

22. Prove that if $\mathbf{A}$ is a skew-symmetric matrix and k is a nonnegative *odd* integer, then $\mathbf{A}^k$ is also skew-symmetric. (Hint: $(-1)^k = 1$ if k is even, but $(-1)^k = -1$ if k is odd.)

23. (This exercise together with the next exercise gives a proof of MM1.) Assume that s is a scalar, $\mathbf{A}$ is an $m \times n$ matrix, and $\mathbf{B}$ is an $n \times p$ matrix. Prove that $s(\mathbf{AB})$, $(s\mathbf{A})\mathbf{B}$, and $\mathbf{A}(s\mathbf{B})$ are all $m \times p$ matrices, and therefore, all have the same size.

24. (In this exercise, we complete the proof of MM1.)
    From the previous exercise, we know that $s(\mathbf{AB})$, $(s\mathbf{A})\mathbf{B}$, and $\mathbf{A}(s\mathbf{B})$ all have the same size. Prove that $s(\mathbf{AB})$, $(s\mathbf{A})\mathbf{B}$, and $\mathbf{A}(s\mathbf{B})$ have equal corresponding entries by deriving a formula for the generic (i,j) entry of each of these matrices, and showing that all three formulas are equal. (Either ellipsis or sigma notation can be used for the sums involved.) (Hint: Setting $\mathbf{C} = s\mathbf{A}$ and $\mathbf{D} = s\mathbf{B}$ helps to organize the notation in the expressions for the (i,j) entries.)

25. (In this exercise, we prove Theorem 3 for the 3×3 matrix case.) Suppose that $\mathbf{A}$ and $\mathbf{B}$ are both 3×3 stochastic matrices. Prove that $\mathbf{AB}$ is also a stochastic matrix, by showing that the entries of each column of $\mathbf{AB}$ are positive or zero, and that the entries in each column of $\mathbf{AB}$ add up to 1. (Hint: First show that each entry of the 1st column of $\mathbf{AB}$ is positive or zero. Note that the entries in the 1st column of $\mathbf{AB}$ are:

$$\begin{aligned} (1,1) \text{ entry of } \mathbf{AB} &= a_{11}b_{11} + a_{12}b_{21} + a_{13}b_{31} \\ (2,1) \text{ entry of } \mathbf{AB} &= a_{21}b_{11} + a_{22}b_{21} + a_{23}b_{31} \\ (3,1) \text{ entry of } \mathbf{AB} &= a_{31}b_{11} + a_{32}b_{21} + a_{33}b_{31} \end{aligned}$$

Use the fact that all of these factors in these sums are positive or zero to explain why each sum is positive or zero. Next, show that the sum of the entries in the 1st column of **AB** equals 1. Notice that, after pulling out common factors, this sum is equal to

$$(a_{11} + a_{21} + a_{31})b_{11} + (a_{12} + a_{22} + a_{32})b_{21} + (a_{13} + a_{23} + a_{33})b_{31}.$$

However, the sums in parentheses all equal 1 (why?), and so this formula reduces to $b_{11} + b_{21} + b_{31}$, which equals 1 (why?). Finally, use similar arguments for the 2nd and 3rd columns of **AB**.])

26. (In this exercise, we prove that STOC is valid for certain cases.) Suppose that **A** is a stochastic matrix. Prove that $\mathbf{A}^2$, $\mathbf{A}^3$, and $\mathbf{A}^4$ are also stochastic matrices. [Hint: First, use Theorem 3 to show that $\mathbf{A}^2 = \mathbf{AA}$ is a stochastic matrix. Then, notice that $\mathbf{A}^3 = \mathbf{A}^2\mathbf{A}$, and apply Theorem 3 again to show that $\mathbf{A}^3$ is stochastic. Finally, note that $\mathbf{A}^4 = \mathbf{A}^3\mathbf{A}$.] (Similar reasoning shows that all positive powers of **A** are stochastic.)

1.8 Matrix Multiplication and Linear Combinations

Multiplying by Single Rows or Single Columns

Sometimes we only need to compute a single row or a single column of a matrix product. The following general principles follow from the definition of matrix multiplication:

If **A** is an $m \times n$ matrix and **B** is an $n \times p$ matrix, then:

MSR1: The ith row of **AB** = (ith row of **A**)(**B**)

MSC1: The jth column of **AB** = (**A**)(jth column of **B**)

In MSR1, the ith row of **A** is treated as a matrix having a single row. In MSC1, the jth column of **B** is treated as a matrix having a single column.

For example, consider

$$\mathbf{A} = \begin{bmatrix} 2 & 8 & -7 & 1 \\ -3 & 2 & 5 & 0 \\ 6 & -1 & 3 & 4 \end{bmatrix}, \quad \mathbf{B} = \begin{bmatrix} 6 & 2 \\ 4 & 9 \\ 3 & 7 \\ 5 & -4 \end{bmatrix}, \quad \text{with } \mathbf{AB} = \begin{bmatrix} 28 & 23 \\ 5 & 47 \\ 61 & 8 \end{bmatrix}.$$

By MSR1,

the 3rd row of **AB** = (3rd row of **A**)(**B**)

$$= \begin{bmatrix} 6 & -1 & 3 & 4 \end{bmatrix} \begin{bmatrix} 6 & 2 \\ 4 & 9 \\ 3 & 7 \\ 5 & -4 \end{bmatrix} = \begin{bmatrix} 61 & 8 \end{bmatrix}.$$

Similarly, by MSC1,

the 1st column of $\mathbf{AB}$ = $(\mathbf{A})$(1st column of $\mathbf{B}$)

$$= \begin{bmatrix} 2 & 8 & -7 & 1 \\ -3 & 2 & 5 & 0 \\ 6 & -1 & 3 & 4 \end{bmatrix} \begin{bmatrix} 6 \\ 4 \\ 3 \\ 5 \end{bmatrix} = \begin{bmatrix} 28 \\ 5 \\ 61 \end{bmatrix}.$$

Linear Combinations of the Rows or Columns of a Matrix

Recall that linear combinations are sums of scalar multiples. Occasionally, it is useful to form linear combinations of the rows or columns of a given matrix. It turns out that these linear combinations can easily be calculated using matrix multiplication!

For example, let $\mathbf{A} = \begin{bmatrix} 2 & 6 & 7 & -1 \\ 0 & 3 & -4 & 2 \\ 1 & 1 & 9 & -6 \end{bmatrix}$. Consider the following linear combination of the rows of this 3×4 matrix:

$$2\begin{bmatrix} 2 & 6 & 7 & -1 \end{bmatrix} - 3\begin{bmatrix} 0 & 3 & -4 & 2 \end{bmatrix} + 5\begin{bmatrix} 1 & 1 & 9 & -6 \end{bmatrix}$$
$$= \begin{bmatrix} 4 & 12 & 14 & -2 \end{bmatrix} - \begin{bmatrix} 0 & 9 & -12 & 6 \end{bmatrix} + \begin{bmatrix} 5 & 5 & 45 & -30 \end{bmatrix}$$
$$= \begin{bmatrix} 9 & 8 & 71 & -38 \end{bmatrix}.$$

Here we multiplied the first row by the scalar 2, the second row by the scalar -3, and the third row by the scalar 5, and then added the results. However, we can actually obtain the same answer using the rules for matrix multiplication, if we put the scalars into a single row matrix, and multiply by this row matrix on the *left* side of $\mathbf{A}$:

$$\begin{bmatrix} 2 & -3 & 5 \end{bmatrix} \begin{bmatrix} 2 & 6 & 7 & -1 \\ 0 & 3 & -4 & 2 \\ 1 & 1 & 9 & -6 \end{bmatrix} = \begin{bmatrix} 9 & 8 & 71 & -38 \end{bmatrix}.$$

This multiplication makes sense, since the first matrix is 1×3 and the second is 3×4; that is, there is one scalar (column) in the first matrix for each row of $\mathbf{A}$, the second matrix. In other words, placing the scalars into a single row and multiplying $\mathbf{A}$ *on the left side* by this row produces the corresponding linear combination of the rows of $\mathbf{A}$.

Similarly, consider the following linear combination of the columns of $\mathbf{A}$:

$$2\begin{bmatrix} 2 \\ 0 \\ 1 \end{bmatrix} - 5\begin{bmatrix} 6 \\ 3 \\ 1 \end{bmatrix} + 1\begin{bmatrix} 7 \\ -4 \\ 9 \end{bmatrix} + 6\begin{bmatrix} -1 \\ 2 \\ -6 \end{bmatrix}$$
$$= \begin{bmatrix} 4 \\ 0 \\ 2 \end{bmatrix} - \begin{bmatrix} 30 \\ 15 \\ 5 \end{bmatrix} + \begin{bmatrix} 7 \\ -4 \\ 9 \end{bmatrix} + \begin{bmatrix} -6 \\ 12 \\ -36 \end{bmatrix} = \begin{bmatrix} -25 \\ -7 \\ -30 \end{bmatrix}.$$

Here we multiplied the first column by the scalar 2, the second column by the scalar -5, the third column by the scalar 1, and the fourth column by the scalar 6, and then added the results. Again, we can obtain the same answer using matrix multiplication, if we put the scalars into a single column matrix, but multiplying instead on the *right* side of $\mathbf{A}$ this time:

$$\begin{bmatrix} 2 & 6 & 7 & -1 \\ 0 & 3 & -4 & 2 \\ 1 & 1 & 9 & -6 \end{bmatrix} \begin{bmatrix} 2 \\ -5 \\ 1 \\ 6 \end{bmatrix} = \begin{bmatrix} -25 \\ -7 \\ -30 \end{bmatrix},$$

This multiplication makes sense, since the first matrix, $\mathbf{A}$, is 3×4 and the second is 4×1; that is, for each column of $\mathbf{A}$, there is one scalar (row) in the second matrix. In other words, placing the scalars into a single column and multiplying $\mathbf{A}$ *on the right side* by this column produces the corresponding linear combination of the columns of $\mathbf{A}$.

Thus, we have the following general principles:

If $\mathbf{A}$ is an $m \times n$ matrix, $\mathbf{R}$ is a $1 \times m$ matrix (single row), and $\mathbf{C}$ is an $n \times 1$ matrix (single column), then

MSR2: The linear combination of the rows of $\mathbf{A}$ which uses the entries in $\mathbf{R}$ as the scalars is equal to $\mathbf{RA}$.

MSC2: The linear combination of the columns of $\mathbf{A}$ which uses the entries in $\mathbf{C}$ as the scalars is equal to $\mathbf{AC}$.

The Special Matrices $\mathbf{e}_i$ and $\mathbf{e}_i^T$

Next, we consider special cases of MSR2 and MSC2. We will use the symbol $\mathbf{e}_i$ to represent a matrix having a single column with the entry 1 in the ith row and all other entries equal to 0. That is,

$$\mathbf{e}_1 = \begin{bmatrix} 1 \\ 0 \\ 0 \\ \vdots \\ 0 \end{bmatrix}, \quad \mathbf{e}_2 = \begin{bmatrix} 0 \\ 1 \\ 0 \\ \vdots \\ 0 \end{bmatrix}, \quad \mathbf{e}_3 = \begin{bmatrix} 0 \\ 0 \\ 1 \\ \vdots \\ 0 \end{bmatrix}, \quad \text{etc.}$$

Notice that we did not specify how many rows $\mathbf{e}_i$ has. That is because we want to use the same notation regardless of the number of rows in the matrix. The number of rows is determined from the context in which we use the symbol. For example, $\mathbf{e}_2$ could represent any of the following, depending upon its use:

$$\begin{bmatrix} 0 \\ 1 \\ 1 \end{bmatrix}, \quad \begin{bmatrix} 0 \\ 1 \\ 0 \end{bmatrix}, \quad \begin{bmatrix} 0 \\ 1 \\ 0 \\ 0 \end{bmatrix}, \quad \begin{bmatrix} 0 \\ 1 \\ 0 \\ 0 \\ 0 \end{bmatrix}, \quad \dots\dots$$

From the definition of e_i, it follows that e_i^T is a matrix having a single *row* with a 1 in the ith column and 0 elsewhere. For example, depending on the situation, e_2^T could represent any of the following:

$$[0, 1], \quad [0, 1, 0], \quad [0, 1, 0, 0], \quad [0, 1, 0, 0, 0], \quad \cdots$$

Now, consider the matrix product:

$$\begin{bmatrix} 2 & 6 & 7 & -1 \\ 0 & 3 & -4 & 2 \\ 1 & 1 & 9 & -6 \end{bmatrix} e_3 = \begin{bmatrix} 2 & 6 & 7 & -1 \\ 0 & 3 & -4 & 2 \\ 1 & 1 & 9 & -6 \end{bmatrix} \begin{bmatrix} 0 \\ 0 \\ 1 \\ 0 \end{bmatrix}.$$

By MSC2, this equals the linear combination

$$0 \begin{bmatrix} 2 \\ 0 \\ 1 \end{bmatrix} + 0 \begin{bmatrix} 6 \\ 3 \\ 1 \end{bmatrix} + 1 \begin{bmatrix} 7 \\ -4 \\ 9 \end{bmatrix} + 0 \begin{bmatrix} -1 \\ 2 \\ -6 \end{bmatrix} = \begin{bmatrix} 7 \\ -4 \\ 9 \end{bmatrix},$$

which is the third column of the given 3×4 matrix. Notice that e_3 is a 4×1 matrix here, since that is the size that makes sense for this matrix product. A similar argument shows that multiplying any matrix *on the right side* by e_3 produces the third column of that matrix.

Similarly, using MSR2 we see that

$$\begin{aligned} e_3^T \begin{bmatrix} 2 & 6 & 7 & -1 \\ 0 & 3 & -4 & 2 \\ 1 & 1 & 9 & -6 \end{bmatrix} &= \begin{bmatrix} 0 & 0 & 1 \end{bmatrix} \begin{bmatrix} 2 & 6 & 7 & -1 \\ 0 & 3 & -4 & 2 \\ 1 & 1 & 9 & -6 \end{bmatrix} \\ &= 0 \begin{bmatrix} 2 & 6 & 7 & -1 \end{bmatrix} + 0 \begin{bmatrix} 0 & 3 & -4 & 2 \end{bmatrix} \\ &\quad + 1 \begin{bmatrix} 1 & 1 & 9 & -6 \end{bmatrix} \\ &= \begin{bmatrix} 1 & 1 & 9 & -6 \end{bmatrix}, \end{aligned}$$

which is the third row of the given 3×4 matrix. Notice that e_3^T is a 1×3 matrix here, since that is the size that makes sense for this matrix product. A similar argument shows that multiplying any matrix *on the left side* by e_3^T produces the third row of that matrix.

These examples illustrate the following general principles:

If **A** is an $m \times n$ matrix, then

MSR3: $e_i^T \mathbf{A}$ equals the ith row of **A**.

MSC3: $\mathbf{A} e_i$ equals the ith column of **A**.

The Identity Matrix

There is another very special type of matrix that we use frequently: the identity matrix. An **identity matrix** is a diagonal (and hence, square) matrix having all 1's along its main diagonal. There is exactly one identity matrix for each square size. We use the symbol $\mathbf{I}_n$ to represent the $n \times n$ identity matrix. Thus,

$$\mathbf{I}_1 = [1], \; \mathbf{I}_2 = \begin{bmatrix} 1 & 0 \\ 0 & 1 \end{bmatrix}, \; \mathbf{I}_3 = \begin{bmatrix} 1 & 0 & 0 \\ 0 & 1 & 0 \\ 0 & 0 & 1 \end{bmatrix}, \; \mathbf{I}_4 = \begin{bmatrix} 1 & 0 & 0 & 0 \\ 0 & 1 & 0 & 0 \\ 0 & 0 & 1 & 0 \\ 0 & 0 & 0 & 1 \end{bmatrix}, \; \text{etc.}$$

Sometimes we use $\mathbf{I}$ without a subscript to indicate an identity matrix whose size is to be determined by context.

The most important property of the identity matrix is that it acts as an identity element for matrix multiplication; that is,

IDM: If $\mathbf{A}$ is an $m \times n$ matrix, then $\mathbf{I}_m \mathbf{A} = \mathbf{A}$ and $\mathbf{A} \mathbf{I}_n = \mathbf{A}$.

For example, if $\mathbf{A} = \begin{bmatrix} 4 & 5 \\ 2 & 7 \\ 6 & 1 \end{bmatrix}$, then

$$\mathbf{I}_3 \mathbf{A} = \begin{bmatrix} 1 & 0 & 0 \\ 0 & 1 & 0 \\ 0 & 0 & 1 \end{bmatrix} \begin{bmatrix} 4 & 5 \\ 2 & 7 \\ 6 & 1 \end{bmatrix} = \begin{bmatrix} 4 & 5 \\ 2 & 7 \\ 6 & 1 \end{bmatrix}$$

and

$$\mathbf{A} \mathbf{I}_2 = \begin{bmatrix} 4 & 5 \\ 2 & 7 \\ 6 & 1 \end{bmatrix} \begin{bmatrix} 1 & 0 \\ 0 & 1 \end{bmatrix} = \begin{bmatrix} 4 & 5 \\ 2 & 7 \\ 6 & 1 \end{bmatrix}.$$

You should try multiplying a few matrices by an appropriately sized identity matrix on each side to convince yourself that IDM is true. However, merely confirming the property with specific examples does not prove it in general. Therefore, we now show that IDM holds using MSR1, MSR3, MRC1, and MRC3:

Proof of IDM: Note that the ith row of $\mathbf{I}_n$ is $\mathbf{e}_i^T$ and the ith column of $\mathbf{I}_n$ is $\mathbf{e}_i$. So, using MSR1 and MSR3, we have

$$(i\text{th row of } \mathbf{I}_m \mathbf{A}) = (i\text{th row of } \mathbf{I}_m)\mathbf{A} = \mathbf{e}_i^T \mathbf{A} = (i\text{th row of } \mathbf{A}).$$

Since this is true for each value of i, every row of $\mathbf{I}_m \mathbf{A}$ equals its corresponding row in $\mathbf{A}$, and so these matrices are equal; that is, $\mathbf{I}_m \mathbf{A} = \mathbf{A}$.

Similarly, using MSC1 and MSC3, we find that

$$(i\text{th column of } \mathbf{A} \mathbf{I}_n) = \mathbf{A}(i\text{th column of } \mathbf{I}_n) = \mathbf{A} \mathbf{e}_i = (i\text{th column of } \mathbf{A}).$$

Again, since this is true for each value of i, every column of $\mathbf{A} \mathbf{I}_n$ equals its corresponding column in $\mathbf{A}$, and so these matrices are equal; that is, $\mathbf{A} \mathbf{I}_n = \mathbf{A}$. This concludes the proof.◆

From IDM, we know that multiplication by $\mathbf{I}_n$ does not change a matrix. Therefore, $(\mathbf{I}_n)^2 = \mathbf{I}_n \mathbf{I}_n = \mathbf{I}_n$, and in general, $(\mathbf{I}_n)^k = \mathbf{I}_n$, for any positive integer k.

Zero Power of a Square Matrix

In the previous section, we introduced positive integer powers of a square matrix. We now consider the zero power of a square matrix: if $\mathbf{A}$ is an $n \times n$ square matrix, we define $\mathbf{A}^0 = \mathbf{I}_n$, the $n \times n$ identity matrix. For example, if $\mathbf{A}$ is any 2×2 matrix, then $\mathbf{A}^0 = \mathbf{I}_2 = \begin{bmatrix} 1 & 0 \\ 0 & 1 \end{bmatrix}$, and if $\mathbf{A}$ is any 3×3 matrix, then $\mathbf{A}^0 = \mathbf{I}_3 = \begin{bmatrix} 1 & 0 & 0 \\ 0 & 1 & 0 \\ 0 & 0 & 1 \end{bmatrix}$. You can easily check that all of the rules for exponents from the previous section (RFE1, RFE2, RFE3, and RFE4) are all still valid when one or both of the exponents is zero.

Glossary

- Identity matrix: A diagonal (and hence, square) matrix having all 1's along its main diagonal. An identity matrix must be square. There is exactly one identity matrix for each square size. We use the symbol $\mathbf{I}_n$ to represent the $n \times n$ identity matrix. The identity matrix is an identity element for matrix multiplication. The identity matrix is also the zero power of any square matrix.

- Zero power of a square matrix: If $\mathbf{A}$ is any $n \times n$ matrix, $\mathbf{A}^0 = \mathbf{I}_n$.

Exercises for Section 1.8

1. Use MSR1 to write a matrix product that will produce only the second row of
$$\begin{bmatrix} 4 & 0 & -2 \\ -1 & 3 & 6 \\ 7 & 1 & 5 \end{bmatrix} \begin{bmatrix} 5 & 2 \\ -4 & 8 \\ 9 & 0 \end{bmatrix}.$$
Then compute that second row.

2. Use MSC1 to write a matrix product that will produce only the third column of
$$\begin{bmatrix} 6 & 1 & -5 \\ 8 & -3 & 4 \end{bmatrix} \begin{bmatrix} 2 & 4 & 3 \\ -4 & 3 & 9 \\ 1 & -2 & 2 \end{bmatrix}.$$
Then compute that third column.

3. For the matrix $\begin{bmatrix} 5 & -6 & 3 & -2 \\ -8 & 2 & -1 & 7 \\ -3 & 0 & 9 & -4 \end{bmatrix}$, use MSR2 to write a matrix product that produces the following linear combination: (3 times the first row) + (-5 times the second row) + (6 times the third row). Then compute this product.

4. For the matrix $\begin{bmatrix} 7 & -3 & 4 \\ -2 & 8 & -6 \\ 0 & 5 & -1 \\ -5 & 2 & -9 \end{bmatrix}$, use MSC2 to write a matrix product

that produces the following linear combination: $(-4$ times the first column$) + (6$ times the second column$) + (-3$ times the third column$)$. Then compute this product.

5. Use MSR2 to express the matrix product

$$\begin{bmatrix} 4 & 7 & -2 \end{bmatrix} \begin{bmatrix} 8 & 3 & -1 & 0 \\ 5 & 6 & 0 & 2 \\ 3 & 1 & -4 & 9 \end{bmatrix}$$

as a linear combination of rows of the second matrix. Then compute that linear combination.

6. Use MSC2 to express the matrix product

$$\begin{bmatrix} 5 & 2 & -3 & 9 \\ 1 & 6 & 0 & -1 \\ -4 & 7 & 8 & 0 \end{bmatrix} \begin{bmatrix} 3 \\ -4 \\ 5 \\ -1 \end{bmatrix}$$

as a linear combination of columns of the first matrix. Then compute that linear combination.

7. Without using a calculator, use MSR3 or MSC3 to compute each of the following:

$$\mathbf{e}_2^T \begin{bmatrix} 6 & 2 & 3 \\ -4 & 5 & 8 \\ 0 & 2 & 1 \end{bmatrix}, \; \mathbf{e}_3^T \begin{bmatrix} 4 & 2 \\ 5 & 9 \\ 4 & -3 \end{bmatrix}, \; \begin{bmatrix} 0 & 6 \\ 5 & -7 \\ 2 & 4 \end{bmatrix} \mathbf{e}_1, \text{ and } \begin{bmatrix} 1 & 2 & 3 & 4 \\ 5 & 6 & 7 & 8 \\ 6 & 5 & 4 & 3 \end{bmatrix} \mathbf{e}_4.$$

8. Without using a calculator, use MSR3 or MSC3 to compute each of the following:

$$\mathbf{e}_1^T \begin{bmatrix} 9 & 1 & 4 \\ 0 & 3 & 2 \\ 5 & 6 & 8 \end{bmatrix}, \; \mathbf{e}_4^T \begin{bmatrix} 2 & 4 \\ 9 & 0 \\ 1 & 3 \\ 6 & 7 \end{bmatrix}, \; \begin{bmatrix} 1 & 2 & 3 \\ 4 & 5 & 6 \\ 9 & 8 & 7 \end{bmatrix} \mathbf{e}_2, \text{ and } \begin{bmatrix} 3 & 9 & 7 & 4 \\ 6 & 4 & 3 & 0 \end{bmatrix} \mathbf{e}_3.$$

9. If $\mathbf{A}$ is a 4×5 matrix, by what matrix should we multiply $\mathbf{A}$ (and in what order) to obtain each of the following?

   (a) the third row of $\mathbf{A}$
   (b) the fourth column of $\mathbf{A}$

10. If **B** is a 5 × 3 matrix, by what matrix should we multiply **B** (and in what order) to obtain each of the following?

    (a) the second column of **B**

    (b) the first row of **B**

11. Show that RFE1 and RFE2 (in Section 1.7) are both valid when the exponent k equals 0.

12. Show that RFE3 and RFE4 (in Section 1.7) are both valid when the exponent k equals 0.

13. Suppose $\mathbf{A} = \begin{bmatrix} 3 & -2 \\ 1 & 5 \end{bmatrix}$. Compute $\mathbf{B} = 3\mathbf{A}^3 - 5\mathbf{A}^2 + 2\mathbf{I}_2$. Then calculate both **AB** and **BA**. What do you notice about your answers? (Note: This exercise gives a particular example of a more general principle discussed in Exercise 16.)

14. Suppose $\mathbf{A} = \begin{bmatrix} 4 & -1 & 2 \\ 3 & 5 & 0 \\ 1 & -2 & 1 \end{bmatrix}$. Compute $\mathbf{B} = 2\mathbf{A}^3 + 3\mathbf{A}^2 - 4\mathbf{I}_3$. Next, find $\mathbf{C} = 6\mathbf{A}^2 - 3\mathbf{A} + 2\mathbf{I}_3$. Finally, calculate both **BC** and **CB**. What do you notice about your answers? (Note: This exercise gives a particular example of a more general principle discussed in Exercise 16.)

15. Suppose **A** and $\mathbf{B}_1, \ldots, \mathbf{B}_k$ are all $n \times n$ matrices such that **A** commutes with each of $\mathbf{B}_1, \ldots, \mathbf{B}_k$; that is, $\mathbf{AB}_1 = \mathbf{B}_1\mathbf{A}, \ldots, \mathbf{AB}_k = \mathbf{B}_k\mathbf{A}$. Suppose $\mathbf{C} = s_1\mathbf{B}_1 + \cdots + s_k\mathbf{B}_k$, a linear combination of the matrices $\mathbf{B}_1, \ldots, \mathbf{B}_k$. Prove that **A** commutes with **C**; that is, show that $\mathbf{AC} = \mathbf{CA}$.

16. If $p(x) = a_n x^n + a_{n-1} x^{n-1} + \cdots + a_1 x + a_0$ is some polynomial, and **A** is an $m \times m$ matrix, we define $p(\mathbf{A})$ as

$$p(\mathbf{A}) = a_n \mathbf{A}^n + a_{n-1} \mathbf{A}^{n-1} + \cdots + a_1 \mathbf{A} + a_0 \mathbf{I}_m.$$

Use the result of the previous exercise to prove that **A** commutes with $p(\mathbf{A})$. (Note: Using similar techniques to that used in this exercise and the previous one, we can prove the more general statement that if $q(x)$ is another polynomial, then $p(\mathbf{A})$ and $q(\mathbf{A})$ must commute. You are not expected to prove this more general result now.)

1.9　Chapter 1 Test

1. In each part, if possible, perform the given matrix operations. If the operations cannot be performed, state why.

(a) $\begin{bmatrix} 6 & 1 & 0 \\ -2 & 3 & 5 \end{bmatrix} 4 - 5 \begin{bmatrix} 4 & 1 \\ 7 & 3 \\ 2 & -8 \end{bmatrix}^T$

(b) $\begin{bmatrix} 3 & 7 & -1 \\ 2 & -2 & 8 \\ 0 & 1 & 5 \end{bmatrix} \begin{bmatrix} 6 & 2 \\ -3 & 7 \\ 8 & 4 \end{bmatrix}^T$

(c) $\begin{bmatrix} 5 & 1 \\ 8 & 3 \\ -1 & 2 \end{bmatrix}^T \begin{bmatrix} 6 & 0 & 4 \\ 5 & 2 & -2 \\ 4 & 2 & 7 \end{bmatrix}$

(d) $7 \begin{bmatrix} 1 & 3 \\ 2 & 4 \end{bmatrix}^3 - 2 \begin{bmatrix} 6 & -1 \\ 4 & 5 \end{bmatrix}^2 + 4\mathbf{I}_2$

(e) $\mathbf{e}_2^T \begin{bmatrix} 1 & -2 & 3 \\ -3 & 4 & 2 \end{bmatrix} \begin{bmatrix} 2 & -1 & 6 \\ 5 & 3 & 4 \\ 0 & 9 & 7 \end{bmatrix} \mathbf{e}_3$

2. As a fund-raiser, the Buoy Scouts hold a yearly sale of small sailboats. These sales take place in San Diego, Daytona Beach, and Cape Cod. They have three sail boat models available: the *Minnow*, the *Dolphin*, and the *Sailfish*. The following matrix gives the number of each type of boat that the Buoy Scouts have at each location:

	Minnow	Dolphin	Sailfish
San Diego	6	2	4
Daytona Beach	3	5	1
Cape Cod	7	8	6

$\mathbf{A} =$ (the above matrix).

(a) The matrix **B** gives the number of orders received for each type of boat in each location:

	Minnow	Dolphin	Sailfish
San Diego	5	0	3
Daytona Beach	3	4	0
Cape Cod	6	2	4

$\mathbf{B} =$ (the above matrix).

Perform a matrix operation to determine the number of each type of boat available in each location after having filled the orders.

(b) The matrix **C** gives the Buoy Scouts' cost and their selling price, in dollars, for each type of boat.

	Minnow	Dolphin	Sailfish
Cost	325	480	650
Selling Price	650	945	1325

$\mathbf{C} =$ (the above matrix).

Indicate matrix operations that we learned in this chapter that will give the total cost of all boats sold and the gross sales in each city, based on the orders from part (a). Then perform those operations, and describe, in words, the meaning of the $(1,2)$ entry of that result.

(c) Find a single row or single column matrix that can be multiplied by your answer to part (b) to give the total profit obtained in each city. (Profit = (Gross Sales) − (Total Cost). Do not include the cost of the unsold boats in your computations.) Then calculate these profits by performing the multiplication. (Hint: A particular linear combination of either the rows or columns of the answer in part (b) is needed: use MSR2 or MSC2.)

3. For each of the following matrices, indicate which of the following categories the matrix is in: upper triangular, lower triangular, diagonal, symmetric, skew-symmetric, stochastic. (A matrix could be in more than one category. List all that apply.)

(a) $\begin{bmatrix} -2 & 6 & -3 \\ -1 & 8 & 2 \end{bmatrix}$

(b) $\begin{bmatrix} 7 & -1 & 3 & -2 \\ 0 & 8 & 6 & -4 \\ 0 & 0 & 1 & -5 \\ 0 & 0 & 0 & 4 \end{bmatrix}$

(c) $\begin{bmatrix} -2 & 0 & 0 \\ 0 & 5 & 0 \\ 0 & 0 & -1 \end{bmatrix}$

(d) $\begin{bmatrix} 2 & -3 & 4 \\ 3 & 6 & -7 \\ -4 & 7 & 8 \end{bmatrix}$

(e) $\begin{bmatrix} 8 & -7 & 4 \\ 0 & 1 & -5 \\ 0 & 0 & -3 \end{bmatrix}^T$

(f) $\begin{bmatrix} 1 & \frac{5}{12} & \frac{1}{8} \\ 0 & \frac{7}{12} & \frac{5}{8} \\ 0 & 0 & \frac{1}{4} \end{bmatrix}$

(g) $\begin{bmatrix} 5 & -9 & 0 & -2 \\ -9 & 7 & 4 & -6 \\ 0 & 4 & -7 & 8 \\ -2 & -6 & 8 & -1 \end{bmatrix}$

(h) $\begin{bmatrix} 0 & -3 & 4 \\ 3 & 0 & -7 \\ -4 & 7 & 0 \end{bmatrix}^T$

4. Explain why the entries on the main diagonal of a skew-symmetric matrix must all equal zero.

5. Decompose the matrix $\mathbf{A} = \begin{bmatrix} 4 & 2 & -3 \\ 6 & 1 & 0 \\ 7 & 8 & -5 \end{bmatrix}$ into the sum of a symmetric matrix and a skew-symmetric matrix. Is your answer the only possible answer?

6. Name the identity element for addition of $n \times n$ matrices and the identity element for multiplication of $n \times n$ matrices.

7. Express the linear combination $c_1 \mathbf{B}_1 + c_2 \mathbf{B}_2 + \cdots + c_k \mathbf{B}_k$ in sigma notation.

8. Suppose $\mathbf{A}_1$ through $\mathbf{A}_{10}$ are matrices of the same size. Write each of the expressions $\sum_{i=2}^{8} \mathbf{A}_i$ and $\sum_{j=4}^{10} \mathbf{A}_j$ in ellipsis notation.

9. Simplify the given expression using various properties from the chapter. At each step, explain which property was used.

$$\left(\mathbf{A}^T \left(5\mathbf{A} - 2\mathbf{B}\right)\right)^T + 3\left(4\mathbf{B} + 7\mathbf{A}\right)^T \mathbf{A}.$$

10. Let $\mathbf{A} = \begin{bmatrix} 4 & 0 & -2 \\ -1 & 3 & 6 \\ 7 & 1 & 5 \end{bmatrix}$ and $\mathbf{B} = \begin{bmatrix} 5 & 2 \\ -4 & 8 \\ 9 & 1 \end{bmatrix}$.

   (a) Write a matrix product that produces only the third row of $\mathbf{AB}$.

   (b) Write a matrix product that produces only the second column of $\mathbf{AB}$.

   (c) Write a matrix product that produces only the second row of $\mathbf{A}$.

   (d) Write a matrix product that produces only the first column of $\mathbf{B}$.

   (e) Write a matrix product that produces the following linear combination: (5 times the first row of $\mathbf{A}$) $-$ (3 times the second row of $\mathbf{A}$) $+$ (4 times the third row of $\mathbf{A}$).

   (f) Write a matrix product that produces the following linear combination: (-6 times the first column of $\mathbf{B}$) $+$ (2 times the second column of $\mathbf{B}$).

11. (a) For the stochastic matrix $\mathbf{A} = \begin{bmatrix} \frac{1}{4} & \frac{1}{2} & \frac{1}{2} \\ \frac{1}{2} & \frac{1}{2} & 0 \\ \frac{1}{4} & 0 & \frac{1}{2} \end{bmatrix}$, calculate $\mathbf{A}^2$ and $\mathbf{A}^3$ and verify that both are stochastic.

   (b) Consider the matrix $\mathbf{B} = \begin{bmatrix} \frac{1}{3} & \frac{1}{3} & \frac{1}{3} \\ \frac{1}{3} & \frac{1}{3} & \frac{1}{3} \\ \frac{1}{3} & \frac{1}{3} & \frac{1}{3} \end{bmatrix}$. Because $\mathbf{B}$ is stochastic, STOC implies that $\mathbf{B}^n$ is stochastic for every positive integer n. Verify computationally that $\mathbf{B}^2$ and $\mathbf{B}^3$ are stochastic. Then use the results of these computations to explain why $\mathbf{B}^n$ is stochastic for every positive integer n.

12. Let $\mathbf{C}$ be an $p \times q$ matrix, and $\mathbf{D}$ be a $q \times r$ matrix. Prove that $(\mathbf{CD})^T$ is an $r \times p$ matrix.

13. For a particular matrix $\mathbf{A}$, the $(1, 2)$ entry of $\mathbf{A}^T \mathbf{A}$ is 5. What is the $(2, 1)$ entry of $\mathbf{A}^T \mathbf{A}$? Why?

14. In each part, indicate whether the given statement is true or whether it is false.

(a) Matrix addition is an associative operation.

(b) Matrix subtraction is an associative operation.

(c) Matrix multiplication is an associative operation.

(d) Matrix addition is a commutative operation.

(e) Matrix subtraction is a commutative operation.

(f) Matrix multiplication is a commutative operation.

(g) A linear combination of matrices is a sum of scalar multiples of the matrices.

(h) Every entry on the main diagonal of a symmetric matrix is nonzero.

(i) If $\mathbf{A}$ is an upper triangular matrix, then $\mathbf{A}^T$ is also upper triangular.

(j) If $\mathbf{A}$ and $\mathbf{B}$ are 2×2 symmetric matrices, then $\mathbf{AB}$ is a symmetric matrix.

(k) If $\mathbf{A}$ and $\mathbf{B}$ are square matrices having the same size, then $(\mathbf{AB})^T = \mathbf{A}^T \mathbf{B}^T$.

(l) If $\mathbf{A}$ is a matrix such that $\mathbf{A}^2$ is defined, then $\mathbf{A}$ must be a square matrix.

(m) If $\mathbf{A}$ is a square matrix, then $\mathbf{A}$ commutes with $\mathbf{A}^7$.

(n) If $\mathbf{A}$ is not a square matrix, then $\mathbf{A}$ cannot commute with any matrix.

(o) If $\mathbf{A}$, $\mathbf{B}$, and $\mathbf{C}$ are 3×3 matrices such that $\mathbf{AB} = \mathbf{AC}$, then $\mathbf{B} = \mathbf{C}$.

(p) If $\mathbf{A}$ and $\mathbf{B}$ are two 3×3 matrices such that $\mathbf{AB} = \mathbf{O}_{33}$, then $\mathbf{A} = \mathbf{O}_{33}$ or $\mathbf{B} = \mathbf{O}_{33}$.

(q) If $\mathbf{A}$, $\mathbf{B}$, and $\mathbf{C}$ are 3×3 matrices such that $\mathbf{B} = \mathbf{C}$, then $\mathbf{AB} = \mathbf{CA}$.

(r) If $\mathbf{A}$ is a symmetric matrix, then $\mathbf{A}^T \mathbf{A} = \mathbf{A}^2$.

(s) If $\mathbf{A}$ is a skew-symmetric matrix, then $\mathbf{AA}^T = \mathbf{A}^2$.

(t) If $\mathbf{A}$ is an upper triangular matrix, then $\mathbf{A}$ is not a lower triangular matrix.

(u) If $\mathbf{A}$ is a skew-symmetric matrix, then $-3\mathbf{A}$ is also skew-symmetric.

(v) If $\mathbf{A}$ and $\mathbf{B}$ are two 2×2 matrices, then $(\mathbf{A} + \mathbf{B})^2 = \mathbf{A}^2 + \mathbf{B}^2$.

(w) If $\mathbf{A}$ is a 3×3 matrix, then $\mathbf{AI}_3 = \mathbf{A}$ and $\mathbf{I}_3 \mathbf{A} = \mathbf{A}$.

(x) The number of rows in the single-column matrix $\mathbf{e}_1$ is typically determined from context.

(y) If $\mathbf{e}_i^T$ has size $1 \times n$, $\mathbf{e}_j$ has size $n \times 1$, and $i \neq j$, then $\mathbf{e}_i^T \mathbf{e}_j = [0]$.

(z) If $\mathbf{e}_i^T$ has size $1 \times n$, $\mathbf{e}_j$ has size $n \times 1$, and $i = j$, then $\mathbf{e}_i^T \mathbf{e}_j = [1]$.

(aa) If $\mathbf{A}$ and $\mathbf{B}$ are two 4×4 matrices, then $5\,(\mathbf{AB}) = (5\mathbf{A})\,(5\mathbf{B})$.

(bb) If $\mathbf{A}$ is a 3×3 matrix, then there is at least one matrix $\mathbf{B}$ such that $\mathbf{AB} = \mathbf{BA}$.

(cc) If $\mathbf{A}$, $\mathbf{B}$, and $\mathbf{C}$ are all 4×4 matrices, then $\mathbf{A}\left(\mathbf{B}^T\mathbf{C}\right) = \left(\mathbf{AB}^T\right)\mathbf{C}$.

(dd) If $\mathbf{A}$, $\mathbf{B}$, and $\mathbf{C}$ are all 4×4 matrices, then
$\mathbf{A}\left(\mathbf{B}^T + \mathbf{C}\right)^T = \left(\mathbf{B} + \mathbf{C}^T\right)\mathbf{A}^T$.

(ee) If $\mathbf{A}$ is a stochastic matrix, then $\mathbf{A}^T$ is stochastic.

Chapter 2

Application: Elementary Graph Theory

This chapter shows how the results we have learned about matrices can be applied to an area of mathematics known as graph theory. Graph theory has a myriad of practical applications, but here we will give only an introduction to this interesting branch of mathematics.

2.1 Graphs and Digraphs: Definitions

Graphs

A graph is a drawing that indicates whether or not certain objects are related to each other in some way. Figure 2.1 illustrates a typical graph, containing five points, A through E, and seven line segments between particular pairs of points. For example, the points here might represent five people, and a line segment between points could indicate that the two people involved know each other.

In general, a **graph** is a collection of objects, called **vertices**, which are usually depicted geometrically as points, together with a set of connections between these vertices, called **edges,** which are often depicted as line segments.

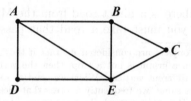

Figure 2.1: A Typical Graph

Note that for easy reference we have given labels to the points representing the vertices. An edge can be represented by a curve instead of a line segment since all that really matters is that the two vertices involved have a connection between them. An edge could also have the same vertex at both ends; that is, the two vertices at the ends of an edge do not have to be distinct. Such an edge is called a **loop.**[1] Some of these ideas are illustrated in Figure 2.2.

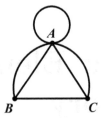

Figure 2.2: A Graph with Curved Edges and a Loop at Vertex A

Example 1. One use for a graph is to indicate roads or trails on a map. The graph is not intended to indicate relative positions of locations, as on a typical map, but merely which locations can be directly accessed from some other location. For example, the graph in Figure 2.3 shows the various roads and trails connecting six features at the Mir Resort.

This graph shows which features have a road or trail directly connecting

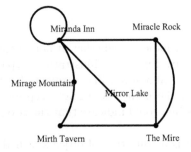

Figure 2.3: Map of the Mir Resort

them. For example, there is a direct road from the Miranda Inn to Mirage Mountain. However, you must take a road that passes at least one other

---

[1]In mathematics, two objects are considered *distinct* if they are not the same object. For example, if an edge in a graph is not a loop, then the vertices at the two ends are distinct; that is, they are different vertices. When we want to say that two objects under consideration could be the same, we frequently express that idea by saying that they are "not necessarily distinct." Hence, in general, the two vertices of an edge are not necessarily distinct. Most of the time, mathematicians generally assume that when two generic objects are chosen, they are not necessarily distinct, and so the phrase "not necessarily distinct" is only used occasionally for added emphasis.

feature to get from the Miranda Inn to the Mirth Tavern. There are two different trails connecting Miracle Rock to The Mire. Notice the nature trail at the Miranda Inn that loops directly back to the Inn. ■

The graph in Figure 2.3 representing the map of the Mir Resort does not indicate distances, nor does it show every twist and turn in the trails.[2] You may have seen similar types of maps on subway or light rail systems, indicating the various stops and connections to other trains. These maps are designed to show the relevant facts about the transit system without getting bogged down by including exact distances between stations or their actual positions. They present just the information we need to get around using the trains.

Example 2. The owners of the Mir Resort decide to build a new road directly from Miracle Rock to Mirage Mountain. The first design for this road has it passing over the road from the Miranda Inn to Mirror Lake with a bridge. We could draw a new graph that includes this new road in two different ways, as shown in Figure 2.4.

The first graph shows the new road crossing the old road from the Miranda Inn to Mirror Lake. However, there is no actual access from one road to the other, because one goes over the other. Hence, there is no vertex representing the intersection. In graphs, when edges are drawn so that one crosses over the other, but no intersection is shown, then we assume there is no "access" from one edge to the other. One can only "get on or off" an edge at the vertex at either end.

The second graph in Figure 2.4, shows a method of depicting the same situation but without using crossing edges. This graph also accurately illustrates the connectivity of the road system at the resort. Both are valid representations for the graph for this collection of roads.

The management of the resort is also considering the second option of having the new road intersect with the old road. To represent this scenario in a graph, we could add a new vertex, Mir Point, at the point of intersection of the edges, turning what were two edges in Figure 2.4 into four edges, as shown in Figure 2.5. ■

We next consider another example of a graph.

Example 3. Ten students are all taking the same linear algebra class. (We will call them $S1, \ldots, S10$.) Each of the ten vertices in the graph in Figure 2.6 represents one of these students. All ten students have *Facespace* accounts. There is an edge between two students in the graph in Figure 2.6 if they are *Facespace* BFFs. In this graph, many edges cross each other without intersecting, but those crossings do not have any effect on the students' BFF status.

---

[2]Some applications of graph theory add a number, or weight, to each edge that could represent the length of road represented by the edge, or, perhaps, the cost to travel from the one vertex to the other. However, we will not consider such weighted graphs in this textbook.

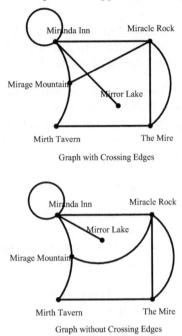

Graph with Crossing Edges

Graph without Crossing Edges

Figure 2.4: New Road Not Intersecting Old Road at the Mir Resort

For the graph in Figure 2.6, it is not immediately clear whether it can be redrawn without using any crossing edges. (In fact, some graphs can not be drawn without edges crossing.[3]) But the graph in Figure 2.7 represents the exact same BFF relationships without having any crossing edges. Both Figures 2.6 and 2.7 are considered representations of the same graph, since all we really care about are the vertices, and which pairs of vertices are connected with edges. ■

Digraphs

A **digraph**, or **directed graph**, is similar to a graph in that it has vertices and connections between vertices. However, in a digraph, edges are replaced by **directed edges**, meaning that the connections only go in one direction, similar to a one-way street. Each directed edge in a digraph still connects two vertices, as in a graph, but one vertex of the directed edge is distinguished as the starting vertex, while the other is the ending vertex. We use arrows on the directed edges in the diagram for a digraph, as in Figure 2.8, to indicate the direction of each edge. As before, we refer to a directed edge as a **loop** if it begins and ends with the same vertex.

---

[3]Graphs that can be drawn without having any crossing edges are called **planar** graphs. For example, Figure 2.7 illustrates that the graph of Figures 2.6 and 2.7 is planar. A graph that cannot be drawn in a plane without having crossing edges is called a nonplanar graph.

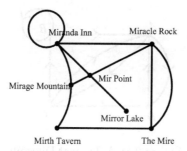

Figure 2.5: New Road Intersects Old Road

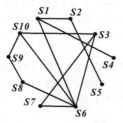

Figure 2.6: A Graph Describing BFF Relationships

Looking at Figure 2.8 we can see that *every* connection between vertices is a directed edge – even the loop! We do not mix directed and non-directed edges. If we want to indicate a "two-way street" in a digraph, we merely put in two directed edges – one in each direction. This is illustrated in Figure 2.8 with the two directed edges between vertices A and C, and again with the two directed edges between vertices A and D.

Example 4. The ten students $S1, \ldots, S10$ from Example 3 taking linear algebra together also take advantage of the POI (Person of Interest) feature on *Facespace*. Some postings on *Facespace* are put there for the general public, while others have various levels of security associated with them that limits

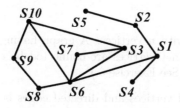

Figure 2.7: The BFF Graph With No Crossing Edges

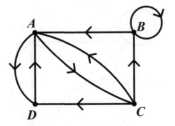

Figure 2.8: A Typical Digraph

access. User X can declare User Y a POI. In this situation, any postings by User Y that are for the general public automatically get posted on the *Facespace* site of User X. In this way User X doesn't have to actually visit User Y's site to check for updates.

The digraph in Figure 2.9 illustrates the POI relationships between the ten students in the linear algebra class. There is a directed edge from student X to student Y if X has declared Y as a POI.

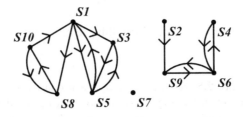

Figure 2.9: *Facespace* POI Relationships

An interesting feature of the digraph in Figure 2.9 is that the vertex $S7$ has no edges connected to it at all! We can see from the digraph that the remaining students are broken into two subgroups, $\{S1, S3, S5, S8, S10\}$ and $\{S2, S4, S6, S9\}$, where no student in either subgroup has any POI relationship with students in the other subgroup. ∎

Glossary

- Complete graph with n vertices: A graph having n vertices such that every pair of distinct vertices are connected by exactly one edge, but having no loops. (See Exercise 6.)

- Digraph: A set of vertices and directed edges between those vertices. Digraphs are typically illustrated geometrically by using points to represent vertices, and using line segments or curves between points with particular directions on them to represent directed edges.

- Directed edge: A connection between vertices that has a specified direction. That is, one of the two vertices is considered the "starting point" of the edge, and the other is considered to be the "ending point."

- Directed graph: A digraph.

- Edge: A connection between two vertices in a graph. The two vertices can be the same, in which case the edge is a loop.

- Graph: A set of vertices and edges between those vertices. Graphs are typically illustrated geometrically by using points to represent vertices, and using line segments or curves between points to represent edges.

- Loop: An edge in a graph, or, a directed edge in a digraph, that connects a vertex to itself.

- Nonplanar graph: A graph that cannot be drawn in a plane without having any of the edges cross.

- Planar graph: A graph that can be drawn in a plane without having any of the edges cross.

- Vertex: An object in a graph or digraph that is usually depicted geometrically as a point.

Exercises for Section 2.1

1. The town of Isolation, Alaska, has four East-West streets, three North-South streets, and one loop street at the northwest corner of the city. Each East-West street intersects with each North-South street. There are no roads coming into or going out of the town. Draw a graph representing the streets in Isolation.

2. Four students, $S1, \ldots, S4$, are comparing their rosters for their Fall classes. Student $S1$ is taking Bio3, Mat2, Eng4, His2, and Psy1. Student $S2$ is scheduled for Mat3, Eng4, Chem2, Psy1, and His3. Student $S3$ has Bio2, Mat2, Soc1, His2, and Psy1. Student $S4$ is rostered for Chem2, Bio2, Mat3, His2, and Psy2. Draw a graph that has an edge connecting two students for each class they have in common. Do not draw any loops.

3. The town of Isolation in Exercise 1 decides to make the four East-West streets one-way streets. The north-most of these streets now runs from west to east. Proceeding southward, the remaining East-West streets alternate in direction. Draw a digraph representing this new street configuration for Isolation.

4. Five students, $S1, \ldots, S5$, are taking a linear algebra course together. Student $S1$ tutors $S2$ and $S3$. Students $S3$, $S4$, and $S5$ work in a study group together, tutoring each other. Student $S2$ also tutors $S4$.

Finally, students $S1$, $S2$, and $S5$ spend time studying on their own. Draw a digraph that indicates which students tutor others. Insert a loop wherever a student studies alone.

5. Consider the graph in Figure 2.10. Redraw the graph so that none of the edges cross each other.

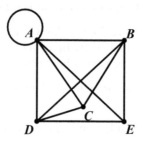

Figure 2.10: Figure for Exercise 5

6. The **complete graph with** n **vertices** is the graph having n vertices such that every pair of distinct vertices are connected by a single edge, but with no loops. Draw the complete graph with two vertices, with three vertices, with four vertices, and with five vertices. Where possible, draw the graph with no crossings.

2.2   The Adjacency Matrix

The Adjacency Matrix For a Graph

Now we want to use what we have learned about matrices in our study of graphs and digraphs.

We have seen that we have the complete information necessary to draw a graph once we know the number of vertices involved and the number of edges that connect each pair of vertices. As long as two drawings of graphs have this same information, they are considered to be drawings of the same graph, even if they do not look alike. We saw this in Figure 2.4 (two versions of the same Mir Resort graph), and in Figures 2.6 and 2.7 (two versions of the same *Facespace* BFF graph).

We can place all of the relevant information for a graph or digraph in table form by using a matrix. This is called the **adjacency matrix** for the graph or digraph. In particular, for a graph having n vertices, $V1, V2, \ldots, Vn$, the adjacency matrix is the $n \times n$ matrix whose (i, j) entry is the number of edges connecting Vi to Vj. For example, consider the graph in Figure 2.11. Since there are five vertices, the adjacency matrix for this graph is a 5×5 matrix. Each row and each column of the matrix corresponds to one of the vertices in the graph. Each entry represents the number of edges connecting the two

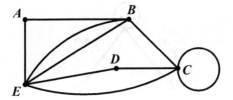

Figure 2.11: Sample Graph with Five Vertices

corresponding vertices. For the graph in Figure 2.11, we get the following
adjacency matrix:

$$\mathbf{M} = \begin{array}{c} \\ A \\ B \\ C \\ D \\ E \end{array} \begin{array}{c} A\ B\ C\ D\ E \\ \left[\begin{array}{ccccc} 0 & 1 & 0 & 0 & 1 \\ 1 & 0 & 1 & 0 & 2 \\ 0 & 1 & 1 & 1 & 1 \\ 0 & 0 & 1 & 0 & 1 \\ 1 & 2 & 1 & 1 & 0 \end{array} \right] \end{array}$$

The $(1, 2)$ entry of $\mathbf{M}$, m_{12}, equals 1 because there is 1 edge connecting vertices
A and B. The $(2, 4)$ entry of $\mathbf{M}$, m_{24}, equals 0 because there is no edge
connecting vertices B and D. The entry $m_{25} = 2$ because there are two edges
connecting B to E.

Naturally, each loop in a graph counts as only one edge from a vertex to
itself. Hence, $m_{33} = 1$ since there is a loop at vertex C. Notice that the main
diagonal of the adjacency matrix indicates the number of loops in the graph.
Also, the adjacency matrix for a graph must be *symmetric*. This is because
the order of the vertices does not matter for any particular edge. For example,
any edge that connects vertex B to vertex E also connects E to B.

All of the relevant information about a graph is in its adjacency matrix. In
fact, given just the adjacency matrix, we can reconstruct the graph. Consider,
for example, the following (symmetric) adjacency matrix:

$$\mathbf{N} = \begin{array}{c} \\ A \\ B \\ C \\ D \\ E \\ F \end{array} \begin{array}{c} A\ B\ C\ D\ E\ F \\ \left[\begin{array}{cccccc} 1 & 0 & 1 & 0 & 1 & 0 \\ 0 & 0 & 2 & 0 & 0 & 1 \\ 1 & 2 & 0 & 3 & 0 & 1 \\ 0 & 0 & 3 & 1 & 0 & 0 \\ 1 & 0 & 0 & 0 & 0 & 2 \\ 0 & 1 & 1 & 0 & 2 & 0 \end{array} \right] \end{array}.$$

We draw the graph corresponding to this adjacency matrix by starting with
six vertices, and then putting in the edges as indicated in the matrix. So,
for example, after considering just the first row of $\mathbf{N}$, we have the following
graph:

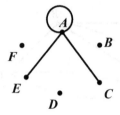

Figure 2.12(a): Partial Graph Using Only the First Row of **N**

Notice that as we proceed, we only need to consider the entries in each row that are *on or above the main diagonal*, because the matrix is symmetric (that is, the entries below the main diagonal simply repeat the information that appears above the main diagonal). Next, we include the three edges indicated in the second row, two of them between B and C, and the other between B and F. This gives:

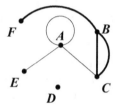

Figure 2.12(b): Partial Graph Using Only the First Two Rows of **N**

Next, we include the edges indicated in the third row. However, we do *not* use the first two entries of the third row since they fall below the main diagonal. (Using these entries would duplicate edges that we have already placed in the graph.) Thus, we only include new edges: those between C and D as well as the edge between C and F:

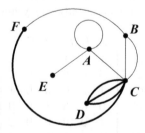

Figure 2.12(c): Partial Graph Using Only the First Three Rows of **N**

The only remaining nonzero entries on or above the main diagonal are $n_{44} = 1$ and $n_{56} = 2$. To complete the graph, we add these last three edges: This is a drawing for the graph corresponding to the adjacency matrix **N**.

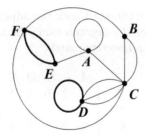

Figure 2.12(d): Final Graph for Adjacency Matrix **N**

Other drawings are possible for this graph, but as long as a drawing has vertices A through F connected with the correct number of edges, it is a valid drawing for the graph.

The Adjacency Matrix for a Digraph

We can create an adjacency matrix for a digraph in a manner similar to that for a graph. In particular, for a digraph having n vertices, the adjacency matrix is the $n \times n$ matrix whose (i, j) entry is the number of directed edges from Vi to Vj. Again, each loop counts as only one directed edge. For example, consider the digraph in Figure 2.13. The adjacency matrix for this

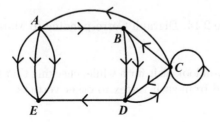

Figure 2.13: Sample Digraph with Five Vertices

digraph is

$$
\mathbf{Q} = \begin{array}{c} \\ A \\ B \\ C \\ D \\ E \end{array} \begin{array}{c} A\ B\ C\ D\ E \\ \left[\begin{array}{ccccc} 0 & 1 & 0 & 0 & 3 \\ 0 & 0 & 0 & 2 & 0 \\ 1 & 1 & 1 & 1 & 0 \\ 0 & 0 & 1 & 0 & 1 \\ 0 & 0 & 0 & 0 & 0 \end{array} \right] \end{array}.
$$

First, notice that the adjacency matrix **Q** is not symmetric. That is typical for a digraph, since there are usually edges in one direction without corresponding edges in the opposite direction.

Just as with a graph, we can start with the adjacency matrix for a digraph and construct a drawing for the digraph using the information in the matrix. For example, consider the adjacency matrix

$$\mathbf{R} = \begin{array}{c} \\ A \\ B \\ C \\ D \\ E \end{array} \begin{array}{c} A\ B\ C\ D\ E \\ \left[\begin{array}{ccccc} 1 & 2 & 0 & 0 & 1 \\ 1 & 0 & 1 & 1 & 0 \\ 1 & 2 & 0 & 1 & 1 \\ 0 & 0 & 1 & 0 & 2 \\ 1 & 0 & 0 & 1 & 1 \end{array} \right] \end{array}.$$

We can tell immediately that this is an adjacency matrix for a digraph rather than a graph because it is not symmetric. The digraph for this matrix is represented by the drawing in Figure 2.14. Now Figure 2.14 is a bit complicated,

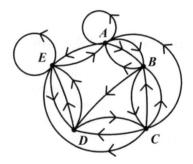

Figure 2.14: Digraph Corresponding to Matrix **R**

so we suggest that you look at it for a while, checking that the correct number of edges are directed from each vertex to other vertices.

Glossary

- Adjacency matrix: For a graph having n vertices, $V1, V2, \ldots, Vn$, the adjacency matrix is the $n \times n$ matrix whose (i, j) entry is the number of distinct edges connecting Vi to Vj. For a digraph having n vertices, the adjacency matrix is the $n \times n$ matrix whose (i, j) entry is the number of distinct edges directed from Vi to Vj. For both a graph and a digraph, a loop counts as only one edge.

Exercises for Section 2.2

1. Determine the adjacency matrix for the graph in Figure 2.15.

2. Determine the adjacency matrix for the graph in Figure 2.16.

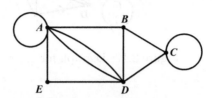

Figure 2.15: Graph for Exercise 1

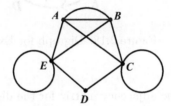

Figure 2.16: Graph for Exercise 2

3. Draw a graph corresponding to the matrix

$$
\mathbf{M} = \begin{array}{c} \\ A \\ B \\ C \\ D \end{array}
\begin{array}{c} A \ B \ C \ D \\
\left[\begin{array}{cccc}
0 & 2 & 0 & 2 \\
2 & 0 & 1 & 1 \\
0 & 1 & 1 & 1 \\
2 & 1 & 1 & 0
\end{array} \right]
\end{array}.
$$

4. Draw a graph corresponding to the matrix

$$
\mathbf{N} = \begin{array}{c} \\ A \\ B \\ C \\ D \\ E \end{array}
\begin{array}{c} A \ B \ C \ D \ E \\
\left[\begin{array}{ccccc}
0 & 0 & 1 & 0 & 1 \\
0 & 0 & 1 & 2 & 0 \\
1 & 1 & 0 & 1 & 0 \\
0 & 2 & 1 & 2 & 2 \\
1 & 0 & 0 & 2 & 1
\end{array} \right]
\end{array}.
$$

5. Determine the adjacency matrix for the digraph in Figure 2.17.

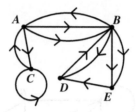

Figure 2.17: Digraph for Exercise 5

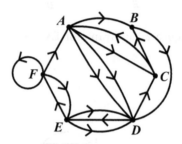

Figure 2.18: Digraph for Exercise 6

6. Determine the adjacency matrix for the digraph in Figure 2.18.

7. Draw a digraph corresponding to the matrix

$$
\begin{array}{c c}
 & \begin{array}{c c c c} A & B & C & D \end{array} \\
\begin{array}{c} A \\ B \\ C \\ D \end{array} &
\left[\begin{array}{c c c c}
0 & 0 & 1 & 2 \\
1 & 0 & 0 & 1 \\
0 & 1 & 1 & 0 \\
1 & 1 & 0 & 1
\end{array}\right].
\end{array}
$$

8. Draw a digraph corresponding to the matrix

$$
\begin{array}{c c}
 & \begin{array}{c c c c c} A & B & C & D & E \end{array} \\
\begin{array}{c} A \\ B \\ C \\ D \\ E \end{array} &
\left[\begin{array}{c c c c c}
1 & 0 & 1 & 1 & 1 \\
1 & 0 & 0 & 2 & 0 \\
1 & 2 & 0 & 1 & 0 \\
0 & 1 & 0 & 1 & 2 \\
1 & 0 & 0 & 1 & 0
\end{array}\right].
\end{array}
$$

9. Find the adjacency matrix for the complete graph with four vertices. (See Exercise 6 in Section 2.1.)

10. Describe the adjacency matrix for the complete graph with n vertices. (See Exercise 6 in Section 2.1.)

2.3 Counting Paths and Connectivity

Paths

A **path** from vertex X to vertex Y in a graph or digraph is a finite sequence of edges such that the first edge starts at vertex X, each edge (except the last) ends where the next edge begins, and the last edge ends at vertex Y. For example, consider the graph in Figure 2.19. One path from vertex $V1$ to

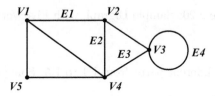

Figure 2.19: Sample Graph with Five Vertices

vertex $V3$ is

$$V1 \xrightarrow{E1} V2 \xrightarrow{E2} V4 \xrightarrow{E3} V3.$$

There are several other paths from $V1$ to $V3$, including

$$V1 \xrightarrow{E1} V2 \xrightarrow{E2} V4 \xrightarrow{E3} V3 \xrightarrow{E4} V3.$$

When describing a path, we frequently leave out the label for the edge used, if there is only one possible choice for that edge. For example, two more paths from $V1$ to $V3$ are:

$$V1 \longrightarrow V4 \longrightarrow V3, \qquad \text{and}$$
$$V1 \longrightarrow V5 \longrightarrow V4 \longrightarrow V2 \longrightarrow V3 \longrightarrow V3.$$

These two paths have length 2 and 5, respectively, where the **length of a path** is the total number of edges used in the path, including repetitions, if any. The lengths of the first two paths we gave (with edges labeled) are 3 and 4, respectively. There are many more possible paths starting at $V1$ and ending at $V3$. Some of these paths might pass through the same vertex more than once. Edges may also be used repeatedly.

Of course, paths in digraphs must use directed edges only in the direction indicated. For example, consider the digraph in Figure 2.20. We will only bother to label the edges in cases where there is more than one edge in the same direction between two particular vertices, such as the two edges from $V2$ to $V4$. These labels are needed so that we can specify which edge is being used in a particular path.

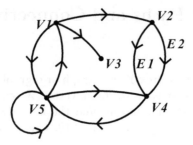

Figure 2.20: Sample Digraph with Five Vertices

There are many different paths from $V1$ to $V5$. Some of them are:

$$V1 \longrightarrow V5,$$
$$V1 \longrightarrow V5 \longrightarrow V4 \longrightarrow V5,$$
$$V1 \longrightarrow V2 \overset{E1}{\longrightarrow} V4 \longrightarrow V5 \longrightarrow V5, \quad \text{and}$$
$$V1 \longrightarrow V2 \overset{E2}{\longrightarrow} V4 \longrightarrow V5 \longrightarrow V5.$$

The lengths of these paths are 1, 3, 4, and 4, respectively.

Notice from Figure 2.20 that we can not have any paths at all starting at $V3$, since there are no directed edges leaving that vertex.

Number of Paths of a Given Length

Given a graph or digraph, our goal is to illustrate a method for computing the number of paths of a particular length going from one vertex to another vertex. We start with an example. Consider the digraph we just studied in Figure 2.20. The adjacency matrix for this digraph is

$$\mathbf{A} = \begin{array}{c} \\ V1 \\ V2 \\ V3 \\ V4 \\ V5 \end{array} \begin{array}{c} \begin{array}{ccccc} V1 & V2 & V3 & V4 & V5 \end{array} \\ \left[\begin{array}{ccccc} 0 & 1 & 1 & 0 & 1 \\ 0 & 0 & 0 & 2 & 0 \\ 0 & 0 & 0 & 0 & 0 \\ 0 & 0 & 0 & 0 & 1 \\ 1 & 0 & 0 & 1 & 1 \end{array} \right] \end{array}.$$

Clearly, the (i, j) entry of $\mathbf{A}$ gives the number of paths of length 1 from Vi to Vj. After all, paths of length 1 are comprised of just a single edge, and the number of such edges is the (i, j) entry of $\mathbf{A}$ by the definition of the adjacency matrix. For example, since the $(2, 4)$ entry of $\mathbf{A}$ equals 2, there are two paths of length 1 from $V2$ to $V4$ (that is, there are two edges from $V2$ to $V4$).

Next, we consider paths of length 2. How many paths of length 2 are there, for example, from $V1$ to $V4$? After studying the diagram in Figure 2.20, we can see that there are three possible paths of length 2 from $V1$ to $V4$. These are:

$$V1 \longrightarrow V2 \xrightarrow{E1} V4,$$

$$V1 \longrightarrow V2 \xrightarrow{E2} V4, \quad \text{and}$$

$$V1 \longrightarrow V5 \longrightarrow V4.$$

But we can imagine how easy it would be to overlook one of the paths if we had a much greater number of vertices and/or edges. However, there is a way to use the adjacency matrix to determine the correct number of paths.

We illustrate this method by working out exactly how many paths of length 2 there are from $V1$ to $V4$. Now, such a path would have to have the form

$$V1 \longrightarrow \text{---} \longrightarrow V4.$$

There are five possible choices for the vertex in the middle. If we place $V1$ in the middle, then we are looking for possible paths of the form $V1 \longrightarrow V1 \longrightarrow V4$. Thus, we need to find the number of $V1 \longrightarrow V1$ edges as well as the number of $V1 \longrightarrow V4$ edges. Now, in mathematics, whenever we have two choices to be made, one following the other, the total number of possible choices is the product of the ways that each individual choice can be made. This means we must calculate

(# of $V1 \longrightarrow V1$ edges) $\times$ (# of $V1 \longrightarrow V4$ edges) $= a_{11}a_{14} = (0)(0) = 0.$

Since this answer is 0, there are actually no paths of the form $V1 \longrightarrow V1 \longrightarrow V4$.

Continuing on, if we next place $V2$ in the middle, then we are looking for possible paths of the form $V1 \longrightarrow V2 \longrightarrow V4$. This means we must calculate

(# of $V1 \longrightarrow V2$ edges) $\times$ (# of $V2 \longrightarrow V4$ edges) $= a_{12}a_{24} = (1)(2) = 2.$

Since this answer is 2, there are two paths of the form $V1 \longrightarrow V2 \longrightarrow V4$. In fact, these are the first two paths of length 2 that we found earlier.

Similarly, if we place each of $V3$, $V4$, and $V5$ in the middle, we obtain:

(# of $V1 \longrightarrow V3$ edges) $\times$ (# of $V3 \longrightarrow V4$ edges) $= a_{13}a_{34} = (1)(0) = 0,$

(# of $V1 \longrightarrow V4$ edges) $\times$ (# of $V4 \longrightarrow V4$ edges) $= a_{14}a_{44} = (0)(0) = 0,$

(# of $V1 \longrightarrow V5$ edges) $\times$ (# of $V5 \longrightarrow V4$ edges) $= a_{15}a_{54} = (1)(1) = 1.$

That is, there is only one other path of length 2 from $V1$ to $V4$, and in fact, this is the path $V1 \longrightarrow V5 \longrightarrow V4$ that we found earlier.

We see now that to get the total number of paths of length 2 from $V1$ to $V4$, we need to add all of these products:

$$a_{11}a_{14} + a_{12}a_{24} + a_{13}a_{34} + a_{14}a_{44} + a_{15}a_{54} = 0 + 2 + 0 + 0 + 1 = 3.$$

But the expression for this sum should look familiar! It is actually what we get when we multiply each entry of the 1st row of $\mathbf{A}$ by its corresponding entry in the 4th column of $\mathbf{A}$. In other words, it is the $(1,4)$ entry of $\mathbf{A}^2$! That is, the

$$(1,4) \text{ entry of } \mathbf{A}^2 = a_{11}a_{14} + a_{12}a_{24} + a_{13}a_{34} + a_{14}a_{44} + a_{15}a_{54}.$$

We can verify this by calculating $\mathbf{A}^2$ and checking that its $(1,4)$ entry does, in fact, equal 3:

$$\mathbf{A}^2 = \begin{array}{c} \\ V1 \\ V2 \\ V3 \\ V4 \\ V5 \end{array} \begin{array}{ccccc} V1 & V2 & V3 & V4 & V5 \\ \left[\begin{array}{ccccc} 1 & 0 & 0 & 3 & 1 \\ 0 & 0 & 0 & 0 & 2 \\ 0 & 0 & 0 & 0 & 0 \\ 1 & 0 & 0 & 1 & 1 \\ 1 & 1 & 1 & 1 & 3 \end{array} \right] \end{array}.$$

Now, a similar analysis shows that, in general, the (i,j) entry of $\mathbf{A}^2$ is the total number of paths of length 2 from Vi to Vj for every possible value of i and j. For example, there is only one path of length 2 from $V5$ to $V4$ because the $(5,4)$ entry of $\mathbf{A}^2$ equals 1. (This path is $V5 \longrightarrow V5 \longrightarrow V4$.) Also, there are exactly three paths of length 2 from $V5$ to $V5$ because the $(5,5)$ entry of $\mathbf{A}^2$ equals 3. (Can you find these three paths?)

We can extend this reasoning to longer paths. Using a similar argument, we can show that the (i,j) entry of $\mathbf{A}^3$ gives the total number of paths of length 3 from Vi to Vj. We first calculate

$$\mathbf{A}^3 = \begin{array}{c} \\ V1 \\ V2 \\ V3 \\ V4 \\ V5 \end{array} \begin{array}{ccccc} V1 & V2 & V3 & V4 & V5 \\ \left[\begin{array}{ccccc} 1 & 1 & 1 & 1 & 5 \\ 2 & 0 & 0 & 2 & 2 \\ 0 & 0 & 0 & 0 & 0 \\ 1 & 1 & 1 & 1 & 3 \\ 3 & 1 & 1 & 5 & 5 \end{array} \right] \end{array}.$$

Then, for example, there is only one path of length 3 from $V1$ to $V4$, because the $(1,4)$ entry of $\mathbf{A}^3$ equals 1. (This path is $V1 \longrightarrow V5 \longrightarrow V5 \longrightarrow V4$.) Also, there are five paths of length 3 from $V5$ to $V4$ since the $(5,4)$ entry of $\mathbf{A}^3$ equals 5. (One of them is $V5 \longrightarrow V4 \longrightarrow V5 \longrightarrow V4$. Can you find the other four? Try it!)

In general, we have the following theorem:

Theorem 1. *If $\mathbf{A}$ is the adjacency matrix for a graph or digraph, and if k is a positive integer, then the (i,j) entry of $\mathbf{A}^k$ equals the number of paths of length k from vertex Vi to vertex Vj.*

Number of Paths Up to a Given Length

For a given graph or digraph having adjacency matrix $\mathbf{A}$, we can also compute the total number of paths between two vertices *up to* a given length by adding together powers of **A**. Now we know that the entries of **A** give us the number of paths of length 1, while the entries of $\mathbf{A}^2$ and $\mathbf{A}^3$ give us the number of paths of lengths 2 and 3, respectively. Therefore, if we add the matrices **A**, $\mathbf{A}^2$, and $\mathbf{A}^3$ together, we obtain the total number of paths of length 1, 2, or 3; that is, the total number of paths *up to* length 3.

In particular, for the digraph in Figure 2.20, we can calculate

$$
\mathbf{A} + \mathbf{A}^2 + \mathbf{A}^3 =
\begin{array}{c}
\\ V1 \\ V2 \\ V3 \\ V4 \\ V5
\end{array}
\begin{array}{c}
\begin{array}{ccccc} V1 & V2 & V3 & V4 & V5 \end{array} \\
\left[
\begin{array}{ccccc}
2 & 2 & 2 & 4 & 7 \\
2 & 0 & 0 & 4 & 4 \\
0 & 0 & 0 & 0 & 0 \\
2 & 1 & 1 & 2 & 5 \\
5 & 2 & 2 & 7 & 9
\end{array}
\right].
\end{array}
$$

From this sum, we find that the $(1,5)$ entry equals 7, so there are seven total paths of length 1, 2, or 3 from $V1$ to $V5$. Similarly, there are no paths of length 3 or less from $V2$ to either $V2$ or $V3$, and, as expected, there are no paths of length 3 or less leaving $V3$.

By looking at various powers of **A**, we can also see that the *shortest* possible path from $V4$ to $V2$ has length 3, because $\mathbf{A}^3$ is the lowest power of **A** that has a nonzero number as its $(4,2)$ entry, since the $(4,2)$ entries of both **A** and $\mathbf{A}^2$ are zero.

Example 1. Consider the BFF graph from Example 3 in Section 2.1. We reproduce Figure 2.7 for this graph in Figure 2.21. Now, *Facespace* allows

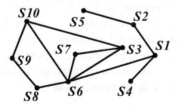

Figure 2.21: BFF Relationships Among Ten Students

several different security settings. One such setting allows a given posting to be seen by both BFFs and BFFs of BFFs. Therefore, the people able to view such a posting are those connected to the sender by a path of either length 1 or length 2.

The adjacency matrix $\mathbf{A}$ for the BFF graph is:

	S1	S2	S3	S4	S5	S6	S7	S8	S9	S10
S1	0	1	0	1	0	1	0	0	0	0
S2	1	0	0	0	1	0	0	0	0	0
S3	0	0	0	0	0	1	1	0	0	1
S4	1	0	0	0	0	0	0	0	0	0
S5	0	1	0	0	0	0	0	0	0	0
S6	1	0	1	0	0	0	1	1	0	1
S7	0	0	1	0	0	1	0	0	0	0
S8	0	0	0	0	0	1	0	0	1	0
S9	0	0	0	0	0	0	0	1	0	1
S10	0	0	1	0	0	1	0	0	1	0

$\mathbf{A} =$ (the matrix above).

After some computation, we find that

	S1	S2	S3	S4	S5	S6	S7	S8	S9	S10
S1	3	1	1	1	1	1	1	1	0	1
S2	1	2	0	1	1	1	0	0	0	0
S3	1	0	3	0	0	3	2	1	1	2
S4	1	1	0	1	0	1	0	0	0	0
S5	1	1	0	0	1	0	0	0	0	0
S6	1	1	3	1	0	5	2	1	2	2
S7	1	0	2	0	0	2	2	1	0	2
S8	1	0	1	0	0	1	1	2	1	2
S9	0	0	1	0	0	2	0	1	2	1
S10	1	0	2	0	0	2	2	2	1	3

$\mathbf{A} + \mathbf{A}^2 =$ (the matrix above).

Any given student can tell which other students can view his or her postings under this security setting by noting which entries in $\mathbf{A} + \mathbf{A}^2$ are nonzero. For example, $S3$'s information cannot be viewed by $S2$, $S4$, or $S5$, because the $(2,3)$, $(4,3)$, and $(5,3)$ entries in $\mathbf{A} + \mathbf{A}^2$ are zero. But, since the rest of column 3 of $\mathbf{A} + \mathbf{A}^2$ is nonzero, everyone else can see $S3$'s postings.

Of course, computing $\mathbf{A} + \mathbf{A}^2 + \mathbf{A}^3$ would provide an analysis of the security setting where BFFs of BFFs of BFFs can read postings. In a similar way, we could perform an analysis of the security settings for the digraph for the POI relationship from Example 4 in Section 2.1. ∎

Example 2. It is widely believed that everyone on Earth is linked by at most six degrees of separation. That is, given two people on Earth, there is a chain (that is, a path) linking them together having at most five other people between them, where each of these people knows both the person behind and ahead of them in the chain. In graph theory terms, if we were to set up a gigantic graph with one vertex for each person, and an edge connecting each pair of people that are acquainted, then this theory says that any pair of distinct vertices of this graph are connected with a path of length at most 6. Therefore, if $\mathbf{A}$ is the (huge) adjacency matrix for this graph, then

$\mathbf{A}+\mathbf{A}^2+\mathbf{A}^3+\mathbf{A}^4+\mathbf{A}^5+\mathbf{A}^6$ would have no zero entries at all! (Notice that the main diagonal entries of this sum are also not zero, because everyone knows at least one other person, and so there is also a path of length 2 starting at each vertex and returning to that vertex.) ■

Connectivity of Graphs

A graph is **connected** if every vertex can be reached from every other distinct vertex by a path of some length. For small graphs, we can easily tell whether or not the graph is connected just by looking at the drawing. In Figure 2.22,

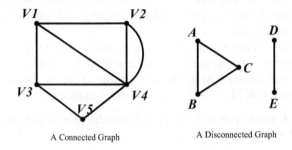

Figure 2.22: Sample Connected and Disconnected Graphs

the graph on the left is clearly connected; that is, we can get from any one vertex to any other vertex using a path of some length. However, the graph on the right is **disconnected** because, for example, there is no path from vertex A to vertex D.

For larger graphs whose diagrams may be more complicated, it is not always so obvious whether or not the graph is connected. However, we can use the adjacency matrix to analyze the connectivity of graphs.

Suppose a graph has n vertices. Notice that if there is a path between two vertices Vi and Vj and there is a repeated vertex Vk in the path, then the path has the form

$$Vi \longrightarrow \ldots \longrightarrow Vk \longrightarrow \ldots \longrightarrow Vk \longrightarrow \ldots \longrightarrow Vj.$$

But in such a case we can find a shorter path from Vi to Vj by eliminating all of the vertices between the Vk's as well as one of the Vk's. In other words, if there is a path between two distinct vertices, then we can find a path between those same vertices that contains *no repeated* vertices. But a path with no repeated vertices has at most $n-1$ edges since we have only n vertices to begin with. This means that we only have to consider paths of length $n - 1$ or less to determine whether there is a path from any vertex to every other vertex. We can summarize this by saying that a graph with n vertices is connected if there is a path of length $n - 1$ or less between every pair of distinct vertices.

Now, for each pair of distinct vertices, we could check the entries of $\mathbf{A}$, $\mathbf{A}^2$, $\mathbf{A}^3$, ..., and $\mathbf{A}^{n-1}$ in turn to see whether there are any paths of length

1, 2, 3, ..., $n-1$ between those vertices. However, this involves a lot of work, especially if n is a large number. Instead, there is a simpler way to determine whether a given graph is connected or not.

Assume we have a graph with n vertices. Suppose we add a loop to every vertex in the graph (whether the vertex already has a loop or not). The new adjacency matrix is then $\mathbf{I}_n + \mathbf{A}$. Notice that whenever there is a path of any length between two vertices in the *original* graph, there is also a path with one more edge between those same two vertices in the *new* graph, because we can add a loop to the beginning (or end) of the original path. Now we can continue to add loops to the beginning (or end) of each path. Therefore, there is a path of length 1 or 2 or 3 or ... $n-1$ between two vertices in the *original* graph, if and only if there is also a path of length $n-1$ between those same vertices in the *new* graph.

In other words, to determine whether a graph is connected, we first add a loop at every vertex, and then check whether there is a path of length $n-1$ between every pair of vertices in the new graph. That is, we need to check whether every entry of $(\mathbf{I}_n + \mathbf{A})^{n-1}$ is a positive number.

Theorem 2. *A graph with n vertices is connected if and only if every entry of $(\mathbf{I}_n + \mathbf{A})^{n-1}$ is a positive integer.*

The following example illustrates this theorem.

Example 3. Consider the connected graph on the left in Figure 2.22. The adjacency matrix for this graph is

$$\mathbf{A} = \begin{array}{c} \\ V1 \\ V2 \\ V3 \\ V4 \\ V5 \end{array} \begin{array}{ccccc} V1 & V2 & V3 & V4 & V5 \\ \left[\begin{array}{ccccc} 0 & 1 & 1 & 1 & 0 \\ 1 & 0 & 0 & 2 & 0 \\ 1 & 0 & 0 & 1 & 1 \\ 1 & 2 & 1 & 0 & 1 \\ 0 & 0 & 1 & 1 & 0 \end{array}\right] \end{array}.$$

Then,

$$(\mathbf{I}_5 + \mathbf{A}) = \begin{array}{c} \\ V1 \\ V2 \\ V3 \\ V4 \\ V5 \end{array} \begin{array}{ccccc} V1 & V2 & V3 & V4 & V5 \\ \left[\begin{array}{ccccc} 1 & 1 & 1 & 1 & 0 \\ 1 & 1 & 0 & 2 & 0 \\ 1 & 0 & 1 & 1 & 1 \\ 1 & 2 & 1 & 1 & 1 \\ 0 & 0 & 1 & 1 & 1 \end{array}\right] \end{array}.$$

Because the graph has five vertices, we raise $(\mathbf{I}_5 + \mathbf{A})$ to the 4th power:

$$(\mathbf{I}_5 + \mathbf{A})^4 = \begin{array}{c} \\ V1 \\ V2 \\ V3 \\ V4 \\ V5 \end{array} \begin{array}{ccccc} V1 & V2 & V3 & V4 & V5 \\ \left[\begin{array}{ccccc} 70 & 78 & 62 & 98 & 46 \\ 78 & 90 & 68 & 108 & 50 \\ 62 & 68 & 59 & 87 & 45 \\ 98 & 108 & 87 & 139 & 65 \\ 46 & 50 & 45 & 65 & 35 \end{array}\right] \end{array}.$$

Because none of the entries of $(\mathbf{I}_5 + \mathbf{A})^4$ equal zero, the graph is connected.

Similarly, consider the disconnected graph on the right in Figure 2.22. The adjacency matrix for this graph is

$$
\mathbf{G} =
\begin{array}{c}
 \\
A \\
B \\
C \\
D \\
E
\end{array}
\begin{array}{c}
A\ B\ C\ D\ E \\
\left[
\begin{array}{ccccc}
0 & 1 & 1 & 0 & 0 \\
1 & 0 & 1 & 0 & 0 \\
1 & 1 & 0 & 0 & 0 \\
0 & 0 & 0 & 0 & 1 \\
0 & 0 & 0 & 1 & 0
\end{array}
\right]
\end{array},
\text{ and } (\mathbf{I}_5 + \mathbf{G}) =
\begin{array}{c}
 \\
A \\
B \\
C \\
D \\
E
\end{array}
\begin{array}{c}
A\ B\ C\ D\ E \\
\left[
\begin{array}{ccccc}
1 & 1 & 1 & 0 & 0 \\
1 & 1 & 1 & 0 & 0 \\
1 & 1 & 1 & 0 & 0 \\
0 & 0 & 0 & 1 & 1 \\
0 & 0 & 0 & 1 & 1
\end{array}
\right]
\end{array}.
$$

Then

$$
(\mathbf{I}_5 + \mathbf{G})^4 =
\begin{array}{c}
 \\
A \\
B \\
C \\
D \\
E
\end{array}
\begin{array}{c}
A\quad B\quad C\ D\ E \\
\left[
\begin{array}{ccccc}
27 & 27 & 27 & 0 & 0 \\
27 & 27 & 27 & 0 & 0 \\
27 & 27 & 27 & 0 & 0 \\
0 & 0 & 0 & 8 & 8 \\
0 & 0 & 0 & 8 & 8
\end{array}
\right]
\end{array}.
$$

Since $(\mathbf{I}_5 + \mathbf{G})^4$ has at least one zero entry, the corresponding graph is disconnected. Notice that the pattern of zeroes in the matrix $(\mathbf{I}_5 + \mathbf{G})^4$ shows that no paths exist from any of the vertices in $\{A, B, C\}$ to the vertices in $\{D, E\}$. ∎

We will not study the connectivity of digraphs in this textbook, since the situation there is much more complicated – too advanced for this brief introduction to graph theory. Two vertices might be connected by a path in one direction, but not the other, and diagrams that might look connected are not, because the directions of the edges are "uncooperative."

Glossary

- Connected graph: A graph in which every vertex can be reached from every other vertex by a path of some length.

- Disconnected graph: A graph that is not connected. That is, there is at least one pair of distinct vertices in the graph having no path of any length connecting them.

- Length of a path: The number of edges used in the path, including repetitions, if any.

- Path: A path from vertex X to vertex Y in a graph or digraph is a finite sequence of edges such that the first edge starts at vertex X, each edge (except the last) ends where the next edge begins, and the last edge ends at vertex Y. For a path in a digraph, directed edges can only be used in their indicated direction.

Exercises for Section 2.3

1. Consider the graph in Figure 2.23.

   (a) Find the adjacency matrix **A** for the graph.

   (b) Use the matrix **A** to determine the number of paths of length 3 from $V1$ to $V3$.

   (c) Write out all of the actual paths of length 3 from $V1$ to $V3$.

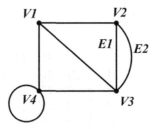

Figure 2.23: Graph for Exercise 1

2. Consider the graph in Figure 2.24.

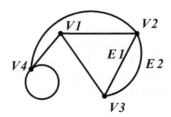

Figure 2.24: Graph for Exercise 2

   (a) Find the adjacency matrix **A** for the graph.

   (b) Use the matrix **A** to determine the number of paths of length 3 from $V2$ to $V4$.

   (c) Write out all of the actual paths of length 3 from $V2$ to $V4$.

3. Consider the digraph in Figure 2.25.

   (a) Find the adjacency matrix **A** for the digraph.

   (b) Use the matrix **A** to determine the number of paths of length 4 from $V2$ to $V3$.

   (c) Write out all of the actual paths of length 4 from $V2$ to $V3$.

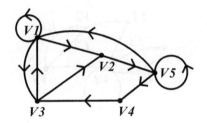

Figure 2.25: Digraph for Exercise 3

4. Consider the digraph in Figure 2.26.

   (a) Find the adjacency matrix **A** for the digraph.

   (b) Use the matrix **A** to determine the number of paths of length 4 from $V3$ to $V2$.

   (c) Write out all of the actual paths of length 4 from $V3$ to $V2$.

5. Consider the graph in Figure 2.27.

   (a) Determine the number of paths of length 7 from $V1$ to $V3$.

   (b) Find the number of paths that have length 7 or less from $V1$ to $V3$.

   (c) Use computations with the adjacency matrix to calculate the length of the shortest path from $V2$ to $V5$.

6. Consider the graph in Figure 2.28.

   (a) Determine the number of paths of length 7 from $V2$ to $V4$.

   (b) Calculate the number of paths that have length 7 or less from $V2$ to $V4$.

   (c) Use computations with the adjacency matrix to calculate the length of the shortest path from $V1$ to $V6$.

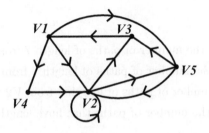

Figure 2.26: Digraph for Exercise 4

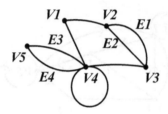

Figure 2.27: Graph for Exercise 5

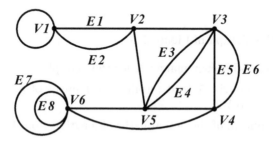

Figure 2.28: Graph for Exercise 6

7. Consider the digraph in Figure 2.29.

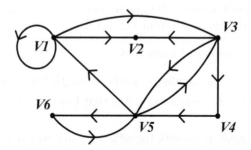

Figure 2.29: Digraph for Exercise 7

(a) Determine the number of paths of length 7 from $V1$ to $V4$.

(b) Compute the number of paths of length 7 from $V4$ to $V1$.

(c) Find the number of paths of length 4 from $V2$ to $V5$.

(d) Calculate the number of paths that have length 7 or less from $V1$ to $V3$.

(e) Use computations with the adjacency matrix to find the length of the shortest path from $V4$ to $V2$.

(f) Ascertain the length of the shortest path from $V5$ back to $V5$.

8. Consider the digraph in Figure 2.30.

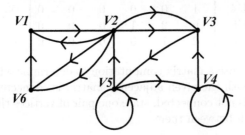

Figure 2.30: Digraph for Exercise 8

(a) Determine the number of paths of length 7 from $V2$ to $V4$.

(b) Compute the number of paths of length 7 from $V4$ to $V2$.

(c) Ascertain the number of paths that have length 7 or less from $V4$ to $V2$.

(d) Find the number of paths of length 4 from $V6$ back to $V6$.

(e) Use computations with the adjacency matrix to calculate the length of the shortest path from $V4$ to $V6$.

(f) Use computations with the adjacency matrix to find the length of the shortest path from $V3$ back to $V3$.

9. In each part, use a matrix computation to determine whether the graph associated with the given adjacency matrix is connected. If the associated graph is not connected, state one pair of vertices that have no path of any length between them.

(a) $\mathbf{A} = $

	$V1$	$V2$	$V3$	$V4$
$V1$	1	0	1	0
$V2$	0	1	0	0
$V3$	1	0	2	1
$V4$	0	0	1	1

(b) $\mathbf{B} = $

	$V1$	$V2$	$V3$	$V4$	$V5$
$V1$	2	0	1	0	1
$V2$	0	2	0	1	0
$V3$	1	0	1	0	2
$V4$	0	1	0	2	0
$V5$	1	0	2	0	1

	V1	V2	V3	V4	V5	V6
V1	0	0	0	1	0	2
V2	0	0	0	0	2	1
(c) $\mathbf{C} =$ V3	0	0	0	0	1	0
V4	1	0	0	0	0	0
V5	0	2	1	0	0	0
V6	2	1	0	0	0	0

10. In each part, use a matrix computation to determine whether the graph associated with the given adjacency matrix is connected. If the associated graph is not connected, state one pair of vertices that have no path of any length between them.

	V1	V2	V3	V4
V1	0	1	1	0
(a) $\mathbf{A} =$ V2	1	0	1	1
V3	1	1	0	1
V4	0	1	1	0

	V1	V2	V3	V4	V5
V1	0	0	2	0	1
V2	0	0	0	0	1
(b) $\mathbf{B} =$ V3	2	0	0	1	0
V4	0	0	1	0	0
V5	1	1	0	0	0

	V1	V2	V3	V4	V5	V6
V1	1	1	0	2	0	1
V2	1	0	0	0	0	0
(c) $\mathbf{C} =$ V3	0	0	0	0	1	0
V4	2	0	0	0	0	1
V5	0	0	1	0	2	0
V6	1	0	0	1	0	0

11. In the parts of Exercise 9 for which the graph is connected, which two distinct vertices are farthest apart from each other? (Hint: If $\mathbf{A}$ is the associated adjacency matrix, examine the largest power of $(\mathbf{I} + \mathbf{A})$ that contains zero entries.)

12. In the parts of Exercise 10 for which the graph is connected, which two distinct vertices are farthest apart from each other? (Hint: If $\mathbf{A}$ is the associated adjacency matrix, examine the largest power of $(\mathbf{I} + \mathbf{A})$ that contains zero entries.)

13. Consider the huge "acquaintance" adjacency matrix $\mathbf{A}$ described in Example 2. According to the theory that the degree of separation between any two people is no larger than six, what is the lowest power of $(\mathbf{I} + \mathbf{A})$ that we should expect to have all nonzero entries? Why?

14. Suppose **A** is the adjacency matrix for a graph having n vertices, where n is some positive *even* number. Also, suppose that the following path exists in the graph:

$$V1 \longrightarrow V2 \longrightarrow V3 \longrightarrow \cdots \longrightarrow Vn \longrightarrow V1.$$

Explain why $(\mathbf{I} + \mathbf{A})^{\frac{n}{2}}$ has no zero entries. (Hint: Explain why there is a path of length $\frac{n}{2}$ or less between any pair of vertices.) (Note: Any path, such as the one in this problem, which begins and ends at the same vertex, and visits every other vertex in the graph exactly once, is called a **Hamiltonian circuit**.)

2.4 Chapter 2 Test

1. Joan gave her phone number to Jack, Janet, and Jason. Jack gave his phone number to Janet, Jennifer, Jason, and Justin. Janet told her phone number to Joan, Jason, Jennifer, and Justin. Jason distributed his phone number to Joan, Jack, Janet, Jennifer, and Justin. Jennifer only told her phone number to Janet. Justin gave his phone number to Joan, Jennifer, and Janet. Of course, everyone knows his or her own phone number. Would a graph or a digraph be more appropriate to describe this given information? Draw such a graph or digraph.

2. The graph in Figure 2.31 shows various transportation connections between six towns in Alaska. ("B" edges are by boat, "P" edges are by plane, and "R" edges are by road.) Give the adjacency matrix for this graph.

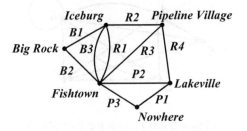

Figure 2.31: Graph for Exercise 2

3. A trucker is in the town of Fishtown, Alaska with his big rig. (See Exercise 2 and Figure 2.31.) He can travel only on roads.

   (a) What is the adjacency matrix for the graph that represents all six towns, but only includes roads as edges?

   (b) Use the matrix from part (a) to show that the graph using only roads is disconnected.

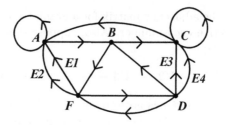

Figure 2.32: Digraph for Exercise 4

4. Consider the digraph in Figure 2.32.

   (a) Find the number of paths of length 6 from vertex B to vertex D.

   (b) Compute the number of paths of length 6 or less from vertex C to vertex B.

   (c) Determine the number of paths of length 4 from vertex D back to vertex D.

   (d) List all paths of length 4 from vertex B to vertex A. Make sure that you have them all!

   (e) Use a computation with the adjacency matrix for the digraph to find the length of the shortest path from vertex A to vertex D.

5. Redraw the graph in Figure 2.33 without having paths crossing over each other.

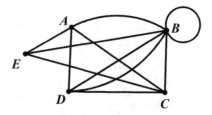

Figure 2.33: Graph for Exercise 5

6. Consider the following adjacency matrix:

$$\mathbf{A} = \begin{bmatrix} 0 & 0 & 0 & 1 & 0 & 0 \\ 0 & 1 & 1 & 0 & 0 & 1 \\ 0 & 1 & 0 & 0 & 1 & 0 \\ 1 & 0 & 0 & 1 & 0 & 0 \\ 0 & 0 & 1 & 0 & 1 & 0 \\ 0 & 1 & 0 & 0 & 0 & 1 \end{bmatrix}.$$

(a) Show that the graph corresponding to this matrix is disconnected.

(b) Use your computation in part (a) to split the set of vertices of the graph corresponding to $\mathbf{A}$ into two subsets of vertices so that there is no path from any vertex in the first subset to any vertex in the second subset, but so that there is a path between any two distinct vertices in the same subset.

7. Draw a graph having seven vertices such that, if $\mathbf{A}$ is the adjacency matrix for the graph, then $(\mathbf{I}_7 + \mathbf{A})^6$ has no zero entries, but $(\mathbf{I}_7 + \mathbf{A})^5$ has at least one zero entry.

8. In each part, indicate whether the given statement is true or whether it is false.

(a) Every edge in a graph connects exactly two, not necessarily distinct, vertices.

(b) Every vertex in a graph must be associated with at least one edge.

(c) Every graph can be drawn in a plane so that none of the edges cross.

(d) The adjacency matrix for a graph must be square.

(e) The adjacency matrix for a graph must be symmetric.

(f) The adjacency matrix for a digraph must be skew-symmetric.

(g) The complete graph with seven vertices is connected.

(h) A graph that has no loops can be thought of as a digraph having twice as many edges.

(i) There are ten (not necessarily distinct) vertices visited in a path of length 10.

(j) Given two distinct vertices in a connected graph having n vertices, the shortest path connecting the two vertices must have length $(n-1)$.

(k) Given two distinct vertices in a connected graph having n vertices, the shortest path connecting the two vertices must have length at most $(n-1)$.

(l) Given two distinct vertices in a connected graph having n vertices, the longest path connecting the two vertices must have length $(n-1)$.

(m) Given two distinct vertices in a connected graph having n vertices, there is at least one path of length $(n-1)$ connecting the two vertices.

(n) Given two distinct vertices in a connected graph having n vertices, there is at least one path of length $(n-1)$ or less connecting the two vertices.

(o) A path that starts and ends at the same vertex is a loop.

(p) The number of paths in a graph from vertex $V1$ to vertex $V2$ is the $(1, 2)$ entry of the adjacency matrix for the graph.

(q) In a graph with adjacency matrix $\mathbf{A}$, the number of paths of length 9 from vertex Vi to vertex Vj is the (i, j) entry of $\mathbf{A}^9$.

(r) In a graph with adjacency matrix $\mathbf{A}$, the number of paths of length 1 to 8, inclusive, from vertex $V1$ back to itself is the $(1, 1)$ entry of $(\mathbf{I} + \mathbf{A})^8$.

(s) In a graph with adjacency matrix $\mathbf{A}$, the number of paths of length 1 to 8, inclusive, from vertex $V1$ back to itself is the $(1, 1)$ entry of $\mathbf{A} + \mathbf{A}^2 + \mathbf{A}^3 + \cdots + \mathbf{A}^8$.

(t) A connected graph must have a loop at each vertex.

(u) Mathematicians have used graph theory to prove that the degree of separation between any two people in the world is at most six.

(v) If $\mathbf{A}$ is the adjacency matrix for a connected graph having n vertices, then $(\mathbf{I}_n + \mathbf{A})^{n-1}$ has no zero entries.

(w) If $\mathbf{A}$ is the adjacency matrix for a connected graph having n vertices, then $(\mathbf{I}_n + \mathbf{A})^{n-2}$ has at least one zero entry.

(x) If the second column of the adjacency matrix for a digraph has every entry equal to zero, then the digraph has no edges going out of vertex $V2$.

(y) The main diagonal entries on the adjacency matrix for a graph indicate the number of loops in the graph.

(z) If $\mathbf{A}$ is the adjacency matrix for a connected graph having n vertices, then $\mathbf{A}^{n-1}$ has no zero entries.

Chapter 3

Vectors

Matrices and vectors are the two most important objects in the subject area of linear algebra. In fact, almost every application of linear algebra involves at least one of these two types of objects, and typically utilizes both of them interacting with each other in some way. We studied some fundamental properties of matrices in Chapter 1. In this chapter, we concentrate on vectors.

3.1 Vector Fundamentals

Definition of a Vector

A **vector** is a matrix that contains only a single column. For example,

$$[3], \quad \begin{bmatrix} 1 \\ 2 \end{bmatrix}, \quad \begin{bmatrix} 4 \\ 5 \\ -6 \end{bmatrix}, \text{ and } \begin{bmatrix} 8 \\ 0 \\ 9 \\ -1 \end{bmatrix}$$

are all vectors. The number of rows in the vector is called the **dimension** of the vector. The vectors given above have dimensions 1, 2, 3, and 4, respectively. In order to save space on the page, we often write vectors as rows instead of as columns, with commas between the entries. So, the four vectors we have seen so far could be written as

$$[3], \quad [1,2], \quad [4,5,-6], \quad \text{and} \quad [8,0,9,-1].$$

However, we do this only for convenience. When it is important in a particular application of vectors, or in a formal proof, we will be careful to express vectors as columns.

Even though vectors are just special types of matrices, we will use *lower case* letters, or variables, in boldface type to represent vectors, instead of the *upper case* letters used for matrices. Another difference is that we call the entries of a vector its **coordinates** instead of its entries. We still use lower case non-bold letters to represent a vector's coordinates (entries), but we will only use a single subscript, representing the row number of the coordinate. For example, we could represent a general vector of dimension n as

$$\mathbf{v} = \begin{bmatrix} v_1 \\ v_2 \\ \vdots \\ v_n \end{bmatrix}, \text{ or, equivalently, to save space, } \mathbf{v} = [v_1, v_2, \ldots, v_n].$$

Notice that we have used the ellipsis here to represent the missing coordinates. We use $\mathbf{0}_n$ to represent the n-dimensional vector having all zeroes as entries. Just $\mathbf{0}$, without the subscript, may be used if the dimension of a particular **zero vector** is understood from context.

In a particular application of vectors, all of the vectors in the problem will generally have the same dimension. Because of this we have a special symbol to represent the *set* of all vectors of a given dimension. The symbol $\mathbb{R}^n$ (pronounced "R n", as for a registered nurse) represents the set of all vectors having dimension n. For example, $\mathbb{R}^3$ represents the *set* of all vectors having three coordinates. Thus,

$$[1, 2, 3], \quad [7, 0, 2], \quad [4, 1, -5], \quad \text{and} \quad \mathbf{0}_3$$

are all vectors in this set. We could not possibly list *all* of the vectors in the set $\mathbb{R}^3$. Notice that $\mathbb{R}^3$ is just the symbol for the collection of all possible 3-dimensional vectors, but it is *not* a vector itself.

We have already seen several important vectors in Chapter 1, namely $\mathbf{e}_1$, $\mathbf{e}_2$, …. These are vectors because they are matrices having a single column each. This explains why we used lower case letters to represent them back in Chapter 1. Recall that the meaning of the symbols $\mathbf{e}_1$, $\mathbf{e}_2$, … is dependent upon the dimension of the vectors, which we determine from context. For example, in the context of $\mathbb{R}^4$, the set of all 4-dimensional vectors, we have

$$\mathbf{e}_1 = [1, 0, 0, 0], \mathbf{e}_2 = [0, 1, 0, 0], \mathbf{e}_3 = [0, 0, 1, 0], \text{ and } \mathbf{e}_4 = [0, 0, 0, 1].$$

Vectors also "inherit" many important properties from the fact that they are just special types of matrices. In particular, we have the operations of vector (matrix) addition and scalar multiplication. The eight ESP properties are all true for vectors, since they are true for all matrices. Thus, we can form linear combinations of vectors and perform basic algebra with vectors just as we did with matrices. We will see how some of the other, more advanced, properties of matrices manifest themselves with vectors later in this chapter.

In particular, the Eight Special Properties play a special role in more advanced courses in linear algebra. Sets of mathematical "objects" with addition and scalar multiplication operations on them that satisfy the Eight Special Properties are called **vector spaces**. While we will not study general vector spaces in this textbook, we note in passing that the set $\mathbb{R}^n$ (for any value of n) is a vector space. Another familiar example of a vector space is the set of all matrices of a common size.

Assignment 1: Many calculators handle matrices and vectors as two different types of objects. Determine how your calculator works with vectors. Learn how to enter vectors and save vectors, how to add them and how to perform scalar multiplication with them. Practice doing this with several vectors. Make sure you try out different dimensions as well. Then, discover what happens on your calculator if you try to multiply a matrix times a vector whose dimension equals the number of *columns* of the matrix. Is the result appropriate? If, instead, the dimension of a vector equals the number of *rows* of the matrix, is there a way to take the product of the matrix and the vector that makes sense?

Application: Markov Chains

In mathematics we regularly encounter a situation in which a given object can move back and forth among several possible "states." For instance, consider a mail carrier that could be assigned to one of three possible routes (Routes A, B, C) each day. It is convenient to create a vector that indicates the probability that the carrier is on a certain route. Suppose the probability on a given week is 40% that the carrier is on Route A, 25% that the carrier is on Route B, and 35% that the carrier is on Route C. We can express this vector as

$$\mathbf{p} = \begin{array}{c} \text{Route A} \\ \text{Route B} \\ \text{Route C} \end{array} \begin{bmatrix} .40 \\ .25 \\ .35 \end{bmatrix}.$$

Notice that the sum of the entries in the vector $\mathbf{p}$ is $1.00 = 100\%$. A vector (single column matrix) whose entries are all positive or zero (since probabilities cannot be negative) and add up to 100% is called a **probability vector**.

Now suppose we are also given the following additional information about how the carrier's route was likely to change from week to week:

- If the carrier is currently on Route A, then there is a 60% chance the carrier will remain on Route A, but a 30% chance the carrier will be switched to Route B, and a 10% chance the carrier will be switched to Route C.

- If the carrier is currently on Route B, then there is a 50% chance the carrier will remain on Route B, but a 20% chance the carrier will be switched to Route A, and a 30% chance the carrier will be switched to Route C.

- If the carrier is currently on Route C, then there is a 80% chance the carrier will remain on Route C, but a 5% chance the carrier will be switched to Route A, and a 15% chance the carrier will be switched to Route B.

Although this is a lot of information, it can be easily summarized using a 3×3 matrix $\mathbf{M}$ as follows:

		Current Week		
		Route A	Route B	Route C
	Route A	.60	.20	.05
$\mathbf{M} =$ Next Week	Route B	.30	.50	.15
	Route C	.10	.30	.80

Notice that each column of $\mathbf{M}$ is actually a probability vector. Thus, $\mathbf{M}$ is also a *stochastic* matrix!

The situation given here is a typical example of a Markov chain. A **Markov chain** is a system of objects occupying several different "states," with known (unchanging) probabilities for moving an object from one state to another. The matrix for a Markov chain gives all of the probabilities for changing states. Here, the matrix $\mathbf{M}$ determines into which route (or "state") the carrier moves from week to week. Notice that it is customary to let the (i, j) entry of this matrix represent the probability of changing from the jth state (the *current* state) to the ith state (the *next* state).[1]

For example, consider the product $\mathbf{Mp}$, which produces a new vector which we call $\mathbf{p}_1$:

		Current Week				Current Week	
		Route A	Route B	Route C			
Next Week	Route A	.60	.20	.05	.40	Route A
	Route B	.30	.50	.15	.25	Route B
	Route C	.10	.30	.80	.35	Route C

			Route A	.3075	
$=$	Next Week	Route B	.2975	$= \mathbf{p}_1.$
		Route C	.3950	

In this matrix product, the column headings for $\mathbf{M}$ agree with the row headings for $\mathbf{p}$, and "cancel each other out," and so the row headings for the product $\mathbf{p}_1 = \mathbf{Mp}$ are the same as the row headings for $\mathbf{M}$. The product $\mathbf{p}_1$ gives us a new set of probabilities for the carrier after one week has elapsed. For example, the $(2, 1)$ entry of $\mathbf{p}_1$ is found by multiplying the probability that the carrier is currently on a particular route by the probability that the carrier switches from that route to Route B, and then adding all these probabilities

---

[1] It might seem odd that the *columns* of $\mathbf{M}$ are used to represent the *current* routes and the *rows* of $\mathbf{M}$ are used to represent the *next* routes instead of the other way around, but this has been done deliberately to make the computations that follow easier.

together:

$$\underbrace{(.30)}_{\substack{\text{probability} \\ \text{carrier} \\ \text{switches} \\ \text{from} \\ \text{Route A} \\ \text{to Route B}}} \underbrace{(.40)}_{\substack{\text{probability} \\ \text{carrier is} \\ \text{currently} \\ \text{on} \\ \text{Route A}}} + \underbrace{(.50)}_{\substack{\text{probability} \\ \text{carrier} \\ \text{stays} \\ \text{on} \\ \text{Route B}}} \underbrace{(.25)}_{\substack{\text{probability} \\ \text{carrier is} \\ \text{currently} \\ \text{on} \\ \text{Route B}}} + \underbrace{(.15)}_{\substack{\text{probability} \\ \text{carrier} \\ \text{switches} \\ \text{from} \\ \text{Route C} \\ \text{to Route B}}} \underbrace{(.35)}_{\substack{\text{probability} \\ \text{carrier is} \\ \text{currently} \\ \text{on} \\ \text{Route C}}}$$

$$= .2975 = 29.75\%,$$

which represents the overall probability that the carrier is on Route B after the first week.

We can now perform a second matrix multiplication $\mathbf{p}_2 = \mathbf{Mp}_1$ to determine the probabilities that the carrier will be on a particular route after the second week.

		Current Week			**Current**	
		Route A	Route B	Route C	**Week**	
Next	Route A	.60	.20	.05	.3075	Route A
Week	Route B	.30	.50	.15	.2975	Route B
	Route C	.10	.30	.80	.3950	Route C

$$= \quad \begin{matrix} \textbf{Next} \\ \textbf{Week} \end{matrix} \quad \begin{matrix} \text{Route A} \\ \text{Route B} \\ \text{Route C} \end{matrix} \begin{bmatrix} .26375 \\ .30025 \\ .43600 \end{bmatrix} = \mathbf{p}_2.$$

Thus, after two weeks, the probabilities that the carrier will be on Routes A, B, and C, respectively, are 26.4%, 30.0%, and 43.6%. This process can be continued indefinitely, as long as the original probabilities of switching from one route to another (given in the matrix $\mathbf{M}$) do not change from week to week. It can be shown that after several more multiplications, the probabilities that the carrier will be on a particular route will eventually stabilize[2] to the "limit probabilities" shown in the vector

$$\mathbf{p}_{\text{lim}} = \begin{matrix} \text{Route A} \\ \text{Route B} \\ \text{Route C} \end{matrix} \begin{bmatrix} .2037 \\ .2778 \\ .5185 \end{bmatrix} = \begin{bmatrix} 20.37\% \\ 27.78\% \\ 51.85\% \end{bmatrix}.$$

Thus, in the long run, the carrier has the greatest chance of being assigned in a particular week to Route C, and the least chance of being assigned in a particular week to Route A.

---

[2] Although we do not prove it here, a Markov chain will always stabilize over time if all entries in the matrix for the Markov chain are positive.

Vector Geometry

Consider a plane with the usual (Cartesian) xy-coordinate system, where the coordinates of points are determined from their position relative to the x-axis and the y-axis, respectively. The center $(0,0)$ of this coordinate system is called the **origin**. We can think of a 2-dimensional vector (a vector in $\mathbb{R}^2$) geometrically as a displacement from one point in this plane to another point. For example, the vector $[2,-3]$ can be thought of as a movement of a point 2 units to the right and 3 units down. Thus, if the starting position, or **initial point**, of the vector $[2,-3]$ is $(1,4)$, then its ending position, or **terminal point**, after the displacement, is $(3,1)$. This is because $(3,1)$ is 2 units to the right and 3 units down from $(1,4)$ (see Figure 3.1).

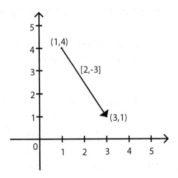

Figure 3.1: The Vector $[2,-3]$ with Initial Point $(1,4)$ and Terminal Point $(3,1)$

Notice that the same vector represents the same displacement even if a different initial point is used. For example, the vector $[2,-3]$ in Figure 3.1 could also be thought of as the displacement from initial point $(-1,2)$ to terminal point $(1,-1)$. Both representations of $[2,-3]$ are shown in Figure 3.2. Notice that the two representations of the vector $[2,-3]$ in Figure 3.2 are parallel to each other and have the same size, since they both represent the very same movement.

Also notice that when we use the origin, $(0,0)$, as the initial point of a vector, the terminal point of the vector has the same coordinates as the vector itself. For example, if the initial point of the vector $[5,-4]$ is the origin, then the terminal point of the vector is $(5,-4)$. In this sense, each point in the plane can be thought of as a 2-dimensional vector starting at the origin. For this reason, we sometimes speak of the points in the plane as being "in $\mathbb{R}^2$," and the xy-plane itself as being "$\mathbb{R}^2$." That is, we can think of the vectors in $\mathbb{R}^2$ starting at the origin as algebraically representing the points in 2-dimensional space. We will return to this idea in later sections.

Similarly, we can consider a 3-dimensional (Cartesian) xyz-coordinate system, with three mutually perpendicular axes; that is, three axes, each of which

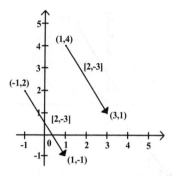

Figure 3.2: Parallel Representations of the Vector $[2, -3]$

is perpendicular to the other two. When representing such a system in two dimensions, as on a page in a book, or on a blackboard, it is common to let the y-axis and z-axis form the plane representing the page of the book, and to let the x-axis be perpendicular to this plane so that its positive direction points toward the reader and its negative direction points away from the reader. However, in order to depict all three dimensions on a flat page, the x-axis is generally "slanted" to give the illusion that it is perpendicular to the y- and z-axes. The center, $(0,0,0)$ of the 3-dimensional coordinate system is called the **origin**.

A point in a Cartesian xyz-coordinate system has three coordinates, corresponding to the x-axis, y-axis, and z-axis, respectively. When we plot a point in an xyz-coordinate system, it is often convenient to use its x- and y-coordinates to form a parallelogram, using part of the x- and y-axes as two of the sides. Then we use the z-coordinate to raise (or lower) the corner of the parallelogram that is opposite the origin. For example, in Figure 3.3, the point $(3,4,7)$ is drawn by first creating a parallelogram whose sides correspond to the values $x = 3$ and $y = 4$. Then, because $z = 7$ here, the furthest corner of that parallelogram from the origin is raised 7 units to obtain the location of the point $(3,4,7)$. However, for the point $(2,0,1)$ in Figure 3.3, the y-coordinate is 0, so no parallelogram needs to be drawn here. Since $z = 1$, we only need to raise the x-coordinate by 1 unit to obtain the location of $(2,0,1)$.

Now, 3-dimensional vectors represent movements or displacements in 3-dimensional space from an initial point to a terminal point. For example, in Figure 3.3, we see the vector $[1,4,6]$ represented as a displacement from the initial point $(2,0,1)$ to the terminal point $(3,4,7)$.

Just as we did earlier with $\mathbb{R}^2$, we can think of each point in $\mathbb{R}^3$ as a vector beginning at the origin, $(0,0,0)$, and ending at that point. For example, if the vector $[3,2,5]$ has its initial point at the origin, then its terminal point is $(3,2,5)$. Hence, we can speak of the points in space as being "in $\mathbb{R}^3$," and 3-dimensional space itself as being "$\mathbb{R}^3$." That is, we can think of the

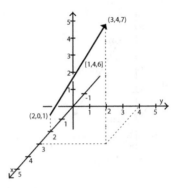

Figure 3.3: The Vector $[1, 4, 6]$ in 3-Dimensional Space

vectors in $\mathbb{R}^3$ starting at the origin as algebraically representing the points in 3-dimensional space.

Now, vectors can be formed in any dimension, not just in two and three dimensions. We can algebraically discuss vectors having four, five, or even twenty-seven dimensions. The dimension of a vector is merely the number of rows in its particular column matrix, and we can create vectors having as many rows as we want! As we will see, mathematicians (and others) use their geometric understanding of vectors in two and three dimensions to imagine the geometry of these higher dimensions, even though they are quite difficult to picture.

Assignment 2: If you have software available that can draw figures involving vectors, learn how to use it. The figures in this chapter were all drawn using The Geometer's Sketchpad, Version 5.01, but other software packages are also able to make such sketches. (The 3-dimensional drawings were much more challenging to create.) Try to draw a few simple vector diagrams in both two and three dimensions.

Glossary

- Coordinates in a vector: The entries in a vector. The ith coordinate of a vector $\mathbf{v}$ is the entry in row i of $\mathbf{v}$, usually denoted by v_i.

- Dimension of a vector: The number of rows in a vector. This also equals the number of coordinates in the vector.

- Initial point of a vector: Geometrically, the starting point for a particular displacement represented by the vector.

- Markov chain: A system of objects occupying several different "states," with known probabilities for moving an object from one state to another.

- Matrix for a Markov chain: A square matrix whose entries are the probabilities for changing states. In particular, the (i, j) entry of this matrix represents the probability of changing from the jth state (the current state) to the ith state (the next state). If $\mathbf{M}$ is the matrix for a Markov chain, and $\mathbf{p}$ is the current probability vector for the states, then $\mathbf{Mp}$ is the probability vector after a single change of states.

- Mutually perpendicular axes: A set of axes, each of which is perpendicular to the others. For example, the x-, y-, and z-axes in 3-dimensional space are mutually perpendicular.

- Origin: The point in a coordinate system all of whose coordinates equal zero.

- Probability vector: A vector (single column matrix) whose entries are all positive or zero, and add up to $100\% = 1$.

- $\mathbb{R}^2$: The set of all vectors having dimension 2.

- $\mathbb{R}^3$: The set of all vectors having dimension 3.

- $\mathbb{R}^n$: The set of all vectors having dimension n.

- Terminal point of a vector: Geometrically, the ending point for a particular displacement represented by the vector.

- Vector: A matrix having only one column.

- Zero vector: A vector in which every coordinate has the value 0. The zero vector having dimension n is denoted by $\mathbf{0}_n$. The symbol $\mathbf{0}$ is often used when the dimension is understood from context.

Exercises for Section 3.1

1. Determine, if possible, the dimension of each of the given vectors:

   (a) $[1, 2, 3]$ (c) $\mathbf{0}_4$

   (b) $[5, 8, -1, -2, 7]$ (d) $\mathbf{e}_4$

2. Determine, if possible, the dimension of each of the given vectors:

   (a) $[2, -9, 0, 0, 3]$ (c) $\mathbf{0}$

   (b) $[0, 0, 0]$ (d) $\mathbf{e}_3$

3. In each part, a vector $\mathbf{v}$ and an initial point P are given. Find the terminal point of the vector.

(a) $\mathbf{v} = [2, -3, 5]$; $P = (1, 3, -7)$

(b) $\mathbf{v} = [1, 0, -1, 0]$; $P = (0, -1, 0, 1)$

(c) $\mathbf{v} = \mathbf{0}$; $P = (3, 1, 2, -5)$

(d) $\mathbf{v} = -2\mathbf{e}_2$; $P = (3, 1, 8)$

4. In each part, a vector $\mathbf{v}$ and an initial point P are given. Find the terminal point of the vector.

(a) $\mathbf{v} = [3, 6, -7]$; $P = (2, -7, 4)$ (c) $\mathbf{v} = [2, 3, 4]$; P is the origin

(b) $\mathbf{v} = \mathbf{0}$; $P = (4, 9, -2, 3, 5)$ (d) $\mathbf{v} = 3\mathbf{e}_3$; $P = (1, -1, 1, 5)$

5. Find the vector having initial point $(2, 6, 3)$ and terminal point $(4, 3, 8)$.

6. What vector has initial point $(4, 2, -3, 7)$ and the origin as its terminal point?

7. What is the initial point of the vector $[3, 7, 5]$ if its terminal point is $(5, 2, 4)$?

8. Find the initial point of the vector $8\mathbf{e}_1$ if its terminal point is $(6, 1, -2)$.

9. In each part, draw the described vector in an appropriate Cartesian coordinate system.

(a) The vector $[3, -1]$ with initial point $(2, 4)$.

(b) The vector $[-2, -3]$ with terminal point $(3, 0)$.

(c) The vector $[2, 3, 6]$ with initial point $(1, 1, -2)$.

(d) The vector $\mathbf{e}_2$ with terminal point $(4, 2, 5)$.

10. In each part, draw the described vector in an appropriate Cartesian coordinate system.

(a) The vector $[-2, 4]$ having initial point $(5, 1)$.

(b) The vector $[-3, -1]$ having its terminal point at the origin.

(c) The vector $-4\mathbf{e}_3$ as a 3-dimensional vector having its terminal point at the origin.

(d) The vector $[1, 3, 2]$ having initial point $(2, 3, 5)$.

11. Suppose in a local park, on a certain day, the probability that a hiker chooses one of two trails is currently as follows: a 70% chance of choosing Grand Walkaround, and a 30% chance of choosing Splendid Vista. Also, suppose we are given the following additional information about how a hiker is likely to switch trails from day to day:

If a hiker currently follows Grand Walkaround, there is a 25% chance the hiker will stay with Grand Walkaround the next day, but a 75% chance the hiker will choose Splendid Vista instead.

If a hiker currently follows Splendid Vista, there is a 60% chance the hiker will stay with Splendid Vista the next day, but a 40% chance the hiker will choose Grand Walkaround instead.

Use all of this information to create a current probability vector $\mathbf{p}$ and Markov chain matrix $\mathbf{M}$ for this situation. Then, use these to calculate the probability vector $\mathbf{p}_1$ for the choice of trails for the next day. Then use $\mathbf{p}_1$ and $\mathbf{M}$ to calculate the probability vector $\mathbf{p}_2$ for the choice of trails on the following day after that. Be sure to label the rows and/or columns of your vectors and matrices appropriately.

12. Suppose a government agency flies two surveillance drones, Seeker and Finder, over a certain region each week. The current probability that Seeker is selected this week is 45%, while the probability that Finder is selected is 55%. Also, suppose we are given the following additional information about the choice of drones from week to week:

If Seeker is currently being used, there is a 40% chance that Seeker will be used the following week, but a 60% chance instead that Finder will be used the following week.

If Finder is currently being used, there is a 65% chance that Finder will be used the following week, but a 35% chance instead that Seeker will be used the following week.

Use all of this information to create a current probability vector $\mathbf{p}$ and Markov chain matrix $\mathbf{M}$ for this situation. Then, use these to calculate the probability vector $\mathbf{p}_1$ for the choice of drones for the next week. Then use $\mathbf{p}_1$ and $\mathbf{M}$ to calculate the probability vector $\mathbf{p}_2$ for the choice of drones in the following week after that. Be sure to label the rows and/or columns of your vectors and matrices appropriately.

13. Suppose in a certain month, the probability that a customer prefers three different types of cleanser is as follows: 25% currently use Brand A, 60% currently use Brand B, and 15% currently use Brand C. Also, suppose we are also given the following additional information about how a typical customer is likely to change brands from month to month:

If a customer currently uses Brand A, there is a 30% chance the customer will stay with Brand A, but a 45% chance the customer will switch to Brand B, and a 25% chance the customer will switch to Brand C.

If a customer currently uses Brand B, there is a 70% chance the customer will stay with Brand B, but a 20% chance the customer will switch to Brand A, and a 10% chance the customer will switch to Brand C.

If a customer currently uses Brand C, there is a 35% chance the customer will stay with Brand C, but a 25% chance the customer will switch to Brand A, and a 40% chance the customer will switch to Brand B.

Use all of this information to create a current probability vector **p** and Markov chain matrix **M** for this situation. Then, use these to calculate the probability vector $\mathbf{p}_1$ for cleanser use for the next month. Then use $\mathbf{p}_1$ and **M** to calculate the probability vector $\mathbf{p}_2$ for cleanser use for the following month after that. Be sure to label the rows and/or columns of your vectors and matrices appropriately.

14. Suppose that a local art museum has its collection displayed in four locations: Room A, Room B, Room C, and Room D. There are two doors between Rooms A and B, one door between Rooms A and C, two doors between Rooms A and D, three doors between Rooms B and C, one door between Rooms B and D, and no door between Rooms C and D. Suppose that a class of students is visiting the museum with orders to study the paintings in one room until a chime sounds, at which point, they are to move to an adjacent room, with each possible exit from the current room an equally likely choice. Suppose that currently 30% of the students are in Room A, 25% are in Room B, 35% are in Room C, and 10% are in Room D.

Use all of this information to create a current probability vector **p** and Markov chain matrix **M** for this situation. Then, use these to calculate the probability vector $\mathbf{p}_1$ for students to be in a particular room after the chime sounds once. Then use $\mathbf{p}_1$ and **M** to calculate the probability vector $\mathbf{p}_2$ for students to be in a particular room after the chime sounds again. Be sure to label the rows and/or columns of your vectors and matrices appropriately.

3.2 Vector Operations and Geometry

Addition of Vectors

Since vectors are just special types of matrices, we add them the same way we add matrices, coordinate by coordinate. Let us see how this works geometrically.

Suppose $\mathbf{v} = [2, 3]$ and $\mathbf{w} = [1, -2]$. Then the sum is $\mathbf{s} = \mathbf{v} + \mathbf{w} = [2, 3] + [1, -2] = [3, 1]$. If we use $(1, 2)$ as the initial point for **s**, then its terminal point is at $(4, 3)$.

Another way we can reach the same terminal point of the sum **s** is as follows: beginning at the same initial point $(1, 2)$ for **v**, we get $(3, 5)$ as the terminal point for **v**. Then using this terminal point of **v** as the initial point for **w**, we calculate that the terminal point for **w** is $(4, 3)$, the same terminal point as the sum **s**. These computations are illustrated in Figure 3.4.

We call this method of representing vector addition geometrically the "Follow Method." This method can be extended to a sum involving any number of vectors. We start with the first vector in the sum, and then place each

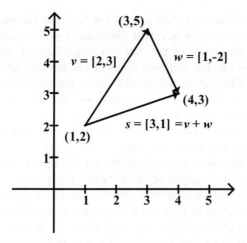

Figure 3.4: The Sum of Two Vectors **v** and **w** Using the Follow Method

subsequent vector so that it starts where the previous vector ended. That is, we reach the desired final destination by simply following the path of the arrows. The "Follow Method" also works in three dimensions, and in higher dimensions as well.

Now we know from ESP1 that vector addition is commutative. Therefore, geometrically, starting from the same initial point, we should reach the same terminal point using the Follow Method if we, instead, draw **w** first and then follow it with **v**. This is shown in Figure 3.5.

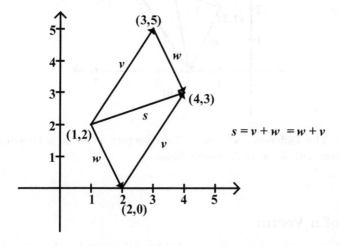

Figure 3.5: The Sum **s** of Vectors **v** and **w** as the Diagonal of a Parallelogram with **v**, **w** as Adjacent Sides

Figure 3.5 also illustrates that if **v** and **w** begin at the same initial point, then the sum **s** = **v** + **w** is a vector (beginning at the same initial point) along one of the diagonals of the parallelogram formed with **v** and **w** as adjacent sides. Notice that opposite sides of the parallelogram are the same vector, but have different initial points. Also notice that the diagonals of the parallelogram are simply line segments, so they do not have a specified direction. Therefore, this same diagonal can be thought of as a vector in the opposite direction, but in that case, this diagonal would instead be represented in vector form as − (**v** + **w**) = −**v** − **w**.

Next, consider the other diagonal of this parallelogram. Let **d** be the vector along that diagonal that begins at the terminal point of **w** (at the bottom of the parallelogram) and ends at the terminal point of **v** (at the top of the parallelogram), as shown in Figure 3.6. Now, when we calculate **w** + **d** by the Follow Method, we get **v** as the sum; that is, **w** + **d** = **v**. But this means **d** is equal to **v** − **w**. In summary, then, if **v** and **w** begin at the same initial point, then the subtraction **v** − **w** represents a vector along another diagonal of the same parallelogram, having its initial point at the terminal point of **w**, and its terminal point at the terminal point of **v**, as in Figure 3.6. Of course, if we think of this diagonal as a vector in the opposite direction, then this diagonal would be represented instead in vector form as − (**v** − **w**) = **w** − **v**.

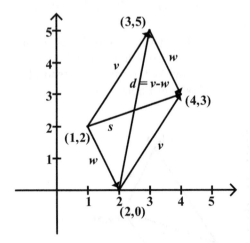

Figure 3.6: The Difference **v** − **w** of Two Vectors **v**, **w** as a Diagonal of a Parallelogram with **v**, **w** as Adjacent Sides

Length of a Vector

The **length** of a vector is the distance between the initial and terminal points of the vector. The length of **v**, written as ‖**v**‖, is also called the **magnitude** of **v**, or the **norm** of **v**. The length of a vector is computed by taking the square root of the sum of the squares of its coordinates; that is,

$$\|\mathbf{v}\| = \|[v_1, v_2, \ldots, v_n]\| = \sqrt{v_1^2 + v_2^2 + \cdots + v_n^2}.$$

For example, the length of $[6, -2, 3]$ is $\|[6, -2, 3]\| = \sqrt{6^2 + (-2)^2 + 3^2}$ $= \sqrt{36 + 4 + 9} = \sqrt{49} = 7$. Also, the norm of $[4, -5, 2, -8]$ is $\|[4, -5, 2, -8]\|$ $= \sqrt{4^2 + (-5)^2 + 2^2 + (-8)^2} = \sqrt{16 + 25 + 4 + 64} = \sqrt{109} \approx 10.44$. (The symbol "$\approx$" means "is approximately equal to.")

In two dimensions, the formula for the length of a vector can easily be derived from the Pythagorean Theorem, which states that, in a right triangle, the square of the length of the hypotenuse equals the sum of the squares of the lengths of the other two sides. This is illustrated in Figure 3.7 for a general vector $\mathbf{v} = [v_1, v_2]$. From the Pythagorean Theorem, we have $\|\mathbf{v}\|^2 = v_1^2 + v_2^2$; that is, $\|\mathbf{v}\| = \sqrt{v_1^2 + v_2^2}$.

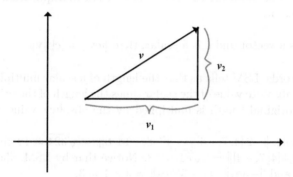

Figure 3.7: $\|\mathbf{v}\|^2 = v_1^2 + v_2^2$

A similar diagram, involving a right triangle in 3-dimensional space, can be used to derive the formula $\|\mathbf{v}\| = \sqrt{v_1^2 + v_2^2 + v_3^2}$ for the norm of $\mathbf{v} = [v_1, v_2, v_3]$ in $\mathbb{R}^3$. In the same manner, we can algebraically apply the Pythagorean Theorem to right triangles in higher-dimensional spaces to justify the formula $\|\mathbf{v}\| = \sqrt{v_1^2 + v_2^2 + \cdots + v_n^2}$ for the length of the general vector $\mathbf{v} = [v_1, v_2, v_3, \ldots, v_n]$ in $\mathbb{R}^n$.

Notice that in the formula for length, the values of all the coordinates are squared. Thus, every nonzero coordinate makes a positive contribution to the overall length of a vector, and no subtraction takes place under the square root. Therefore, in each dimension, the zero vector is the only vector having magnitude 0. That is,

LZV: $\|\mathbf{v}\| = 0$ if and only if $\mathbf{v} = \mathbf{0}$.

The use of the phrase "if and only if" here is shorthand for saying:

If $\|\mathbf{v}\| = 0$, then $\mathbf{v} = \mathbf{0}$,

and
$$\text{If } \mathbf{v} = \mathbf{0}, \text{ then } \|\mathbf{v}\| = 0.$$

In any dimension, the length of $\mathbf{e}_i$, for every i, is equal to 1, because $\sqrt{0^2 + \cdots + 0^2 + 1^2 + 0^2 + \cdots + 0^2} = \sqrt{1} = 1$. Any vector having length 1 is called a **unit vector**. Therefore, each $\mathbf{e}_i$ is an example of a unit vector.

Assignment 3: Learn how to compute the length of a vector on your calculator. Practice by finding the lengths of several different vectors in various dimensions, including at least one zero vector.

Scalar Multiplication of Vectors

Because vectors are just special types of matrices, scalar multiplication is performed simply by multiplying the value of each coordinate of the vector by the scalar. The following rule shows how scalar multiplication changes the length of a vector:

LSM: If $\mathbf{v}$ is a vector and c is a scalar, then $\|c\mathbf{v}\| = |c|\,\|\mathbf{v}\|$.

In plain words, LSM tells us that the length of a scalar multiple of a vector is merely the absolute value of the scalar times the length of the original vector. That is, the original length is multiplied by the absolute value of the scalar involved.

For example, $\|-2\,[4, 7, -4]\| = \|[-8, -14, 8]\| = \sqrt{324} = 18$ gives the same result as $|-2|\,\|[4, 7, -4]\| = 2\sqrt{81} = 18$. Notice that by LSM, $\|5\mathbf{e}_2\| = 5\,\|\mathbf{e}_2\| = 5 \times 1 = 5$, and $\|-3\mathbf{e}_4\| = |-3|\,\|\mathbf{e}_4\| = 3 \times 1 = 3$.

Now, LSM indicates how scalar multiplication alters the magnitude of a vector. But what does scalar multiplication do to the direction of a vector? The answer is simple. A *positive* scalar multiple of a nonzero vector goes in the *same* direction as the original vector. A *negative* scalar multiple of a nonzero vector points in the *opposite* direction of the original vector. (Of course, the zero scalar multiple of any vector is a zero vector, which has no direction.) Either way, nonzero scalar multiples of nonzero vectors are always parallel to the original vectors. (See Figure 3.8.)

As we can see in Figure 3.8, if the scalar c has absolute value greater than 1, then $c\mathbf{v}$ is parallel to $\mathbf{v}$, but is *longer* than $\mathbf{v}$. Such a transformation of the vector is called a **dilation**. If c has absolute value less than 1 (but nonzero), then $c\mathbf{v}$ is parallel to $\mathbf{v}$, but *shorter* than $\mathbf{v}$. This type of transformation of the vector is called a **contraction**. Finally, notice that $(-1)\mathbf{v} = -\mathbf{v}$ has the same length as $\mathbf{v}$, but goes in the opposite direction as $\mathbf{v}$.

Normalization of a Nonzero Vector

Unit vectors play a fundamental role in linear algebra. Every nonzero vector is a scalar multiple of some unit vector. For example, if a given vector $\mathbf{v}$ has length 10, then it is 10 times as large as a unit vector. Therefore, if $\mathbf{u}$ is

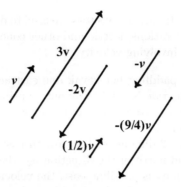

Figure 3.8: Various Positive and Negative Scalar Multiples of a Vector **v**

the unit vector in the same direction as **v**, then **v** = 10**u**. Similarly, to find a vector of length 7 in the direction of the positive x-axis in $\mathbb{R}^2$, we merely multiply $\mathbf{e}_1$ by 7, producing $7\mathbf{e}_1 = 7\,[1, 0] = [7, 0]$.

Now, given a nonzero vector **v**, how do we find the unit vector in the same direction as **v**? We know that any positive scalar multiple of **v** will be in the same direction as **v**, so we just need to find the right scalar, c, so that $\mathbf{u} = c\mathbf{v}$ has length 1. The scalar we need is $c = \frac{1}{\|\mathbf{v}\|}$; that is,

$$\mathbf{u} = \left(\frac{1}{\|\mathbf{v}\|}\right)\mathbf{v}.$$

We can check this using LSM, since

$$\|\mathbf{u}\| = \left\|\left(\frac{1}{\|\mathbf{v}\|}\right)\mathbf{v}\right\| = \left|\frac{1}{\|\mathbf{v}\|}\right| \|\mathbf{v}\| = \frac{1}{\|\mathbf{v}\|}\|\mathbf{v}\| \left(\text{because } \frac{1}{\|\mathbf{v}\|} > 0\right) = 1,$$

since the $\|\mathbf{v}\|$'s cancel. This process of multiplying a nonzero vector **v** by $\frac{1}{\|\mathbf{v}\|}$ is called **normalizing** the vector **v**, and produces the unit vector in the same direction as **v**.

For example, suppose $\mathbf{v} = [9, -12]$. Now, $\|\mathbf{v}\| = \sqrt{81 + 144} = \sqrt{225} = 15$. Then, by the normalization process, the unit vector, **u**, in the same direction as **v** is $\mathbf{u} = \left(\frac{1}{\|\mathbf{v}\|}\right)\mathbf{v} = \frac{1}{15}\,[9, -12] = \left[\frac{3}{5}, -\frac{4}{5}\right]$. Now that we have found the unit vector in in the same direction as **v**, we can easily compute a vector of any length in that direction. For example, a vector having length 55 in the same direction as **v** is just $55\mathbf{u}$, which equals $55\left[\frac{3}{5}, -\frac{4}{5}\right] = [33, -44]$. Also notice that a vector having length 4 in the *opposite* direction as **v** would simply be $(-4)\mathbf{u} = (-4)\left[\frac{3}{5}, -\frac{4}{5}\right] = \left[-\frac{12}{5}, \frac{16}{5}\right]$.

Velocity

A vector is uniquely determined once we know both its length and direction. For this reason, vectors are useful when both the magnitude and direction of a quantity are important. In physics, vectors are used to designate movements of objects, velocities, accelerations, forces, and other concepts. We will consider some specific examples involving velocity.

Example 1. Cindy is paddling her kayak due east across the bay at 3 feet per second (ft/sec). The tide is on its way out, causing the water in the bay to move 5 ft/sec due south. We will compute the overall velocity of the kayak from these two movements.

Consider a traditional 2-dimensional map with east in the direction of e_1 (the positive x-axis) and north in the direction e_2 (the positive y-axis). (See Figure 3.9.) Because Cindy is paddling east, the velocity of Cindy's paddling can be expressed by a vector in the direction of e_1, having magnitude 3. Because e_1 is a unit vector, we see that $3e_1$ represents Cindy's paddling effort in ft/sec. The water in the bay is moving due south. A unit vector pointing south (opposite of north), is $-e_2$. Thus, a vector pointing south having magnitude 5 is $5(-e_2) = -5e_2$. This represents the velocity of the water in ft/sec.

Cindy's overall velocity, v, therefore, is the sum of these two vectors; that is,

$$v = 3e_1 + (-5e_2) = [3, 0] + [0, -5] = [3, -5] \text{ ft/sec.}$$

The length of v tells us how fast Cindy is going:

$$\text{Cindy's speed } = \|v\| = \sqrt{9 + 25} = \sqrt{34} \approx 5.83 \text{ ft/sec.}$$

The direction of v is the direction in which Cindy is actually moving, despite her efforts to go due east. ∎

When two or more velocities are applied simultaneously to the same object, the sum of these velocities is called the **resultant velocity**; that is, the overall result of all the velocities involved. In Example 1, the resultant velocity $[3, -5]$ ft/sec of Cindy's kayak represents the combination of her paddling together with the flow of the tide. The magnitude of the resultant velocity is called the **resultant speed**. In Example 1, we found the resultant speed of Cindy's kayak to be about 5.83 ft/sec.

Example 2. The *Sea Lion*, a small research submarine, is travelling under water in a current that is moving in the direction $[4, -1, 2]$ at 3 km/hr, as in Figure 3.10. (We are assuming here that the positive x-axis points east, the positive y-axis points north, and the positive z-axis points upward to the surface of the water.) The *Sea Lion*'s propeller is pushing the submarine 7.2 km/hr in the direction $[2, 3, 1]$. Also, the air in the submarine's ballast tanks is making the boat float straight upward at 4.25 km/hr. We will compute the resultant velocity of the *Sea Lion*.

First, we need to find a velocity vector, v_1, for the current. To do this, we compute a unit vector, u_1, in the direction of the current, and then multiply that by 3, which is the speed of the current. The vector u_1 is calculated by

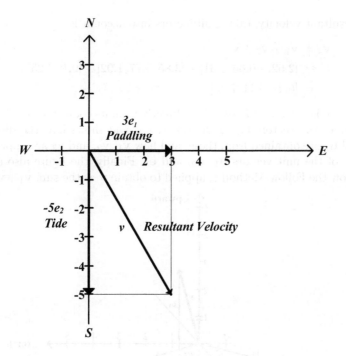

Figure 3.9: Cindy's Resultant Velocity $\mathbf{v} = 3\mathbf{e}_1 + (-5\mathbf{e}_2)$

normalizing the given direction vector $[4, -1, 2]$. That is,

$$\mathbf{u}_1 = \frac{1}{\|[4, -1, 2]\|}[4, -1, 2] = \frac{1}{\sqrt{21}}[4, -1, 2].$$

Hence,

$$\mathbf{v}_1 = 3\mathbf{u}_1 = \frac{3}{\sqrt{21}}[4, -1, 2] \approx [2.62, -0.66, 1.31].$$

Next, we determine the velocity vector, $\mathbf{v}_2$, corresponding to the action of the propeller. A unit vector, $\mathbf{u}_2$, in the direction the propeller is pushing the sub is found by normalizing the given direction vector $[2, 3, 1]$. Therefore,

$$\mathbf{u}_2 = \frac{1}{\|[2, 3, 1]\|}[2, 3, 1] = \frac{1}{\sqrt{14}}[2, 3, 1].$$

Since the speed resulting from the propeller is 7.2 km/hr, we multiply this unit vector by 7.2 to calculate the velocity vector due to the propeller:

$$\mathbf{v}_2 = 7.2\mathbf{u}_2 = \frac{7.2}{\sqrt{14}}[2, 3, 1] \approx [3.85, 5.77, 1.92].$$

The submarine is moving straight upward at a speed of 4.25 km/hr. Since $\mathbf{e}_3$ is a unit vector pointing straight up, the velocity vector due to the ballast tanks is

$$\mathbf{v}_3 = 4.25\mathbf{e}_3 = [0, 0, 4.25].$$

The resultant velocity, taking all factors into account is

$$\begin{aligned}
\mathbf{v} &= \mathbf{v}_1 + \mathbf{v}_2 + \mathbf{v}_3 \\
&\approx [2.62, -0.66, 1.31] + [3.85, 5.77, 1.92] + [0, 0, 4.25] \\
&= [6.47, 5.11, 7.48].
\end{aligned}$$

The vector $\mathbf{v}$ is plotted in Figure 3.10 directly from its coordinates. The figure also includes the vectors $[4, -1, 2]$ and $[2, 3, 1]$ and shows how the directions of $\mathbf{u}_1$ and $\mathbf{u}_2$ are obtained from these, and how $\mathbf{v}_1$, $\mathbf{v}_2$, and $\mathbf{v}_3$ are appropriate multiples of the unit vectors $\mathbf{u}_1$, $\mathbf{u}_2$, and $\mathbf{e}_3$. Finally, the figure also demonstrates how the Follow Method is applied to obtain $\mathbf{v}$ as the sum $\mathbf{v}_1 + \mathbf{v}_2 + \mathbf{v}_3$.

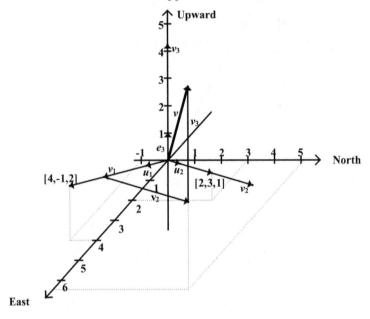

Figure 3.10: Resultant Velocity $\mathbf{v} = \mathbf{v}_1 + \mathbf{v}_2 + \mathbf{v}_3$ of the *Sea Lion*

The resultant speed of the submarine is the magnitude of the velocity vector $\mathbf{v}$, which is

$$\|\mathbf{v}\| \approx \sqrt{(6.47)^2 + (5.11)^2 + (7.48)^2} \approx \sqrt{123.92} \approx 11.13 \text{ km/hr.} \ \blacksquare$$

The Triangle Inequality

One important formula regarding vectors that is easily seen to be true geometrically is known as the Triangle Inequality.

Theorem 1. *(**Triangle Inequality**) Suppose that $\mathbf{v}$ and $\mathbf{w}$ are two n-dimensional vectors. Then,*

$$\|\mathbf{v} + \mathbf{w}\| \leq \|\mathbf{v}\| + \|\mathbf{w}\|.$$

The Triangle Inequality states that the *magnitude of a sum* is always less than or equal to the *sum of the magnitudes* of the individual vectors in the sum. Figure 3.11 contains a "picture proof" of the Triangle Inequality in $\mathbb{R}^2$. In the figure, we have used the Follow Method to illustrate the sum of the

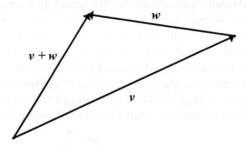

Figure 3.11: The Triangle Inequality

vectors **v** and **w**. Theorem 1 merely states that the length of the **v** + **w** side of the triangle is less than or equal to the sum of the lengths of the other two sides. This is obvious! Since the shortest distance between two points is a straight line, any side of any triangle is always shorter than the sum of the lengths of the other two sides. The only case in which the rule in Theorem 1 is an equality, that is, $\|\mathbf{v} + \mathbf{w}\| = \|\mathbf{v}\| + \|\mathbf{w}\|$, is when **v** and **w** have the same direction. In this case, the triangle "collapses," and has no interior. (See Figure 3.12.)

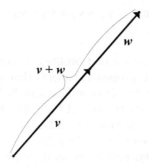

Figure 3.12: Case of the Triangle Inequality for which $\|\mathbf{v} + \mathbf{w}\| = \|\mathbf{v}\| + \|\mathbf{w}\|$

To illustrate the Triangle Inequality, recall Example 1, where we computed the resultant velocity of Cindy's kayak. There, we found that 5.83 ft/sec, the *magnitude of the sum* of the two given velocity vectors (that is, the magnitude of the resultant velocity) was indeed less than the *sum of the magnitudes* of those two velocity vectors: 3 ft/sec + 5 ft/sec = 8 ft/sec.

Theorem 1 can be generalized to a sum of many vectors:[3]

---

[3] A corollary is a theorem whose proof follows immediately from a previous theorem.

Corollary 2. *Suppose that* $\mathbf{v}_1, \ldots, \mathbf{v}_k$ *are n-dimensional vectors. Then,*

$$\|\mathbf{v}_1 + \mathbf{v}_2 + \cdots + \mathbf{v}_k\| \leq \|\mathbf{v}_1\| + \|\mathbf{v}_2\| + \cdots + \|\mathbf{v}_k\|.$$

Geometrically, Corollary 2 again says that the shortest distance between two points is a straight line. Consider the example in Figure 3.13, which shows a sum of six vectors using the Follow Method, with A as the initial point of the sum and B as the terminal point. By the Follow Method, this sum is equivalent to a single vector $\mathbf{v}$ that starts at A and ends at B. In fact, $\mathbf{v}$ represents the shortest path from A to B. We can see from the figure that $\|\mathbf{v}\|$, the magnitude of the sum $\mathbf{v}$, is less than the sum $\|\mathbf{v}_1\| + \|\mathbf{v}_2\| + \|\mathbf{v}_3\| + \|\mathbf{v}_4\| + \|\mathbf{v}_5\| + \|\mathbf{v}_6\|$ of the magnitudes of the six individual vectors $\mathbf{v}_1$ through $\mathbf{v}_6$. That is precisely what Corollary 2 asserts in this situation.

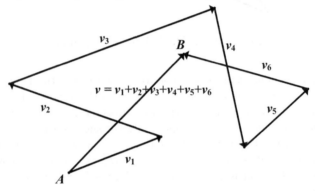

Figure 3.13: Shortest Path $\mathbf{v} = \mathbf{v}_1 + \mathbf{v}_2 + \mathbf{v}_3 + \mathbf{v}_4 + \mathbf{v}_5 + \mathbf{v}_6$ from Point A to Point B

To give another illustration of Corollary 2, consider the three velocity vectors $\mathbf{v}_1$, $\mathbf{v}_2$, $\mathbf{v}_3$ (current, propeller, ballast) that we added to calculate the resultant velocity $\mathbf{v}$ of the *Sea Lion* in Example 2. We found the magnitude $\|\mathbf{v}\|$ of the resultant velocity to be 11.13 km/hr. This is less than the sum $\|\mathbf{v}_1\| + \|\mathbf{v}_2\| + \|\mathbf{v}_3\|$ of the magnitudes: 3 km/hr + 7.2 km/hr + 4.25 km/hr = 14.45 km/hr, as predicted by Corollary 2.

The Triangle Inequality, and its generalization in Corollary 2, are used frequently for finding upper estimates for various sums. Mathematicians also find these very helpful in proving statements in calculus and in other areas of mathematics.

Glossary

- Contraction: An operation on a vector whose result is in the same or opposite direction from the original vector and whose length is shorter than the original vector. A contraction of a nonzero vector is obtained by taking a scalar multiple of the vector in which the nonzero scalar has absolute value less than 1.

- Corollary: A theorem whose proof follows immediately from a previous theorem.

- Dilation: An operation on a vector whose result is in the same or opposite direction from the original vector and whose length is longer than the original vector. A dilation of a nonzero vector is obtained by taking a scalar multiple of the vector in which the scalar has absolute value greater than 1.

- Follow Method: A method of adding vectors geometrically, in which we start with the first vector in the sum, and then place each subsequent vector so that it starts where the previous vector ended, as illustrated in Figure 3.4. (That is, each new vector has its initial point at the terminal point of the previous vector.) We reach the desired final destination by simply following the path of the arrows.

- Length of a vector: The distance between the initial and terminal points of the vector. For a vector $\mathbf{v}$, the length of $\mathbf{v}$ is written as $\|\mathbf{v}\|$, and is computed using the formula $\|\mathbf{v}\| = \|[v_1, v_2, \ldots, v_n]\| = \sqrt{v_1^2 + v_2^2 + \cdots + v_n^2}$. The length of $\mathbf{v}$ is also called the magnitude of $\mathbf{v}$, or the norm of $\mathbf{v}$.

- Magnitude of a vector: Another name for the length of a vector.

- Norm of a vector: Another name for the length of a vector.

- Normalizing a vector: The process of multiplying a nonzero vector $\mathbf{v}$ by $\frac{1}{\|\mathbf{v}\|}$. The result is the vector $\mathbf{u} = \left(\frac{1}{\|\mathbf{v}\|}\right)\mathbf{v}$, which is a unit vector in the same direction as $\mathbf{v}$.

- Opposite directions: Two nonzero vectors are in opposite directions if and only if each is a negative scalar multiple of the other.

- Parallel vectors: Two nonzero vectors that are nonzero scalar multiples of each other. From a geometric point of view, two vectors are parallel if and only if they are in the same direction or opposite directions.

- Resultant speed: The magnitude of the resultant velocity of an object.

- Resultant velocity: The overall velocity of an object due to the combined effect of two or more simultaneous velocities. The resultant velocity is calculated by finding the sum of these simultaneous velocities.

- Same direction: Two nonzero vectors are in the same direction if and only if each is a positive scalar multiple of the other.

- Unit vector: A vector having length 1.

Exercises for Section 3.2

1. Use the Follow Method to illustrate each of the following vector expressions using the given initial point. What is the terminal point?

    (a) $[3, 1] + [-2, 4]$; initial point: $(0, 0)$
    (b) $[-1, 5] + [4, -4]$; initial point: $(2, 1)$
    (c) $2[3, -1] - [2, 4]$; initial point: $(-1, 5)$
    (d) $[2, 4, 3] + [1, -1, 2]$; initial point: $(1, 2, -4)$

2. Use the Follow Method to illustrate each of the following vector expressions using the given initial point. What is the terminal point?

    (a) $[5, 2] + [-4, 3]$; initial point: $(0, 0)$
    (b) $[4, 3] - [2, 2]$; initial point: $(-1, 2)$
    (c) $2[2, 4] + 3[1, -2]$; initial point: $(1, -2)$
    (d) $[4, -1, 2] + [1, 4, 3]$; initial point: $(1, -2, -1)$

3. Draw a parallelogram having two of its adjacent sides represented by the vectors $[1, 4]$ and $[3, 2]$ with their initial points at the origin. Then draw the two diagonals, and label each one as an appropriate linear combination of the sides.

4. Draw a parallelogram having two of its adjacent sides represented by the vectors $[-1, 3]$ and $[5, -2]$ with their initial points at the origin. Then draw the two diagonals, and label each one as an appropriate linear combination of the sides.

5. In each part, compute the length of the given vector. Round to two places after the decimal point.

    (a) $[2, -6, 3]$ (b) $[3, -3, -3, 3]$ (c) $[7, 2, 5, 1, 2]$

6. In each part, calculate the length of the given vector. Round to two places after the decimal point.

    (a) $[4, -7, -4]$ (b) $[3, -12, 0, 4]$ (c) $[8, -3, 0, 2, -2, 1]$

7. In each part, find a unit vector in the same direction as the given vector. If possible, express the coordinates of the unit vector as fractions in lowest terms. Otherwise, round to three places after the decimal point.

    (a) $[11, -2, 10]$ **(c)** $[1, 1]$

    (b) $[-6, 0, 4, 5, -2]$ **(d)** $[8, 0, 1, -1, 5]$

8. In each part, find a unit vector in the same direction as the given vector. If possible, express the coordinates of the unit vector as fractions in lowest terms. Otherwise, round to three places after the decimal point.

    (a) $[12, 8, -9]$ **(b)** $[7, -1, 0, 8, -5, 2, 1]$

                                  (c) $[5, -2]$

                                  (d) $[-6, 1, 3, 0, -9]$

9. Suppose $\|\mathbf{v}\| = 18$, and $c = -4$. What value(s) can $\|c\mathbf{v}\|$ have?

10. If $\|c\mathbf{v}\| = 112$ and $\|\mathbf{v}\| = 14$, what value(s) can c have?

11. Ralph is riding on a parade float that is travelling northwest on 19th Street at 3 mph. Ralph walks across the float in the northeast direction at 1 mph. Determine a vector representing Ralph's resultant velocity. How fast is Ralph moving? Round your answers to two places after the decimal point.

12. The submarine *Sea Lion* is out in a river that is flowing Southwest at 5 km/hr. There is no vertical movement of the water. The *Sea Lion*'s propeller is pushing the submarine in the direction $[12, 12, -1]$ at 8 km/hr. The ballast tanks are being filled with water, making the *Sea Lion* go straight down at 1 km/hr. Compute the resultant velocity of the *Sea Lion*. How fast is the submarine moving? Round your answers to two places after the decimal point.

13. Which two of the Properties ESP1 through ESP8 for vectors are illustrated in Figure 3.14? Explain.

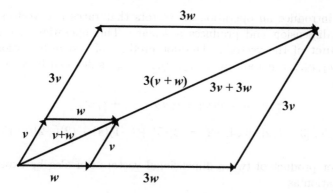

Figure 3.14: Figure for Exercise 13

14. Which one of the Properties ESP1 through ESP8 for vectors is illustrated in Figure 3.15? Explain.

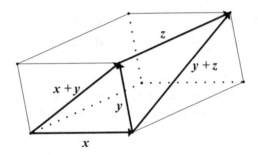

Figure 3.15: Figure for Exercise 14

15. Use Corollary 2 to find an upper estimate for the length of $[2, -9, 6]$ $+ [28, 10, -4] + [0, -5, 12]$.

16. The Triangle Inequality is also true for 1-dimensional vectors.

    (a) Explain why the norm of a 1-dimensional vector equals the absolute value of its single coordinate.

    (b) Restate the Triangle Inequality for 1-dimensional vectors in terms of absolute values of numbers instead of norms of vectors.

3.3   The Dot Product of Vectors

Definition of the Dot Product

Next, we introduce an operation on vectors that takes two vectors having the same dimension and produces a scalar. This operation is called the **dot product** of the vectors. The dot product of two n-dimensional vectors $\mathbf{v} = [v_1, v_2, \ldots, v_n]$ and $\mathbf{w} = [w_1, w_2, \ldots, w_n]$ is denoted by $\mathbf{v} \cdot \mathbf{w}$, and is defined by

$$\mathbf{v} \cdot \mathbf{w} = v_1 w_1 + v_2 w_2 + \cdots + v_n w_n.$$

For example, $[3, -1, 2] \cdot [5, 4, -2] = (3)(5) + (-1)(4) + (2)(-2) = 15 - 4 - 4 = 7$.

The dot product of two n-dimensional vectors can also be expressed in sigma notation as

$$\mathbf{v} \cdot \mathbf{w} = \sum_{i=1}^{n} v_i w_i.$$

A third way to express the dot product is to take advantage of the fact that vectors are just special types of matrices, and use matrix multiplication to express the dot product $\mathbf{v} \cdot \mathbf{w}$ of two vectors as the single entry of

$$\mathbf{v}^T\mathbf{w} = \underbrace{[v_1, v_2, \ldots, v_n]}_{1 \times n \text{ matrix}} \underbrace{\begin{bmatrix} w_1 \\ w_2 \\ \vdots \\ w_n \end{bmatrix}}_{n \times 1 \text{ matrix}} = \underbrace{[v_1 w_1 + v_2 w_2 + \cdots + v_n w_n]}_{1 \times 1 \text{ matrix}} = [\mathbf{v} \cdot \mathbf{w}].$$

That is, $\mathbf{v}^T\mathbf{w}$ is a 1×1 matrix whose single entry has the value $v_1 w_1 + v_2 w_2 + \cdots + v_n w_n$, which equals $\mathbf{v} \cdot \mathbf{w}$. However, this is one of those instances in which we can essentially think of a 1×1 matrix as being a scalar, since it only has one entry. For example, if $\mathbf{v} = [2, -3, 0, 1]$ and $\mathbf{w} = [-4, 2, 5, 7]$, then

$$\mathbf{v}^T\mathbf{w} = [2, -3, 0, 1] \begin{bmatrix} -4 \\ 2 \\ 5 \\ 7 \end{bmatrix}$$

$$= [(2)(-4) + (-3)(2) + (0)(5) + (1)(7)] = [-7],$$

and therefore $\mathbf{v} \cdot \mathbf{w}$ is -7, the single entry of this product.

In fact, we can actually describe matrix multiplication in terms of the dot product: if $\mathbf{A}$ and $\mathbf{B}$ are two compatible matrices, then the formula for the (i, j) entry of $\mathbf{AB}$ actually equals the dot product of the ith row of $\mathbf{A}$ with the jth column of $\mathbf{B}$, where the rows of $\mathbf{A}$ and the columns of $\mathbf{B}$ are thought of as vectors.

Assignment 4: Learn how to compute the dot product of two vectors having the same dimension on your calculator. Practice this with a few examples of your own. Then see how your calculator responds if the two vectors have different dimensions.

Properties of the Dot Product

Just as each of matrix addition, scalar multiplication, matrix transpose, and matrix multiplication has a list of properties which allow us to simplify expressions involving these operations, the dot product also has its own list of fundamental properties.

Suppose that $\mathbf{v}$, $\mathbf{w}$, and $\mathbf{z}$ are n-dimensional vectors and c is a scalar. Then,

DP1: $\mathbf{v} \cdot \mathbf{v} = \|\mathbf{v}\|^2$ (**The Dot Product and the Norm**)

DP2: $\mathbf{v} \cdot \mathbf{w} = \mathbf{w} \cdot \mathbf{v}$ (**Commutative Property of the Dot Product**)

DP3: $c(\mathbf{v} \cdot \mathbf{w}) = (c\mathbf{v}) \cdot \mathbf{w} = \mathbf{v} \cdot (c\mathbf{w})$ (**Associative Property of the Dot Product and Scalar Multiplication**)

DP4: $\mathbf{v} \cdot (\mathbf{w} \pm \mathbf{z}) = (\mathbf{v} \cdot \mathbf{w}) \pm (\mathbf{v} \cdot \mathbf{z})$ (**First Distributive Property of the Dot Product**)

DP5: $(\mathbf{v} \pm \mathbf{w}) \cdot \mathbf{z} = (\mathbf{v} \cdot \mathbf{z}) \pm (\mathbf{w} \cdot \mathbf{z})$ (**Second Distributive Property of the Dot Product**)

These five properties can all be easily proved just by writing out what the various dot products represent using ellipsis notation. We will not do that here. However, if we think of $\mathbf{v} \cdot \mathbf{w}$ as the (single entry of the) matrix product $\mathbf{v}^T\mathbf{w}$, DP2 through DP5 can all be proved using the analogous properties of matrix multiplication and properties of the transpose, with which we are already familiar. For example, for DP2, using that fact that a 1×1 matrix must equal its own transpose (at the second equal sign, below), we see that

$$\left[\mathbf{v} \cdot \mathbf{w}\right] = \mathbf{v}^T\mathbf{w} = \left(\mathbf{v}^T\mathbf{w}\right)^T = \mathbf{w}^T \left(\mathbf{v}^T\right)^T = \mathbf{w}^T\mathbf{v} = \left[\mathbf{w} \cdot \mathbf{v}\right],$$

and so, $\mathbf{v} \cdot \mathbf{w} = \mathbf{w} \cdot \mathbf{v}$. Notice that DP2 tells us that the dot product is always commutative, even though, in general, matrix multiplication is *not* commutative.

In DP1, we take the dot product of a vector with itself! Suppose $\mathbf{v} = [4, -1, 3, 7]$. Then $\mathbf{v} \cdot \mathbf{v} = [4, -1, 3, 7] \cdot [4, -1, 3, 7] = (4)(4) + (-1)(-1) + (3)(3) + (7)(7) = 16 + 1 + 9 + 49 = 75$, while $\|\mathbf{v}\|^2 = \left(\sqrt{4^2 + (-1)^2 + 3^2 + 7^2}\right)^2 = \left(\sqrt{16 + 1 + 9 + 49}\right)^2 = \left(\sqrt{75}\right)^2 = 75$. You can see from this example that the dot product of a vector $\mathbf{v}$ with itself is equal to the sum of the squares of its coordinates. But that is precisely the square of the length of the vector $\mathbf{v}$. That is why $\mathbf{v} \cdot \mathbf{v}$ is always equal to $\|\mathbf{v}\|^2$.

DP1 has two important consequences, which we will frequently use. First is that, for any vector $\mathbf{v}$,

$$\mathbf{v} \cdot \mathbf{v} \geq 0.$$

This is because $\mathbf{v} \cdot \mathbf{v}$ is the square of the nonnegative number $\|\mathbf{v}\|$.

Second, combining DP1 with LZV, we see that $\mathbf{v} \cdot \mathbf{v} = 0$ if and only if $\|\mathbf{v}\|^2 = 0$ (by DP1) if and only if $\|\mathbf{v}\| = 0$ if and only if $\mathbf{v} = \mathbf{0}$ (by LZV). Therefore,

$$\mathbf{v} \cdot \mathbf{v} = 0 \text{ if and only if } \mathbf{v} = 0.$$

Thus, to prove that a particular vector is the zero vector, it is enough to show that the dot product of the vector with itself equals zero.

DP1 through DP5 are frequently used to expand and simplify various expressions involving the dot product and linear combinations of vectors, just as we do in algebra when multiplying out polynomial expressions.

Example 1. If **x** and **y** are vectors having the same dimension, then

$$
\begin{aligned}
(3\mathbf{x} - 4\mathbf{y}) \cdot (2\mathbf{x} + 5\mathbf{y}) &= (3\mathbf{x} - 4\mathbf{y}) \cdot (2\mathbf{x}) + (3\mathbf{x} - 4\mathbf{y}) \cdot (5\mathbf{y}) \quad \text{by DP4} \\
&= (3\mathbf{x}) \cdot (2\mathbf{x}) + (-4\mathbf{y}) \cdot (2\mathbf{x}) + (3\mathbf{x}) \cdot (5\mathbf{y}) \\
&\qquad + (-4\mathbf{y}) \cdot (5\mathbf{y}) \qquad\qquad\qquad\quad \text{by DP5} \\
&= 6(\mathbf{x} \cdot \mathbf{x}) - 8(\mathbf{y} \cdot \mathbf{x}) + 15(\mathbf{x} \cdot \mathbf{y}) - 20(\mathbf{y} \cdot \mathbf{y}) \quad \text{by DP3} \\
&= 6(\mathbf{x} \cdot \mathbf{x}) - 8(\mathbf{x} \cdot \mathbf{y}) + 15(\mathbf{x} \cdot \mathbf{y}) - 20(\mathbf{y} \cdot \mathbf{y}) \quad \text{by DP2} \\
&= 6(\mathbf{x} \cdot \mathbf{x}) + 7(\mathbf{x} \cdot \mathbf{y}) - 20(\mathbf{y} \cdot \mathbf{y}) \\
&\qquad\qquad\qquad\qquad\qquad \text{by combining like terms} \\
&= 6 \|\mathbf{x}\|^2 + 7(\mathbf{x} \cdot \mathbf{y}) - 20 \|\mathbf{y}\|^2 \quad \text{by DP1.} \quad \blacksquare
\end{aligned}
$$

FOILing

Rather than taking many steps to expand a dot product of binomials as in Example 1, it is more efficient to have a general property for use in such situations. The FOIL method for the product of binomials in algebra applies to the dot product as well. Recall that FOIL is an acronym for "First, Outer, Inner, Last," which represents the order in which we multiply together the respective terms of the binomials. In particular, for the dot product of a linear combination of two vectors, we have:

DP-FOIL1: $(a\mathbf{x} + b\mathbf{y}) \cdot (c\mathbf{x} + d\mathbf{y}) = ac(\mathbf{x} \cdot \mathbf{x}) + (ad + bc)(\mathbf{x} \cdot \mathbf{y}) + bd(\mathbf{y} \cdot \mathbf{y})$
$$= ac\|\mathbf{x}\|^2 + (ad + bc)(\mathbf{x} \cdot \mathbf{y}) + bd\|\mathbf{y}\|^2.$$

The first part of DP-FOIL1 uses DP2 through DP5 as was done in the first five steps in Example 1. The second part extends this result using DP1, as we did in the last step in Example 1. Hence, using DP-FOIL1, we could have done the computation in Example 1 in just one step.

More generally, the vectors in the two linear combinations involved could be different. In such a case, we have:

DP-FOIL2: $(a\mathbf{w} + b\mathbf{x}) \cdot (c\mathbf{y} + d\mathbf{z}) = ac(\mathbf{w} \cdot \mathbf{y}) + ad(\mathbf{w} \cdot \mathbf{z}) + bc(\mathbf{x} \cdot \mathbf{y}) + bd(\mathbf{x} \cdot \mathbf{z}).$

"FOILing" can be extended to the dot product of linear combinations involving more than two vectors, just as in algebra where there are processes analogous to FOIL for multiplying two polynomials, each having any number of terms.

The Polarization Identity

The next result, known as the Polarization Identity, allows us to express the dot product of two vectors solely in terms of the norms of those vectors.

Theorem 3. *(Polarization Identity)* *Suppose* **v** *and* **w** *are two vectors having the same dimension. Then,*

$$\mathbf{v} \cdot \mathbf{w} = \frac{1}{4}\left(\|\mathbf{v} + \mathbf{w}\|^2 - \|\mathbf{v} - \mathbf{w}\|^2\right).$$

Example 2. Suppose $\mathbf{v} = [3, -2, 1]$ and $\mathbf{w} = [4, 2, 7]$. Then,

$$\mathbf{v} \cdot \mathbf{w} = 12 - 4 + 7 = 15.$$

Also,

$$\frac{1}{4}\left(\|\mathbf{v}+\mathbf{w}\|^2 - \|\mathbf{v}-\mathbf{w}\|^2\right) = \frac{1}{4}\left(\|[7,0,8]\|^2 - \|[-1,-4,-6]\|^2\right)$$
$$= \frac{1}{4}\left((49+0+64) - (1+16+36)\right)$$
$$= \frac{1}{4}(113-53) = \frac{1}{4}(60) = 15,$$

which equals $\mathbf{v} \cdot \mathbf{w}$. ∎

Proof of the Polarization Identity: We prove this formula by starting with $\frac{1}{4}\left(\|\mathbf{v}+\mathbf{w}\|^2 - \|\mathbf{v}-\mathbf{w}\|^2\right)$. We then use the DP properties to expand and simplify that expression until we obtain $\mathbf{v} \cdot \mathbf{w}$.

$$\frac{1}{4}\left(\|\mathbf{v}+\mathbf{w}\|^2 - \|\mathbf{v}-\mathbf{w}\|^2\right)$$
$$= \frac{1}{4}((\mathbf{v}+\mathbf{w})\cdot(\mathbf{v}+\mathbf{w}) - (\mathbf{v}-\mathbf{w})\cdot(\mathbf{v}-\mathbf{w})) \quad \text{by DP1}$$
$$= \frac{1}{4}\left(\begin{array}{l}((\mathbf{v}\cdot\mathbf{v})+2(\mathbf{v}\cdot\mathbf{w})+(\mathbf{w}\cdot\mathbf{w}))\\ -((\mathbf{v}\cdot\mathbf{v})-2(\mathbf{v}\cdot\mathbf{w})+(\mathbf{w}\cdot\mathbf{w}))\end{array}\right) \quad \text{by DP-FOIL1}$$
$$= \frac{1}{4}\left(\begin{array}{l}(\mathbf{v}\cdot\mathbf{v})+2(\mathbf{v}\cdot\mathbf{w})+(\mathbf{w}\cdot\mathbf{w})\\ -(\mathbf{v}\cdot\mathbf{v})+2(\mathbf{v}\cdot\mathbf{w})-(\mathbf{w}\cdot\mathbf{w})\end{array}\right) \quad \text{distributing the minus sign}$$
$$= \frac{1}{4}(2(\mathbf{v}\cdot\mathbf{w})+2(\mathbf{v}\cdot\mathbf{w})) \quad \text{cancelling like terms}$$
$$= \frac{1}{4}(4(\mathbf{v}\cdot\mathbf{w})) \quad \text{combining like terms}$$
$$= \mathbf{v}\cdot\mathbf{w}. \quad \blacklozenge$$

The Cauchy-Schwarz Inequality

The DP properties can also be used to prove another important result – the Cauchy-Schwarz Inequality. We ask you to prove this inequality in the exercises, using the Triangle Inequality, in order to gain practice in using properties of the dot product and the norm.

Theorem 4. *(Cauchy-Schwarz Inequality) If $\mathbf{v}$ and $\mathbf{w}$ are two vectors having the same dimension, then*

$$|\mathbf{v}\cdot\mathbf{w}| \le \|\mathbf{v}\|\,\|\mathbf{w}\|.$$

Just as the Triangle Inequality is used to find an upper estimate for the norm of the sum of two vectors, the Cauchy-Schwarz Inequality provides an upper estimate for the absolute value of the dot product of two vectors. That is, it gives us a limit on how large the dot product of two given vectors can become, since it says that the absolute value of the dot product of $\mathbf{v}$ and $\mathbf{w}$ cannot be any greater than the product of the magnitudes of $\mathbf{v}$ and $\mathbf{w}$.

For example, if $\mathbf{v} = [3, 5, -2]$ and $\mathbf{w} = [-6, 1, 4]$, then

$$|\mathbf{v} \cdot \mathbf{w}| = |[3, 5, -2] \cdot [-6, 1, 4]| = |-21| = 21.$$

But,

$$\|\mathbf{v}\| \, \|\mathbf{w}\| = \|[3, 5, -2]\| \, \|[-6, 1, 4]\| = \sqrt{38}\sqrt{53} \approx 44.878.$$

In this case, $\|\mathbf{v}\| \, \|\mathbf{w}\|$ is actually more than twice the value of $|\mathbf{v} \cdot \mathbf{w}|$.

However, there are vectors $\mathbf{v}$ and $\mathbf{w}$ for which $|\mathbf{v} \cdot \mathbf{w}| = \|\mathbf{v}\| \, \|\mathbf{w}\|$. In particular, if $\mathbf{v} = \mathbf{w}$, we know from DP1 that $\mathbf{v} \cdot \mathbf{v} = \|\mathbf{v}\|^2$; that is, $|\mathbf{v} \cdot \mathbf{v}| = \|\mathbf{v}\| \, \|\mathbf{v}\|$. In fact, if $\mathbf{v}$ and $\mathbf{w}$ are any two nonzero parallel vectors, in either the same or opposite directions, then $|\mathbf{v} \cdot \mathbf{w}|$ will be equal to $\|\mathbf{v}\| \, \|\mathbf{w}\|$.

The Cauchy-Schwarz Inequality can be written without using the absolute value on the dot product as

$$- \|\mathbf{v}\| \, \|\mathbf{w}\| \leq \mathbf{v} \cdot \mathbf{w} \leq \|\mathbf{v}\| \, \|\mathbf{w}\|.$$

That is, the actual value of the dot product of $\mathbf{v}$ and $\mathbf{w}$ is between $\pm \|\mathbf{v}\| \, \|\mathbf{w}\|$. Hence, the Cauchy-Schwarz Inequality is also giving a *lower* estimate for the dot product. In other words, $\mathbf{v} \cdot \mathbf{w}$ cannot be *less than* $- \|\mathbf{v}\| \, \|\mathbf{w}\|$.

We will see an application of these upper and lower estimates for the dot product in the next section, in which we discuss the angle between two vectors.

Glossary

- Dot Product: The dot product of two n-dimensional vectors $\mathbf{v}$ and $\mathbf{w}$ is denoted by $\mathbf{v} \cdot \mathbf{w}$, and is defined by the formula $\mathbf{v} \cdot \mathbf{w} = v_1 w_1 + v_2 w_2 + \cdots + v_n w_n$. Expressed in sigma notation, $\mathbf{v} \cdot \mathbf{w} = \sum_{i=1}^{n} v_i w_i$. We can also essentially express the dot product as a matrix product as follows: $\mathbf{v} \cdot \mathbf{w} =$ the single entry of the 1×1 matrix $\mathbf{v}^T \mathbf{w}$.

- FOIL: An acronym for "First, Outer, Inner, Last." A process for expanding the dot product of two linear combinations of vectors, analogous to expanding a product of two binomials in algebra.

Exercises for Section 3.3

1. In each part, perform the indicated operations.

   (a) $[1, 5, -3] \cdot [4, -2, 6]$ (b) $[5, 2, 1, 0] \cdot [-3, 7, 1, 8]$

2. In each part, compute the dot product of the two given vectors.

(a) $[3, -4, 5, 0]$; $[2, 0, 3, -7]$ (b) $[2, 3, -4, 5]$; $[3, 4, 7, 2]$

3. Compute the expressions on each side of DP4 for the vectors $\mathbf{v} = [4, 1, -2]$, $\mathbf{w} = [6, 8, 3]$ and $\mathbf{z} = [-1, 5, 7]$ to verify that DP4 is true in this case.

4. Compute the expressions on each side of DP5 for the vectors $\mathbf{v} = [3, 0, -5, 9]$, $\mathbf{w} = [4, -2, 6, 1]$ and $\mathbf{z} = [-7, 1, -2, 8]$ to verify that DP5 is true in this case.

5. Suppose $\|\mathbf{v}\| = 7$, $\|\mathbf{w}\| = 8$, and $\mathbf{v} \cdot \mathbf{w} = -5$.

   (a) Calculate $(\mathbf{v} + 3\mathbf{w}) \cdot (6\mathbf{v})$.
   (b) Find $(2\mathbf{v} - 5\mathbf{w}) \cdot (3\mathbf{v} + 4\mathbf{w})$.

6. Suppose $\|\mathbf{v}\| = 6$, $\|\mathbf{w}\| = 5$, and $\mathbf{v} \cdot \mathbf{w} = 7$.

   (a) Determine $(5\mathbf{v} - \mathbf{w}) \cdot (4\mathbf{w})$.
   (b) Ascertain the value of $(4\mathbf{v} + 9\mathbf{w}) \cdot (2\mathbf{w} - \mathbf{v})$.

7. If $\|\mathbf{v}\| = 3$, $\|\mathbf{w}\| = 5$, give the possible range of values for $\mathbf{v} \cdot \mathbf{w}$.

8. If $\|\mathbf{w}\| = 6$ and $\mathbf{v} \cdot \mathbf{w} = -2$, what is the smallest possible value for $\|\mathbf{v}\|$?

9. The two diagonals of a certain parallelogram have lengths 7 units and 5 units. Compute the absolute value of the dot product of the two vectors representing the sides of the parallelogram. Also, explain why we cannot determine from the given information whether the dot product is positive or negative. (Hint: Use the Polarization Identity.)

10. The two diagonals of a certain parallelogram have lengths 12 units and 15 units. Compute the absolute value of the dot product of the two vectors representing the sides of the parallelogram. (Hint: We cannot determine from the given information whether the dot product is positive or negative.)

11. In each part, use the two given vectors to compute both sides of the Cauchy-Schwarz Inequality, verifying that they are equal in these particular special cases.

    (a) $\mathbf{v} = [1, 2, -5]$; $\mathbf{w} = [1, 2, -5]$. (Here, $\mathbf{w} = \mathbf{v}$.)
    (b) $\mathbf{v} = [2, -1, 8]$; $\mathbf{w} = [-2, 1, -8]$. (Here, $\mathbf{w} = -\mathbf{v}$.)
    (c) $\mathbf{v} = [-4, 6, 7]$; $\mathbf{w} = [-12, 18, 21]$. (Here, $\mathbf{v}$ and $\mathbf{w}$ are in the same direction.)
    (d) $\mathbf{v} = [0, 0, 0]$; $\mathbf{w} = [8, 11, -13]$.

12. In each part, use the two given vectors to compute both sides of the Cauchy-Schwarz Inequality, verifying that they are equal in these particular special cases.

   (a) $\mathbf{v} = [4, 3, -1, 6]$; $\mathbf{w} = [4, 3, -1, 6]$. (Here, $\mathbf{w} = \mathbf{v}$.)
   (b) $\mathbf{v} = [-3, 2, 9, 5]$; $\mathbf{w} = [3, -2, -9, -5]$. (Here, $\mathbf{w} = -\mathbf{v}$.)
   (c) $\mathbf{v} = [6, -2, 10, -8]$; $\mathbf{w} = [-3, 1, -5, 4]$. (Here, $\mathbf{v}$ and $\mathbf{w}$ are parallel, but in opposite directions.)
   (d) $\mathbf{v} = [-5, 3, 7, -9]$; $\mathbf{w} = [0, 0, 0, 0]$.

13. Use the Cauchy-Schwarz Inequality to give both an upper estimate and a lower estimate for $[7, -4, 4] \cdot [2, 3, -6]$.

14. Use the Cauchy-Schwarz Inequality to give both an upper estimate and a lower estimate for $[1, 3, -7, 1, -6, 5] \cdot [5, -6, 0, 2, -4, 0]$.

15. Explain why the dot product is not an associative operation. That is, explain why it is *not* true that $\mathbf{x} \cdot (\mathbf{y} \cdot \mathbf{z}) = (\mathbf{x} \cdot \mathbf{y}) \cdot \mathbf{z}$ for three vectors $\mathbf{x}$, $\mathbf{y}$ and $\mathbf{z}$ having the same dimension.

16. Explain why the following "distributive law" for the dot product and scalar addition is true:

   If a and b are scalars, and $\mathbf{v}$ and $\mathbf{w}$ are vectors having the same dimension, then

   $$(a + b)(\mathbf{v} \cdot \mathbf{w}) = a(\mathbf{v} \cdot \mathbf{w}) + b(\mathbf{v} \cdot \mathbf{w}).$$

   (We used this fact in the proof of the Polarization Identity, saying that we were "combining like terms.")

17. Suppose that $\mathbf{v}$ and $\mathbf{w}$ are two vectors having the same dimension. The Parallelogram Law states that

   $$\|\mathbf{v}\|^2 + \|\mathbf{w}\|^2 = \frac{1}{2}\left(\|\mathbf{v} + \mathbf{w}\|^2 + \|\mathbf{v} - \mathbf{w}\|^2\right).$$

   Consider a parallelogram having adjacent sides determined by $\mathbf{v}$ and $\mathbf{w}$. Explain what the Parallelogram Law is saying from a geometric point of view. (For a particular example, turn back and consider Figure 3.6.)

18. Prove the Parallelogram Law, given in the previous exercise, using the DP properties. (Hint: The proof is similar to that of the Polarization Identity given in the text.)

19. Suppose that $\mathbf{v}$ and $\mathbf{w}$ are two vectors having the same dimension.

   (a) Use the DP properties to show that

   $$\|\mathbf{v} + \mathbf{w}\|^2 = \|\mathbf{v}\|^2 + 2\mathbf{v} \cdot \mathbf{w} + \|\mathbf{w}\|^2.$$

(b) Use ordinary algebra to prove that

$$\left(\|\mathbf{v}\| + \|\mathbf{w}\|\right)^2 = \|\mathbf{v}\|^2 + 2\,\|\mathbf{v}\|\,\|\mathbf{w}\| + \|\mathbf{w}\|^2.$$

(c) Consider the Triangle Inequality. Square both sides. Then use parts (a) and (b) of this exercise to prove that $\mathbf{v} \cdot \mathbf{w} \le \|\mathbf{v}\|\,\|\mathbf{w}\|$.

20. Suppose that $\mathbf{v}$ and $\mathbf{w}$ are two vectors having the same dimension. Assume that the results in the previous problem have been proved.

(a) Substitute "$-\mathbf{y}$" for $\mathbf{w}$ in the formula proved in part (c) of the previous problem. Use this to prove that $-\mathbf{v} \cdot \mathbf{y} \le \|\mathbf{v}\|\,\|\mathbf{y}\|$.

(b) Substitute "$\mathbf{w}$" back in for "$\mathbf{y}$" in the formula you proved in part (b). What do you get?

(c) Explain why the final results of part (c) of the previous problem and part (b) of this problem, taken together, prove the Cauchy-Schwarz Inequality.

3.4 The Angle Between Vectors

Two Dimensions

When a pair of 2-dimensional vectors $\mathbf{v}$ and $\mathbf{w}$ share the same initial point, there are two angles formed by $\mathbf{v}$ and $\mathbf{w}$, one "inside" and one "outside," as shown in Figure 3.16.

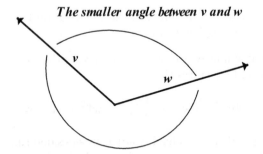

The smaller angle between v and w

The larger angle between v and w

Figure 3.16: The Two Angles Formed By a Pair of Vectors Having the Same Initial Point

When we speak of the "angle" between two vectors, we generally mean the smaller of the two angles involved, which always has a measure from $0°$ to $180°$. A formula involving the measure, θ, of the (smaller) angle[4] between

---

[4] Mathematicians like to use lower-case Greek letters to represent the measures of angles. Theta (θ) is the "favorite," but alpha (α), beta (β), phi (ϕ), and psi (ψ) are also frequently used.

vectors **v** and **w** is

$$\cos(\theta) = \frac{\mathbf{v} \cdot \mathbf{w}}{\|\mathbf{v}\| \, \|\mathbf{w}\|},$$

where $\cos(\theta)$ is the value of the "cosine" trigonometric function of the angle θ. Since we are not assuming a background in trigonometry in this text, we will not prove here that this formula is correct. However, such a proof can be given using the Law of Cosines from trigonometry.

For an example, consider the vectors $\mathbf{v} = [3,2]$ and $\mathbf{w} = [-1,4]$, both starting from the same initial point, as in Figure 3.17. Using the given formula,

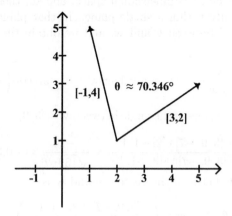

Figure 3.17: The (Smaller) Angle Between $[3,2]$ and $[-1,4]$

we get

$$\cos(\theta) = \frac{[3,2] \cdot [-1,4]}{\|[3,2]\| \, \|[-1,4]\|} = \frac{5}{\sqrt{13}\sqrt{17}} \approx 0.3363364.$$

To find θ, we use the "arccosine" function, also known as the "inverse cosine." You should be able to find this function on your calculator as either "arccos" or as "$\cos^{-1}$," depending upon what calculator you have. Then,

$$\theta = \arccos\left(\frac{[3,2] \cdot [-1,4]}{\|[3,2]\| \, \|[-1,4]\|}\right) \approx \arccos(0.3363364) \approx 70.346°.$$

Your calculator might give the angle measure in radians instead of degrees. If so, your answer for the preceding example would have been about 1.228 radians. Throughout this text, we will use degrees rather than radians. However, your instructor may prefer radians instead. Make sure you know which mode your calculator is currently using!

Assignment 5: Find the "arccos" or "$\cos^{-1}$" function on your calculator and learn how to use it. Figure out how to change the calculator's output mode from degrees to radians, and vice versa. Go through the preceding example on your calculator, making sure that you can produce both answers: 70.346° and 1.228 radians. (These answers were rounded to 3 digits after the decimal point *after* the arccosine calculation.)

Three Dimensions

Suppose we have two 3-dimensional vectors $\mathbf{v}$ and $\mathbf{w}$ that share the same initial point. Then, that common initial point and the terminal points of the two vectors represent three points in 3-dimensional space. Any three points lie in a common plane in 3-dimensional space, and so, these vectors can be depicted geometrically within a single plane. In that plane, we can picture the (smaller) angle θ between $\mathbf{v}$ and $\mathbf{w}$, and we obtain the same formula as before:

$$\cos{(\theta)} = \frac{\mathbf{v} \cdot \mathbf{w}}{\|\mathbf{v}\| \, \|\mathbf{w}\|}, \quad \text{or, equivalently,} \quad \theta = \arccos\left(\frac{\mathbf{v} \cdot \mathbf{w}}{\|\mathbf{v}\| \, \|\mathbf{w}\|}\right).$$

For example, the cosine of the angle between $\mathbf{v} = [3, 9, -7]$ and $\mathbf{w} = [2, 1, 1]$ is

$$\cos{(\theta)} = \frac{[3, 9, -7] \cdot [2, -1, 1]}{\|[3, 9, -7]\| \, \|[2, -1, 1]\|} = \frac{-10}{\sqrt{139}\sqrt{6}} \approx -0.3462717.$$

Therefore, the angle between the vectors $\mathbf{v}$ and $\mathbf{w}$ is

$$\theta = \arccos\left(\frac{[3, 9, -7] \cdot [2, -1, 1]}{\|[3, 9, -7]\| \, \|[2, -1, 1]\|}\right)$$
$$\approx \arccos{(-0.3462717)} \approx 110.259°.$$

Higher Dimensions

More generally, suppose $\mathbf{v}$ and $\mathbf{w}$ are two n-dimensional vectors that share the same initial point. That common initial point and the terminal points of the two vectors give three points in n-dimensional space. Analogous to the 3-dimensional case, these three points lie in a common plane in n-dimensional space, and so, again, geometrically, these two vectors can be depicted within a single plane. Therefore, we can imagine "drawing" the two vectors in a plane, even though they exist in higher-dimensional space. No matter what dimension the vectors have, we can also imagine the (smaller) angle formed between $\mathbf{v}$ and $\mathbf{w}$. Hence, we again obtain the same formula for the angle θ between $\mathbf{v}$ and $\mathbf{w}$. That is,

$$\cos{(\theta)} = \frac{\mathbf{v} \cdot \mathbf{w}}{\|\mathbf{v}\| \, \|\mathbf{w}\|}, \quad \text{or, equivalently,} \quad \theta = \arccos\left(\frac{\mathbf{v} \cdot \mathbf{w}}{\|\mathbf{v}\| \, \|\mathbf{w}\|}\right).$$

For example, in dimension 5, the cosine of the angle between $\mathbf{v} = [4, -3, 5, -6, 2]$ and $\mathbf{w} = [7, 9, 1, 3, -4]$ is

$$\cos{(\theta)} = \frac{[4, -3, 5, -6, 2] \cdot [7, 9, 1, 3, -4]}{\|[4, -3, 5, -6, 2]\| \, \|[7, 9, 1, 3, -4]\|} = \frac{-20}{\sqrt{90}\sqrt{156}} \approx -0.16878989.$$

Therefore, the angle between **v** and **w** is

$$\theta = \arccos\left(\frac{[4,-3,5,-6,2]\cdot[7,9,1,3,-4]}{\|[4,-3,5,-6,2]\|\,\|[7,9,1,3,-4]\|}\right)$$
$$\approx \arccos(-0.16878989) \approx 99.718°.$$

Notice that the cosine of any angle measure is a number between -1 and 1, so the arccosine needs for its *input* a number in that same range: between -1 and 1. But for nonzero vectors **v** and **w**, the Cauchy-Schwarz Inequality tells us that

$$-\|\mathbf{v}\|\,\|\mathbf{w}\| \le \mathbf{v}\cdot\mathbf{w} \le \|\mathbf{v}\|\,\|\mathbf{w}\|.$$

Dividing both sides by $\|\mathbf{v}\|\,\|\mathbf{w}\|$, we obtain

$$-1 \le \frac{\mathbf{v}\cdot\mathbf{w}}{\|\mathbf{v}\|\,\|\mathbf{w}\|} \le 1,$$

which shows that the input for the arccosine is indeed always in the correct range!

Notice also that the arccosine always produces an *output* in the range from $0°$ to $180°$. That is why the formula $\theta = \arccos\left(\frac{\mathbf{v}\cdot\mathbf{w}}{\|\mathbf{v}\|\|\mathbf{w}\|}\right)$ always gives the "smaller" of the two possible angles between **v** and **w**, as shown earlier in Figure 3.16.

Acute, Obtuse, or Perpendicular?

If the angle between two nonzero vectors is $90°$, then the two vectors are at right angles; that is, they are perpendicular to each other. If the angle is less than $90°$, then the vectors form an acute angle. Similarly, if the angle is greater than $90°$, then the vectors form an obtuse angle.

Now the cosine of an acute angle is positive, the cosine of an obtuse angle is negative, and the cosine of a right angle equals 0. Therefore, the arccosine will produce an acute angle if $\frac{\mathbf{v}\cdot\mathbf{w}}{\|\mathbf{v}\|\|\mathbf{w}\|}$ is positive, an obtuse angle if $\frac{\mathbf{v}\cdot\mathbf{w}}{\|\mathbf{v}\|\|\mathbf{w}\|}$ is negative, and a right angle if $\frac{\mathbf{v}\cdot\mathbf{w}}{\|\mathbf{v}\|\|\mathbf{w}\|} = 0$. But, for two nonzero vectors, the denominator of $\frac{\mathbf{v}\cdot\mathbf{w}}{\|\mathbf{v}\|\|\mathbf{w}\|}$ must be a positive number! Dividing $\mathbf{v}\cdot\mathbf{w}$ by this positive number does not change the sign of $\mathbf{v}\cdot\mathbf{w}$, and so the expression $\frac{\mathbf{v}\cdot\mathbf{w}}{\|\mathbf{v}\|\|\mathbf{w}\|}$ always has the same *sign* as its numerator, $\mathbf{v}\cdot\mathbf{w}$. Hence we can tell just by looking at the sign of $\mathbf{v}\cdot\mathbf{w}$ whether the angle between **v** and **w** is acute, obtuse, or a right angle! Specifically, we have the following three "angle rules:"

If **v** and **w** are two nonzero vectors having the same dimension, then:

AR1: The angle θ between **v** and **w** is acute ($0° \le \theta < 90°$) if and only if $\mathbf{v}\cdot\mathbf{w} > 0$.

AR2: The angle θ between **v** and **w** is a right angle ($\theta = 90°$) if and only if $\mathbf{v}\cdot\mathbf{w} = 0$.

AR3: The angle θ between $\mathbf{v}$ and $\mathbf{w}$ is obtuse ($90° < \theta \leq 180°$) if and only if $\mathbf{v} \cdot \mathbf{w} < 0$.

For example, from AR1, we see that the angle between $[2, 3, 4]$ and $[7, 2, -3]$ is an acute angle, because $[2, 3, 4] \cdot [7, 2, -3] = 8$, which is positive.

AR2 tells us that $[3, 6, 1, 0]$ and $[5, -2, -3, 8]$ are perpendicular to each other, since $[3, 6, 1, 0] \cdot [5, -2, -3, 8] = 0$.

Using AR3, we can determine that the angle between $[4, 7]$ and $[-5, 2]$ is obtuse, because $[4, 7] \cdot [-5, 2] = -6$, which is negative.

Recognizing when the dot product of two vectors is zero is often useful in mathematics. Any two vectors $\mathbf{v}$ and $\mathbf{w}$ such that $\mathbf{v} \cdot \mathbf{w} = 0$ are said to be **orthogonal** vectors, or to be **orthogonal to each other**. Thus, AR2 tells us that the term "orthogonal" means the same thing as the term "perpendicular" for nonzero vectors. However, the term "orthogonal" also allows zero vectors. In fact, a zero vector is orthogonal to every other vector having the same dimension. On the other hand, the term "perpendicular" is not used when a zero vector is involved, because zero vectors do not have a direction. Hence, the term "orthogonal" describes an *algebraic* relationship between two vectors, using the dot product, while the term "perpendicular" describes a *geometric* relationship between two nonzero vectors, based on the right angle between them. These algebraic and geometric concepts are linked by AR2.

Mutually Orthogonal Sets of Vectors

A *set* of vectors $\{\mathbf{v}_1, \ldots, \mathbf{v}_k\}$ is **mutually orthogonal** if $\mathbf{v}_i \cdot \mathbf{v}_j = 0$ whenever $i \neq j$. This means that any pair of distinct vectors within the set are orthogonal to each other. The most important set of mutually orthogonal n-dimensional vectors is $\{\mathbf{e}_1, \ldots, \mathbf{e}_n\}$. Specifically, in $\mathbb{R}^2$, the set $\{\mathbf{e}_1, \mathbf{e}_2\} = \{[1, 0], [0, 1]\}$ is a mutually orthogonal set. These are unit vectors in the direction of the positive x-axis and the positive y-axis, respectively. In 3-dimensions, $\{\mathbf{e}_1, \mathbf{e}_2, \mathbf{e}_3\}$ is the mutually orthogonal set of unit vectors in the direction of the positive x-axis, the positive y-axis, and the positive z-axis, respectively. When a mutually orthogonal set contains only unit vectors, such as $\{\mathbf{e}_1, \ldots, \mathbf{e}_n\}$, the set is said to be **orthonormal**.

The set $\{\mathbf{e}_1, \ldots, \mathbf{e}_n\}$ is not the only mutually orthogonal set in $\mathbb{R}^n$. In particular, $\{[3, -4], [4, 3]\}$ is a mutually orthogonal set in $\mathbb{R}^2$, because $[3, -4] \cdot [4, 3] = 12 - 12 = 0$. Similarly, $\{[4, 2, -1], [0, 0, 0], [3, -5, 2]\}$ is a mutually orthogonal set in $\mathbb{R}^3$, since

$$[4, 2, -1] \cdot [0, 0, 0] = 0 + 0 + 0 = 0,$$
$$[4, 2, -1] \cdot [3, -5, 2] = 12 - 10 - 2 = 0,$$
$$\text{and} \quad [0, 0, 0] \cdot [3, -5, 2] = 0 + 0 + 0 = 0.$$

If we have a mutually orthogonal set of nonzero vectors, we can easily turn it into an orthonormal set, merely by normalizing each of the vectors. The dot products of distinct vectors in the set will still all be zero, which is easy to see using DP3. For example, consider the mutually orthogonal set $\{[3, -4], [4, 3]\}$ mentioned earlier. Because both vectors in the set have length 5, we can multiply each vector by $\frac{1}{5}$ to turn them into unit vectors, producing the orthonormal set $\{[\frac{3}{5}, -\frac{4}{5}], [\frac{4}{5}, \frac{3}{5}]\}$.

Similarly, consider the set $\{[4, -1, 2, 6], [3, 4, -4, 0], [8, -2, 4, -7]\}$. These vectors are mutually orthogonal since

$$
\begin{aligned}
[4, -1, 2, 6] \cdot [3, 4, -4, 0] &= 12 - 4 - 8 + 0 = 0, \\
[4, -1, 2, 6] \cdot [8, -2, 4, -7] &= 32 + 2 + 8 - 42 = 0, \\
\text{and} \quad [3, 4, -4, 0] \cdot [8, -2, 4, -7] &= 24 - 8 - 16 + 0 = 0.
\end{aligned}
$$

Therefore, we can obtain an orthonormal set of vectors by normalizing each of these vectors. We have

$$
\begin{aligned}
\|[4, -1, 2, 6]\| &= \sqrt{4^2 + (-1)^2 + 2^2 + 6^2} = \sqrt{57}, \\
\|[3, 4, -4, 0]\| &= \sqrt{3^2 + 4^2 + (-4)^2 + 0^2} = \sqrt{41}, \\
\text{and} \quad \|[8, -2, 4, -7]\| &= \sqrt{8^2 + (-2)^2 + 4^2 + (-7)^2} = \sqrt{133}.
\end{aligned}
$$

Dividing each vector by its norm, we obtain the following orthonormal set:

$$
\left\{ \left[\frac{4}{\sqrt{57}}, -\frac{1}{\sqrt{57}}, \frac{2}{\sqrt{57}}, \frac{6}{\sqrt{57}} \right], \left[\frac{3}{\sqrt{41}}, \frac{4}{\sqrt{41}}, -\frac{4}{\sqrt{41}}, 0 \right], \right.
$$
$$
\left. \left[\frac{8}{\sqrt{133}}, -\frac{2}{\sqrt{133}}, \frac{4}{\sqrt{133}}, -\frac{7}{\sqrt{133}} \right] \right\}.
$$

We will study mutually orthogonal sets of vectors more deeply later in this chapter.

Glossary

- **Angle between two vectors:** The smaller of the two angles formed between the vectors when they begin at the same initial point.

- **Arccosine function:** The inverse function for the cosine function. Also called "inverse cosine," and often written as "arccos" or "$\cos^{-1}$."

- **Acute angle:** An angle θ for which $0° \leq \theta < 90°$.

- **Inverse cosine function:** See "arccosine function."

- **Mutually orthogonal:** A *set* of vectors $\{\mathbf{v}_1, \ldots, \mathbf{v}_k\}$ is mutually orthogonal if $\mathbf{v}_i \cdot \mathbf{v}_j = 0$ whenever $i \neq j$. This means that any pair of distinct vectors within the set are orthogonal to each other.

- **Obtuse angle:** An angle θ for which $90° < \theta \leq 180°$.

- **Orthogonal vectors:** Any two vectors $\mathbf{v}$ and $\mathbf{w}$ such that $\mathbf{v} \cdot \mathbf{w} = 0$.

- Orthonormal set of vectors: A mutually orthogonal set of unit vectors.

- Perpendicular vectors: Two nonzero vectors having the same dimension such that the angle between them measures 90°.

- Right angle: An angle whose measure equals 90°.

Exercises for Section 3.4

1. In each part, use the dot product of the given vectors to determine whether the angle between the vectors is acute, obtuse, or a right angle.

   (a) $\mathbf{v} = [5, 7]$; $\mathbf{w} = [14, -10]$

   (b) $\mathbf{v} = [4, 8, -3]$; $\mathbf{w} = [-6, 9, 5]$

   (c) $\mathbf{v} = [7, -5, 6]$; $\mathbf{w} = [-2, 4, 3]$

   (d) $\mathbf{v} = [8, 3, -4, 2, 5]$; $\mathbf{w} = [-7, 6, 1, 4, 9]$

2. In each part, use the dot product of the given vectors to determine whether the angle between the vectors is acute, obtuse, or a right angle.

   (a) $\mathbf{v} = [7, 12]$; $\mathbf{w} = [-8, 5]$

   (b) $\mathbf{v} = [1, -5, 4]$; $\mathbf{w} = [2, 6, 7]$

   (c) $\mathbf{v} = [9, -3, 5, 8]$; $\mathbf{w} = [1, 7, 6, -4]$

   (d) $\mathbf{v} = [3, -8, 5, 11, -2]$; $\mathbf{w} = [-1, 5, 9, 4, 1]$

3. In each part, find the angle between the two given vectors. Give your answer in degrees, rounded to three places after the decimal point.

   (a) $\mathbf{v} = [6, -6, 7]$; $\mathbf{w} = [11, 2, -10]$

   (b) $\mathbf{v} = [-6, 6, 3]$; $\mathbf{w} = [-12, 9, -8]$

   (c) $\mathbf{v} = [9, 3, -1, 2]$; $\mathbf{w} = [2, 4, -3, 5]$

   (d) $\mathbf{v} = [6, -1, 0, 3, 5]$; $\mathbf{w} = [7, -1, 9, -6, -5]$

4. In each part, find the angle between the two given vectors. Give your answer in degrees, rounded to three places after the decimal point.

   (a) $\mathbf{v} = [12, -1, 12]$; $\mathbf{w} = [2, -2, 1]$

   (b) $\mathbf{v} = [4, 2, -6]$; $\mathbf{w} = [-6, -3, 9]$

   (c) $\mathbf{v} = [6, -5, 1, 4]$; $\mathbf{w} = [5, 8, -2, 3]$

   (d) $\mathbf{v} = [8, 0, -3, 7, 2]$; $\mathbf{w} = [-2, 1, 7, 3, 4]$

5. In each part, verify that the given set of vectors is mutually orthogonal.

   (a) $\{[6, 1, -4], [5, 2, 8], [0, 0, 0], [16, -68, 7]\}$

(b) $\{[4, 1, 2, -3], [2, 19, 0, 9], [1, -2, 5, 4]\}$

6. In each part, verify that the given set of vectors is mutually orthogonal.

   (a) $\{[-17, 38, 23], [0, 0, 0], [4, 3, -2], [5, -2, 7]\}$
   (b) $\{[1, 2, -1, 2], [-5, 0, 3, 4], [2, -1, 2, 1], [0, 5, 4, -3]\}$

7. Normalize the vectors in the mutually orthogonal set

$$\{[2, 10, 11], [1, 2, -2], [14, -5, 2]\}$$

   to produce an orthonormal set of vectors. Use fractions in lowest terms in your answers for the coordinates of the vectors instead of using decimals.

8. Normalize the vectors in the mutually orthogonal set

$$\{[5, 3, -1, 0, 1], [2, -4, 0, 1, 2], [1, 1, 5, 8, -3]\}$$

   to produce an orthonormal set of vectors. Use fractions in lowest terms in your answers for the coordinates of the vectors instead of using decimals.

9. In the parallelograms in each of parts (a) and (b) of Figure 3.18, the lengths of two adjacent sides and the lengths of the diagonals are given. In each parallelogram, use the Polarization Identity to find the dot product of the vectors (having a common initial point) along the two adjacent sides whose lengths are given, and then use this answer to find the angle between those vectors. Then, use this result to calculate the remaining angles of the parallelogram. Give your final answers in degrees, rounded to the nearest whole number, but do not round any of your intermediate computations.

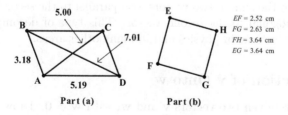

Figure 3.18: Figures for Exercise 9

10. In the parallelograms in each of parts (a) and (b) of Figure 3.19, the lengths of two adjacent sides and the lengths of the diagonals are given. In each parallelogram, use the Polarization Identity to find the dot product of the vectors (having a common initial point) along the two adjacent sides whose lengths are given, and then use this answer to find the angle between those vectors. Then, use this result to calculate the remaining angles of the parallelogram. Give your final answers in degrees, rounded to 1 place after the decimal point, but do not round any of your intermediate computations.

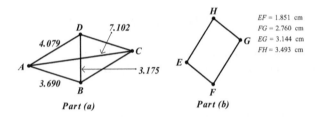

Figure 3.19: Figures for Exercise 10

11. If $\mathbf{v}$ and $\mathbf{w}$ are nonzero vectors and $\mathbf{v} \cdot \mathbf{w} = \|\mathbf{v}\| \|\mathbf{w}\|$, how are $\mathbf{v}$ and $\mathbf{w}$ related geometrically? (Hint: For $0° \leq \theta \leq 180°$, the value of $\cos(\theta)$ equals 1 only when $\theta = 0°$.)

12. If $\mathbf{v}$ and $\mathbf{w}$ are nonzero vectors and $\mathbf{v} \cdot \mathbf{w} = -\|\mathbf{v}\| \|\mathbf{w}\|$, how are $\mathbf{v}$ and $\mathbf{w}$ related geometrically? (Hint: For $0° \leq \theta \leq 180°$, the value of $\cos(\theta)$ equals -1 only when $\theta = 180°$.)

3.5 Projection Vectors

The main result of this section is another decomposition theorem, called the Projection Theorem. Given two vectors, we will show how to decompose the first vector into the sum of two vectors, one parallel to the second vector and the other orthogonal to the second vector. This type of decomposition has important applications in physics and in many other areas.

The Projection of v Onto w

Suppose we are given two vectors $\mathbf{v}$ and $\mathbf{w}$, with $\mathbf{w} \neq \mathbf{0}$. Draw $\mathbf{v}$ and $\mathbf{w}$ so that they have the same initial point. Then, drop a perpendicular line from the terminal point of $\mathbf{v}$ to the line l through $\mathbf{w}$, as shown in Figure 3.20. Consider the vector $\mathbf{p}$ that begins at the common initial point of $\mathbf{v}$ and $\mathbf{w}$ and ends where the perpendicular line meets l. This vector $\mathbf{p}$ is called the

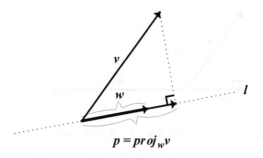

Figure 3.20: Projection of **v** Onto **w**

projection of v onto w, and is usually written as $\mathbf{proj_w v}$. (The vector $\mathbf{proj_w v}$ is sometimes referred to as the **component of v parallel to w**.)

From trigonometry, we can derive a formula for $\mathbf{proj_w v}$ using the angle between **v** and **w**. Then, we can use the formula for the angle between two vectors to simplify the trigonometric formula in order to obtain the following:

$$\mathbf{proj_w v} = \left(\frac{\mathbf{v} \cdot \mathbf{w}}{\|\mathbf{w}\|^2} \right) \mathbf{w}.$$

Notice that the part of this formula in parentheses is just a scalar, since it is a dot product divided by the square of a norm. This confirms algebraically that $\mathbf{proj_w v}$ is simply a scalar multiple of the vector **w**, and therefore parallel to **w** (unless, of course, $\mathbf{proj_w v}$ equals the zero vector).

For example, suppose $\mathbf{v} = [4, -1, 5]$ and $\mathbf{w} = [6, 2, -3]$. Then

$$
\begin{aligned}
\mathbf{proj_w v} &= \left(\frac{\mathbf{v} \cdot \mathbf{w}}{\|\mathbf{w}\|^2} \right) \mathbf{w} = \left(\frac{[4, -1, 5] \cdot [6, 2, -3]}{\|[6, 2, -3]\|^2} \right) [6, 2, -3] \\
&= \left(\frac{7}{49} \right) [6, 2, -3] = \left(\frac{1}{7} \right) [6, 2, -3] = \left[\frac{6}{7}, \frac{2}{7}, -\frac{3}{7} \right].
\end{aligned}
$$

In this case, $\mathbf{proj_w v} = \frac{1}{7}\mathbf{w}$, a scalar multiple of **w**.

Take some time to familiarize yourself with this new notation. A common error that students often make when asked to compute $\mathbf{proj_w v}$ is accidentally reversing the roles of **v** and **w**. Remember that $\mathbf{proj_w v}$ is always parallel to **w**, the "subscript" vector.

When the angle between **v** and **w** is obtuse, the dropped perpendicular intersects the line l "behind" **w**, as illustrated in Figure 3.21, so that $\mathbf{proj_w v}$ is actually in the opposite direction as **w**. However, $\mathbf{proj_w v}$ is still parallel to **w** here since it has the opposite direction as **w**. The formula for $\mathbf{proj_w v}$ given above is also valid in the obtuse case.

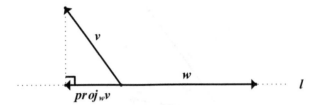

Figure 3.21: Projection of **v** Onto **w** When the Angle Between **v** and **w** is Obtuse

For example, suppose $\mathbf{v} = [1, -2, 3]$ and $\mathbf{w} = [2, 5, -1]$. Then

$$
\begin{aligned}
\mathbf{proj_w v} &= \left(\frac{\mathbf{v} \cdot \mathbf{w}}{\|\mathbf{w}\|^2} \right) \mathbf{w} = \left(\frac{[1, -2, 3] \cdot [2, 5, -1]}{\|[2, 5, -1]\|^2} \right) [2, 5, -1] \\
&= \left(\frac{-11}{30} \right) [2, 5, -1] = \left[-\frac{11}{15}, -\frac{11}{6}, \frac{11}{30} \right].
\end{aligned}
$$

Notice in this example that $\mathbf{v} \cdot \mathbf{w}$ is negative, and so, by AR3, the angle between **v** and **w** is obtuse. The projection vector $\mathbf{proj_w v} = -\frac{11}{30}\mathbf{w}$, a negative scalar multiple of **w**, and so in this case, $\mathbf{proj_w v}$ has the opposite direction as **w**.

The Projection Theorem

Consider Figure 3.22 below, containing vectors **v**, **w**, and $\mathbf{p} = \mathbf{proj_w v}$. Let **r** be the vector whose initial point is at the terminal point of $\mathbf{p} = \mathbf{proj_w v}$, and whose terminal point is the same as the terminal point of **v**. From the Follow Theorem, we know that $\mathbf{v} = \mathbf{p} + \mathbf{r}$. That is, $\mathbf{r} = \mathbf{v} - \mathbf{p} = \mathbf{v} - \mathbf{proj_w v}$.

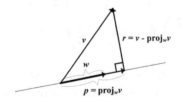

Figure 3.22: The Projection Theorem

To summarize, we have $\mathbf{v} = \mathbf{p} + \mathbf{r} = \mathbf{proj_w v} + (\mathbf{v} - \mathbf{proj_w v})$. This is a way of decomposing **v** into the sum of two component vectors, one (assuming it is nonzero) parallel to **w**, and the other orthogonal to **w**. The next theorem states that this is the only possible way to decompose **v** in such a manner.

Theorem 5. (*Projection Theorem*) *Suppose* $\mathbf{v}$ *and* $\mathbf{w}$ *are two vectors having the same dimension with* $\mathbf{w} \neq \mathbf{0}$. *Then, assuming that* $\mathbf{proj_w v} \neq \mathbf{0}$, *there is one and only one way to express* $\mathbf{v}$ *as the sum of two vectors so that one of the vectors in the sum is parallel to* $\mathbf{w}$ *and the other is orthogonal to* $\mathbf{w}$. *This unique decomposition of* $\mathbf{v}$ *is given by* $\mathbf{v} = \mathbf{proj_w v} + (\mathbf{v} - \mathbf{proj_w v})$.

(Although the proof that this type of decomposition of $\mathbf{v}$ is unique is not very difficult, we will not present it here.)

Proof of the Projection Theorem: Suppose we are given vectors $\mathbf{v}$ and $\mathbf{w}$ and want to compute the Projection Theorem decomposition of $\mathbf{v}$. Now, we already know the formula for one of the vectors in the sum for $\mathbf{v}$; namely, $\mathbf{p} = \mathbf{proj_w v} = \left(\frac{\mathbf{v} \cdot \mathbf{w}}{\|\mathbf{w}\|^2} \right) \mathbf{w}$. That is, $\mathbf{proj_w v}$ is a scalar multiple, $c\mathbf{w}$, of $\mathbf{w}$, where $c = \left(\frac{\mathbf{v} \cdot \mathbf{w}}{\|\mathbf{w}\|^2} \right)$. Once we have calculated $\mathbf{proj_w v}$, we can use it to find $\mathbf{r} = \mathbf{v} - \mathbf{proj_w v}$, the other vector in the sum for $\mathbf{v}$.

For example, consider $\mathbf{v} = [3, 2, 5, -4]$ and $\mathbf{w} = [-1, 3, 2, 1]$. Then

$$
\begin{aligned}
\mathbf{p} = \mathbf{proj_w v} &= \left(\frac{\mathbf{v} \cdot \mathbf{w}}{\|\mathbf{w}\|^2} \right) \mathbf{w} = \left(\frac{[3,2,5,-4] \cdot [-1,3,2,1]}{\|[-1,3,2,1]\|^2} \right) [-1,3,2,1] \\
&= \frac{9}{15} [-1,3,2,1] = \frac{3}{5} [-1,3,2,1] = \left[-\frac{3}{5}, \frac{9}{5}, \frac{6}{5}, \frac{3}{5} \right] = \frac{3}{5} \mathbf{w}.
\end{aligned}
$$

Then, we can use this result to calculate $\mathbf{r}$, as follows:

$$
\begin{aligned}
\mathbf{r} = \mathbf{v} - \mathbf{proj_w v} &= [3,2,5,-4] - \left[-\frac{3}{5}, \frac{9}{5}, \frac{6}{5}, \frac{3}{5} \right] \\
&= \left[\frac{15}{5}, \frac{10}{5}, \frac{25}{5}, -\frac{20}{5} \right] - \left[-\frac{3}{5}, \frac{9}{5}, \frac{6}{5}, \frac{3}{5} \right] = \left[\frac{18}{5}, \frac{1}{5}, \frac{19}{5}, -\frac{23}{5} \right].
\end{aligned}
$$

As a check on our work, we can easily verify that $\mathbf{p} + \mathbf{r} = \mathbf{v}$, since

$$
\begin{aligned}
\mathbf{p} + \mathbf{r} &= \left[-\frac{3}{5}, \frac{9}{5}, \frac{6}{5}, \frac{3}{5} \right] + \left[\frac{18}{5}, \frac{1}{5}, \frac{19}{5}, -\frac{23}{5} \right] \\
&= \left[\frac{15}{5}, \frac{10}{5}, \frac{25}{5}, -\frac{20}{5} \right] = [3, 2, 5, -4] = \mathbf{v},
\end{aligned}
$$

and that $\mathbf{r}$ is orthogonal to $\mathbf{w}$ because

$$
\begin{aligned}
\mathbf{w} \cdot \mathbf{r} &= [-1,3,2,1] \cdot \left[\frac{18}{5}, \frac{1}{5}, \frac{19}{5}, -\frac{23}{5} \right] \\
&= (-1) \left(\frac{18}{5} \right) + (3) \left(\frac{1}{5} \right) + (2) \left(\frac{19}{5} \right) + (1) \left(-\frac{23}{5} \right) \\
&= -\frac{18}{5} + \frac{3}{5} + \frac{38}{5} - \frac{23}{5} = 0.
\end{aligned}
$$

In fact, we can easily prove algebraically that, in general, **w** and **r** are orthogonal, as follows:

$$
\begin{aligned}
\mathbf{w} \cdot \mathbf{r} &= \mathbf{w} \cdot (\mathbf{v} - \mathbf{proj_w v}) & \text{definition of } \mathbf{r} \\
&= \mathbf{w} \cdot \mathbf{v} - \mathbf{w} \cdot \mathbf{proj_w v} & \text{by DP4} \\
&= \mathbf{w} \cdot \mathbf{v} - \mathbf{w} \cdot \left(\left(\frac{\mathbf{v} \cdot \mathbf{w}}{\|\mathbf{w}\|^2} \right) \mathbf{w} \right) & \text{formula for } \mathbf{proj_w v} \\
&= \mathbf{w} \cdot \mathbf{v} - \left(\frac{\mathbf{v} \cdot \mathbf{w}}{\|\mathbf{w}\|^2} \right) (\mathbf{w} \cdot \mathbf{w}) & \text{by DP3} \\
&= \mathbf{w} \cdot \mathbf{v} - \left(\frac{\mathbf{v} \cdot \mathbf{w}}{\|\mathbf{w}\|^2} \right) \|\mathbf{w}\|^2 & \text{by DP1} \\
&= \mathbf{w} \cdot \mathbf{v} - \mathbf{v} \cdot \mathbf{w} & \text{cancelling the } \|\mathbf{w}\|^2 \text{ factors} \\
&= 0 \quad \text{by DP2. } \blacklozenge
\end{aligned}
$$

A Box on a Ramp

Let us consider a simple example from physics in which the projection vector plays an important role.

Example 1. Suppose a box weighing 12 pounds is sitting on a ramp. The base of the ramp is 18 feet long. The lower end of the ramp is 8 feet below the higher end of the ramp. The force of gravity is pulling downward on this box. We want to compute the component of the gravitational force that is pushing the box along the ramp, as well as the component of the gravitational force that is pushing the box against the ramp. This situation is illustrated in Figure 3.23.

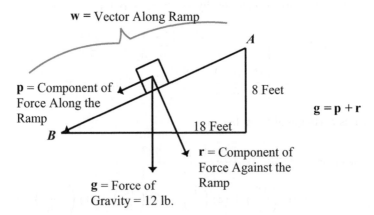

Figure 3.23: Components of Gravitational Force Acting on a Box on a Ramp

Imagine this ramp in a 2-dimensional coordinate system, with the origin at point B and the base of the ramp along the x-axis. Notice that the vector from A to B lies along the ramp; let us call this vector $\mathbf{w}$. The initial point of $\mathbf{w}$ is $A = (18,8)$ and the terminal point of $\mathbf{w}$ is $B = (0,0)$, and so $\mathbf{w} = [-18,-8]$. The vector, $\mathbf{g}$, representing the force of gravity, points vertically downward and has magnitude 12, representing 12 lbs. of force in that direction. Hence, $\mathbf{g} = [0,-12]$.

Here, the vector $\mathbf{g}$ is going to play the role of $\mathbf{v}$ in the Projection Theorem. We will decompose $\mathbf{g}$ into two forces, one force, $\mathbf{p}$, that is parallel to $\mathbf{w}$ (that is, parallel to the ramp), and a second force, $\mathbf{r}$, that is orthogonal to $\mathbf{w}$ (that is, orthogonal to the ramp).

Now, the component of $\mathbf{g}$ parallel to $\mathbf{w}$ is

$$\mathbf{p} = \mathbf{proj_w g} = \left(\frac{\mathbf{g} \cdot \mathbf{w}}{\|\mathbf{w}\|^2}\right)\mathbf{w} = \left(\frac{[0,-12] \cdot [-18,-8]}{\|[-18,-8]\|^2}\right)[-18,-8]$$

$$= \frac{96}{388}[-18,-8] = \left[-\frac{432}{97}, -\frac{192}{97}\right] \approx [-4.45, -1.98].$$

Then, the magnitude of this force $\mathbf{p}$ pushing the box along the ramp is $\|\mathbf{p}\| \approx \|[-4.45,-1.98]\| \approx 4.87$ lbs.

Also, the component of $\mathbf{g}$ orthogonal to $\mathbf{w}$ is

$$\mathbf{r} = \mathbf{g} - \mathbf{proj_w g} = [0,-12] - \left[-\frac{432}{97}, -\frac{192}{97}\right]$$

$$= \left[\frac{432}{97}, -\frac{972}{97}\right] \approx [4.45, -10.02].$$

Then, the magnitude of this force $\mathbf{r}$ pushing the box against the ramp is $\|\mathbf{r}\| \approx \|[4.45,-10.02]\| \approx 10.96$ lbs. (The value of $\|\mathbf{r}\|$ is one of the factors that contributes to the force of friction that slows the progress of the box as it moves along the ramp. In this type of situation, the force of friction is often approximately proportional to $\|\mathbf{r}\|$.) ∎

Mutually Orthogonal Vectors and the Gram-Schmidt Process

Occasionally, we will be presented with a set of vectors and want to replace them with a related set of mutually orthogonal vectors. For a set $\{\mathbf{x},\mathbf{y}\}$ of two vectors, the Projection Theorem makes this process relatively easy. We can "adjust" $\mathbf{y}$ by replacing it with $\mathbf{z} = \mathbf{y} - \mathbf{proj_x y}$, which we know is orthogonal to $\mathbf{x}$. Then, $\{\mathbf{x},\mathbf{z}\}$ is a mutually orthogonal set of vectors related to the original set $\{\mathbf{x},\mathbf{y}\}$. What we have actually done here is use subtraction to "throw away" the component of $\mathbf{y}$ that is parallel to $\mathbf{x}$, thereby retaining only the component of $\mathbf{y}$ that is orthogonal to $\mathbf{x}$.

We can carry out a similar process with three or more vectors, as shown in the following example.

Example 2. Suppose $\mathbf{v}_1 = [4, 1, -3]$, $\mathbf{v}_2 = [-7, -3, 7]$, and $\mathbf{v}_3 = [-6, -1, 9]$. We will leave the first vector as is, and then "adjust" the other two vectors to form a mutually orthogonal set $\{\mathbf{x}_1, \mathbf{x}_2, \mathbf{x}_3\}$. Let $\mathbf{x}_1 = \mathbf{v}_1 = [4, 1, -3]$, the first vector. Next, we use the Projection Theorem with $\mathbf{v}_2$ to find a second vector, $\mathbf{x}_2$, that is orthogonal to $\mathbf{x}_1$, as follows:

$$
\begin{aligned}
\mathbf{x}_2 &= \mathbf{v}_2 - \left(\mathbf{proj}_{\mathbf{x}_1} \mathbf{v}_2\right) \\
&= \mathbf{v}_2 - \left(\frac{\mathbf{v}_2 \cdot \mathbf{x}_1}{\|\mathbf{x}_1\|^2}\right) \mathbf{x}_1 \\
&= [-7, -3, 7] - \left(\frac{[-7, -3, 7] \cdot [4, 1, -3]}{\|[4, 1, -3]\|^2}\right) [4, 1, -3] \\
&= [-7, -3, 7] - \left(\frac{-52}{26}\right) [4, 1, -3] \\
&= [-7, -3, 7] + 2[4, 1, -3] = [-7, -3, 7] + [8, 2, -6] = [1, -1, 1].
\end{aligned}
$$

Notice that $\mathbf{x}_1 \cdot \mathbf{x}_2 = [4, 1, -3] \cdot [1, -1, 1] = 0$, confirming that $\mathbf{x}_1$ and $\mathbf{x}_2$ are orthogonal to each other. (We accomplished this by subtracting away the component of $\mathbf{v}_2$ that was parallel to $\mathbf{x}_1$, which left only the component orthogonal to $\mathbf{x}_1$.)

To finish the process, we use $\mathbf{v}_3$ to find a vector $\mathbf{x}_3$ that is orthogonal to both $\mathbf{x}_1$ and $\mathbf{x}_2$. To do this, we start with $\mathbf{v}_3$ and subtract off both the component of $\mathbf{v}_3$ in the $\mathbf{x}_1$ direction *as well as* the component of $\mathbf{v}_3$ in the $\mathbf{x}_2$ direction. This will leave only the component of $\mathbf{v}_3$ that is orthogonal to both $\mathbf{x}_1$ and $\mathbf{x}_2$. Thus,

$$
\begin{aligned}
\mathbf{x}_3 &= \mathbf{v}_3 - \left(\mathbf{proj}_{\mathbf{x}_1} \mathbf{v}_3\right) - \left(\mathbf{proj}_{\mathbf{x}_2} \mathbf{v}_3\right) \\
&= \mathbf{v}_3 - \left(\frac{\mathbf{v}_3 \cdot \mathbf{x}_1}{\|\mathbf{x}_1\|^2}\right) \mathbf{x}_1 - \left(\frac{\mathbf{v}_3 \cdot \mathbf{x}_2}{\|\mathbf{x}_2\|^2}\right) \mathbf{x}_2 \\
&= [-6, -1, 9] - \left(\frac{[-6, -1, 9] \cdot [4, 1, -3]}{\|[4, 1, -3]\|^2}\right) [4, 1, -3] \\
&\qquad - \left(\frac{[-6, -1, 9] \cdot [1, -1, 1]}{\|[1, -1, 1]\|^2}\right) [1, -1, 1] \\
&= [-6, -1, 9] - \left(\frac{-52}{26}\right) [4, 1, -3] - \left(\frac{4}{3}\right) [1, -1, 1] \\
&= [-6, -1, 9] + 2[4, 1, -3] - \left[\frac{4}{3}, -\frac{4}{3}, \frac{4}{3}\right] = \left[\frac{2}{3}, \frac{7}{3}, \frac{5}{3}\right].
\end{aligned}
$$

Since we already know that $\mathbf{x}_1 \cdot \mathbf{x}_2 = 0$, we can confirm that $\{\mathbf{x}_1, \mathbf{x}_2, \mathbf{x}_3\}$ is a mutually orthogonal set by checking that $\mathbf{x}_1 \cdot \mathbf{x}_3 = [4, 1, -3] \cdot [\frac{2}{3}, \frac{7}{3}, \frac{5}{3}] = 0$ and $\mathbf{x}_2 \cdot \mathbf{x}_3 = [1, -1, 1] \cdot [\frac{2}{3}, \frac{7}{3}, \frac{5}{3}] = 0$. Thus, we have created a mutually

orthogonal set $\{x_1, x_2, x_3\}$ of vectors related to the original set $\{v_1, v_2, v_3\}$.
∎

The method illustrated in the last example is called the **Gram-Schmidt Process**. It takes a given set of vectors and "adjusts" them to form a related mutually orthogonal set. This method can be extended to any number of vectors. For each successive vector, we simply subtract away its projection onto each of the previously "adjusted" vectors.

More formally, given a set $\{v_1, v_2, \ldots, v_n\}$ of vectors, we create a related set $\{x_1, x_2, \ldots, x_n\}$ of mutually orthogonal vectors as follows:

Step 1: Let $x_1 = v_1$.
Step 2: Let $x_2 = v_2 - (\mathbf{proj}_{x_1} v_2)$.
Step 3: Let $x_3 = v_3 - (\mathbf{proj}_{x_1} v_3) - (\mathbf{proj}_{x_2} v_3)$.
Step 4: Let $x_4 = v_4 - (\mathbf{proj}_{x_1} v_4) - (\mathbf{proj}_{x_2} v_4) - (\mathbf{proj}_{x_3} v_4)$, etc.

Unfortunately, the computations involved in the Gram-Schmidt Process often produce ugly and complicated fractions. One method that is often used to make the results more manageable is to substitute an appropriate positive scalar multiple for any of the x_i vectors in order to eliminate denominators. For instance, in the previous example, we could replace $x_3 = \left[\frac{2}{3}, \frac{7}{3}, \frac{5}{3}\right]$ with $3\left[\frac{2}{3}, \frac{7}{3}, \frac{5}{3}\right] = [2, 7, 5]$ to obtain the simpler-looking mutually orthogonal set $\{[4, 1, -3], [1, -1, 1], [2, 7, 5]\}$ instead. (Of course, replacing any of these vectors with a positive scalar multiple will not alter the fact that they are mutually orthogonal. This is because multiplying a nonzero vector by a positive scalar will not change the direction of that vector. Hence, the nonzero vectors in the set will still be perpendicular to each other.) Another advantage to replacing a vector with a simpler-looking multiple is that it often makes the calculations for any remaining steps much easier. However, we must always take special care *never* to perform this type of substitution midway through one of the steps of the Gram-Schmidt Process listed above, but *only between steps*. This is illustrated in the next example, in which two vector replacements are performed.

Example 3. Suppose $v_1 = [-1, 0, 5, 2]$, $v_2 = [0, 0, 2, 1]$, $v_3 = [6, 1, -7, -2]$, and $v_4 = [-1, -1, 7, 12]$. We will use the Gram-Schmidt Process to create a related set of four mutually orthogonal vectors $\{x_1, x_2, x_3, x_4\}$. We leave the first vector as is; that is, we let $x_1 = v_1 = [-1, 0, 5, 2]$.

Next, we calculate x_2:

$$
\begin{aligned}
x_2 &= v_2 - (\mathbf{proj}_{x_1} v_2) \\
&= v_2 - \left(\frac{v_2 \cdot x_1}{\|x_1\|^2}\right) x_1 \\
&= [0, 0, 2, 1] - \left(\frac{[0, 0, 2, 1] \cdot [-1, 0, 5, 2]}{\|[-1, 0, 5, 2]\|^2}\right)[-1, 0, 5, 2]
\end{aligned}
$$

$$= [0,0,2,1] - \left(\frac{-12}{30}\right) [-1,0,5,2]$$

$$= [0,0,2,1] - \frac{2}{5}[-1,0,5,2] = \left[\frac{2}{5},0,0,\frac{1}{5}\right].$$

In order to remove fractions, and to simplify further calculations, we multiply this final vector by 5 and use $\mathbf{x}_2 = [2,0,0,1]$ in all that follows.

Next, we calculate $\mathbf{x}_3$:

$$\mathbf{x}_3 = \mathbf{v}_3 - \left(\mathbf{proj}_{\mathbf{x}_1}\mathbf{v}_3\right) - \left(\mathbf{proj}_{\mathbf{x}_2}\mathbf{v}_3\right)$$

$$= \mathbf{v}_3 - \left(\frac{\mathbf{v}_3 \cdot \mathbf{x}_1}{\|\mathbf{x}_1\|^2}\right)\mathbf{x}_1 - \left(\frac{\mathbf{v}_3 \cdot \mathbf{x}_2}{\|\mathbf{x}_2\|^2}\right)\mathbf{x}_2$$

$$= [6,1,-7,-2] - \left(\frac{[6,1,-7,-2]\cdot[-1,0,5,2]}{\|[-1,0,5,2]\|^2}\right)[-1,0,5,2]$$

$$\qquad - \left(\frac{[6,1,-7,-2]\cdot[2,0,0,1]}{\|[2,0,0,1]\|^2}\right)[2,0,0,1]$$

$$= [6,1,-7,-2] - \left(\frac{-45}{30}\right)[-1,0,5,2] - \left(\frac{10}{5}\right)[2,0,0,1]$$

$$= [6,1,-7,-2] + \frac{3}{2}[-1,0,5,2] - 2[2,0,0,1] = \left[\frac{1}{2},1,\frac{1}{2},-1\right].$$

We multiply this final vector by 2 to simplify further calculations, and use $\mathbf{x}_3 = [1,2,1,-2]$ in all that follows.

Finally, we calculate $\mathbf{x}_4$:

$$\mathbf{x}_4 = \mathbf{v}_4 - \left(\mathbf{proj}_{\mathbf{x}_1}\mathbf{v}_4\right) - \left(\mathbf{proj}_{\mathbf{x}_2}\mathbf{v}_4\right) - \left(\mathbf{proj}_{\mathbf{x}_3}\mathbf{v}_4\right)$$

$$= \mathbf{v}_4 - \left(\frac{\mathbf{v}_4 \cdot \mathbf{x}_1}{\|\mathbf{x}_1\|^2}\right)\mathbf{x}_1 - \left(\frac{\mathbf{v}_4 \cdot \mathbf{x}_2}{\|\mathbf{x}_2\|^2}\right)\mathbf{x}_2 - \left(\frac{\mathbf{v}_4 \cdot \mathbf{x}_3}{\|\mathbf{x}_3\|^2}\right)\mathbf{x}_3$$

$$= [-1,-1,7,12] - \left(\frac{[-1,-1,7,12]\cdot[-1,0,5,2]}{\|[-1,0,5,2]\|^2}\right)[-1,0,5,2]$$

$$\qquad - \left(\frac{[-1,-1,7,12]\cdot[2,0,0,1]}{\|[2,0,0,1]\|^2}\right)[2,0,0,1]$$

$$\qquad - \left(\frac{[-1,-1,7,12]\cdot[1,2,1,-2]}{\|[1,2,1,-2]\|^2}\right)[1,2,1,-2]$$

$$= [-1,-1,7,12] - \left(\frac{60}{30}\right)[-1,0,5,2] - \left(\frac{10}{5}\right)[2,0,0,1]$$

$$\qquad - \left(\frac{-20}{10}\right)[1,2,1,-2]$$

$$= [-1, -1, 7, 12] - 2[-1, 0, 5, 2] - 2[2, 0, 0, 1] + 2[1, 2, 1, -2]$$
$$= [-1, 3, -1, 2].$$

You can easily verify that $\{x_1, x_2, x_3, x_4\} = \{[-1, 0, 5, 2], [2, 0, 0, 1],$ $[1, 2, 1, -2], [-1, 3, -1, 2]\}$ is a mutually orthogonal set of vectors. ∎

Occasionally, when performing the Gram-Schmidt Process, one of the "adjusted" x_i vectors could turn out to be the zero vector.[5] However, this is not a problem because the zero vector is automatically orthogonal to all the others. Also, this simplifies any further computations, because there is no need to compute any projections of other vectors onto the zero vector (since those projections would also be zero vectors).

Glossary

- Component of v parallel to w: The vector $\mathbf{proj_w v}$; that is, the projection of v onto w (assuming v is not orthogonal to w).

- Component of v orthogonal to w: The vector $v - (\mathbf{proj_w v})$.

- Gram-Schmidt Process: A method for "adjusting" a given set of vectors to form a related mutually orthogonal set of vectors. Each subsequent vector v_i in the set is replaced by a vector x_i that is obtained by subtracting away any components of v_i that are parallel to each of the previously "adjusted" vectors in the set.

- Projection of v onto w: Symbolically, $\mathbf{proj_w v}$. Given v and w, with $w \neq 0$, the vector $\mathbf{proj_w v} = \left(\frac{v \cdot w}{\|w\|^2} \right) w$. If $\mathbf{proj_w v} \neq 0$, then $\mathbf{proj_w v}$ is the component of v parallel to w.

Exercises for Section 3.5

1. In each part, compute $\mathbf{proj_w v}$. Round your answer to two places after the decimal point.

   (a) $v = [4, -1, 5]$; $w = [5, 2, -1]$
   (b) $v = [1, 7, -3, 4]$; $w = [2, 0, -2, 1]$

2. In each part, compute $\mathbf{proj_w v}$. Round your answer to two places after the decimal point.

   (a) $v = [5, 2, -1]$; $w = [4, -1, 5]$
   (b) $v = [8, -3, 4, -2]$; $w = [4, -2, 2, 1]$

---

[5] In fact, more than one of the x_i's could be the zero vector. In such a case, the number of distinct vectors in the final mutually orthogonal set will be fewer than the original number of vectors.

3. In each part, decompose the vector **v** into the sum **p** + **r** of two vectors, so that **p** is parallel to **w**, and **r** is orthogonal to **w**. Round your answers to one place after the decimal point. Then verify that **r** is orthogonal to **w** by computing an appropriate dot product.

   (a) **v** = $[17, -5, 16]$; **w** = $[6, -4, 9]$

   (b) **v** = $[7, 9, -3, 6]$; **w** = $[3, -2, 1, -4]$

4. In each part, decompose the vector **v** into the sum **p** + **r** of two vectors, so that **p** is parallel to **w**, and **r** is orthogonal to **w**. Round your answers to two places after the decimal point. Then verify that **r** is orthogonal to **w** by computing an appropriate dot product. (Note: This dot product may not actually equal zero due to roundoff error, but should be very close to zero.)

   (a) **v** = $[12, -5, 6]$; **w** = $[-3, 2, -8]$

   (b) **v** = $[-9, 16, -3, 10]$; **w** = $[2, -5, 5, -3]$

5. In each part, decompose the vector **x** into the sum **s** + **t** of two vectors, so that **s** is parallel to **y**, and **t** is orthogonal to **y**. Round your answers to one place after the decimal point. Then verify that **t** is orthogonal to **y** by computing an appropriate dot product. (Note: This dot product may not actually equal zero due to roundoff error, but should be very close to zero.)

   (a) **x** = $[3, 1, -8]$; **y** = $[-2, 4, 5]$

   (b) **x** = $[-2, 6, 4, 5]$; **y** = $[-4, 0, -2, 3]$

6. In each part, decompose the vector **x** into the sum **s** + **t** of two vectors, so that **s** is parallel to **y**, and **t** is orthogonal to **y**. Round your answers to two places after the decimal point. Then verify that **t** is orthogonal to **y** by computing an appropriate dot product. (Note: This dot product may not actually equal zero due to roundoff error, but should be very close to zero.)

   (a) **x** = $[8, 2, -5]$; **y** = $[-6, 5, 3]$

   (b) **x** = $[-6, 4, 2, 1]$; **y** = $[1, 7, 4, -5]$

7. In each part, given vectors **y** and **z**, find a scalar c and vector **r** such that **z** = c**y** + **r**, and **y** · **r** = 0. Round your answers to two places after the decimal point. Also, state whether or not your answers are the only possible answers.

   (a) **y** = $[6, 1, 5]$; **z** = $[9, -2, 3]$

   (b) **y** = $[6, -2, 4]$; **z** = $[-9, 3, -6]$

   (c) **y** = $[3, 2, -1, 5]$; **z** = $[6, -5, -2, -2]$

8. In each part, given vectors **y** and **z**, find a scalar c and vector **r** such that $\mathbf{z} = c\mathbf{y} + \mathbf{r}$, and $\mathbf{y} \cdot \mathbf{r} = 0$. Round your answers to two places after the decimal point. Also, state whether or not your answers are the only possible answers.

   (a) $\mathbf{y} = [7, 1, 3]$; $\mathbf{z} = [1, -4, -1]$
   (b) $\mathbf{y} = [0, 0, 0]$; $\mathbf{z} = [-6, 7, 8]$
   (c) $\mathbf{y} = [5, 1, -1, 2]$; $\mathbf{z} = [3, 2, 9, -6]$
   (d) $\mathbf{y} = [2, 8, -6, -4]$; $\mathbf{z} = [-5, -20, 15, 10]$

9. A box weighing 25 pounds is sitting on a ramp that descends from right to left. The base of the ramp is 20 feet long. The top of the ramp is 6 feet higher than the bottom of the ramp. Compute the gravitational force that is pushing the box along the ramp and the gravitational force that is pushing the box against the ramp. State the forces in vector form, and then compute the magnitudes of these forces. Round your answers to two places after the decimal point.

10. A pendulum, as pictured in Figure 3.24, is swinging downward. The ball at the end of the pendulum weighs 10 lbs. (Assume that the string has negligible weight.) Also assume that the vector **w** along the string is currently $[-4, -3]$ as shown. Find the vector force, **p**, that gravity is exerting in the direction parallel to **w** (that is, along the string). This force is applying tension to the string. The vector component **r** of the force of gravity orthogonal to **w** (that is, orthogonal to the string) causes the ball to swing in a circular arc. Since the vector **r** is perpendicular to **w**, it follows that **r** is always tangent to the circular arc, as shown. Solve for **r**. Then, find the magnitudes of both **p** and **r**. Round your answers to two places after the decimal point.

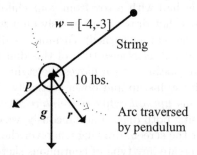

Figure 3.24: Swinging Pendulum for Exercise 10

11. In each part, use the Gram-Schmidt Process to "adjust" the given set of vectors to create a related mutually orthogonal set of vectors. If

necessary, replace each "adjusted" vector with an appropriate scalar multiple so that the vectors in the final mutually orthogonal set contain only whole numbers.

(a) $\{[1, 3, 2], [4, 8, 7], [7, -9, -4]\}$

(b) $\{[1, 3, 1, 2], [4, 1, -5, 2], [23, 14, -22, 16], [5, -7, 0, 2]\}$

12. In each part, use the Gram-Schmidt Process to "adjust" the given set of vectors to create a related mutually orthogonal set of vectors. If necessary, replace each "adjusted" vector with an appropriate scalar multiple so that the vectors in the final mutually orthogonal set contain only whole numbers.

(a) $\{[5, 7, 4], [7, 15, 10], [27, 17, 4], [2, 13, -14]\}$

(b) $\{[1, -1, 1, -1], [2, 3, 1, 4], [3, -2, 1, 2], [10, -2, 3, -2]\}$

13. Use the answer to part (a) of Exercise 11 (which can be found in Appendix A) to create an orthonormal set of vectors. Round your answers to two places after the decimal point.

14. Use the answer to part (b) of Exercise 11 (which can be found in Appendix A) to create an orthonormal set of vectors. (Eliminate the zero vector first.) Round your answers to two places after the decimal point.

3.6   Linear Combinations with Mutually Orthogonal Vectors

The Etch-a-Sketch^{TR}

Recall the great fun you had with a toy from your childhood – the Etch-a-Sketch^{TR}. It has a stylus that draws continuously on a screen. The stylus is controlled by two dials that can be turned. Turning one dial moves the stylus left and right horizontally. A *clockwise* turn of this dial moves the stylus to the *right*, while a *counterclockwise* turn moves the stylus to the *left*. Turning the other dial moves the stylus up and down vertically. A *clockwise* turn of this dial moves the stylus *upward*, while a *counterclockwise* turn moves the stylus *downward*. By using both dials simultaneously, we can move the stylus in any direction all over the screen, drawing wherever the stylus goes. In this way we can essentially create any type of continuous sketch on the screen.

Now consider the dial that moves the stylus horizontally. Define one unit of distance as the amount the stylus moves during one complete rotation of this dial (assuming the stylus does not reach an edge of the screen). Then, if we turn this dial *clockwise* through one complete rotation, we have moved the stylus from its initial point to some terminal point one unit to the right. The vector between these two points is $\mathbf{e}_1$, the unit vector in the horizontal direction. On the other hand, if we merely turn this dial clockwise halfway around, the movement of the stylus would be represented instead by the vector $\frac{1}{2}\mathbf{e}_1$. Similarly, if we turn this dial *counterclockwise* two times, the movement would be represented by the vector $-2\mathbf{e}_1$. Therefore, using just this dial, we can move the stylus wherever we want, left or right, along a horizontal line; that is, a line parallel to the vector $\mathbf{e}_1$.

A completely analogous situation occurs when we turn the other dial, except that turning it clockwise one time represents $\mathbf{e}_2$ instead of $\mathbf{e}_1$. This other dial allows us to move the stylus any desired scalar multiple of $\mathbf{e}_2$ (at least until an edge of the screen is reached). That is, this other dial moves the stylus wherever we want, up or down, along a vertical line; that is, a line parallel to the vector $\mathbf{e}_2$.

Now suppose we turn both dials together so that the horizontal dial is turned three times clockwise and the vertical dial is turned twice counterclockwise. If, instead, we had turned the dials one at a time, the picture created on the screen would look different, but the commutative, associative, and distributive ESP properties assure us that we arrive at the same final destination regardless of the manner in which we combine these vectors. That is, the final position of the stylus is the same whether the dials are turned simultaneously or not. Thus we know in this particular case that the vector representing the overall movement of the stylus is $3\mathbf{e}_1 - 2\mathbf{e}_2$, a linear combination of the vectors $\mathbf{e}_1$ and $\mathbf{e}_2$ that represent single dial turns. (Recall that a linear combination of vectors is merely a sum of scalar multiples of those vectors.) In this way, we can form any linear combination of $\mathbf{e}_1$ and $\mathbf{e}_2$ by merely making the appropriate number of turns of each dial (unless, of course, we run out of space on the screen).

The Vector Etch-a-Sketch

Now, consider a more general kind of Etch-a-Sketch[TR] – the Vector Etch-a-Sketch! Imagine that this device has a "screen" that is not limited to two dimensions, but can have as many dimensions as we want. Imagine also that we have as many dials as we want, and that the dials are not limited to moving only horizontally and vertically, but that we can set up each dial so that a clockwise turn of that dial corresponds to some vector of our own selection (not necessarily $\mathbf{e}_1$ or $\mathbf{e}_2$). However, all of the dials must correspond to vectors of the same dimension as the "screen." Finally, imagine also that the "screen" extends infinitely in every direction, so we never have the problem of running out of room for a drawing. Of course, constructing such a device is not physically possible, but we can *imagine* it!

For example, we could have a 3-dimensional Vector Etch-a-Sketch with several dials. The screen represents $\mathbb{R}^3$; that is, 3-dimensional space. Suppose the first dial corresponds to the vector $[1, 2, 3]$. That is, turning this dial one full turn in the clockwise direction moves the stylus from its initial point to a terminal point so that it traverses the vector $[1, 2, 3]$. Turning the dial halfway clockwise results in the movement $\frac{1}{2}[1,2,3]$. Turning the dial three times counterclockwise moves the stylus along the vector $-3[1,2,3]$. Using just this dial, we can get any scalar multiple of $[1,2,3]$. But, just as with the regular Etch-a-Sketch^{TR}, using this one dial merely allows us to move back and forth along a line parallel to $[1,2,3]$. The movement from this single dial alone is very limited.

Now suppose a second dial corresponds to the vector $[4,5,6]$. Because $[4,5,6]$ is not a scalar multiple of $[1,2,3]$, the second dial moves in a new direction that is not parallel to $[1,2,3]$. Each turn of this dial moves the stylus a scalar multiple of $[4,5,6]$. That is, using the second dial moves the stylus back and forth along a line parallel to $[4,5,6]$.

Now imagine turning both dials (either simultaneously or separately). The resulting movement of the stylus will be some linear combination of $[1,2,3]$ and $[4,5,6]$. For example, if we turn the first dial three times clockwise and the second dial $\frac{1}{4}$ of a turn counterclockwise, the total movement corresponds to the vector $3[1,2,3] - \frac{1}{4}[4,5,6] = \left[2, \frac{19}{4}, \frac{15}{2}\right]$. A little thought will convince you that because these two dials allow us to move in two separate directions, we can move anywhere in an entire plane. While this is an improvement over merely being able to move along a single line, a plane is still not all of our 3-dimensional screen. There are places above or below that plane that we still cannot reach yet.

Imagine now that we add a third dial. We set up the third dial to correspond to the vector $[7,8,9]$. By turning all three dials, we can form linear combinations using all three of our dial vectors, such as $-5[1,2,3]+3[4,5,6]-2[7,8,9] = [-7,-11,-15]$. We might naturally expect to reach new destinations with the help of the third dial, since $[7,8,9]$ is not parallel to either $[1,2,3]$ or $[4,5,6]$. But, interestingly, it turns out that the third dial does not take us out of the plane determined by the first two dials!

What is the problem here? Well, our "new" vector $[7,8,9]$ isn't really all that new, because $[7,8,9]$ is, in fact, a linear combination of $[1,2,3]$ and $[4,5,6]$. In particular, $[7,8,9] = (-1)[1,2,3] + 2[4,5,6]$. In other words, we can get to the same destination either by turning the third dial once clockwise or by simultaneously turning the first dial once counterclockwise and the second dial twice clockwise. More generally, we can multiply both sides of the equation $[7,8,9] = (-1)[1,2,3] + 2[4,5,6]$ by any scalar to show that any amount of turns we give to the third dial can be replicated by turning the first two dials an appropriate amount instead. For example,

$$\begin{aligned} 4[7,8,9] &= 4\left((-1)[1,2,3] + 2[4,5,6]\right) \\ &= (-4)[1,2,3] + 8[4,5,6]. \end{aligned}$$

In a similar manner, any linear combination of all three dial vectors can be expressed as a linear combination of the first two vectors alone. For example, for the linear combination of all three dial vectors given above, we have:

$$-5\,[1,2,3] + 3\,[4,5,6] - 2\,[7,8,9]$$
$$= -5\,[1,2,3] + 3\,[4,5,6] - 2((-1)[1,2,3] + 2[4,5,6])$$
$$= -5[1,2,3] + 3[4,5,6] + 2[1,2,3] + 4[4,5,6]$$
$$= (-5+2)[1,2,3] + (3+4)[4,5,6]$$
$$= (-3)[1,2,3] + 7[4,5,6],$$

which is a linear combination of the first two dial vectors alone.

In retrospect, we made a poor choice for the third dial vector. The third dial vector really brings us nowhere "new," only to those locations that we were already able to reach using the first two dial vectors together. (There are other choices we can make for the third dial vector that would allow us to break out of the plane, however.)

More generally, the Vector Etch-a-Sketch gives us the ability to work with linear combinations of vectors in a graphical manner – even if only in our imagination. In general, we can have as many dials as we want, and work in any dimension, as long as all of the dials correspond to vectors having the same dimension. If the dials of the Vector Etch-a-Sketch correspond to the vectors $\mathbf{v}_1, \ldots, \mathbf{v}_k$ respectively, then by turning the dials appropriately, we can obtain any linear combination of the form $c_1\mathbf{v}_1 + \cdots + c_k\mathbf{v}_k$. The values of the scalars $c_1, \ldots, c_k$ in this linear combination represent exactly how far we need to turn each dial, and the sign of each scalar indicates whether the dial is turned clockwise or counterclockwise. We will use the Vector Etch-a-Sketch to help us to understand the meaning and proof of the next theorem.

Mutually Orthogonal Sets of Vectors

In the next theorem, we examine linear combinations that are created from a mutually orthogonal set of vectors.

Theorem 6. *(Linear Combinations of Mutually Orthogonal Vectors) (LCMOV) Suppose that $\{\mathbf{v}_1, \ldots, \mathbf{v}_k\}$ is a mutually orthogonal set of nonzero vectors in $\mathbb{R}^n$. Then, no vector in this set can be expressed as a linear combination of the other vectors in the set.*

Let us take some time to fully understand the statement of the Linear Combinations of Mutually Orthogonal Vectors (LCMOV) Theorem before proceeding to the proof.

First, we are dealing with a set of k distinct vectors. Also, the vectors in the set are all in $\mathbb{R}^n$, which means they all have the same dimension, namely n. Next, they are all nonzero. Finally, they are mutually orthogonal. Recall that this means that the dot product of any two distinct vectors in the set equals zero. Symbolically, $\mathbf{v}_i \cdot \mathbf{v}_j = 0$ whenever $i \neq j$. Thus, what we have here is a set of nonzero vectors, all of which are perpendicular to each other.

Next, what is the conclusion of the theorem? It says that no vector in such a set can be expressed as a linear combination of the other vectors.[6] Let us look at this statement from the Vector Etch-a-Sketch point of view. Suppose we set up an n-dimensional Vector Etch-a-Sketch with k different dials, where dial number i corresponds to vector $\mathbf{v}_i$. Now, let us use just dials 1 through $k-1$; that is, all but the last dial. We can turn those $k-1$ dials any way we want, and each different setting of the dials produces a different linear combination of $\mathbf{v}_1, \ldots, \mathbf{v}_{k-1}$. As a result, the stylus moves to many different positions on the n-dimensional "screen" of the Vector Etch-a-Sketch. Now, imagine turning the dial for the last vector $\mathbf{v}_k$. Does this take us in a new direction, or are we only revisiting places that we can already reach with the first $k-1$ dials?

This is analogous to the situation we saw earlier with the vectors $[1, 2, 3]$, $[4, 5, 6]$, and $[7, 8, 9]$ in which $[7, 8, 9]$ did not take us anywhere new. Remember that this happened because $[7, 8, 9]$ is a linear combination of $[1, 2, 3]$ and $[4, 5, 6]$, and so any effect we produced by turning the $[7, 8, 9]$ dial could be accomplished by turning the $[1, 2, 3]$ and $[4, 5, 6]$ dials instead. The question here is whether a similar result is possible with our vectors $\mathbf{v}_1, \ldots, \mathbf{v}_k$. Could any turning of the $\mathbf{v}_k$ dial be accomplished instead by manipulating the $\mathbf{v}_1, \ldots, \mathbf{v}_{k-1}$ dials in some manner?

The answer is *no*! In fact, the LCMOV Theorem tells us that the $\mathbf{v}_k$ direction is, indeed, a "new" direction. Turning the $\mathbf{v}_k$ dial does take the stylus to new regions on the screen. Of course, this theorem also claims a similar property for each of the other vectors in the set, not simply the last vector, $\mathbf{v}_k$. That is, it claims that every one of the dials moves the stylus in a genuinely new direction that cannot be duplicated merely by using some combination of the other dials.

A specific example may help to make this more obvious. For example, consider the familiar mutually orthogonal set $\{\mathbf{e}_1, \mathbf{e}_2, \mathbf{e}_3\}$ in $\mathbb{R}^3$. This is a set consisting of unit vectors in the directions of the x-, y-, and z-axes in three dimensions. The LCMOV Theorem claims that for $\{\mathbf{e}_1, \mathbf{e}_2, \mathbf{e}_3\}$, the z-axis points in a fundamentally different direction than the x- and y-axes. That is, merely moving in directions parallel to the x- and y-axes will never allow us to move in the direction of the z-axis. Similarly, moving only in the x- and z-directions will never allow us to move in the y-direction. These mutually orthogonal directions are all "different." The LCMOV Theorem asserts that a similar concept is true for any mutually orthogonal set in any dimension, not just in those dimensions we can visualize geometrically.

---

[6]Recall from Chapter 1 that a linear combination of a single matrix is simply a scalar multiple of that matrix. Therefore, a linear combination of a single vector is simply a scalar multiple of that vector. For this reason, if a set contains exactly two distinct nonzero vectors, then one of these vectors is a linear combination of the other if and only if it is a scalar multiple of the other.

Proof of the Linear Combinations of Mutually Orthogonal Vectors Theorem

To prove the LCMOV Theorem, we will use a method called **proof by contradiction**. In this method of reasoning, we take advantage of the Law of the Excluded Middle from logic,[7] which tells us that a given statement must be either true or false. That is, it cannot be neither of these, and it also cannot be both of these at the same time. In other words, a statement must take on exactly one of these two possible "truth values." Therefore, to prove a given statement using the method of proof by contradiction, we temporarily *assume* that the statement is *false*, and then show that such an assumption leads to a logical contradiction. We are then forced to conclude that the given statement must, in fact, be *true*.

Our goal here is to prove the following statement:

Suppose that $\{\mathbf{v}_1, \ldots, \mathbf{v}_k\}$ is a mutually orthogonal set of nonzero vectors in $\mathbb{R}^n$. Then, no vector in this set can be expressed as a linear combination of the other vectors in the set.

Proof: We use the method of proof by contradiction. We assume that one of the vectors $\mathbf{v}_1, \ldots, \mathbf{v}_k$ actually *can* be expressed as a linear combination of the other vectors in the set.

Suppose, to begin with, that the last vector $\mathbf{v}_k$ in the set can be written as a linear combination of the others. That is, we are *assuming* there are some scalars $c_1, \ldots, c_{k-1}$ such that

$$\mathbf{v}_k = c_1\mathbf{v}_1 + \cdots + c_{k-1}\mathbf{v}_{k-1}.$$

This means that any movement using the $\mathbf{v}_k$ dial on the Vector Etch-a-Sketch can be replicated by using the other $k - 1$ dials.

Now, take the dot product of both sides of the previous equation (our *assumption*) with the vector $\mathbf{v}_k$. We get

$$
\begin{aligned}
\mathbf{v}_k \cdot \mathbf{v}_k &= (c_1\mathbf{v}_1 + \cdots + c_{k-1}\mathbf{v}_{k-1}) \cdot \mathbf{v}_k \\
&= c_1(\mathbf{v}_1 \cdot \mathbf{v}_k) + \cdots + c_{k-1}(\mathbf{v}_{k-1} \cdot \mathbf{v}_k) \quad \text{by DP3 and DP5} \\
&= c_1(0) + \cdots + c_{k-1}(0) \quad \text{since } \mathbf{v}_k \text{ is orthogonal to each } \mathbf{v}_i \\
&= 0.
\end{aligned}
$$

Now, we noted earlier in the chapter that whenever the dot product of a vector with itself is zero, that vector must be the zero vector, so $\mathbf{v}_k = \mathbf{0}$ here. But this gives us a logical contradiction, because one of the conditions in the LCMOV Theorem is that all of the given vectors are nonzero. Therefore, the assumption that $\mathbf{v}_k = c_1\mathbf{v}_1 + \cdots + c_{k-1}\mathbf{v}_{k-1}$ must be false, and so the original given statement is true: that is, the vector $\mathbf{v}_k$ *cannot* be expressed as a linear combination of $\mathbf{v}_1, \ldots, \mathbf{v}_{k-1}$.

---

[7]See the Logic Supplement for this textbook for more details on the Law of the Excluded Middle, as well as other topics in logic. The supplement can be downloaded at the URL given in the Preface for the Instructor at the beginning of this book.

Thus far, our proof has only shown that the last vector $\mathbf{v}_k$ cannot be expressed as a linear combination of the other vectors $\mathbf{v}_1, \ldots, \mathbf{v}_{k-1}$. But a similar argument shows that no matter which vector we start with, that vector cannot be expressed as linear combination of the other vectors. One easy way to see this is simply to imagine "relabelling" all of the vector dials on the Vector Etch-a-Sketch so that any dial of our choice corresponds to $\mathbf{v}_k$, while the other dials correspond to $\mathbf{v}_1$ through $\mathbf{v}_{k-1}$. The exact same proof above then works for that dial! This argument shows that the LCMOV Theorem is true for *all* of the vectors in the mutually orthogonal set $\{\mathbf{v}_1, \ldots, \mathbf{v}_k\}$. ◆

At the end of the proof of the LCMOV Theorem we claimed that proving the theorem for $\mathbf{v}_k$ essentially proved the theorem for the other vectors as well. Mathematicians frequently employ such a technique. However, rather than give such a long argument as we presented here, a mathematician would simply *start* the proof by saying "Without loss of generality, assume that $\mathbf{v}_k$ can be expressed as a linear combination of $\mathbf{v}_1, \ldots, \mathbf{v}_{k-1}$." The phrase "without loss of generality" means that the objects involved in the proof (in our case, the vectors $\mathbf{v}_1, \ldots, \mathbf{v}_k$) have no special distinguishing characteristics other than their labels. Therefore, anything we prove about one of the objects could just as easily be proved about every other object in the list just by switching the names of the objects in the proof. Thus, we simply do the proof for whichever object makes the proof easiest to write. In our case, it was easiest to choose $\mathbf{v}_k$ because it was the last one in the list of vectors. (If we had chosen some $\mathbf{v}_i$ in the middle of the list instead, we would have had to use ellipsis notation both before and after $\mathbf{v}_i$ when writing a list of the remaining vectors: $\mathbf{v}_1, \ldots, \mathbf{v}_{i-1}, \mathbf{v}_{i+1}, \ldots, \mathbf{v}_k$.) Mathematicians prefer using phrases such as "without loss of generality" to make proofs shorter, and therefore, easier to understand.

Assignment 6: Try to explain the concepts of "proof by contradiction" and "without loss of generality" to a willing friend or classmate. You can frequently get a better understanding of a concept by trying to explain it to someone else. In doing so, you quickly realize what aspects of the concept you may not fully understand yourself.

Glossary

- Etch-a-Sketch^{TR}: A truly ingenious toy that provided the authors with many hours of enjoyment during their childhoods. A stylus is used to create a continuous drawing on the device's screen by turning two dials (either separately or together), one moving the stylus horizontally, the other moving the stylus vertically.

- Proof by contradiction: A type of proof in which, instead of proving a given statement is true directly, we assume that the statement is false, and then show that this assumption leads to a logical contradiction.

Therefore, the given statement cannot actually be false as it was assumed to be, and must, instead, be true.

- Vector Etch-a-Sketch: An imaginary gadget, based on the principles of the Etch-a-Sketch^{TR}, that graphically depicts linear combinations of particular n-dimensional vectors by turning dials that correspond to those vectors.

- Without loss of generality: A phrase used in writing proofs to indicate that the objects involved in the proof have no special distinguishing characteristics other than the names or labels given to them. Therefore, if an object needs to be singled out in the proof, we can choose whichever is the most convenient to use, and the proof will also apply to all of the other objects as well.

Exercises for Section 3.6

1. Suppose the first two dials of a 3-dimensional Vector Etch-a-Sketch correspond to the vectors $[2, 1, -1]$ and $[1, 2, 4]$, respectively.

   (a) Explain why these two dials will take the stylus in different directions.

   (b) Suppose a third dial corresponds to the vector $[5, -2, -16]$, which equals $4[2, 1, -1] - 3[1, 2, 4]$. Explain what turns can be made on the first two dials in order to accomplish the same overall movement as turning the third dial twice counterclockwise.

   (c) Explain why the first and third dials take the stylus in different directions.

   (d) Explain what turns of the first and third dials can be made in order to accomplish the same overall movement as turning the second dial once clockwise.

   (e) Suppose a fourth dial corresponds to the vector $[2, -3, 1]$. Can the movement of the fourth dial be replicated using only the first two dials? Why or why not? (Hint: Consider the LCMOV Theorem.)

2. Suppose the first two dials of a 4-dimensional Vector Etch-a-Sketch correspond to the vectors $[4, 1, 3, -2]$ and $[2, 1, -1, 3]$, respectively.

   (a) Explain why these two dials will take the stylus in different directions.

   (b) Suppose a third dial corresponds to the vector $[-1, -1, 3, 2]$. Can the movement of the third dial be replicated using only the first two dials? Why or why not? (Hint: Consider the LCMOV Theorem.)

   (c) Suppose a fourth dial corresponds to the vector $[2, -5, -1, 0]$. Can the movement of the fourth dial be replicated using only the first three dials? Why or why not?

(d) Suppose a fifth dial corresponds to the vector $[4, 2, -2, 6]$. Can the movement of the fifth dial be replicated using the first four dials? Why or why not?

(e) Suppose a sixth dial corresponds to the vector $[-1, 20, 18, 5]$, which equals

$$2[4, 1, 3, -2] + [2, 1, -1, 3] + 3[-1, -1, 3, 2] - 4[2, -5, -1, 0].$$

Explain what turns can be made on the first five dials in order to produce the same overall movement as turning the sixth dial once around clockwise.

(f) Explain what turns can be made on dials 2 through 6 in order to produce the same overall movement as turning the first dial once clockwise.

3. In this exercise, we discover a way to move out of the plane determined by the vectors $[1, 2, 3]$ and $[4, 5, 6]$.

(a) Beginning with the set $\{[1, 2, 3], [4, 5, 6]\}$, use the Gram-Schmidt Process to form an orthogonal set $\{\mathbf{v}_1, \mathbf{v}_2\}$ whose vectors have integer entries.

(b) Imagine resetting the two dials for $[1, 2, 3]$ and $[4, 5, 6]$ on the Vector Etch-a-Sketch so that they correspond to the vectors $\mathbf{v}_1$ and $\mathbf{v}_2$ from part (a). Explain why all of the vectors that can be produced by manipulating these two dials still remain in the same plane as before.

(c) Use the Gram-Schmidt Process with the set $\{\mathbf{v}_1, \mathbf{v}_2, \mathbf{e}_1\}$ to find a vector with integer entries that is orthogonal to both $\mathbf{v}_1$ and $\mathbf{v}_2$. Call this vector $\mathbf{v}_3$.

(d) Explain why resetting the third dial of the Vector Etch-a-Sketch to correspond to the vector $\mathbf{v}_3$ will produce vectors that move out of the plane determined by $[1, 2, 3]$ and $[4, 5, 6]$. (Hint: Consider the LCMOV Theorem.)

4. Repeat Exercise 3, except use the vectors $[4, 1, 2, -1]$ and $[6, 5, 6, -3]$ in $\mathbb{R}^4$ instead of $[1, 2, 3]$ and $[4, 5, 6]$.

5. This exercise assumes some prior knowledge of elementary logic, including familiarity with existential and universal quantifiers.[8]

(a) Rewrite the LCMOV Theorem as a conditional statement (that is, as a statement of the form "If...then").

---

[8] A summary of the fundamental principles of elementary logic, including rules for negating quantifiers, can be found in the Logic Supplement for this textbook. This supplement can be downloaded from the URL given in the Preface for the Instructor at the beginning of this book.

(b) What is the negation of the conditional statement in part (a)? Does this negation contain any universal quantifiers, or does it contain only existential quantifiers? (This negation is the "assumption" we make in the proof by contradiction of the LCMOV Theorem.)

3.7   Chapter 3 Test

1. Suppose $\mathbf{v} = [3, -2, 5]$ and $\mathbf{w} = [6, 1, -7]$. Determine the terminal point of the vector $(6\mathbf{v} - 5\mathbf{w})$, if the initial point is $(4, 2, 3)$.

2. Consider $\mathbf{v} = [5, 1]$ and $\mathbf{w} = [-2, 7]$. Provide a single drawing that illustrates the vectors $\mathbf{v}$, $\mathbf{w}$, $(\mathbf{v} + \mathbf{w})$, and $(\mathbf{v} - \mathbf{w})$. Be sure to label all four vectors in your drawing.

3. Suppose $\|\mathbf{v}\| = 12$, $\|\mathbf{w}\| = 7$, and $\mathbf{v} \cdot \mathbf{w} = 35$. Determine the value of $(4\mathbf{v} - 2\mathbf{w}) \cdot (5\mathbf{w} + 3\mathbf{v})$.

4. Assume that $\mathbf{u}$, $\mathbf{v}$, and $\mathbf{w}$ are three vectors having the same dimension, and that $\mathbf{u}$ is a unit vector. Prove that

$$|\mathbf{u} \cdot (\mathbf{v} + \mathbf{w})| \le \|\mathbf{v}\| + \|\mathbf{w}\|.$$

5. Compute the angle between the vectors $\mathbf{v} = [5, -2, 7, 9]$ and $\mathbf{w} = [3, 1, -1, 4]$. Give your final answer in degrees, rounding to two places after the decimal point. (Do not round in any intermediate steps.)

6. Consider the vector $\mathbf{v} = [3, -7, 4, -5]$. Find vectors $\mathbf{x}$ and $\mathbf{y}$, both having positive integers in every coordinate, such that the angle between $\mathbf{v}$ and $\mathbf{x}$ is acute, while the angle between $\mathbf{v}$ and $\mathbf{y}$ is obtuse. (There are many possible answers here.)

7. Decompose $\mathbf{v} = [4, -8, 3, 1]$ into the sum of two vectors, one of which is parallel to $\mathbf{w} = [2, -1, 1, -2]$, and a second which is perpendicular to $\mathbf{w}$.

8. Use the Gram-Schmidt Process to adjust the set $\{[2, -1, 1]$, $[5, 3, 5]$, $[8, 7, 9]$, $[0, 7, 0]\}$ to obtain a mutually orthogonal set of vectors, each having all integer entries.

9. A certain Vector Etch-a-Sketch has three dials set up so that turning the first dial twice clockwise and the second dial once counterclockwise produces the same overall movement as turning the third dial three times clockwise. Explain what turns can be made on the second and third dials in order to produce the same overall movement as turning the first dial once clockwise.

10. Suppose that $\mathbf{x} = [11, 68, -69]$, $\mathbf{y} = [-6, 3, 2]$, and $\mathbf{z} = [7, 8, 9]$.

    (a) Show that $\{\mathbf{x}, \mathbf{y}, \mathbf{z}\}$ is a mutually orthogonal set of vectors.

(b) Find a unit vector in the direction of $\mathbf{y}$.

(c) If possible, find scalars c_1 and c_2 such that $\mathbf{x} = c_1\mathbf{y} + c_2\mathbf{z}$. If it is not possible, explain why not.

11. Suppose that a large collection of robots is programmed to wander among four rooms in a maze: Room A, Room B, Room C, and Room D. There is one door between Rooms A and B, two doors between Rooms A and C, three doors between Rooms A and D, two doors between Rooms B and C, no door between Rooms B and D, and one door between Rooms C and D. Every hour a coin is flipped. If the coin shows "Heads," then each robot remains in its current room. If the coin shows "Tails," then each robot moves to a different room, with each possible exit from the current room an equally likely choice. Suppose that currently 20% of the robots are in Room A, 15% are in Room B, 35% are in Room C, and 30% are in Room D.

    Use all of this information to create a current probability vector $\mathbf{p}$ and Markov chain matrix $\mathbf{M}$ for this situation. Then, use these to calculate the probability vector $\mathbf{p}_1$ for a robot to be in a particular room after one hour. Then use $\mathbf{p}_1$ and $\mathbf{M}$ to calculate the probability vector $\mathbf{p}_2$ for a robot to be in a particular room after two hours. Be sure to label the rows and/or columns of your vectors and matrices appropriately.

12. In each part, indicate whether the given statement is true or whether it is false.

    (a) A vector having dimension n is a matrix having n columns, but only one row.

    (b) If two nonzero vectors are parallel, then they are in the same direction.

    (c) If two nonzero vectors are in the same direction, then they are parallel.

    (d) The Follow Method is a technique for adding vectors geometrically in which each subsequent term in a sum has its initial point at the terminal point of the previous vector.

    (e) If $\mathbf{A}$ and $\mathbf{B}$ are two matrices having compatible sizes, then the formula in the definition of matrix multiplication for the (i, j) entry of $\mathbf{AB}$ is equal to the dot product of the ith row of $\mathbf{A}$ with the jth column of $\mathbf{B}$, where the ith row of $\mathbf{A}$ and the jth column of $\mathbf{B}$ are thought of as vectors.

    (f) If $\mathbf{v} \cdot \mathbf{w} = 0$, then $\mathbf{v} = \mathbf{0}$ or $\mathbf{w} = \mathbf{0}$.

    (g) If $\mathbf{u}_1$ and $\mathbf{u}_2$ are two unit vectors having the same dimension, then $\mathbf{u}_1 \cdot \mathbf{u}_2 \leq 1$.

    (h) If $\mathbf{v}$ is an n-dimensional vector, and $\mathbf{A}$ is an $m \times n$ matrix, then $\mathbf{v}^T \mathbf{A} \mathbf{e}_i$ equals the dot product of $\mathbf{v}$ with the ith column of $\mathbf{A}$.

(i) If $\mathbf{v}$, $\mathbf{w}$, and $\mathbf{y}$ are three vectors having the same dimension such that $\mathbf{v} \cdot \mathbf{w} = \mathbf{v} \cdot \mathbf{y}$, then $\mathbf{w} = \mathbf{y}$.

(j) To find a unit vector in the same direction as a nonzero vector $\mathbf{v}$, merely multiply $\mathbf{v}$ by its magnitude.

(k) The zero n-dimensional vector is orthogonal to every other n-dimensional vector.

(l) The zero n-dimensional vector is perpendicular to every other n-dimensional vector.

(m) It is impossible to find two 4-dimensional vectors $\mathbf{v}$ and $\mathbf{w}$ such that $\mathbf{v} \cdot \mathbf{w} = 15$, $\|\mathbf{v}\| = 3$, and $\|\mathbf{w}\| = 4$.

(n) A linear combination of vectors involves adding together a collection of vectors that all lie along the same line.

(o) The process of taking the projection of one vector onto another is called "normalizing" the vector.

(p) The associative law holds for the dot product of vectors; that is, $\mathbf{x} \cdot (\mathbf{y} \cdot \mathbf{z}) = (\mathbf{x} \cdot \mathbf{y}) \cdot \mathbf{z}$, for all $\mathbf{x}$, $\mathbf{y}$, $\mathbf{z} \in \mathbb{R}^n$.

(q) Proof by contradiction is used when one is trying to prove that a given statement is false.

(r) $(5\mathbf{v}) \cdot (5\mathbf{w}) = 5\,(\mathbf{v} \cdot \mathbf{w})$ for all vectors $\mathbf{v}$ and $\mathbf{w}$ in $\mathbb{R}^n$.

(s) The unit vector $\mathbf{e}_3$ in $\mathbb{R}^5$ is the vector $[0,0,1,0,0]^T$.

(t) If the angle between two nonzero vectors $\mathbf{v}$ and $\mathbf{w}$ is acute, then $\mathbf{v} \cdot \mathbf{w}$ is positive.

(u) If $\mathbf{v}$ and $\mathbf{w}$ are two nonzero, nonparallel vectors in $\mathbb{R}^2$, then the vectors $(\mathbf{v} + \mathbf{w})$ and $(\mathbf{v} - \mathbf{w})$ lie along distinct diagonals of the parallelogram formed by using the vectors $\mathbf{v}$ and $\mathbf{w}$ as adjacent sides.

(v) If the measure of the angle between two nonzero vectors $\mathbf{v}$ and $\mathbf{w}$ equals $0°$, then $\mathbf{w} = c\mathbf{v}$ for some nonzero scalar c.

(w) If $\mathbf{v}$ and $\mathbf{w}$ are two nonzero vectors such that $\mathbf{w} = c\mathbf{v}$ for some nonzero scalar c, then the measure of the angle between $\mathbf{v}$ and $\mathbf{w}$ equals $0°$.

(x) Given a mutually orthogonal set of nonzero vectors, we can create an orthonormal set of vectors by normalizing each of the vectors in the given set.

(y) Multiplying a nonzero vector by a scalar c with $|c| < 0.2$ performs a dilation on the vector.

(z) If a dial on a Vector Etch-a-Sketch corresponds to the zero vector, then turning the dial any amount in either direction will not move the stylus at all.

(aa) If the dials on a Vector Etch-a-Sketch correspond to distinct vectors from an orthonormal set of vectors, then there is no dial whose movement can be replicated by turning any combination of the other dials instead.

(bb) If the dials on a Vector Etch-a-Sketch correspond to nonzero vectors, then turning any particular dial one complete turn clockwise will move the stylus a distance of exactly one unit.

(cc) If the dials on a Vector Etch-a-Sketch correspond to vectors from an orthonormal set of vectors, then turning any particular dial one complete turn counterclockwise will move the stylus a distance of exactly one unit.

(dd) For a nonzero vector $\mathbf{v}$ in $\mathbb{R}^n$, $\mathbf{v} \cdot \mathbf{v}$ cannot equal zero.

(ee) If $\mathbf{A}$ is a matrix whose columns form an orthonormal set of vectors, then $\mathbf{A}^T \mathbf{A}$ is an identity matrix.

(ff) The Polarization Identity can be used to prove that whenever the two diagonals of a parallelogram have the same length, the parallelogram is a rectangle.

(gg) If $\mathbf{x}$, $\mathbf{y}$ and $\mathbf{z}$ are 3-dimensional vectors, then $\|2\mathbf{x} - 3\mathbf{y} + 4\mathbf{z}\| \leq 2\|\mathbf{x}\| + 3\|\mathbf{y}\| + 4\|\mathbf{z}\|$.

(hh) Arccosine and inverse cosine are two different names for the same function.

(ii) The Triangle Inequality does not apply to 1-dimensional vectors.

(jj) If $(\mathbf{v} + \mathbf{w}) \cdot (\mathbf{v} + \mathbf{w}) = 0$, then $\mathbf{w} = -\mathbf{v}$.

(kk) The Cauchy-Schwarz Inequality is not true if one of the vectors is a zero vector.

(ll) The term "orthogonal" describes an algebraic relationship between two vectors, while the term "perpendicular" describes a geometric relationship between two vectors.

(mm) If $\mathbf{u}$ is a unit vector and $\mathbf{v}$ is a vector having the same dimension as $\mathbf{u}$, then $\mathbf{u} \cdot \mathbf{v} = \mathbf{v}$.

(nn) The dot product of vectors is a commutative operation.

(oo) If the vector $\mathbf{w}$ is the result of normalizing the vector $\mathbf{v}$, then $\mathbf{v}$ and $\mathbf{w}$ are in the same direction.

(pp) The zero vector cannot be normalized.

(qq) If the norm of a vector is zero, then all of the coordinates of the vector equal zero.

(rr) If the length of a vector $\mathbf{v}$ is 8, then the magnitude of $-0.375\mathbf{v}$ is 3.

(ss) If $\mathbf{v}$ and $\mathbf{w}$ are two vectors in $\mathbb{R}^3$, then the value of the dot product $\mathbf{v} \cdot \mathbf{w}$ equals an entry in the matrix $\mathbf{v}\mathbf{w}^T$.

(**tt**) The set of all polynomials is an example of a vector space.

(**uu**) Every row of a Markov chain matrix is a probability vector.

Chapter 4

Solving Systems of Linear Equations

One of the most important reasons to study linear algebra is its usefulness for solving systems of linear equations. In this chapter we introduce systems of linear equations and show how we can find their solutions using operations on matrices.

4.1 Systems of Linear Equations

Introduction to Systems of Linear Equations

A **linear equation** is an equation having one or more variables in which none of the variables are multiplied together or raised to powers (other than the first power). For example, $5x - 3y = 7$ is a linear equation since neither x nor y are raised to a power different than the first power, and since the x and y variables are not multiplied together. On the other hand, neither $5xy + 6y = 4$ nor $9y^3 - 2x = 8$ is a linear equation. In the first of these, the variables are multiplied together in the expression $5xy$, while in the second, the variable y is cubed. In this chapter, we are not interested in equations like the last two above, but will focus only on linear equations.

The variables in a linear equation are sometimes called **unknowns**. A linear equation can have any number of unknowns (variables). When there are only two unknowns, we generally label them as x and y. We frequently use additional letters of the alphabet when more variables are involved, as in $3x + 4y - 5z + 2w = 10$. We also sometimes refer to the unknowns as x_1, x_2, x_3, etc. (Notice we are using subscripts rather than superscripts here.) Thus, for example, $5x_1 - 2x_2 - 4x_3 + x_4 - 6x_5 = -3$ is a linear equation having 5 different unknowns (variables) x_1 through x_5.

When we have a set of two or more linear equations involving the same unknowns, we refer to the set as a **system of linear equations** (or, a **linear**

system, for short). We usually place a brace in front of the equations to make it clear that they are part of a common system. For example,

$$\begin{cases} 5x & + & 2y & = & 14 \\ 4x & + & 3y & = & 7 \end{cases}$$

is a system of two linear equations having two unknowns x and y. On the other hand,

$$\begin{cases} 3x_1 & - & 5x_2 & - & 7x_3 & = & 25 \\ 6x_1 & + & x_2 & - & 3x_3 & = & 17 \\ x_1 & - & 2x_2 & + & 5x_3 & = & -52 \\ -2x_1 & - & 3x_2 & - & 9x_3 & = & 61 \end{cases}$$

is a system of four linear equations having three unknowns x_1, x_2, and x_3.

Solutions of Linear Systems

When a system of linear equations is given, we naturally want to find values of the unknowns that make each equation in the system true. Such a set of values for the unknowns is called a **solution** to the system. For example, the first system above has $x = 4$ and $y = -3$ as a solution. You should check this by plugging these values into each equation to make sure the equations are satisfied. When two unknowns are involved, we often express a solution as an **ordered pair**, such as $(4, -3)$. Here, we use the word "ordered" to imply that the first value in the pair corresponds to the first unknown (x, in this case) and the second value in the pair corresponds to the second unknown (y, in this case). Another term for ordered pair is an **ordered 2-tuple**.

The second system above has $x_1 = -2$, $x_2 = 5$, and $x_3 = -8$ as a solution. You should check this for yourself! Here we would express this solution as an **ordered triple** $(-2, 5, -8)$, where the numbers are placed in the same order as their corresponding unknowns. Another term for ordered triple is **ordered 3-tuple**.

In general, we say that a **solution of a linear system** with n unknowns x_1, x_2, ..., x_n is an **ordered n-tuple**, a collection of n numbers in the same order as the unknowns, so that when each number is substituted for its corresponding unknown, every equation in the system is satisfied.

Number of Solutions of a Linear System

We will be able to show later that both of the systems of linear equations given above have only a single solution: $(4, -3)$ for the first system, $(-2, 5, -8)$ for the second system. However, some linear systems actually have more than one solution, or no solution at all. For instance,

$$\begin{cases} 4x & + & 3y & = & 32 \\ 8x & + & 6y & = & 20 \end{cases}$$

has no solution for x and y that make both equations true simultaneously. A quick way to see this is that the left side of the second equation is 2 times the left side of the first equation, but the right sides of each equation are not related in this same way. Formally, we express the solution set of this system as { } (the empty set), or ϕ.

We can also see why this system has no solution from a geometric point of view. Each of the equations in the system is the equation for a line in the coordinate plane. A solution for the system would be a common point on both lines, since only then would it satisfy both equations. But in this case, the two lines are parallel, as shown in Figure 4.1. Since there is no common point on the two lines, the solution set for the system is empty.

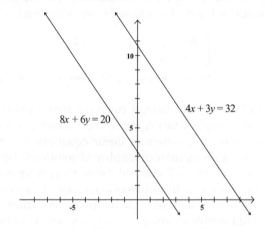

Figure 4.1: Parallel Lines: No Solution

On the other hand,

$$\begin{cases} 4x & + & 3y & = & 32 \\ 8x & + & 6y & = & 64 \end{cases}$$

has an infinite number of solutions. The second equation is merely the first equation multiplied by 2. For this system, both equations represent the *same* line in the plane: the line on the right in Figure 4.1. Hence, all of the points on that line are solutions to the system. We can solve for y in terms of x in either equation to obtain $y = -\frac{4}{3}x + \frac{32}{3}$. Therefore every solution of the second system has the general form $\left(x, -\frac{4}{3}x + \frac{32}{3}\right)$. Choosing a particular value for x would produce one of these solutions. For example, if $x = 2$, $y = -\frac{4}{3}(2) + \frac{32}{3} = 8$, giving us the particular solution $(2, 8)$, which is one of the points on the line.

When additional equations are involved, the linear systems become more complicated geometrically than the situation pictured in Figure 4.1. (See Exercises 3 and 4 below.) Also, when additional variables are involved, we need to use higher dimensions to picture the linear equations. Although these larger systems are often cumbersome to graph, we will see that they can actually be solved efficiently using matrices.

The Complete Solution Set for a Linear System

We can formally express the complete solution set of the last system above as $\left\{ \left(x, -\frac{4}{3}x + \frac{32}{3} \right) \right\}$, where the number of solutions is infinite, because a different solution is obtained for every distinct choice for x. The complete solution set can also be written formally as $\left\{ (x, y) \ \middle| \ y = -\frac{4}{3}x + \frac{32}{3} \right\}$. This translates into English as "the set of all ordered pairs (x, y) such that the condition $y = -\frac{4}{3}x + \frac{32}{3}$ is true."

When a linear system has only a single solution, the complete solution set contains only a single n-tuple. For example, we noted earlier that the only solution for

$$\begin{cases} 5x & + & 2y & = & 14 \\ 4x & + & 3y & = & 7 \end{cases}$$

is $x = 4$ and $y = -3$. Thus, the complete solution set for this system is $\{(4, -3)\}$.

Although we will not prove it here, it turns out that every system of linear equations has either a single solution, an infinite number of solutions, or no solution. In other words, if a system of linear equations has more than one solution, then it must have an infinite number of solutions. (It can be shown that if a linear system has two different solutions, then every point on the line connecting these two solutions in n-dimensional space is also a solution, and so the number of solutions is infinite.) Therefore, the **complete solution set** (that is, the set of all possible solutions) for a given system of linear equations with n unknowns is exactly one of the following: a single ordered n-tuple, an infinite number of n-tuples, or the empty set (if there are no solutions).

Example 1. Recall the system of linear equations

$$\begin{cases} 5x & + & 2y & = & 14 \\ 4x & + & 3y & = & 7 \end{cases}$$

introduced earlier. Suppose x represents the rise or fall of Narnes & Boble stock (in dollars), while y represents the rise or fall of Boundaries stock (also in dollars) on a single day. Jennifer owns five shares of Narnes & Boble stock and two shares of Boundaries stock. Peter's portfolio contains four shares of Narnes & Boble and three shares of Boundaries. Jennifer's portfolio increased $14 in value on this day, while Peter's increased $7 in value.

The first equation represents the change in Jennifer's portfolio. The left side adds up the change in value of her five shares of Narnes & Boble stock plus the change in value of her two shares of Boundaries stock. The right side of the first equation is the total change in value of her entire portfolio. Similarly, the second equation represents the change in Peter's portfolio.

Therefore, finding the solution set for this system would indicate how much each stock actually rose or fell that day. We noted earlier that the complete solution set for this system is $\{(4, -3)\}$; that is, the only possible solution is $x = 4$ and $y = -3$. Thus, the Narnes & Boble stock went *up* $4, while the Boundaries stock went *down* $3. ∎

Linear Systems and Word Problems

As we saw in Example 1, systems of linear equations can arise in real world situations. We will describe a procedure for analyzing a word problem and extracting from it a linear system whose solution will solve the word problem.

The Five-Step Method for Analyzing Word Problems Involving Linear Systems

Step 1: Read the problem carefully in its entirety.

Step 2: Find the question asked by the problem. Use it to determine the values or quantities the problem is asking you to find.

Step 3: Assign variables to the various values determined in Step 2. Write out the meaning of these variables in detail.

Step 4: Determine any restrictions or conditions given in the problem.

Step 5: Write an equation for each restriction found in Step 4.

Let us consider a couple of examples which illustrate this method.

Example 2. Recall Example 2 from Section 1.6 in which a nutritionist is designing meals at a nursing home. In that example, we were given a matrix **A** that describes the vitamin content of three different foods in appropriate units per serving:

$$\mathbf{A} = \begin{matrix} & \text{Vitamin A} & \text{Vitamin C} & \text{Vitamin D} \\ \text{Food 1} & 20 & 35 & 0 \\ \text{Food 2} & 10 & 50 & 40 \\ \text{Food 3} & 25 & 5 & 12 \end{matrix} .$$

Suppose Fred is a resident of the nursing home. The nutritionist wants to adjust Fred's diet so that he receives 80 units of vitamin A, 215 units of vitamin C, and 152 units of vitamin D. How many servings of each food should Fred be given to exactly meet these needs?

We will use our Five-Step Method:

Step 1: At this point, go back and reread the entire problem to be solved. This step is important because it allows you to make connections among the given pieces of information. This also prevents you from making missteps by trying to start to write a system of equations before you really understand the problem.

Step 2: Determine the question being asked. In this case, the question is "How many servings of each food should Fred be given to exactly meet these needs?" Hence, we need to solve for the number of servings of each type of food.

Step 3: Assign variables. In most cases, the variables will represent the unknown quantities that the word problem is asking us to find. Since we are asked to determine the number of servings of each food that should be given to Fred, and there are three types of foods, we will have three variables; a different variable for each type of food. Hence, we let

$$x = \text{the number of servings of Food 1 given to Fred,}$$
$$y = \text{the number of servings of Food 2 given to Fred, and}$$
$$z = \text{the number of servings of Food 3 given to Fred.}$$

As part of your solution, you should always write out a detailed description like this of each variable in the problem, specifying its exact meaning. This will help you keep organized later on, and also make it easier for other people to understand your work.

Step 4: Determine any restrictions. In this problem, the restrictions are that the nutritionist has specific goals for the quantity of various vitamins that Fred receives in his diet. In particular, the nutritionist wants Fred to receive 80 units of vitamin A, 215 units of vitamin C, and 152 units of vitamin D. Hence, the constraints are the quantity of each vitamin Fred gets in his diet.

Step 5: List equations for the restrictions in Step 4. Because we have one restriction for each of the three types of vitamin, we will get one equation for each vitamin.

Now Fred must get exactly 80 units of vitamin A. So, we need a vitamin A equation to ensure this. If Fred is to eat x servings of Food 1, then he will get $20x$ units of vitamin A from these x servings, since there are 20 units of vitamin A per serving. Similarly, Fred will get $10y$ units of vitamin A from his y servings of Food 2. Also, he receives $25z$ units of vitamin A from his z servings of Food 3. We must set this equal to the nutritionist's vitamin A requirement for Fred. This produces the equation

$$20x + 10y + 25z = 80. \quad \text{(vitamin A)}$$

The right side of the equation is the total quantity desired (in this case, of vitamin A), while the left side of the equation is a calculation of the amount received using the (unknown) values of the variables from Step 3.

Notice that the coefficients for x, y and z came from the first column of the matrix **A**. However, that will not always be the case. The data could have been organized differently, with, say, the vitamin A data going across a row instead of down a column. In real life, data comes to us in many forms. You must learn to examine the information you are given, and use it appropriately.

Next, Fred must get exactly 215 units of vitamin C. Using an analysis similar to that for vitamin A, we get the equation

$$35x + 50y + 5z = 215. \quad \text{(vitamin C)}$$

Finally, Fred needs 152 units of vitamin D. This requirement provides us with the equation

$$0x + 40y + 12z = 152, \text{ or just } 40y + 12z = 152. \quad \text{(vitamin D)}$$

Therefore, to determine Fred's diet, the nutritionist must solve the following system of linear equations:

$$\begin{cases} 20x + 10y + 25z = 80 & \text{(vitamin A)} \\ 35x + 50y + 5z = 215 \ . & \text{(vitamin C)} \\ 40y + 12z = 152 & \text{(vitamin D)} \end{cases}$$

Later in this chapter, we will see how to find the complete solution set for such a system. Once a solution is found, the values of the variables x, y, and z in that solution can be used to answer the question asked in the word problem. ■

Example 3. In Exercise 14 in Section 1.6, we discuss the various mixes of grass seed distributed by Green Lawns, Inc. In particular, Matrix $\mathbf{A}$ indicates the number of pounds of bluegrass, rye, and fescue seed used in a single package of each mixture:

$$\mathbf{A} = \begin{array}{c} \text{Bluegrass} \\ \text{Rye} \\ \text{Fescue} \end{array} \overset{\begin{array}{ccc} \text{Standard Mix} & \text{Shade Mix} & \text{Dry Area Mix} \end{array}}{\left[\begin{array}{ccc} 15 & 3 & 3 \\ 7 & 11 & 0 \\ 3 & 11 & 22 \end{array} \right]} .$$

Now, suppose that Green Lawns has available exactly 53,085 lbs. of Bluegrass seed, 38,508 lbs. of Rye seed, and 38,582 lbs. of Fescue seed at its Des Moines plant. The company needs to package all of this seed to clean out their inventory before stocking up on new seed for the next season. How many packages of Standard Mix, Shade Mix, and Dry Area Mix should they produce in Des Moines to exactly use up all of the old seed in their inventory?

Again, we use the Five-Step Method.

Step 1: We first go back and carefully reread the word problem in its entirety.

Step 2: Determine the question being asked. The question is: "How many packages of Standard Mix, Shade Mix, and Dry Area Mix should they produce in Des Moines to exactly use up all of the old seed in their inventory?"

Step 3: Assign variables. In Step 2, the question asks for the number of packages of each mix, so we use one variable to represent each seed mix, as follows:

$x =$ the number of packages of Standard Mix to be produced,

$y =$ the number of packages of Shade Mix to be produced, and

$z =$ the number of packages of Dry Mix to be produced.

Step 4: Determine any restrictions. In this case, we are told exactly how many pounds of each kind of grass seed to use in total. That is, we want to use exactly 53,085 lbs. of Bluegrass seed, 38,508 lbs. of Rye seed, and 38,582 lbs. of Fescue seed.

Step 5: List equations for the restrictions in Step 4. Therefore, we need an equation for each type of grass seed: one for Bluegrass, one for Rye, and one for Fescue. For each type of seed, we calculate the amount needed to create x Standard Mix packages, y Shade Mix packages, and z Dry Mix packages and set the sum of these equal to the total amount available of that seed type. Doing this yields the following linear system:

$$\begin{cases} 15x + 3y + 3z = 53085 & \text{(Bluegrass)} \\ 7x + 11y = 38508 \; . & \text{(Rye)} \\ 3x + 11y + 22z = 38582 & \text{(Fescue)} \end{cases}$$

Solving this linear system for x, y, and z will tell us how many packages of each type of mix that Green Lawns should make in Des Moines. ∎

Notice that in Example 3, the coefficients for the equations were obtained by going across the *rows* of the matrix in which the data is given, rather than by going down the columns as in Example 2. Now, in these last two examples, the data was given to us already arranged in matrix form. However, the data can be presented in many different forms, such as in a paragraph format or in a table that is arranged differently than the way it will ultimately appear in the linear system that we are creating. You must consider carefully what information you are actually given and express it correctly, rather than trying to memorize a single template with which to set up each problem.

Limit Vectors for Markov Chains

Recall that in Section 3.1, we introduced Markov chains. A Markov chain is a system of objects occupying several different "states," with given probabilities for moving an object from one state to another. These probabilities are listed in a matrix $\mathbf{M}$ for the Markov chain, where we let the (i, j) entry of $\mathbf{M}$ represent the probability of an object changing from the jth state (the *current* state) to the ith state (the *next* state).

The example given in Section 3.1 involved a carrier serving on three different routes, A, B, and C. The initial probabilities that the carrier was on a particular route were given in the following probability vector:

$$\mathbf{p} = \begin{array}{c} \text{Route A} \\ \text{Route B} \\ \text{Route C} \end{array} \begin{bmatrix} .40 \\ .25 \\ .35 \end{bmatrix} .$$

We were also given the following Markov chain matrix for this situation:

			Current Week		
			Route A	Route B	Route C
$\mathbf{M} =$	**Next Week**	Route A	.60	.20	.05
		Route B	.30	.50	.15
		Route C	.10	.30	.80

In that example, we calculated the products $\mathbf{p}_1 = \mathbf{Mp}$ and $\mathbf{p}_2 = \mathbf{Mp}_1$ which respectively represent the probabilities that the carrier is on a particular route after one week and two weeks:

$$\mathbf{p}_1 = \begin{array}{c} \textbf{After 1} \\ \textbf{Week} \end{array} \begin{array}{c} \text{Route A} \\ \text{Route B} \\ \text{Route C} \end{array} \begin{bmatrix} .3075 \\ .2975 \\ .3950 \end{bmatrix} = \begin{bmatrix} 30.75\% \\ 29.75\% \\ 39.50\% \end{bmatrix},$$

$$\mathbf{p}_2 = \begin{array}{c} \textbf{After 2} \\ \textbf{Weeks} \end{array} \begin{array}{c} \text{Route A} \\ \text{Route B} \\ \text{Route C} \end{array} \begin{bmatrix} .26375 \\ .30025 \\ .43600 \end{bmatrix} = \begin{bmatrix} 26.38\% \\ 30.03\% \\ 43.60\% \end{bmatrix}.$$

However, this process can be continued indefinitely, as long as the original probabilities given in $\mathbf{M}$ of switching from one route to another do not change from week to week. We noted in that example that after several more multiplications, the probabilities that the carrier is on a particular route will eventually converge to those shown in the vector

$$\mathbf{p}_{\text{lim}} = \begin{array}{c} \text{Route A} \\ \text{Route B} \\ \text{Route C} \end{array} \begin{bmatrix} .2037 \\ .2778 \\ .5185 \end{bmatrix} = \begin{bmatrix} 20.37\% \\ 27.78\% \\ 51.85\% \end{bmatrix}.$$

Thus, in the long run, the carrier has the greatest chance of being assigned in a particular week to Route C, and the least chance to Route A. The vector $\mathbf{p}_{\text{lim}}$ is often referred to as the **limit vector** for the successive vectors $\mathbf{p}$, $\mathbf{p}_1$, $\mathbf{p}_2$, ..., etc. of the given Markov chain.

Unfortunately, not every Markov chain has a limit vector for a given initial probability vector $\mathbf{p}$. However, if a limit vector does exist for a given Markov chain, it would be helpful to determine that limit vector without having to perform a long string of matrix multiplications to determine $\mathbf{p}_1$, $\mathbf{p}_2$, $\mathbf{p}_3$, ..., etc. Now, notice that once the limit vector is reached, further multiplication by the Markov chain matrix has no effect. That is, the limit vector $\mathbf{p}_{\text{lim}}$ has the property that $\mathbf{Mp}_{\text{lim}} = \mathbf{p}_{\text{lim}}$. Mathematically, a vector $\mathbf{x}$ with the property that $\mathbf{Mx} = \mathbf{x}$ is called a **steady-state vector** for the matrix $\mathbf{M}$. Although we will not prove it here, every limit vector for a Markov chain must be a steady-state vector.

Thus, we can determine limit vectors by solving for steady-state vectors instead. This is done by solving an appropriate linear system of equations. In particular, we want to find a vector solution $\mathbf{x}$ for the linear system $\mathbf{Mx} = \mathbf{x}$ where $\mathbf{x}$ is a probability vector – that is, where the sum of the entries of the vector $\mathbf{x}$ equals 1.

For the particular Markov chain given here, the corresponding linear system would be

$$\begin{cases} .60x + .20y + .05z = x & \text{(Route A)} \\ .30x + .50y + .15z = y & \text{(Route B)}, \\ .10x + .30y + .80z = z & \text{(Route C)} \end{cases}$$

under the condition that $x + y + z = 1$. We can incorporate this condition into the linear system by including a fourth row as follows:

$$\begin{cases} .60x + .20y + .05z = x & \text{(Route A)} \\ .30x + .50y + .15z = y & \text{(Route B)} \\ .10x + .30y + .80z = z & \text{(Route C)} \\ 1.00x + 1.00y + 1.00z = 1 & \text{(Sum Condition)} \end{cases}$$

When solving systems later in this chapter, we will see that it is helpful to bring all variables to the left-hand side of each equation and to combine like terms. Therefore, we can write this system in a slightly simpler form as follows:

$$\begin{cases} -.40x + .20y + .05z = 0 & \text{(Route A)} \\ .30x - .50y + .15z = 0 & \text{(Route B)} \\ .10x + .30y - .20z = 0 & \text{(Route C)} \\ x + y + z = 1 & \text{(Sum Condition)} \end{cases}$$

You can check that the vector

$$\begin{bmatrix} x \\ y \\ z \end{bmatrix} = \begin{bmatrix} 20.37\% \\ 27.78\% \\ 51.85\% \end{bmatrix}$$

is a solution of this linear system. Therefore, it is a steady-state vector for the Markov chain matrix $\mathbf{M}$. Later in this chapter we will be able to verify that it is, in fact, the *only* solution – in other words, it is the *only* steady-state vector for this system.

Although we will not prove it here, it can be shown that if a Markov chain matrix $\mathbf{M}$ (or some positive power of $\mathbf{M}$) has all of its entries nonzero, then a limit vector will exist for $\mathbf{M}$ for *any* given initial probability vector $\mathbf{p}$. Now, in the particular example above, every entry of $\mathbf{M}$ is nonzero, so we are guaranteed to have a limit vector $\mathbf{p}_{\lim}$ for $\mathbf{M}$ for any given initial probability vector $\mathbf{p}$. But, since each $\mathbf{p}_{\lim}$ must be a steady-state vector for $\mathbf{M}$, and since we discovered above that there is only one steady-state vector for $\mathbf{M}$, this effectively means that the vector $\mathbf{p}_{\lim}$ that we calculated above is actually the limit vector for $\mathbf{M}$ for *every* initial probability vector $\mathbf{p}$! That is, regardless of the initial probability vector $\mathbf{p}$ that is given, the given Markov chain ultimately results in the same final vector

$$\mathbf{p}_{\lim} = \begin{bmatrix} 20.37\% \\ 27.78\% \\ 51.85\% \end{bmatrix}.$$

Glossary

- Complete solution set of a linear system: The collection of all possible solutions to the system. For example, if a linear system has three unknowns, then any solutions are ordered triples. A linear system with three unknowns has either a unique solution (a single ordered triple), an infinite number of solutions (an infinite number of ordered triples), or no solution. In general, any solutions to a linear system with n unknowns are ordered n-tuples, and the system either has a unique solution (a single ordered n-tuple), an infinite number of solutions (an infinite number of ordered n-tuples), or no solution.

- Limit vector for a Markov chain: The vector $\mathbf{p}_{\text{lim}}$ that the probability vectors $\mathbf{p}$, $\mathbf{p}_1$, $\mathbf{p}_2$, $\mathbf{p}_3$, ..., converge to for a given Markov chain matrix $\mathbf{M}$, if such a vector exists. (If a limit vector $\mathbf{p}_{\text{lim}}$ for a Markov chain exists, then it is a steady-state vector for the Markov chain matrix $\mathbf{M}$. That is, $\mathbf{M}\mathbf{p}_{\text{lim}} = \mathbf{p}_{\text{lim}}$.)

- Linear equation: An equation in which none of the variables are multiplied together, and every variable (unknown) appears to the first power. For example, $3x + 5y - 6z = 12$ is a linear equation. Most equations are not linear; for example, any equation containing an "xy" term or an "x^2" term or a "y^3" term would *not* be linear.

- Linear system: An alternate expression for "system of linear equations."

- Ordered n-tuple: A collection of n numbers listed in a particular order, surrounded by parentheses. For example, $(6, -3, 7, -2)$ is an ordered 4-tuple.

- Ordered pair: Two numbers listed in a particular order. An ordered pair of numbers is surrounded by parentheses; for example, $(4, -3)$. An ordered pair of numbers is also referred to as an ordered 2-tuple.

- Ordered triple: Three numbers listed in a particular order. An ordered triple of numbers is surrounded by parentheses; for example, $(-5, 2, -1)$. An ordered triple of numbers is also referred to as an ordered 3-tuple.

- Solution of a linear system: An ordered n-tuple whose values satisfy every equation in the linear system when they are substituted for the corresponding unknowns in that system.

- Steady-state vector for a Markov chain: A probability vector $\mathbf{x}$ having the property that $\mathbf{M}\mathbf{x} = \mathbf{x}$, if $\mathbf{M}$ is the matrix for the Markov chain. (If a limit vector $\mathbf{p}_{\text{lim}}$ exists for a Markov chain matrix $\mathbf{M}$ with initial probability vector $\mathbf{p}$, then $\mathbf{p}_{\text{lim}}$ is a steady-state vector for that Markov chain.)

- System of linear equations: Two or more linear equations considered simultaneously that have the same unknowns.

- Unknowns: The variables in a particular system of linear equations. For example, the unknowns for the linear system
$$\begin{cases} 3x - 5y = 27 \\ -2x + 4y = -20 \end{cases}$$
are x and y.

Exercises for Section 4.1

1. Determine in each case whether the given ordered pairs or triples are solutions of the indicated system of linear equations.

    (a) $(24, -15)$ for $\begin{cases} 3x & - & 5y & = & 147 \\ -2x & + & 3y & = & -93 \end{cases}$

    (b) $(-18, 23, -13)$ for $\begin{cases} 2x_1 & + & 5x_2 & - & x_3 & = & 92 \\ -3x_1 & - & x_2 & + & 2x_3 & = & 5 \\ x_1 & - & 2x_2 & + & 5x_3 & = & 115 \end{cases}$

2. Determine in each case whether the given ordered n-tuples are solutions of the indicated system of linear equations.

    (a) $(13, -9, 16)$ for $\begin{cases} 5x & - & 6y & + & z & = & 135 \\ -3x & - & y & + & 2z & = & -9 \\ -2x & - & 5y & + & 8z & = & 147 \end{cases}$

    (b) $(7, -11, 8, -9)$ for

    $$\begin{cases} 4x_1 & + & 2x_2 & - & 5x_3 & - & 7x_4 & = & 29 \\ -2x_1 & + & 3x_2 & - & x_3 & + & 6x_4 & = & -109 \\ 3x_1 & - & 5x_2 & - & 2x_3 & + & x_4 & = & 51 \end{cases}$$

3. The three lines in Figure 4.2 represent the equations in a linear system having three equations and two unknowns. Determine whether this system of linear equations has no solution, exactly one solution, or an infinite number of solutions. Explain how you made this determination.

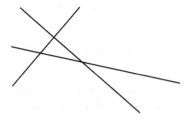

Figure 4.2: Figure for Exercise 3

4. The three lines in Figure 4.3 represent the equations in a linear system having three equations and two unknowns. Determine whether this system of linear equations has no solution, exactly one solution, or an infinite number of solutions. Explain how you made this determination.

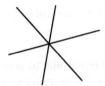

Figure 4.3: Figure for Exercise 4

5. Write a system of linear equations to represent the given situation. Be sure to explicitly explain the meaning of each of the variables you use.

   Assume the price (per gallon) of both regular gasoline and premium gasoline at a particular service station has not changed in the last month. Suppose during this time that a motorist bought 12 gallons of regular gasoline and 9 gallons of premium gasoline and spent $67.50, but the following week bought 11 gallons of regular gasoline and 10 gallons of premium gasoline and spent $68.00. What were the prices (per gallon) of regular gasoline and premium gasoline? (You do not need to solve the linear system.)

6. Write a system of linear equations to represent the given situation. Be sure to explicitly explain the meaning of each of the variables you use.

   Eye Was Framed, Inc. makes three different styles of frames for eyeglasses. Each Pilot model uses $27 in materials, 0.31 hours of manufacturing time, and 0.12 hours of assembly time. Each Classic model uses $25 in materials, 0.26 hours of manufacturing time, and 0.11 hours of assembly time. Each Xtreme model uses $34 in materials, 0.41 hours of manufacturing time, and 0.15 hours of assembly time. In the next month, the company has $31,488 to invest in materials, has 356.3 hours of manufacturing time available, and has 139.26 hours of assembly time available. How many of each type of frame should the company make in the next month so that it spends all of its available capital on materials and uses all available manufacturing and assembly time? (You do not need to solve the linear system.)

7. Write a system of linear equations to represent the given situation. Be sure to explicitly explain the meaning of each of the variables you use.

   Grey Eyes, Inc. makes three different types of security cameras. Last month, they made and sold a total of 1200 cameras, costing them $68,584 to make, and making them a profit of $94,224. Each indoor Nanny-Cam costs them $38 to make, and makes them a profit of $57. Each indoor Shoplift-Stopper costs them $62 to make, and makes them a profit of $91. Each Weather-All outdoor camera costs them $85 to make, and makes them a profit of $104. How many of each type of camera did they make and sell last month? (Hint: Do not overlook the constraint involving the total number of cameras.) (You do not need to solve the linear system.)

8. Write a system of linear equations to represent the given situation. Be sure to explicitly explain the meaning of each of the variables you use.

    A company manufactured a total of 1000 refrigerators and sent them to three different retail chains. The costs of shipping a single refrigerator to the first, second, and third chains were $10, $12, and $8, respectively. The profits for selling a single refrigerator to the first, second, and third chains were $80, $75, and $85, respectively. The manufacturing company's total shipping cost was $10, 100, while its total profit was $79, 750. How many refrigerators did the manufacturing company send to each chain? (Hint: Do not overlook the constraint involving the total number of refrigerators.) (You do not need to solve the linear system.)

9. Write a system of linear equations to represent the given situation. Be sure to explicitly explain the meaning of each of the variables you use.

    Storage Systems, Inc. makes three types of steel shelving units. Each HomeSpace model costs them $10 in materials and takes 1.2 hours of labor to make. Each Hobbyist model costs them $13 in materials and takes 1.4 hours of labor to make. Each Warehouser model costs them $17 in materials and takes 1.7 hours of labor to make. They also want to make exactly twice as many HomeSpace units as the number of Hobbyist units made. In the next month, the company has $12, 997 to invest in materials and has 1363.2 hours of labor available. How many units of each type should they manufacture in the next month in order to use all of their labor and all of the material capital that is available, and satisfy the condition regarding the number of HomeSpace units to make? (You do not need to solve the linear system.)

10. Write a system of linear equations to represent the given situation. Be sure to explicitly explain the meaning of each of the variables you use.

    Mario's Pipes, Inc. makes PVC pipe in ten-foot sections in three different sizes. One section of the $\frac{1}{4}$ inch pipe costs them $0.24 in materials and takes 1.2 minutes of labor to make. One section of the $\frac{3}{8}$ inch pipe costs them $0.35 in materials and takes 1.3 minutes of labor to make. One section of the $\frac{1}{2}$ inch pipe costs them $0.48 in materials and takes 1.5 minutes of labor to make. Mario needs exactly as many sections of $\frac{1}{2}$ pipe as the number of sections of the other two types combined. In the next month, he has $15, 979.11 available to spend on materials and he has 57800.1 minutes of labor available. How many sections of each type of pipe should Mario make in the next month to use up all available money for materials, use all available labor, and satisfy the condition regarding the amount of $\frac{1}{2}$ inch pipe to make? (You do not need to solve the linear system.)

11. Suppose the matrix for a given Markov chain is

$$\mathbf{M} = \begin{bmatrix} .45 & .10 & .15 \\ .20 & .65 & .45 \\ .35 & .25 & .40 \end{bmatrix}.$$

Set up an appropriate system of linear equations (with all variables placed on the left-hand side of each equation, and with like terms combined) whose solution is a steady-state vector for $\mathbf{M}$. Then check that the vector

$$\begin{bmatrix} 17.81\% \\ 50.68\% \\ 31.51\% \end{bmatrix}$$

is a steady-state vector for $\mathbf{M}$ by showing that it satisfies all of the equations in this system (after rounding to two places after the decimal point). (It can be shown that this vector is the *only* steady-state vector for this system. Since all entries of $\mathbf{M}$ are nonzero, this steady-state vector is actually $\mathbf{p}_{\lim}$ for *any* initial probability vector $\mathbf{p}$ for this Markov chain.)

12. Suppose the matrix for a given Markov chain is

$$\mathbf{M} = \begin{bmatrix} .18 & .26 & .32 \\ .48 & .52 & .26 \\ .34 & .22 & .42 \end{bmatrix}.$$

Set up an appropriate system of linear equations (with all variables placed on the left-hand side of each equation, and with like terms combined) whose solution is a steady-state vector for $\mathbf{M}$. Then check that the vector

$$\begin{bmatrix} 25.82\% \\ 42.81\% \\ 31.37\% \end{bmatrix}$$

is a steady-state vector for $\mathbf{M}$ by showing that it satisfies all of the equations in this system (after rounding to two places after the decimal point). (It can be shown that this vector is the *only* steady-state vector for this system. Since all entries of $\mathbf{M}$ are nonzero, this steady-state vector is actually $\mathbf{p}_{\lim}$ for *any* initial probability vector $\mathbf{p}$ for this Markov chain.)

4.2 Row Operations on Augmented Matrices

In this section, we show how to translate every system of linear equations into a matrix multiplication, and how to perform certain useful operations on the rows of corresponding matrices. In the next two sections, we will see how these tools enable us to find the complete solution set for a given linear system.

Expressing Systems of Linear Equations using Matrix Multiplication

Recall the linear system

$$\begin{cases} 3x_1 & - & 5x_2 & - & 7x_3 & = & 25 \\ 6x_1 & + & x_2 & - & 3x_3 & = & 17 \\ x_1 & - & 2x_2 & + & 5x_3 & = & -52 \\ -2x_1 & - & 3x_2 & - & 9x_3 & = & 61 \end{cases}$$

introduced in the previous section. Since this system has four equations and three unknowns, we create a 4×3 matrix $\mathbf{A}$ that contains the coefficients of the unknowns as follows:

$$\mathbf{A} = \begin{bmatrix} 3 & -5 & -7 \\ 6 & 1 & -3 \\ 1 & -2 & 5 \\ -2 & -3 & -9 \end{bmatrix}.$$

Notice that the *rows* of $\mathbf{A}$ represent the left side of the *equations* in the linear system, and that the *columns* of $\mathbf{A}$ correspond to the coefficients of the *unknowns*. We also create a 3×1 (column) matrix $\mathbf{X}$ containing the three unknowns, and a 4×1 (column) matrix $\mathbf{B}$ containing the four numbers appearing on the right-hand side of the equations in the system:

$$\mathbf{X} = \begin{bmatrix} x_1 \\ x_2 \\ x_3 \end{bmatrix}, \quad \text{and} \quad \mathbf{B} = \begin{bmatrix} 25 \\ 17 \\ -52 \\ 61 \end{bmatrix}.$$

We next show that the given system of linear equations is equivalent to the single matrix equation $\mathbf{AX} = \mathbf{B}$! That is,

$$\begin{bmatrix} 3 & -5 & -7 \\ 6 & 1 & -3 \\ 1 & -2 & 5 \\ -2 & -3 & -9 \end{bmatrix} \begin{bmatrix} x_1 \\ x_2 \\ x_3 \end{bmatrix} = \begin{bmatrix} 25 \\ 17 \\ -52 \\ 61 \end{bmatrix}.$$

First notice that in carrying out the matrix multiplication on the left side of this matrix equation, the result is a 4×1 matrix, so its size matches that of $\mathbf{B}$ on the right side of the equation. Now we compare the entries. Each row entry of $\mathbf{AX}$ is obtained by multiplying the coefficients in a row of $\mathbf{A}$ by their corresponding unknowns in $\mathbf{X}$, and then adding the results. For example, the first entry of $\mathbf{AX}$ is found by multiplying each entry of the first row of $\mathbf{A}$ by its corresponding unknown in $\mathbf{X}$, and then adding the results together to get $3x_1 - 5x_2 - 7x_3$. This must equal the first row of $\mathbf{B}$, which is the number 25. But setting these equal yields precisely the first equation in the linear system above. Similarly, the second row of $\mathbf{AX}$ is found by multiplying each entry of the second row of $\mathbf{A}$ by the corresponding unknowns in $\mathbf{X}$ and then adding the results to obtain $6x_1 + x_2 - 3x_3$. Setting this equal to the entry in the second row of $\mathbf{B}$ produces the second equation, $6x_1 + x_2 - 3x_3 = 17$. Similarly, setting the results from the third and fourth row entries of $\mathbf{AX}$ equal to the corresponding entries in $\mathbf{B}$ establish the third and fourth equations.

In a similar manner, any system of linear equations can be expressed as a matrix multiplication. In general, if there are m equations and n unknowns, then the coefficient matrix $\mathbf{A}$ is an $m \times n$ matrix, the matrix $\mathbf{X}$ of unknowns is an $n \times 1$ (column) matrix, and the matrix $\mathbf{B}$ of values from the right side of the equations is an $m \times 1$ (column) matrix. Multiplying the matrix $\mathbf{A}$ of coefficients by the column matrix $\mathbf{X}$ of unknowns produces the column matrix $\mathbf{B}$; that is, $\mathbf{AX} = \mathbf{B}$.

Now, since $\mathbf{X}$ and $\mathbf{B}$ are matrices having a single column, we can also think of them as vectors instead of as matrices. Whenever we do this, we will use lower case letters for these column matrices and write the system of equations as $\mathbf{Ax} = \mathbf{b}$. In this case, since the coordinates of the vector $\mathbf{x}$ are the variables, we will also express the solution set for the system as a set of vectors rather than as a set of n-tuples; that is, we will use square brackets "[ ]" around the individual solutions instead of round parentheses, "( )." This is related to the idea from Chapter 3 of melding the concepts of points in n-dimensional space and n-dimensional vectors. Generally, we will use whichever notation is most convenient in a particular situation.

Representing Linear Systems Using Augmented Matrices

To save space, we often "eliminate" the unknowns by creating a single matrix containing all of the numbers in the linear system. For the system above, this matrix has the form

$$[\mathbf{A} \mid \mathbf{B}] = \left[\begin{array}{ccc|c} 3 & -5 & -7 & 25 \\ 6 & 1 & -3 & 17 \\ 1 & -2 & 5 & -52 \\ -2 & -3 & -9 & 61 \end{array}\right].$$

Matrices of this form, which are constructed by joining two matrices together with a vertical bar between them, are called **augmented matrices**. In this particular case, the matrix $\mathbf{A}$ is "augmented" (that is, enlarged) by introducing an additional single column matrix $\mathbf{B}$ after the bar. The *rows* of this augmented matrix correspond to the *equations* in the linear system. The *columns to the left of the bar* correspond to the *unknowns*. The column to the right of the bar corresponds to the constants on the right side of the equations. Therefore, this system of four equations and three unknowns is represented by a 4×4 augmented matrix. In general, a linear system with m equations and n unknowns is represented by an $m \times (n + 1)$ augmented matrix.

Example 1. Consider the system of linear equations

$$\begin{cases} 4x_1 + 2x_2 - 5x_3 - 7x_4 = 29 \\ -2x_1 + 3x_2 - x_3 + 6x_4 = -109 \\ 3x_1 - 5x_2 - 2x_3 + x_4 = 51 \end{cases}$$

from part (b) of Exercise 2 in the previous section. Let

$$\mathbf{A} = \begin{bmatrix} 4 & 2 & -5 & -7 \\ -2 & 3 & -1 & 6 \\ 3 & -5 & -2 & 1 \end{bmatrix} \text{ be the } 3 \times 4 \text{ matrix of coefficients, } \mathbf{X} = \begin{bmatrix} x_1 \\ x_2 \\ x_3 \\ x_4 \end{bmatrix}$$

be the 4×1 matrix of unknowns, and $\mathbf{B} = \begin{bmatrix} 29 \\ -109 \\ 51 \end{bmatrix}$ be the 3×1 matrix

containing the values on the right side of the given equations. Then the given linear system can be expressed using matrix multiplication as $\mathbf{AX} = \mathbf{B}$, or

$$\begin{bmatrix} 4 & 2 & -5 & -7 \\ -2 & 3 & -1 & 6 \\ 3 & -5 & -2 & 1 \end{bmatrix} \begin{bmatrix} x_1 \\ x_2 \\ x_3 \\ x_4 \end{bmatrix} = \begin{bmatrix} 29 \\ -109 \\ 51 \end{bmatrix}.$$

The 3×5 augmented matrix corresponding to this linear system is

$$[\mathbf{A}|\,\mathbf{B}] = \left[\begin{array}{cccc|c} 4 & 2 & -5 & -7 & 29 \\ -2 & 3 & -1 & 6 & -109 \\ 3 & -5 & -2 & 1 & 51 \end{array} \right].$$

Notice that, in both forms, the rows of the matrices correspond to the equations in the linear system, and the columns of $\mathbf{A}$ correspond to the unknowns in the system. ∎

You may not be able to create the vertical bar when entering an augmented matrix into your calculator. Don't panic! Just enter the matrix without the bar and proceed through any computations *imagining* that the bar is there.

Row Operations on Augmented Matrices

There are three fundamental operations that we can perform on the rows of an augmented matrix that do not alter the solution set of its corresponding linear system. To illustrate how these row operations work, we give an example of each type.

Type (I) Row Operation: This operation allows us to multiply a row by a *nonzero* number. For example, consider the augmented matrix

$$\mathbf{M} = \left[\begin{array}{ccc|c} 5 & -10 & 10 & 75 \\ -3 & 1 & -6 & -35 \\ 4 & -3 & 2 & 26 \end{array} \right].$$

We can substitute $\frac{1}{5}$ times the first row of the augmented matrix $\mathbf{M}$ above in place of the original first row in order to obtain

$$\begin{bmatrix} 1 & -2 & 2 & \bigg| & 15 \\ -3 & 1 & -6 & \bigg| & -35 \\ 4 & -3 & 2 & \bigg| & 26 \end{bmatrix}.$$

Notice that only the first row has been changed, while the other two rows remained the same. Since the first *row* of $\mathbf{M}$ corresponds to the first *equation* in a linear system, multiplying row 1 by $\frac{1}{5}$ corresponds to multiplying both sides of the first equation by $\frac{1}{5}$.

We often use shorthand notation to describe the row operations. For instance, we express this particular type (I) row operation as $\frac{1}{5} < 1 > \implies$ $< 1 >$.

In general, if c is a nonzero number and we substitute c times the ith row of an augmented matrix in place of the original ith row, we express this operation formally as:

$$(\text{I}): c < i > \implies < i >.$$

When illustrating the effect of a row operation, we often display the row operation together with the matrix that results from that row operation. For example, the row operation above that substitutes $\frac{1}{5}$ times the first row of $\mathbf{M}$ in place of the original first row is displayed as follows:

$$(\text{I}): \frac{1}{5} < 1 > \implies < 1 > \qquad \begin{bmatrix} 1 & -2 & 2 & \bigg| & 15 \\ -3 & 1 & -6 & \bigg| & -35 \\ 4 & -3 & 2 & \bigg| & 26 \end{bmatrix}.$$

A little thought will convince you that any type (I) row operation does not affect the solution set of the corresponding linear system. The reason is that by multiplying a row by a nonzero number, you are actually multiplying both sides of one of the equations in the linear system by that number. We know from algebra that multiplying both sides of an equation by the same nonzero number does not change the set of solutions to that equation.

Type (II) Row Operation: This operation allows us to add a multiple of one row to another row. For example, consider the augmented matrix

$$\mathbf{N} = \begin{bmatrix} 1 & -5 & 3 & \bigg| & 21 \\ -4 & 2 & -1 & \bigg| & -20 \\ 5 & -3 & -2 & \bigg| & 13 \end{bmatrix}.$$

We can add 4 times the first row of $\mathbf{N}$ to the second row and then substitute this result in place of the second row. The resulting computation looks like this:

$4 \times$ (row 1)	4	-20	12	84
(row 2)	-4	2	-1	-20
(sum = new row 2)	0	-18	11	64

Even though two rows were involved in calculating this sum, we only alter one of them when writing out the result. In particular, the row that is multiplied by a number is *not* altered in the final result. Only the "other" row gets changed. That is, the matrix after this row operation is:

$$\begin{bmatrix} 1 & -5 & 3 & 21 \\ 0 & -18 & 11 & 64 \\ 5 & -3 & -2 & 13 \end{bmatrix}.$$

Only the second row has been changed here, while the other two rows have not been altered. This particular type (II) row operation is described as $4 < 1 > + < 2 > \implies < 2 >$.

In general, if we substitute the sum of c times the ith row and the jth row for the original jth row, we express this operation formally as:

$$\text{(II)}: c < i > \ + \ < j > \ \implies \ < j > .$$

Hence, the row operation above which adds 4 times the first row to the second row of (the original) augmented matrix $\mathbf{N}$ is displayed as follows:

$$\text{(II)}: 4 < 1 > \ + \ < 2 > \ \implies \ < 2 > \qquad \begin{bmatrix} 1 & -5 & 3 & 21 \\ 0 & -18 & 11 & 64 \\ 5 & -3 & -2 & 13 \end{bmatrix}.$$

It turns out that any type (II) row operation does not affect the solution set of the corresponding linear system. Informally, the reason this works is that a type (II) operation adds equal quantities to both sides of one of the equations in the system. However, we will not provide a formal proof here that this actually leaves the solution set unchanged.

Type (III) Row Operation: This operation allows us to switch the entries of two rows with each other. For example, consider the augmented matrix

$$\mathbf{P} = \begin{bmatrix} 0 & -5 & 4 & 26 \\ -2 & -6 & 1 & 10 \\ 5 & -2 & 1 & 23 \end{bmatrix}.$$

We can switch the first and second rows of $\mathbf{P}$ above to obtain

$$\begin{bmatrix} -2 & -6 & 1 & 10 \\ 0 & -5 & 4 & 26 \\ 5 & -2 & 1 & 23 \end{bmatrix}.$$

Notice that this changes *both* the first and second rows of $\mathbf{P}$, while the third row remained the same. This is the *only* type of row operation in which we allow two rows to be changed simultaneously when we compute the resulting matrix. We express this particular type (III) row operation as $< 1 > \iff < 2 >$ (or, $< 2 > \iff < 1 >$). This notation means that the entries of the first row are replaced by those in the second row, and vice versa.

In general, if we switch the ith and jth rows of an augmented matrix, we express this operation formally as:

$$(\text{III}): \quad <i> \quad \Longleftrightarrow \quad <j> \, .$$

That is, the entries of the ith row are replaced with those in the jth row, and vice versa. Therefore, the row operation above which swaps the first and second rows of the (original) augmented matrix **P** above is displayed as follows:

$$(\text{III}): \quad <1> \quad \Longleftrightarrow \quad <2> \qquad \left[\begin{array}{rrr|r} -2 & -6 & 1 & 10 \\ 0 & -5 & 4 & 26 \\ 5 & -2 & 1 & 23 \end{array} \right].$$

A little thought will convince you that any type (III) row operation does not affect the solution set of the corresponding linear system. The reason is that only the *order* of the rows has been changed, so the actual equations that they represent have not been altered. We have only changed the order in which the equations have been listed. Any values for the unknowns that make every equation in a linear system true must also make every equation in the system true after a row operation of this type because precisely the same equations are involved.

Performing a Sequence of Row Operations

We often want to perform several row operations in turn, with each row operation in the sequence acting on the most recent result obtained.

Example 2. We perform the following row operations in succession:

$$(\text{III}): \quad <1> \quad \Longleftrightarrow \quad <2>$$

$$(\text{I}): \left(-\frac{1}{3}\right) <1> \quad \Longrightarrow \quad <1>$$

$$(\text{II}): (-5)<1> \; + \; <3> \quad \Longrightarrow \quad <3>$$

starting with the augmented matrix

$$\left[\begin{array}{rrr|r} 0 & -3 & 2 & -19 \\ -3 & -9 & 6 & -15 \\ 5 & -2 & -4 & -6 \\ -4 & 3 & 10 & -25 \end{array} \right].$$

After the first row operation, $<1> \Longleftrightarrow <2>$, we obtain:

$$(\text{III}): \quad <1> \quad \Longleftrightarrow \quad <2> \qquad \left[\begin{array}{rrr|r} -3 & -9 & 6 & -15 \\ 0 & -3 & 2 & -19 \\ 5 & -2 & -4 & -6 \\ -4 & 3 & 10 & -25 \end{array} \right].$$

For the second row operation, $\left(-\frac{1}{3}\right) <1> \implies <1>$, we have

$$(\text{I}): \left(-\frac{1}{3}\right) <1> \implies <1> \qquad \begin{bmatrix} 1 & 3 & -2 & 5 \\ 0 & -3 & 2 & -19 \\ 5 & -2 & -4 & -6 \\ -4 & 3 & 10 & -25 \end{bmatrix}.$$

For the third row operation, $(-5) <1> + <3> \implies <3>$, the corresponding calculations are:

$(-5) \times$ (row 1)	-5	-15	10	-25
(row 3)	5	-2	-4	-6
(sum = new row 3)	0	-17	6	-31

and therefore, the final result is

$$(\text{II}): (-5) <1> + <3> \implies <3> \qquad \begin{bmatrix} 1 & 3 & -2 & 5 \\ 0 & -3 & 2 & -19 \\ 0 & -17 & 6 & -31 \\ -4 & 3 & 10 & -25 \end{bmatrix}. \quad \blacksquare$$

Glossary

- **Augmented matrix:** Matrices constructed by joining two matrices together with a vertical bar between them. In general, we represent a linear system with m equations and n unknowns using an $m \times (n+1)$ augmented matrix. The rows of the matrix correspond to equations in the linear system. The columns to the left of the bar correspond to unknowns in the system. The column to the right of the bar corresponds to the constants on the right side of the equations.

- **Row operations:** Three operations (of types (I), (II), (III), respectively) on an (augmented) matrix. These row operations do not affect the solution set of the linear system represented by the matrix.

- **Sequence of row operations:** Two or more row operations performed in sequence; that is, each row operation is performed on the matrix resulting from the previous row operation.

- **Type (I) row operation:** A row operation which multiplies a particular row of a matrix by a nonzero number. If the ith row of a matrix is multiplied by a nonzero number c, then the operation is represented symbolically as: (I): $c <i> \implies <i>$.

- **Type (II) row operation:** A row operation which adds a multiple of one row of a matrix to another row. If c times the ith row of a matrix is added to the jth row, then the operation is represented symbolically as: (II): $c <i> + <j> \implies <j>$.

- Type (III) row operation: A row operation which swaps two rows of a matrix with each other. If the entries of the ith and jth rows switch places, then the operation is represented symbolically as: (III): $< i > \iff < j >$.

Exercises for Section 4.2

1. Express the following system of linear equations in two ways: as a matrix multiplication, and also as an augmented matrix.

$$\begin{cases} 5x & + & 3y & - & 2z & + & 4w & = & -36 \\ -2x & + & 6y & - & z & - & 5w & = & 28 \\ 7x & - & y & + & 3z & - & 2w & = & -1 \end{cases}$$

2. Express the following system of linear equations as a matrix multiplication, and also state the augmented matrix for this system.

$$\begin{cases} 2x_1 & - & 4x_2 & - & 5x_3 & = & 36 \\ -x_1 & + & 3x_2 & - & 6x_3 & = & -6 \\ 3x_1 & - & 5x_2 & + & 2x_3 & = & 30 \\ 4x_1 & + & x_2 & - & 8x_3 & = & 23 \end{cases}$$

3. For each of the following row operations, indicate the type of operation, and express the operation in symbolic form:

   (a) Multiply the third row by (-7).

   (b) Add 4 times the second row to the first row.

   (c) Switch the second row and the third row.

   (d) Add (-6) times the fourth row to the third row.

   (e) Switch the first row and the fourth row.

   (f) Multiply the second row by $\frac{1}{3}$.

4. For each of the following row operations, indicate the type of operation, and express the operation in symbolic form:

   (a) Switch the third row and the fourth row.

   (b) Multiply the first row by $(-\frac{1}{2})$.

   (c) Add 5 times the second row to the fourth row.

   (d) Multiply the fourth row by 3.

   (e) Add 5 times the third row to the first row.

   (f) Switch the second row and the fourth row.

5. In each case, perform the indicated row operation on the given augmented matrix, and state the resulting matrix obtained.

(a) (I): $(-\frac{1}{4}) < 2 > \Longrightarrow < 2 >$ on $\begin{bmatrix} 1 & -3 & 4 & 2 & | & -16 \\ 0 & -4 & -2 & -8 & | & 24 \\ 0 & -2 & 3 & -5 & | & -15 \end{bmatrix}$.

(b) (III): $< 2 > \Longleftrightarrow < 3 >$ on $\begin{bmatrix} 1 & 2 & -6 & | & -5 \\ 0 & 0 & 5 & | & 12 \\ 0 & -3 & 2 & | & -22 \end{bmatrix}$.

(c) (II): $(-5) < 3 > + < 2 > \Longrightarrow < 2 >$ on $\begin{bmatrix} 1 & 0 & -3 & | & -8 \\ 0 & 1 & 5 & | & -24 \\ 0 & 0 & 1 & | & -6 \\ 0 & 0 & -2 & | & -9 \end{bmatrix}$.

(d) (III): $< 2 > \Longleftrightarrow < 4 >$ on $\begin{bmatrix} 1 & 6 & -3 & 4 & | & -11 \\ 0 & 0 & -2 & -8 & | & 21 \\ 0 & 0 & 3 & -5 & | & -13 \\ 0 & -5 & 12 & -4 & | & -8 \end{bmatrix}$.

(e) (II): $4 < 2 > + < 3 > \Longrightarrow < 3 >$ on
$\begin{bmatrix} 1 & -1 & -2 & 8 & -3 & | & -18 \\ 0 & 1 & 5 & -4 & 12 & | & 2 \\ 0 & -4 & -18 & 13 & -36 & | & -7 \end{bmatrix}$.

(f) (I): $(-\frac{1}{4}) < 1 > \Longrightarrow < 1 >$ on $\begin{bmatrix} -4 & 2 & -6 & | & 8 \\ -7 & -3 & 9 & | & -13 \\ 3 & -8 & -2 & | & -6 \end{bmatrix}$.

6. In each case, perform the indicated row operation on the given augmented matrix, and state the resulting matrix obtained.

(a) (III): $< 3 > \Longleftrightarrow < 4 >$ on $\begin{bmatrix} 8 & -5 & 2 & | & -12 \\ 0 & -9 & -2 & | & 20 \\ 0 & 0 & 3 & | & -15 \\ 0 & 6 & -9 & | & 14 \\ 0 & 5 & -3 & | & 7 \end{bmatrix}$.

(b) (II): $(-4) < 1 > + < 3 > \Longrightarrow < 3 >$ on $\begin{bmatrix} 1 & -7 & -3 & -2 & | & -4 \\ 0 & 10 & -5 & 3 & | & 18 \\ 4 & -8 & -2 & -1 & | & -13 \end{bmatrix}$.

(c) (I): $(-5) < 1 > \Longrightarrow < 1 >$ on $\begin{bmatrix} -\frac{1}{5} & \frac{4}{5} & -\frac{9}{5} & \frac{7}{5} & -\frac{2}{5} & | & -\frac{11}{5} \\ \frac{1}{2} & 0 & -\frac{5}{2} & -\frac{3}{2} & -1 & | & -2 \\ \frac{1}{3} & -\frac{5}{3} & 0 & -\frac{2}{3} & \frac{4}{3} & | & -\frac{8}{3} \end{bmatrix}$.

(d) (II): $5 < 2 > + < 4 > \Longrightarrow < 4 >$ on $\begin{bmatrix} 1 & 0 & -5 & -4 & | & -13 \\ 0 & 1 & -6 & -3 & | & 6 \\ 0 & 0 & 3 & -2 & | & -11 \\ 0 & -5 & 9 & 19 & | & -18 \end{bmatrix}$.

(e) (I): $4 < 3 > \Longrightarrow < 3 >$ on $\begin{bmatrix} 1 & 0 & -8 & | & -14 \\ 0 & 1 & 5 & | & 22 \\ 0 & 0 & \frac{1}{4} & | & -\frac{5}{4} \end{bmatrix}$.

(f) (III): $< 1 > \Longleftrightarrow < 4 >$ on $\begin{bmatrix} 0 & 2 & -12 & \bigm| & 5 \\ 0 & 4 & -2 & \bigm| & 24 \\ 0 & -3 & 3 & \bigm| & -15 \\ -6 & -4 & -5 & \bigm| & 9 \end{bmatrix}$.

7. For the augmented matrix

$$\begin{bmatrix} 1 & 5 & 2 & \bigm| & -27 \\ 0 & 0 & -4 & \bigm| & 12 \\ 0 & 0 & -5 & \bigm| & 15 \\ 0 & -3 & 9 & \bigm| & -12 \end{bmatrix},$$

perform the following sequence of row operations, showing the intermediate steps involved, and state the final matrix obtained.

(III): $< 2 > \Longleftrightarrow < 4 >$
(I): $(-\frac{1}{3}) < 2 > \Longrightarrow < 2 >$
(II): $(-5) < 2 > + < 1 > \Longrightarrow < 1 >$.

8. For the augmented matrix

$$\begin{bmatrix} 0 & 4 & -6 & 5 & -3 & \bigm| & -11 \\ 0 & -3 & 1 & -8 & -5 & \bigm| & 35 \\ -4 & -8 & 2 & 8 & -2 & \bigm| & -12 \\ 2 & 5 & -3 & -4 & -1 & \bigm| & 9 \end{bmatrix},$$

perform the following sequence of row operations, showing the intermediate steps involved, and state the final matrix obtained.

(III): $< 1 > \Longleftrightarrow < 3 >$
(I): $(-\frac{1}{4}) < 1 > \Longrightarrow < 1 >$
(II): $(-2) < 1 > + < 4 > \Longrightarrow < 4 >$

9. For the augmented matrix

$$\begin{bmatrix} 1 & -7 & 7 & -3 & \bigm| & 12 \\ 0 & -1 & 5 & -2 & \bigm| & -9 \\ 4 & -2 & 6 & -5 & \bigm| & -1 \end{bmatrix},$$

state a single type (II) row operation that converts the $(3,1)$ entry to zero, and give the resulting matrix after this row operation is performed.

10. For the augmented matrix

$$\begin{bmatrix} 1 & 0 & 3 & -5 & -3 & \bigm| & 7 \\ 0 & 1 & 2 & -1 & -7 & \bigm| & -3 \\ 0 & -6 & 5 & -2 & -1 & \bigm| & -18 \end{bmatrix},$$

state a single type (II) row operation that would convert the $(3,2)$ entry to zero, and give the resulting matrix after this row operation is performed.

4.3 Introduction to Gauss-Jordan Row Reduction

In the previous section we introduced three types of row operations that do not affect the solution set of a system of linear equations. In this section and the next, we discover how to use these operations to determine the complete solution set of a linear system. This solution technique is called the **Gauss-Jordan Row Reduction Method**. (To save space, we sometimes refer to this as the "Gauss-Jordan Method," or simply, "row reduction.")

A Simple Example

Suppose we want to find the complete solution set for the system of linear equations represented by the augmented matrix

$$\mathbf{Q} = \left[\begin{array}{ccc|c} 1 & 0 & 0 & 2 \\ 0 & 1 & 0 & 3 \\ 0 & 0 & 1 & 4 \\ 0 & 0 & 0 & 0 \end{array}\right].$$

First, because there are four rows in $\mathbf{Q}$, there are four equations in the linear system. Since there are three columns to the left of the bar, there are three variables, say, x, y, and z. Therefore, the corresponding system of linear equations is

$$\begin{cases} 1x & + & 0y & + & 0z & = & 2 \\ 0x & + & 1y & + & 0z & = & 3 \\ 0x & + & 0y & + & 1z & = & 4 \\ 0x & + & 0y & + & 0z & = & 0 \end{cases}, \text{ or, more simply, } \begin{cases} x & = & 2 \\ y & = & 3 \\ z & = & 4 \\ 0 & = & 0 \end{cases}.$$

The last equation, $0 = 0$, is true no matter what values x, y, and z have, and so it contributes no information about the solution set. Whenever we encounter the equation $0 = 0$, we just ignore it as irrelevant. However, the other three equations tell us that the unique solution for this linear system is $x = 2$, $y = 3$, $z = 4$. That is, the complete solution set is $\{(2, 3, 4)\}$.

In this example, the augmented matrix was in the form

$$\left[\begin{array}{c|c} \mathbf{I} & \mathbf{B} \\ \hline \mathbf{O} & \mathbf{O} \end{array}\right],$$

where $\mathbf{I}$ is an identity matrix, $\mathbf{B}$ is some column matrix having the same number of rows as $\mathbf{I}$, and the $\mathbf{O}$'s are zero matrices of compatible sizes. Whenever the augmented matrix for a system is in this form, the $\mathbf{O}$'s at the bottom correspond to equations of the form $0 = 0$, and are irrelevant. The $[\mathbf{I} \mid \mathbf{B}]$ portion at the top corresponds to a linear system as we saw in the example above, in which each equation in the system indicates a unique value for one of the variables. Such an augmented matrix always corresponds to a system having a unique solution, and that solution is easily read off from the entries of **B**. For example, the complete solution set for the linear system corresponding to

$$\begin{bmatrix} 1 & 0 & 17 \\ 0 & 1 & 19 \\ 0 & 0 & 0 \\ 0 & 0 & 0 \end{bmatrix} \quad \text{is } \{(17, 19)\}.$$

Basics of the Gauss-Jordan Method

Now we consider some more complicated linear systems. Suppose, for example, that we want to find the complete solution set for the linear system:

$$\begin{cases} 2x_1 & - & 4x_2 & + & 3x_3 & = & 26 \\ 3x_1 & - & 8x_2 & - & 2x_3 & = & 17 \\ -2x_1 & - & 4x_2 & + & x_3 & = & 6 \end{cases}.$$

We begin by creating the augmented matrix for this system:

$$\mathbf{R} = \begin{bmatrix} 2 & -4 & 3 & 26 \\ 3 & -8 & -2 & 17 \\ -2 & -4 & 1 & 6 \end{bmatrix}.$$

Our goal is to use a sequence of row operations to put **R** into the form

$$\begin{bmatrix} \mathbf{I} & \mathbf{B} \\ \mathbf{O} & \mathbf{O} \end{bmatrix},$$

if possible. If we can accomplish this, the complete solution set will be apparent, as we saw in the previous example.

In performing these row operations, we use the following basic rules:

- We work on one column at a time, proceeding through the columns from left to right, until we reach the bar.

- In each column, we begin by placing the "1" in its proper spot, and then convert the remaining entries of that column into "0."

- To convert a nonzero entry to "1:"

  − We label that entry as the current **pivot entry**, and the row containing that entry as the current **pivot row**. The column containing the pivot entry is the current **pivot column**.

  − We use the type (I) row operation

  $$\left(\frac{1}{\text{pivot entry}} \right) < \text{pivot row} > \Longrightarrow < \text{pivot row} >.$$

  That is, we multiply the pivot row by the reciprocal of the pivot entry.

- To convert an entry to "0:"

    - We label that entry as the **target entry**, and the row containing that entry as the **target row**.
    - We use the type (II) row operation

$$- \text{(target entry)} < \text{pivot row} > + < \text{target row} >$$
$$\Longrightarrow < \text{target row} > .$$

    That is, we add a multiple of the pivot row to the target row. The number we multiply by is the *negative* of the target entry. We often use **target** as a verb; that is, to "target" an entry means that we use the type (II) row operation above to convert that particular entry into a zero.

You should always use the specific type (I) and type (II) row operations given above to turn a nonzero pivot into a "1" and to change a target entry into a "0." We illustrate these rules in the next example by performing the Gauss-Jordan Row Reduction Method on the augmented matrix **R** above. For easier reference, we place a circle around each former and current pivot entry that has been converted to "1" during this process, and henceforth refer to each of these as a **circled pivot entry**.

Example 1. We find the complete solution set for the linear system

$$\begin{cases} 2x_1 & - & 4x_2 & + & 3x_3 & = & 26 \\ 3x_1 & - & 8x_2 & - & 2x_3 & = & 17 \\ -2x_1 & - & 4x_2 & + & x_3 & = & 6 \end{cases}$$

by following the Gauss-Jordan Method. The augmented matrix for this system is

$$\left[\begin{array}{ccc|c} 2 & -4 & 3 & 26 \\ 3 & -8 & -2 & 17 \\ -2 & -4 & 1 & 6 \end{array} \right].$$

We begin with the first column as the *pivot column*. Our first goal is to convert the $(1,1)$ entry into "1." Thus, we make the $(1,1)$ entry the current *pivot entry*, and the first row our current *pivot row*. We use a type (I) row operation in which we multiply the pivot row by the reciprocal of the pivot entry. The result is:

$$(\text{I}): \frac{1}{2} <1> \implies <1> \qquad \left[\begin{array}{ccc|c} ① & -2 & \frac{3}{2} & 13 \\ 3 & -8 & -2 & 17 \\ -2 & -4 & 1 & 6 \end{array} \right].$$

Next we want to target the remaining entries in the first column, converting them to "0." We do this using type (II) row operations. We begin with the $(2,1)$ entry (that is, "3") as our *target entry*, and the second row as our *target row*. From the general method above, we see that the required row operation is

$$(-3)<1> + <2> \implies <2>.$$

That is, we add (-3) times the pivot row to the target row. Performing this computation, we get

$(-3) \times$ (row 1)	-3	6	$-\frac{9}{2}$	-39
(row 2)	3	-8	-2	17
(sum = new row 2)	0	-2	$-\frac{13}{2}$	-22

which results in the next matrix in the sequence:

$$(\text{II}): (-3)<1> + <2> \implies <2> \qquad \begin{bmatrix} ① & -2 & \frac{3}{2} & 13 \\ 0 & -2 & -\frac{13}{2} & -22 \\ -2 & -4 & 1 & 6 \end{bmatrix}.$$

To complete our work on the first column, we target the $(3,1)$ entry, converting it to "0." Thus, the $(3,1)$ entry becomes our new *target entry*, and the third row becomes our new *target row*. Again, we use a type (II) row operation, this time adding 2 times the pivot row to the target row:

$2 \times$ (row 1)	2	-4	3	26
(row 3)	-2	-4	1	6
(sum = new row 3)	0	-8	4	32

$$(\text{II}): 2<1> + <3> \implies <3> \qquad \begin{bmatrix} ① & -2 & \frac{3}{2} & 13 \\ 0 & -2 & -\frac{13}{2} & -22 \\ 0 & -8 & 4 & 32 \end{bmatrix}.$$

Now our work on the first column is finished, so we begin work on the second column, making that the current *pivot column*. Our first goal is to place a "1" in the $(2,2)$ position. Thus, the $(2,2)$ entry becomes our new *pivot entry*, and the second row becomes our new *pivot row*. We use a type (I) row operation in which we multiply the pivot row by the reciprocal of the pivot entry:

$$(\text{I}): \left(-\frac{1}{2}\right)<2> \implies <2> \qquad \begin{bmatrix} ① & -2 & \frac{3}{2} & 13 \\ 0 & ① & \frac{13}{4} & 11 \\ 0 & -8 & 4 & 32 \end{bmatrix}.$$

Next we target the remaining entries of the second column, converting them to "0." We begin with the $(1,2)$ entry. Thus, the $(1,2)$ entry is our current *target entry*, and the first row is our current *target row*. To produce a "0" in the target entry, we add 2 times the pivot row to the target row:

$2 \times$ (row 2)	0	2	$\frac{13}{2}$	22
(row 1)	1	-2	$\frac{3}{2}$	13
(sum = new row 1)	1	0	8	35

$$(\text{II}): 2 < 2 > \; + \; < 1 > \;\; \Longrightarrow \;\; < 1 > \qquad \begin{bmatrix} ① & 0 & 8 & \bigm| & 35 \\ 0 & ① & \frac{13}{4} & \bigm| & 11 \\ 0 & -8 & 4 & \bigm| & 32 \end{bmatrix}.$$

Next we target the $(3, 2)$ entry. Thus, the $(3, 2)$ entry is our current *target entry* and the third row is our current *target row*. To produce a "0" in the target entry, we add 8 times the pivot row to the target row:

$8 \times (\text{row } 2)$	0	8	26	88
$(\text{row } 3)$	0	-8	4	32
$(\text{sum} = \text{new row } 3)$	0	0	30	120

$$(\text{II}): 8 < 2 > \; + \; < 3 > \;\; \Longrightarrow \;\; < 3 > \qquad \begin{bmatrix} ① & 0 & 8 & \bigm| & 35 \\ 0 & ① & \frac{13}{4} & \bigm| & 11 \\ 0 & 0 & 30 & \bigm| & 120 \end{bmatrix}.$$

This completes our work on the second column. We now turn to the third column as our new *pivot column*. Our first goal is to place a "1" in the $(3, 3)$ position. The $(3, 3)$ entry becomes our new *pivot entry*, and the third row is our new *pivot row*. We use a type (I) row operation to convert the pivot entry to "1" by multiplying the pivot row by the reciprocal of the pivot entry.

$$(\text{I}): \frac{1}{30} < 3 > \;\; \Longrightarrow \;\; < 3 > \qquad \begin{bmatrix} ① & 0 & 8 & \bigm| & 35 \\ 0 & ① & \frac{13}{4} & \bigm| & 11 \\ 0 & 0 & ① & \bigm| & 4 \end{bmatrix}.$$

Our next task is to target the other two positions in the third column, making them "0." First, the $(1, 3)$ entry becomes our new *target entry*, and the first row becomes our new *target row*. To convert the target entry to "0," we use a type (II) row operation to add (-8) times the pivot row to the target row:

$(-8) \times (\text{row } 3)$	0	0	-8	-32
$(\text{row } 1)$	1	0	8	35
$(\text{sum} = \text{new row } 1)$	1	0	0	3

$$(\text{II}): (-8) < 3 > \; + \; < 1 > \;\; \Longrightarrow \;\; < 1 > \qquad \begin{bmatrix} ① & 0 & 0 & \bigm| & 3 \\ 0 & ① & \frac{13}{4} & \bigm| & 11 \\ 0 & 0 & ① & \bigm| & 4 \end{bmatrix}.$$

Finally, we target the $(2, 3)$ entry. We let this entry be our new *target entry*, and the second row be our new *target row*. We convert the target entry to "0" by adding $\left(-\frac{13}{4}\right)$ times the pivot row to the target row:

$\left(-\frac{13}{4}\right) \times (\text{row } 3)$	0	0	$-\frac{13}{4}$	-13
$(\text{row } 2)$	0	1	$\frac{13}{4}$	11
$(\text{sum} = \text{new row } 2)$	0	1	0	-2

(II): $\left(-\dfrac{13}{4}\right) < 3 > \ + \ < 2 > \ \Longrightarrow \ < 2 > \qquad \begin{bmatrix} ① & 0 & 0 & 3 \\ 0 & ① & 0 & -2 \\ 0 & 0 & ① & 4 \end{bmatrix}.$

At this point, no further row operations are needed since the augmented matrix is now in the desired form. This augmented matrix represents the linear system

$$\begin{cases} 1x_1 + 0x_2 + 0x_3 = 3 \\ 0x_1 + 1x_2 + 0x_3 = -2 \\ 0x_1 + 0x_2 + 1x_3 = 4 \end{cases}, \text{ or more simply, } \begin{cases} x_1 = 3 \\ x_2 = -2 \\ x_3 = 4 \end{cases}.$$

The complete solution set of the latter system is $\{(3, -2, 4)\}$. But we know that the row operations which we performed do not affect the solution set. Therefore, the complete solution of the original system is also $\{(3, -2, 4)\}$. ∎

Notice that in Example 1, the final augmented matrix did not contain any rows of the form $[0\ 0\ 0\ |\ 0]$ for us to ignore. In fact, the final augmented matrix actually had the simpler form

$$[\mathbf{I}_3 \mid \mathbf{B}]$$

where the values of $\mathbf{B}$ represent the unique solution of the linear system.

Type (III) Row Operations to the Rescue

Occasionally, we encounter a linear system in which a "0" is occupying the position where we want the next pivot entry to be. Using a type (I) row operation will not help us in such a case because the reciprocal of 0 does not exist – there is no number with which we can multiply the row to change that "0" into "1." When this dilemma occurs, we turn to a type (III) row operation instead, as in the next example.

Example 2. Consider the following system of five linear equations with four unknowns:

$$\begin{cases} 5x_1 - 5x_2 + 10x_3 - 15x_4 = -25 \\ 2x_1 - 2x_2 + 4x_3 - 6x_4 = -10 \\ 6x_1 - 4x_2 + 20x_3 - x_4 = 5 \\ -3x_1 + 8x_2 + 11x_3 + 36x_4 = 59 \\ 4x_1 - x_2 + 20x_3 + 6x_4 = 10 \end{cases}.$$

The augmented matrix for this linear system is

$$\begin{bmatrix} 5 & -5 & 10 & -15 & -25 \\ 2 & -2 & 4 & -6 & -10 \\ 6 & -4 & 20 & -1 & 5 \\ -3 & 8 & 11 & 36 & 59 \\ 4 & -1 & 20 & 6 & 10 \end{bmatrix}.$$

We use the Gauss-Jordan Method to find the complete solution set for this linear system. Our goal is to use row operations to put the augmented matrix in the form

$$\left[\begin{array}{c|c} \mathbf{I} & \mathbf{B} \\ \hline \mathbf{O} & \mathbf{O} \end{array}\right],$$

for some column matrix $\mathbf{B}$, after which the solution set will be apparent.

We begin with the first column as our *pivot column*. We want to convert the $(1,1)$ entry into "1." Thus, our current *pivot entry* is the $(1,1)$ entry, and our current *pivot row* is the first row. We multiply the pivot row by the reciprocal of the pivot entry:

$$(\mathrm{I}):\ \left(\tfrac{1}{5}\right) <1> \implies <1> \qquad \left[\begin{array}{cccc|c} ① & -1 & 2 & -3 & -5 \\ 2 & -2 & 4 & -6 & -10 \\ 6 & -4 & 20 & -1 & 5 \\ -3 & 8 & 11 & 36 & 59 \\ 4 & -1 & 20 & 6 & 10 \end{array}\right].$$

Next we target all of the remaining entries in the first column, converting them to "0." These are the $(2,1)$, $(3,1)$, $(4,1)$, and $(5,1)$ entries. Each of these becomes a *target entry* in turn, with its corresponding row as the *target row*. In all cases, we add the appropriate multiple of the pivot row to the target row to convert the target entry to "0." The cumulative effect of these four type (II) row operations is:

$$\begin{array}{l}
(\mathrm{II}):\ (-2)<1> + <2> \implies <2> \\
(\mathrm{II}):\ (-6)<1> + <3> \implies <3> \\
(\mathrm{II}):\quad 3<1> + <4> \implies <4> \\
(\mathrm{II}):\ (-4)<1> + <5> \implies <5>
\end{array} \qquad \left[\begin{array}{cccc|c} ① & -1 & 2 & -3 & -5 \\ 0 & 0 & 0 & 0 & 0 \\ 0 & 2 & 8 & 17 & 35 \\ 0 & 5 & 17 & 27 & 44 \\ 0 & 3 & 12 & 18 & 30 \end{array}\right].$$

This completes our work on the first column, so we move on to the second column, which becomes our new *pivot column*. The first task is to place a "1" in the $(2,2)$ position. Unfortunately, the $(2,2)$ entry is "0," so there is no number by which we can multiply the second row in order to place a "1" in that position.

At this point, a type (III) row operation comes to our aid. We could switch the second row with any of the rows *below* it. However, since the second row is a row of zeroes, it cannot contain any future pivots, and will ultimately end up at the bottom of the matrix. Therefore, it is advantageous to switch the second row with the fifth row:

$$(\mathrm{III}):\ <2> \iff <5> \qquad \left[\begin{array}{cccc|c} ① & -1 & 2 & -3 & -5 \\ 0 & 3 & 12 & 18 & 30 \\ 0 & 2 & 8 & 17 & 35 \\ 0 & 5 & 17 & 27 & 44 \\ 0 & 0 & 0 & 0 & 0 \end{array}\right].$$

Now we have a nonzero number in the $(2,2)$ position, so we can pivot there. The $(2,2)$ entry becomes our current *pivot entry* and the second row becomes our current *pivot row*. We use type (I) row operation to convert the pivot entry to "1" by multiplying the pivot row by the reciprocal of the pivot entry:

(I): $\left(\dfrac{1}{3}\right) < 2 > \implies < 2 >$
$$\begin{bmatrix} ① & -1 & 2 & -3 & -5 \\ 0 & ① & 4 & 6 & 10 \\ 0 & 2 & 8 & 17 & 35 \\ 0 & 5 & 17 & 27 & 44 \\ 0 & 0 & 0 & 0 & 0 \end{bmatrix}.$$

Next we target the remaining nonzero entries in the second column to convert them to "0." These are the $(1,2)$, $(3,2)$, and $(4,2)$ entries, and each in turn becomes the *target entry* with its corresponding row as the *target row*. In all cases, we add the appropriate multiple of the pivot row to the target row to convert the target entry to "0." The cumulative effect of these three type (II) row operations is:

(II): $\quad 1 < 2 > + < 1 > \implies < 1 >$
(II): $(-2) < 2 > + < 3 > \implies < 3 >$
(II): $(-5) < 2 > + < 4 > \implies < 4 >$
$$\begin{bmatrix} ① & 0 & 6 & 3 & 5 \\ 0 & ① & 4 & 6 & 10 \\ 0 & 0 & 0 & 5 & 15 \\ 0 & 0 & -3 & -3 & -6 \\ 0 & 0 & 0 & 0 & 0 \end{bmatrix}.$$

Our work on the second column is completed, and so we move on to the third column, which becomes our new *pivot column*. We would like to convert the $(3,3)$ entry to "1," but there is a "0" in that position. Once again we use a type (III) row operation, switching the third row with a row below it. The only row below the third row with a nonzero entry in the third column is the fourth row. Therefore, we switch the third row with the fourth row, and obtain:

(III): $< 3 > \iff < 4 >$
$$\begin{bmatrix} ① & 0 & 6 & 3 & 5 \\ 0 & ① & 4 & 6 & 10 \\ 0 & 0 & -3 & -3 & -6 \\ 0 & 0 & 0 & 5 & 15 \\ 0 & 0 & 0 & 0 & 0 \end{bmatrix}.$$

Now the $(3,3)$ entry is nonzero, so we can pivot on this entry. Hence, the $(3,3)$ entry becomes our current *pivot entry*, and the third row becomes our current *pivot row*. We use a type (I) row operation to convert the pivot entry to "1" by multiplying the pivot row by the reciprocal of the pivot entry:

(I): $\left(-\dfrac{1}{3}\right) < 3 > \implies < 3 >$
$$\begin{bmatrix} ① & 0 & 6 & 3 & 5 \\ 0 & ① & 4 & 6 & 10 \\ 0 & 0 & ① & 1 & 2 \\ 0 & 0 & 0 & 5 & 15 \\ 0 & 0 & 0 & 0 & 0 \end{bmatrix}.$$

Next, we target all of the remaining nonzero entries in the third column to convert them to "0." These are the $(1,3)$ and $(2,3)$ entries. Each of these becomes a *target entry* in turn, with its corresponding row as the *target row*. In both cases, we add the appropriate multiple of the pivot row to the target row to convert the target entry to "0." The cumulative effect of these two type (II) row operations is:

$$\text{(II): } (-6) < 3 > \; + \; < 1 > \implies \; < 1 >$$
$$\text{(II): } (-4) < 3 > \; + \; < 2 > \implies \; < 2 >$$

$$\left[\begin{array}{cccc|c} ① & 0 & 0 & -3 & -7 \\ 0 & ① & 0 & 2 & 2 \\ 0 & 0 & ① & 1 & 2 \\ 0 & 0 & 0 & 5 & 15 \\ 0 & 0 & 0 & 0 & 0 \end{array}\right].$$

Our work on the third column is now completed, and finally, we turn to the fourth column, which becomes our new *pivot column*. We want to convert the $(4,4)$ entry to "1." Thus, the $(4,4)$ entry becomes our current *pivot entry*, and the fourth row becomes our current *pivot row*. We use a type (I) row operation to multiply the fourth row by the reciprocal of the pivot entry:

$$\text{(I): } \left(-\frac{1}{5}\right) < 4 > \implies \; < 4 >$$

$$\left[\begin{array}{cccc|c} ① & 0 & 0 & -3 & -7 \\ 0 & ① & 0 & 2 & 2 \\ 0 & 0 & ① & 1 & 2 \\ 0 & 0 & 0 & ① & 3 \\ 0 & 0 & 0 & 0 & 0 \end{array}\right].$$

We now target all of the remaining entries in the fourth column to convert them to "0." These are the $(1,4)$, $(2,4)$, and $(3,4)$ entries. Each of these becomes a *target entry* in turn, with its corresponding row as the *target row*. In all cases, we add the appropriate multiple of the pivot row to the target row to convert the target entry to "0." The cumulative effect of these three type (II) row operations is:

$$\text{(II): } \quad\;\; 3 < 4 > \; + \; < 1 > \implies \; < 1 >$$
$$\text{(II): } (-2) < 4 > \; + \; < 2 > \implies \; < 2 >$$
$$\text{(II): } (-1) < 4 > \; + \; < 3 > \implies \; < 3 >$$

$$\left[\begin{array}{cccc|c} ① & 0 & 0 & 0 & 2 \\ 0 & ① & 0 & 0 & -4 \\ 0 & 0 & ① & 0 & -1 \\ 0 & 0 & 0 & ① & 3 \\ 0 & 0 & 0 & 0 & 0 \end{array}\right].$$

At this point, no further row operations are needed since the augmented matrix is in the desired form. The last row can be ignored since it contributes no information about the solution set. Therefore, this augmented matrix represents the linear system

$$\begin{cases} 1x_1 & + & 0x_2 & + & 0x_3 & + & 0x_4 & = & 2 \\ 0x_1 & + & 1x_2 & + & 0x_3 & + & 0x_4 & = & -4 \\ 0x_1 & + & 0x_2 & + & 1x_3 & + & 0x_4 & = & -1 \\ 0x_1 & + & 0x_2 & + & 0x_3 & + & 1x_4 & = & 3 \end{cases}, \text{ or, } \begin{cases} x_1 & = & 2 \\ x_2 & = & -4 \\ x_3 & = & -1 \\ x_4 & = & 3 \end{cases}.$$

The complete solution set of this is $\{(2,-4,-1,3)\}$. But we know that the row operations which we performed do not affect the solution set. Therefore, the complete solution of the original system is also $\{(2,-4,-1,3)\}$. ∎

The previous example illustrates an additional rule for the Gauss-Jordan Method:

- To convert an entry into "1" when that entry is currently "0:"

  - First use a type (III) row operation to switch the row containing that entry with a *later* row so that a nonzero number is placed in the desired position. (We take care not to switch with an earlier row so that we do not destroy the pattern of pivot entries that have already been created.)

  - Then use a type (I) row operation as usual to convert that (nonzero) entry into "1."

The linear systems that we solved thus far using the Gauss-Jordan Method have unique solutions. However, as we know, some linear systems have an infinite number of solutions or no solutions instead. We will consider those types of systems in the next section, and explain how the Gauss-Jordan Method handles those cases.

Glossary

- Circled pivot entry: A pivot entry that has been converted to "1" during the Gauss-Jordan Row Reduction Method, and circled for easier reference.

- Gauss-Jordan Row Reduction Method: A procedure for determining the complete solution set of a linear system by using the three types of row operations as necessary. This method simplifies the appearance of a linear system by converting pivot entries to "1" and target entries to "0," so that the solution set is readily apparent. (To save space, we sometimes refer to this as the "Gauss-Jordan Method," or simply, "row reduction.")

- Pivot column: The column containing the current pivot entry.

- Pivot entry: The most recent entry to be converted to "1" during Gauss-Jordan row reduction. Once we have converted a pivot entry to "1," we circle that entry for easier reference.

- Pivot row: The row containing the current pivot entry during Gauss-Jordan row reduction.

- Row reduction: Shorthand for "Gauss-Jordan Row Reduction Method."

- Target: A verb, meaning to perform a type (II) row operation that converts a target entry to "0."

- Target entry: The most recent entry to be converted to "0" during Gauss-Jordan row reduction. Target entries lie above and below pivot entries.

- Target row: The row containing the current target entry during Gauss-Jordan row reduction.

Exercises for Section 4.3

1. For each of the following augmented matrices, what is the next row operation that should be performed according to the Gauss-Jordan Row Reduction Method? What augmented matrix is produced by that row operation?

(a)
$$\left[\begin{array}{rrr|r} 1 & -3 & 5 & -16 \\ 0 & 2 & -7 & 13 \\ 4 & -2 & -5 & -9 \end{array}\right]$$

(b)
$$\left[\begin{array}{rrr|r} 1 & 2 & -6 & 40 \\ 0 & 0 & 3 & -15 \\ 0 & -5 & -1 & -10 \end{array}\right]$$

(c)
$$\left[\begin{array}{rrrr|r} 1 & -3 & 2 & -5 & 17 \\ 0 & -4 & -2 & -2 & 6 \\ 0 & 2 & -4 & 1 & -8 \\ 0 & 5 & 3 & 0 & -2 \end{array}\right]$$

2. For each of the following augmented matrices, what is the next row operation that should be performed according to the Gauss-Jordan Row Reduction Method?

(a)
$$\left[\begin{array}{rrr|r} 5 & -4 & 3 & 25 \\ -1 & 3 & -2 & -10 \\ -4 & 1 & 3 & -7 \end{array}\right]$$

(b)
$$\left[\begin{array}{rrr|r} 1 & -6 & 5 & -31 \\ 0 & 1 & -4 & 8 \\ 0 & 3 & 2 & 10 \end{array}\right]$$

(c)
$$\left[\begin{array}{rrrr|r} 1 & -2 & 4 & 3 & 17 \\ 0 & 0 & -1 & 2 & 9 \\ 0 & 0 & 2 & -1 & -6 \\ 0 & -3 & -1 & 4 & 23 \end{array}\right]$$

3. Find the solution set for each of the following systems of linear equations using the Gauss-Jordan Row Reduction Method. (Note: Only type (I) and (II) row operations are needed in these problems.)

(a)
$$\begin{cases} 5x & - & 7y & = & -80 \\ 3x & + & 5y & = & -2 \end{cases}$$

(b)
$$\begin{cases} 2x & + & 2y & + & 3z & = & 4 \\ 3x & + & y & + & 2z & = & 11 \\ x & - & 3y & + & 2z & = & 23 \end{cases}$$

(c)
$$\begin{cases} 2x_1 & - & 6x_2 & - & 12x_3 & + & 21x_4 & = & 42 \\ 2x_1 & & & - & 3x_3 & + & 3x_4 & = & 9 \\ -4x_1 & + & 10x_2 & + & 24x_3 & - & 54x_4 & = & -124 \\ -3x_1 & + & 15x_2 & + & 33x_3 & - & 84x_4 & = & -195 \end{cases}$$

(d)
$$\begin{cases} 3x & - & 9y & - & 30z & = & 72 \\ 2x & - & y & & & = & 18 \\ 2x & - & 4y & - & 8z & = & 32 \\ 5x & & & + & 5z & = & 35 \end{cases}$$

4. Find the solution set for each of the following systems of linear equations using the Gauss-Jordan Row Reduction Method. (Note: Only type (I) and (II) row operations are needed in these problems.)

(a) $\begin{cases} 6x & + & 5y & = & -30 \\ 4x & + & 7y & = & 24 \end{cases}$

(b) $\begin{cases} 2x_1 & - & 12x_2 & - & 13x_3 & = & -59 \\ 3x_1 & - & 20x_2 & - & 20x_3 & = & -110 \\ -x_1 & & & + & 6x_3 & = & -44 \end{cases}$

(c) $\begin{cases} -x_1 & - & x_2 & + & 8x_3 & + & 25x_4 & = & 442 \\ -2x_1 & - & 3x_2 & + & 4x_3 & + & 12x_4 & = & 238 \\ -x_1 & & & + & 16x_3 & + & 51x_4 & = & 880 \\ -2x_1 & - & 4x_2 & & & - & x_4 & = & 28 \end{cases}$

5. Find the solution set for each of the following systems of linear equations using the Gauss-Jordan Row Reduction Method. (Note: Row operation (III) is needed in these problems.)

(a) $\begin{cases} 2x & + & 8y & - & 6z & = & -96 \\ 3x & + & 12y & - & 5z & = & -96 \\ -4x & - & 19y & + & 17z & = & 267 \end{cases}$

(b) $\begin{cases} x_1 & - & 2x_2 & + & 2x_3 & + & 3x_4 & = & 28 \\ -4x_1 & + & 8x_2 & - & 8x_3 & - & 17x_4 & = & -162 \\ 5x_1 & - & 7x_2 & + & 4x_3 & + & 17x_4 & = & 115 \\ -2x_1 & - & x_2 & + & 5x_3 & - & 2x_4 & = & 47 \end{cases}$

6. Find the solution set for each of the following systems of linear equations using the Gauss-Jordan Row Reduction Method. (Note: Row operation (III) is needed in these problems.)

(a) $\begin{cases} 3x_1 & - & 9x_2 & - & 15x_3 & = & 21 \\ -4x_1 & + & 12x_2 & - & 17x_3 & = & -287 \\ 3x_1 & - & 14x_2 & + & 17x_3 & = & 300 \end{cases}$

(b) $\begin{cases} 2x & - & 10y & + & 4z & - & 6w & = & -16 \\ -5x & + & 25y & - & 10z & + & 19w & = & 8 \\ 2x & - & 7y & + & 10z & - & 8w & = & 45 \\ -3x & + & 9y & - & 13z & + & 8w & = & -38 \end{cases}$

(c) $\begin{cases} 4a & + & 8b & - & 4c & = & 44 \\ 3a & + & 10b & - & 11c & = & 49 \\ -2a & + & 4b & - & 14c & = & 10 \\ 2a & + & 7b & - & 2c & = & 28 \end{cases}$

7. A candy maker produces two kinds of fudge – vanilla and chocolate. Each pound of vanilla fudge requires 3 cups of sugar and 2 cups of butter. Each pound of chocolate fudge requires 2.5 cups of sugar and 3

cups of butter (as well as lots of chocolate, which is in plentiful supply). The company has available 123 cups of sugar and 114 cups of butter.

(a) Create a system of linear equations whose solution will tell the candy maker how many pounds of each type of fudge to make in order to exactly use up all of the sugar and butter available. Indicate what your variables represent.

(b) Use the Gauss-Jordan Method on the augmented matrix for the linear system in part (a) to determine the number of pounds of each kind of fudge that should be made so that all of the available sugar and butter is used.

8. Three brands of gerbil food are available at a certain store. Each bag of Brand A contains 25 units of fiber and 35 units of protein. Each bag of Brand B contains 50 units of fiber and 30 units of protein. Each bag of Brand C contains 60 units of fiber and 20 units of protein. Suppose we want to buy twice as many bags of Brand C as Brand A, and we also want a total of 440 units of fiber and 240 units of protein.

(a) Create a system of linear equations whose solution will tell us how many bags of each gerbil food to buy in order to exactly meet all of the given requirements. Indicate what your variables represent.

(b) Use the Gauss-Jordan Method on the augmented matrix for the linear system in part (a) to determine how many bags of Brands A, B, and C should be bought to satisfy the given requirements.

9. Young MacDonald needs to put 4 tons of nitrogen, 3.2 tons of phosphate, and 2 tons of potash onto a field. Mr. MacDonald plans to combine three different types of fertilizer: *SureGrow* fertilizer of which 10% is nitrogen, 5% is phosphate, and 3% is potash, and *QuickGrow* fertilizer of which 20% is nitrogen, 10% is phosphate, and 14% is potash, and *EvenGrow* fertilizer, of which 5% is nitrogen, 10% is phosphate, and 2% is potash.

(a) Create a system of linear equations whose solution will tell Young MacDonald how many tons of each type of fertilizer he should put on his field. Indicate what your variables represent.

(b) Use the Gauss-Jordan Method on the augmented matrix for the linear system in part (a) to determine how many tons of each type of fertilizer Mr. MacDonald should use to satisfy the given requirements. (This problem may need a type (III) row operation.)

10. A citrus farmer is making up fruit baskets for sale. The *Small Pleasures* basket contains 2 navel oranges, 4 Valencia oranges, and 3 tangerines. The *Big Mix* contains 12 navel oranges, 24 Valencia oranges, and 10 tangerines. The *Crate O' Fruit* contains 14 navel oranges, 30 Valencia

oranges, and 24 tangerines. The farmer has 10, 922 navel oranges, 22, 668 Valencia oranges, and 14, 523 tangerines available.

(a) Create a system of linear equations whose solution tells the farmer the number of baskets of each type to make up in order to exactly use up all of the oranges available. Indicate what your variables represent.

(b) Use the Gauss-Jordan Method on the augmented matrix for the linear system in part (a) to determine how many of each type of basket the farmer should make.

4.4 The Remaining Cases of the Gauss-Jordan Method

In the previous section, we illustrated how to use the Gauss-Jordan Row Reduction Method to find the solution set for linear systems with unique solutions. In this section, we show how this method can also be used to find the solution set for linear systems having no solutions or an infinite number of solutions.

Linear Systems with an Infinite Number of Solutions

During the Gauss-Jordan Method, we may encounter a situation in which the current pivot entry is "0," but there is no lower row in the matrix having a nonzero number directly below that "0." In such a case, there is no type (III) row operation that can be used to place a nonzero number in the pivot entry position. Whenever such a situation occurs, we *skip over* that column. We move instead to the next column to the right, and label that column as our newest pivot column. Now, even though we may have to skip over columns occasionally, we *never skip over a row*. That is, each additional pivot entry is always located in the row immediately below the most recent circled pivot entry. The next example illustrates this procedure.

Example 1. Consider the following system of linear equations:

$$\begin{cases} x - 3y - 4z - 7w = -51 \\ -3x + 9y + 14z + 23w = 169 \\ -2x + 6y + 13z + 14w = 117 \\ 2x - 6y - 10z - 13w = -103 \end{cases}$$

The augmented matrix for this linear system is:

$$\left[\begin{array}{cccc|c} 1 & -3 & -4 & -7 & -51 \\ -3 & 9 & 14 & 23 & 169 \\ -2 & 6 & 13 & 14 & 117 \\ 2 & -6 & -10 & -13 & -103 \end{array}\right].$$

We use the Gauss-Jordan Method to row reduce this augmented matrix, beginning with the first column as our *pivot column*, the $(1,1)$ entry as our *pivot entry*, and the first row as our *pivot row*. Since the $(1,1)$ entry is already equal to "1," we proceed to target the remaining entries in the first column:

$$
\begin{array}{rlcl}
\text{(II):} & 3<1> + <2> & \Longrightarrow & <2> \\
\text{(II):} & 2<1> + <3> & \Longrightarrow & <3> \\
\text{(II): } (-2)<1> + <4> & & \Longrightarrow & <4>
\end{array}
\qquad
\left[\begin{array}{cccc|c}
① & -3 & -4 & -7 & -51 \\
0 & 0 & 2 & 2 & 16 \\
0 & 0 & 5 & 0 & 15 \\
0 & 0 & -2 & 1 & -1
\end{array}\right].
$$

We now move to the second column as our next *pivot column*, and the second row as our next *pivot row*. However, the $(2,2)$ entry is zero, and all entries below that entry are also zero. No type (III) operation switching the second row with a later row would help here. Therefore, we move instead to the third column as our next *pivot column, while keeping the second row as the current pivot row*. The $(2,3)$ entry becomes our current *pivot entry*, and we convert this entry to "1" as follows:

$$
\text{(I): } \left(\tfrac{1}{2}\right)<2> \;\Longrightarrow\; <2>
\qquad
\left[\begin{array}{cccc|c}
① & -3 & -4 & -7 & -51 \\
0 & 0 & ① & 1 & 8 \\
0 & 0 & 5 & 0 & 15 \\
0 & 0 & -2 & 1 & -1
\end{array}\right].
$$

Targeting the remaining entries in the third column gives:

$$
\begin{array}{rlcl}
\text{(II):} & 4<2> + <1> & \Longrightarrow & <1> \\
\text{(II): } (-5)<2> + <3> & & \Longrightarrow & <3> \\
\text{(II):} & 2<2> + <4> & \Longrightarrow & <4>
\end{array}
\qquad
\left[\begin{array}{cccc|c}
① & -3 & 0 & -3 & -19 \\
0 & 0 & ① & 1 & 8 \\
0 & 0 & 0 & -5 & -25 \\
0 & 0 & 0 & 3 & 15
\end{array}\right].
$$

We move on by choosing the fourth column as our next *pivot column*, and the next row down (that is, the third row) as our next *pivot row*. Therefore, the $(3,4)$ entry becomes the current *pivot entry*. Converting this entry to "1," we have:

$$
\text{(I): } \left(-\tfrac{1}{5}\right)<3> \;\Longrightarrow\; <3>
\qquad
\left[\begin{array}{cccc|c}
① & -3 & 0 & -3 & -19 \\
0 & 0 & ① & 1 & 8 \\
0 & 0 & 0 & ① & 5 \\
0 & 0 & 0 & 3 & 15
\end{array}\right].
$$

Finally, targeting the remaining entries in the fourth column, we obtain:

$$
\begin{array}{rlcl}
\text{(II):} & 3<3> + <1> & \Longrightarrow & <1> \\
\text{(II): } (-1)<3> + <2> & & \Longrightarrow & <2> \\
\text{(II): } (-3)<3> + <4> & & \Longrightarrow & <4>
\end{array}
\qquad
\left[\begin{array}{cccc|c}
① & -3 & 0 & 0 & -4 \\
0 & 0 & ① & 0 & 3 \\
0 & 0 & 0 & ① & 5 \\
0 & 0 & 0 & 0 & 0
\end{array}\right].
$$

The last row of the augmented matrix represents the equation $0x + 0y + 0z + 0w = 0$, which is true no matter what values the unknowns have. This equation therefore gives us no information whatsoever about the solution set, and we can ignore it. Therefore, the linear system represented by the final augmented matrix is:

$$\begin{cases} x \;-\; 3y & = & -4 \\ & z & = & 3 \\ & w & = & 5 \end{cases}.$$

We can see that z must equal 3, and w must equal 5. However, there is an infinite number of values for x and y that make the first equation true, since this equation represents the straight line $x - 3y = -4$. Therefore the original linear system also has an infinite number of solutions. ■

In cases where the solution set for a system of linear equations is infinite, we often want to write out the complete solution set in a formal manner. This process is streamlined once we realize that the final columns having no circled pivot entries represent unknowns that can take on any value. That is, *we can choose arbitrary values for those particular unknowns.* The values for the remaining unknowns can then be determined from those.

We illustrate this procedure for the final linear system

$$\begin{cases} x \;-\; 3y & = & -4 \\ & z & = & 3 \\ & w & = & 5 \end{cases}$$

from the previous example. The only column without a circled pivot entry is the second column; that is, the "y" column. Therefore, we allow y to represent any arbitrary value. Then from the first equation $x - 3y = -4$, we can solve for the value of x in terms of y, obtaining $x = 3y - 4$. That is, the value of x is determined once we have chosen a value for y. We have already seen that the values for z and w are 3 and 5, respectively. Hence, we can now represent the infinite solution set for this linear system as

$$\{(3y - 4, \; y, \; 3, \; 5)\}.$$

Since there is an infinite number of possible choices for y, we have an infinite number of solutions overall. For instance, if we let $y = 1$, then $x = 3y - 4 = 3(1) - 4 = -1$, so one particular solution is $(x, y, z, w) = (-1, 1, 3, 5)$. Similarly, we can let $y = 4$, and then $x = 3y - 4 = 3(4) - 4 = 8$, giving us another particular solution $(x, y, z, w) = (8, 4, 3, 5)$. You should check that these two particular solutions really do satisfy all of the equations in the original linear system.

We summarize these additional new rules for the Gauss-Jordan Method as follows:

- If the current pivot entry is "0," and there is no lower row having a nonzero number directly below that "0," then skip over that column.

Move instead to the next column to the right, and label that column as the newest pivot column, and label the entry of that column in the *current* pivot row as the new pivot entry. (This ensures that we never skip a row when creating circled pivots.)

- To determine the complete solution set when a linear system has an infinite number of solutions:

  - Write out the linear system corresponding to the final augmented matrix.
  - Assume that the unknowns representing the columns *without* circled pivots have been given any arbitrary values.
  - Use the nonzero rows to solve, where necessary, for the remaining unknowns in terms of those. (That is, solve for the unknowns for columns *with* circled pivots in terms of the unknowns for columns *without* circled pivots.)

Example 2. Suppose that the Gauss-Jordan Method has been performed for a linear system with five equations and five unknowns x_1, x_2, x_3, x_4, x_5, and the final augmented matrix obtained has the form:

$$
\begin{array}{ccccc}
x_1 & x_2 & x_3 & x_4 & x_5 \\
\end{array}
$$

$$
\left[\begin{array}{ccccc|c}
① & 5 & 0 & -4 & 0 & 8 \\
0 & 0 & ① & 6 & 0 & -2 \\
0 & 0 & 0 & 0 & ① & -3 \\
0 & 0 & 0 & 0 & 0 & 0 \\
0 & 0 & 0 & 0 & 0 & 0
\end{array} \right].
$$

We see that the second and fourth columns do not have a circled pivot entry. The unknowns for these columns are x_2 and x_4, respectively. Therefore, we allow these unknowns to take on any arbitrary values, and determine the values of the remaining unknowns, where necessary, by solving for their values in terms of x_2 and x_4.

The first row indicates that $x_1 + 5x_2 - 4x_4 = 8$. Solving for x_1 in terms of x_2 and x_4 gives $x_1 = -5x_2 + 4x_4 + 8$.

The second row indicates that $x_3 + 6x_4 = -2$. Solving for x_3 in terms of x_4 gives $x_3 = -6x_4 - 2$.

The final nonzero row is the third row, which tells us that $x_5 = -3$. (The value of x_5 in this case does not actually depend on x_2 or x_4.)

Combining all of this information gives the infinite solution set

$$\{(-5x_2 + 4x_4 + 8, \ x_2, \ -6x_4 - 2, \ x_4, \ -3)\}.$$

For instance, we can let $x_2 = 1$ and $x_4 = 0$ to obtain the particular solution $(x_1, x_2, x_3, x_4, x_5) = (3, 1, -2, 0, -3)$. Similarly, we can let $x_2 = 0$ and $x_4 = 1$ to obtain the particular solution $(x_1, x_2, x_3, x_4, x_5) = (12, 0, -8, 1, -3)$. Of course, there is an infinite number of other choices for the values of x_2 and x_4, leading to an infinite number of solutions overall. ∎

Example 3. Suppose that the Gauss-Jordan Method has been performed for a linear system with four equations and six unknowns x_1, x_2, x_3, x_4, x_5, x_6, and the final augmented matrix obtained has the form:

$$
\begin{array}{cccccc}
x_1 & x_2 & x_3 & x_4 & x_5 & x_6
\end{array}
$$
$$
\left[
\begin{array}{cccccc|c}
① & 0 & 3 & -7 & 0 & -5 & -1 \\
0 & ① & -2 & 6 & 0 & 2 & 8 \\
0 & 0 & 0 & 0 & ① & -9 & 5 \\
0 & 0 & 0 & 0 & 0 & 0 & 0
\end{array}
\right].
$$

Here, the third, fourth, and sixth columns do not have a circled pivot entry. The unknowns for these columns are x_3, x_4, and x_6, respectively. We allow these unknowns to take on any arbitrary values, and determine the values of the remaining unknowns by solving for them in terms of x_3, x_4, and x_6.

The first row indicates that $x_1 + 3x_3 - 7x_4 - 5x_6 = -1$. Solving for x_1 in terms of x_3, x_4, and x_6 gives $x_1 = -3x_3 + 7x_4 + 5x_6 - 1$.

The second row tells us that $x_2 - 2x_3 + 6x_4 + 2x_6 = 8$. Solving for x_2 in terms of x_3, x_4, and x_6 gives $x_2 = 2x_3 - 6x_4 - 2x_6 + 8$.

The final nonzero row is the third row, which indicates that $x_5 - 9x_6 = 5$. Solving for x_5 in terms of x_6 gives $x_5 = 9x_6 + 5$.

Putting all of this information together gives the infinite solution set

$$\{(-3x_3 + 7x_4 + 5x_6 - 1, \ 2x_3 - 6x_4 - 2x_6 + 8, \ x_3, \ x_4, \ 9x_6 + 5, \ x_6)\}.$$

For instance, we can let $x_3 = 1$ and $x_4 = 0$ and $x_6 = 0$ to obtain the particular solution $(x_1, x_2, x_3, x_4, x_5, x_6) = (-4, 10, 1, 0, 5, 0)$. Similarly, we can let $x_3 = 0$ and $x_4 = 1$ and $x_6 = 0$ to obtain the particular solution $(x_1, x_2, x_3, x_4, x_5, x_6) = (6, 2, 0, 1, 5, 0)$. Of course, there is an infinite number of other choices for the values of x_3, x_4, and x_6, leading to an infinite number of solutions overall. ∎

Linear Systems with No Solution

The next example involves a system of linear equations that has no solutions.

Example 4. Suppose that after applying the Gauss-Jordan Method to the first three columns of the augmented matrix for a linear system with unknowns x_1, x_2, x_3, and x_4, we obtain:

$$
\left[
\begin{array}{cccc|c}
① & 0 & 0 & -3 & 6 \\
0 & ① & 0 & 5 & -2 \\
0 & 0 & ① & -2 & 4 \\
0 & 0 & 0 & 0 & 7
\end{array}
\right].
$$

The fourth row represents the equation $0x_1 + 0x_2 + 0x_3 + 0x_4 = 7$. However, no matter what values x_1, x_2, x_3, and x_4 have, the sum on the left side of the equation is zero. But the right side of this equation is nonzero. Therefore the left side of this equation can never equal the right side, and so this one equation cannot have any possible solutions. For this reason, there are no values of the unknowns that can satisfy *every* equation in the corresponding linear system for this augmented matrix. The row operations that produced this augmented matrix did not affect the solution set, so the original linear system also has no solutions. That is, the solution set of the original system is the empty set: $\{\ \} = \phi$. ∎

In general, when solving a system of linear equations using the Gauss-Jordan Method, if at any point during the row reduction, a row of the associated augmented matrix has the form

$$[0,\ 0,\ 0,\ ...,\ 0\ |\ *],$$

where all of the values in the row to the left of the bar are zeros, and where the "$*$" symbol represents a nonzero number, then the system of linear equations has no solutions. The reason is that such a row represents an equation of the form $0x_1 + 0x_2 + 0x_3 + \cdots + 0x_n = *$. No matter what values are chosen for the variables $x_1, x_2, x_3, \ldots, x_n$, the left side of this equation is zero, and therefore cannot equal the nonzero number $*$ on the other side of the equal sign. Whenever we encounter a row of this form during the row reduction process, we know immediately that no solutions can exist for this particular equation, and therefore there can be no solution for the linear system as a whole. A linear system having no solution is said to be **inconsistent**.

We summarize this additional new rule for the Gauss-Jordan Method as follows:

- If at any point during the Gauss-Jordan Method, the augmented matrix contains a row of the form

$$[0,\ 0,\ 0,\ ...,\ 0\ |\ *],$$

  where the "$*$" symbol represents a nonzero number, then the original system of linear equations has *no solutions*. It is inconsistent.

Note that linear systems that *do* have at least one solution are said to be **consistent**.

Example 5. Consider the linear system

$$\begin{cases} -2x_1 + 8x_2 - 28x_3 = 20 \\ 3x_1 - 11x_2 + 39x_3 = -23 \\ -1x_1 + 5x_2 - 17x_3 = 14 \end{cases}.$$

We apply the Gauss-Jordan Method to the augmented matrix for this system, which is:

$$\begin{bmatrix} -2 & 8 & -28 & 20 \\ 3 & -11 & 39 & -23 \\ -1 & 5 & -17 & 14 \end{bmatrix}.$$

We begin with the first column as our *pivot column*, the $(1,1)$ entry as our *pivot entry*, and the first row as our *pivot row*. Converting the pivot entry to "1," we obtain:

$$(\text{I}): \left(-\frac{1}{2}\right) < 1 > \implies < 1 > \qquad \begin{bmatrix} ① & -4 & 14 & -10 \\ 3 & -11 & 39 & -23 \\ -1 & 5 & -17 & 14 \end{bmatrix}.$$

Next, we target the remaining entries in the first column:

$$\begin{array}{l} (\text{II}): (-3) < 1 > + < 2 > \implies < 2 > \\ (\text{II}): 1 < 1 > + < 3 > \implies < 3 > \end{array} \qquad \begin{bmatrix} ① & -4 & 14 & -10 \\ 0 & 1 & -3 & 7 \\ 0 & 1 & -3 & 4 \end{bmatrix}.$$

Next we move to the second column as our *pivot column*, the $(2,2)$ entry as our *pivot entry*, and the second row as our *pivot row*. But since the $(2,2)$ entry is already equal to 1, we proceed to target the remaining entries in the second column:

$$\begin{array}{l} (\text{II}): 4 < 2 > + < 1 > \implies < 1 > \\ (\text{II}): (-1) < 2 > + < 3 > \implies < 3 > \end{array} \qquad \begin{bmatrix} ① & 0 & 2 & 18 \\ 0 & ① & -3 & 7 \\ 0 & 0 & 0 & -3 \end{bmatrix}.$$

Before moving on to the third column, we notice that the final row of the augmented matrix has all zeroes to the left of the bar, and a *nonzero* number to the right of the bar. This signals that the equation corresponding to this row (as well as the linear system as a whole) has no solution. Since the row operations that we performed did not affect the solution set, the original linear system also has no solutions. That is, the solution set of the original system is the empty set: $\{\ \} = \phi$. ∎

Number of Solutions

We can combine the principles from this section and the previous section to determine the number of solutions for a given system of linear equations:

- If at any point during the Gauss-Jordan Method, the augmented matrix contains a row of the form

$$[0, \ 0, \ 0, \ ..., \ 0 \mid *],$$

  where the "$*$" symbol represents a nonzero number, then the original system of linear equations has *no solutions*. It is inconsistent.

- Otherwise, the linear system is consistent, and:

  - If there is a circled pivot in every column before the bar, then the original system of linear equations has a *unique solution*. (The number after the bar in each row represents the value of the unknown corresponding to the circled pivot in that row.)

— If there is at least one column before the bar without a circled pivot, then the original system of linear equations has an *infinite number of solutions*. (The unknowns corresponding to columns without a circled pivot can take on any arbitrary value, and the values of the remaining unknowns can be determined from those values.)

Glossary

- Consistent system of linear equations: A linear system that has at least one solution. Therefore, systems that have either a unique solution or an infinite number of solutions are consistent.

- Inconsistent system of linear equations: A linear system that has no solutions.

Exercises for Section 4.4

1. For the given final augmented matrix after Gauss-Jordan row reduction, write out the complete solution set for the corresponding system of linear equations. If the solution set is infinite, state two different particular solutions.

(a)
$$\begin{array}{ccc} x & y & z \end{array}$$
$$\left[\begin{array}{ccc|c} ① & 0 & -5 & 3 \\ 0 & ① & 6 & -2 \\ 0 & 0 & 0 & 0 \end{array}\right]$$

(b)
$$\begin{array}{cccc} x & y & z & w \end{array}$$
$$\left[\begin{array}{cccc|c} ① & 0 & 0 & 6 & 9 \\ 0 & ① & 0 & -1 & -4 \\ 0 & 0 & ① & 5 & 3 \\ 0 & 0 & 0 & 0 & -2 \end{array}\right]$$

(c)
$$\begin{array}{cccc} x_1 & x_2 & x_3 & x_4 \end{array}$$
$$\left[\begin{array}{cccc|c} ① & 0 & 12 & 0 & -15 \\ 0 & ① & -9 & 0 & 14 \\ 0 & 0 & 0 & ① & -8 \\ 0 & 0 & 0 & 0 & 0 \end{array}\right]$$

(d)
$$\begin{array}{ccccc} x_1 & x_2 & x_3 & x_4 & x_5 \end{array}$$
$$\left[\begin{array}{ccccc|c} ① & 0 & -7 & 0 & 4 & 13 \\ 0 & ① & 5 & 0 & -11 & -3 \\ 0 & 0 & 0 & ① & -6 & 19 \\ 0 & 0 & 0 & 0 & 0 & 0 \\ 0 & 0 & 0 & 0 & 0 & 0 \end{array}\right]$$

(e)
$$\begin{array}{ccccc} x_1 & x_2 & x_3 & x_4 & x_5 \end{array}$$
$$\left[\begin{array}{ccccc|c} ① & -16 & 0 & 0 & 12 & -6 \\ 0 & 0 & ① & 0 & -10 & 12 \\ 0 & 0 & 0 & ① & 17 & -11 \\ 0 & 0 & 0 & 0 & ① & -2 \\ 0 & 0 & 0 & 0 & 0 & 4 \end{array}\right]$$

2. For the given final augmented matrix after Gauss-Jordan row reduction, write out the complete solution set for the corresponding system of linear

equations. If the solution set is infinite, state two different particular solutions.

(a)
$$\begin{array}{ccc} x & y & z \end{array}$$
$$\left[\begin{array}{ccc|c} ① & 0 & -7 & 4 \\ 0 & ① & 5 & -9 \\ 0 & 0 & 0 & -3 \end{array}\right]$$

(c)
$$\begin{array}{cccc} x_1 & x_2 & x_3 & x_4 \end{array}$$
$$\left[\begin{array}{cccc|c} ① & -8 & 0 & 0 & 26 \\ 0 & 0 & ① & 0 & -15 \\ 0 & 0 & 0 & ① & 23 \\ 0 & 0 & 0 & 0 & -17 \end{array}\right]$$

(b)
$$\begin{array}{cccc} x & y & z & w \end{array}$$
$$\left[\begin{array}{cccc|c} ① & 0 & -24 & 0 & 16 \\ 0 & ① & 18 & 0 & -13 \\ 0 & 0 & 0 & ① & 14 \\ 0 & 0 & 0 & 0 & 0 \end{array}\right]$$

(d)
$$\begin{array}{ccccc} x_1 & x_2 & x_3 & x_4 & x_5 \end{array}$$
$$\left[\begin{array}{ccccc|c} ① & 4 & 0 & 0 & -6 & 17 \\ 0 & 0 & ① & 0 & 10 & -5 \\ 0 & 0 & 0 & ① & 7 & -20 \\ 0 & 0 & 0 & 0 & 0 & 0 \\ 0 & 0 & 0 & 0 & 0 & 0 \end{array}\right]$$

(e)
$$\begin{array}{cccccc} x_1 & x_2 & x_3 & x_4 & x_5 & x_6 \end{array}$$
$$\left[\begin{array}{cccccc|c} ① & -2 & 3 & 0 & 0 & -11 & 15 \\ 0 & 0 & 0 & ① & 0 & 12 & -19 \\ 0 & 0 & 0 & 0 & ① & 4 & -18 \\ 0 & 0 & 0 & 0 & 0 & 0 & 0 \end{array}\right]$$

3. Use the Gauss-Jordan Method to find the complete solution set for each of the following systems of linear equations.

(a)
$$\begin{array}{ccc} x & y & z \end{array}$$
$$\left[\begin{array}{ccc|c} 2 & 4 & 24 & 6 \\ 4 & 5 & 39 & 0 \\ -5 & -8 & -54 & -10 \end{array}\right]$$

(c)
$$\begin{array}{cccc} x_1 & x_2 & x_3 & x_4 \end{array}$$
$$\left[\begin{array}{cccc|c} -2 & 12 & -3 & 7 & -12 \\ 1 & -6 & 2 & -6 & 5 \\ 5 & -30 & 11 & -35 & 23 \\ -2 & 12 & -7 & 27 & -4 \end{array}\right]$$

(b)
$$\begin{array}{cccc} x & y & z & w \end{array}$$
$$\left[\begin{array}{cccc|c} 2 & 2 & 8 & 2 & 32 \\ -2 & -1 & -1 & -3 & -14 \\ -1 & 1 & 10 & -2 & 15 \\ 3 & 0 & -9 & 3 & 9 \end{array}\right]$$

(d)
$$\begin{array}{cccc} x_1 & x_2 & x_3 & x_4 \end{array}$$
$$\left[\begin{array}{cccc|c} 1 & 2 & -5 & -28 & -7 \\ 1 & 1 & -3 & -13 & -8 \\ 2 & 0 & -8 & -32 & -48 \\ -1 & -1 & 7 & 37 & 24 \end{array}\right]$$

(e)
$$\begin{array}{ccccc} x_1 & x_2 & x_3 & x_4 & x_5 \end{array}$$
$$\left[\begin{array}{ccccc|c} 4 & -4 & -32 & 44 & 6 & -96 \\ -2 & 6 & 28 & -38 & -1 & 88 \\ 4 & -1 & -23 & 32 & 4 & -38 \\ -4 & 0 & 20 & -28 & -7 & 48 \end{array}\right]$$

4. Use the Gauss-Jordan Method to find the complete solution set for each of the following systems of linear equations.

(a)
$$\begin{array}{ccc} x & y & z \\ \end{array}$$
$$\left[\begin{array}{ccc|c} 2 & 6 & -40 & 24 \\ -2 & -3 & 16 & -3 \\ 1 & -1 & 12 & -16 \end{array}\right]$$

(c)
$$\begin{array}{cccc} x_1 & x_2 & x_3 & x_4 \\ \end{array}$$
$$\left[\begin{array}{cccc|c} 1 & 3 & -11 & -2 & 10 \\ -1 & -1 & -1 & 1 & 1 \\ 4 & 11 & -38 & -9 & 21 \\ 3 & 6 & -15 & -4 & 10 \end{array}\right]$$

(b)
$$\begin{array}{cccc} x & y & z & w \\ \end{array}$$
$$\left[\begin{array}{cccc|c} 2 & -8 & 0 & 22 & -16 \\ 1 & -4 & 4 & -9 & 48 \\ -2 & 8 & -5 & 3 & -54 \\ -2 & 8 & 4 & -42 & 60 \end{array}\right]$$

(d)
$$\begin{array}{ccccc} x_1 & x_2 & x_3 & x_4 & x_5 \\ \end{array}$$
$$\left[\begin{array}{ccccc|c} 2 & 8 & -2 & 0 & -42 & -18 \\ 2 & 9 & -3 & -5 & -47 & -13 \\ -2 & -6 & 1 & -2 & 30 & 15 \\ 3 & 9 & -2 & -3 & -44 & -16 \end{array}\right]$$

(e)
$$\begin{array}{cccccc} x_1 & x_2 & x_3 & x_4 & x_5 & x_6 \\ \end{array}$$
$$\left[\begin{array}{cccccc|c} 1 & 0 & 5 & 6 & 6 & 10 & 7 \\ 3 & 1 & 7 & 13 & 12 & 34 & 14 \\ 1 & -1 & 13 & 10 & 12 & 0 & 9 \\ -4 & -1 & -12 & -17 & -15 & -44 & -26 \\ -1 & -2 & 11 & 1 & 4 & -28 & 2 \end{array}\right]$$

4.5   Reduced Row Echelon Form and Rank of a Matrix

In this section we examine more closely the matrices that result from the Gauss-Jordan row reduction process.

Row Reducing After the Augmentation Bar

Recall that in Example 5 in Section 4.4, we obtained the following final augmented matrix:

$$\left[\begin{array}{ccc|c} ① & 0 & 2 & 18 \\ 0 & ① & -3 & 7 \\ 0 & 0 & 0 & -3 \end{array}\right],$$

which indicated that the corresponding linear system has no solutions. Notice, however, that we could have actually continued the row reduction process a little further. Skipping over the last column before the bar and placing an additional pivot in the column after the bar, we obtain:

$$\text{(I): } \left(-\frac{1}{3}\right) < 3 > \implies < 3 > \quad \left[\begin{array}{ccc|c} ① & 0 & 2 & 18 \\ 0 & ① & -3 & 7 \\ 0 & 0 & 0 & ① \end{array}\right].$$

Then targeting the remaining entries in the column after the bar, we have:

(II): $(-18) < 3 > \; + \; < 1 > \; \Longrightarrow \; < 1 >$

(II): $(-7) < 3 > \; + \; < 2 > \; \Longrightarrow \; < 2 >$

$$\left[\begin{array}{ccc|c} ① & 0 & 2 & 0 \\ 0 & ① & -3 & 0 \\ 0 & 0 & 0 & ① \end{array}\right].$$

The final row has the form $0x_1 + 0x_2 + 0x_3 + 0x_4 = 1$, which can have no possible solutions, and thereby gives another verification that the original linear system has no solutions.

Similarly, we can row reduce beyond the augmentation bar in the augmented matrix of Example 4 from Section 4.4 to obtain the following final augmented matrix:

$$\left[\begin{array}{cccc|c} ① & 0 & 0 & -3 & 0 \\ 0 & ① & 0 & 5 & 0 \\ 0 & 0 & ① & -2 & 0 \\ 0 & 0 & 0 & 0 & ① \end{array}\right],$$

(Try it!) The final row has the form $0x_1 + 0x_2 + 0x_3 + 0x_4 = 1$, which can have no possible solutions, and confirms once again that the corresponding linear system has no solutions.

Reduced Row Echelon Form

The final augmented matrices we obtained in previous examples after completing the row reduction process have a number of properties in common. These properties are summarized in the following definition:

A (possibly augmented) matrix is in **reduced row echelon form** if and only if all of the following conditions hold:

(1) The first nonzero entry in each row is "1" (the unique circled pivot entry for that row).

(2) Each successive row has its circled pivot in a later column.

(3) All entries above and below each circled pivot are " 0."

(4) Any rows consisting entirely of zeros are the final rows.

For example, the final augmented matrix of Example 3 in Section 4.4 has the form:

$$\left[\begin{array}{cccccc|c} ① & 0 & 3 & -7 & 0 & -5 & -1 \\ 0 & ① & -2 & 6 & 0 & 2 & 8 \\ 0 & 0 & 0 & 0 & ① & -9 & 5 \\ 0 & 0 & 0 & 0 & 0 & 0 & 0 \end{array}\right].$$

Notice that the first nonzero entry in each of the first three rows is the circled pivot, and each new circled pivot occurs in a later column. Also, each entry above and below these circled pivots has been targeted (i.e., converted to "0"), and the (only) row of zeros is the final row of the matrix.

You should go back and verify that all of these conditions really do hold for all of the other final augmented matrices in the examples of this section and the previous section – taking special care in cases where the linear system has no solution to continue the row reduction process into the column after the bar, as we did above for Examples 4 and 5 from Section 4.4.[1]

In the final augmented matrix from Example 3 in Section 4.4 shown above, we have drawn in a descending "staircase" which drops down to a new step only when the next circled pivot is encountered. Notice that not all of the steps on the staircase are the same width, since some of the steps include columns where there are no circled pivots. In general, properties (1), (2), and (4) in the definition of reduced row echelon form above assure us that when a Gauss-Jordan row reduction is carried out completely, the circled pivots must always form a staircase pattern in the final augmented matrix. In particular, property (1) tells us that all of the nonzero rows contain a circled pivot, and property (4) tells us that these rows are bunched together at the top of the matrix. Property (2) then assures us that among the rows containing circled pivots, each new circled pivot is one "step" lower on the staircase than the previous circled pivot.

Uniqueness of Reduced Row Echelon Form

Whenever we have to solve a given system of linear equations, the equations can really be stated in any order. Consequently, there are different possible initial augmented matrices for the same linear system. Also, when we are applying a type (III) row operation to switch a certain row with a lower row, we sometimes have different choices for which lower row to use. In addition, some people may use a non-standard sequence of row operations in order to obtain the final reduced row echelon form matrix – although we do not recommend this practice. A natural question then arises about whether it is possible to obtain different final augmented matrices for the same linear system. However, the following theorem, which we state without proof, assures us that after Gauss-Jordan row reduction is carried out completely on all columns (including the column after the bar in cases having no solution), then there is only *one* possible final augmented matrix for any given linear system.

Theorem 1. *Every (possibly augmented) matrix reduces, by the use of row operations, to a unique corresponding (possibly augmented) matrix in reduced row echelon form.*

As a consequence of this theorem, we know that for a given linear system, we always obtain the same final reduced row echelon form augmented matrix for that system when the row reduction is carried out completely, even if we change the order of the equations in a given linear system, or even if we use a non-standard sequence of row operations to row reduce the augmented matrix. In fact, many calculators and computers have been programmed to calculate the unique reduced row echelon form matrix for a given linear system once

---

[1]There is no need to be concerned about this when the linear system has a unique solution or an infinite number of solutions, for in those cases there is no way to get an additional nonzero pivot entry in the column after the bar.

the initial augmented matrix for the system has been entered. Calculators and computer programs often use the function name "**rref**" for the process that finds the unique reduced row echelon form of the initial matrix.

Caveats When Using Calculators or Computers to Perform Row Reduction

When using a calculator or computer to calculate the reduced row echelon form of a matrix, we should be aware of the following caveats:

- When finding the solution set of a given system, we do not need to proceed past the bar when row reducing by hand. But calculators and computers generally do – that is, they row reduce completely on all possible columns.

- The "rref" command on some calculators and computers does not work if the matrix has more rows than columns, or, sometimes, more generally, if the matrix is not square. If your calculator gives you an error in such a case, simply add rows or columns of zeros as necessary to make the initial matrix square. Doing so will not affect the solution set. Just ignore the added row or column of zeros in your final result.

- Calculators and computers frequently have to round the results of their calculations (because they can use only a finite number of decimal places for each number due to memory limitations). Thus, the results of calculations are not always exact, and subsequent calculations which depend on previously rounded numbers may not be totally accurate.

- In particular, the result of a calculation that should equal "0" when done by hand may instead be represented in the memory of the calculator or computer as a very tiny number. If such a number is in a pivot entry position, then as the row reduction process continues, the calculator or computer would incorrectly proceed to convert that entry to "1," and the ultimate solution set obtained would be incorrect.

A mathematician normally must consider the results obtained with calculators and computers carefully to interpret the final augmented matrix correctly, and use more sophisticated software where necessary. However, in this textbook, we have taken care to create the examples and exercises to work out relatively "nicely" and avoid problems that would result in significant roundoff error.

Assignment 1: Learn how to perform the "rref" command on your calculator. Enter the initial augmented matrices for some of the matrices in Examples 1 and 2 in Section 4.3, as well as for Examples 1 and 5 in Section 4.4, and verify that the calculator does indeed produce the same final augmented matrix given in each example. Finally, rearrange some of the rows in the initial augmented matrix in some of these examples and again verify that the same final augmented matrix is obtained.

Rank of a Matrix

From Theorem 1 above, we know that when the Gauss-Jordan Method is applied to all possible columns of a given matrix, only one unique reduced row echelon form matrix can be obtained. In particular, the Gauss-Jordan Method always produces the same final number of circled pivots for a given initial matrix. We refer to this final number of circled pivots as the **rank** of the initial matrix. For a matrix $\mathbf{A}$, we say that $\text{rank}(\mathbf{A}) = k$ if there are exactly k circled pivots after a complete row reduction is performed on $\mathbf{A}$, including any pivots to the right of the bar.

Example 1. Consider the initial augmented matrix

$$\mathbf{A} = \left[\begin{array}{rrrr|r} 5 & -5 & 10 & -15 & -25 \\ 2 & -2 & 4 & -6 & -10 \\ 6 & -4 & 20 & -1 & 5 \\ -3 & 8 & 11 & 36 & 59 \\ 4 & -1 & 20 & 6 & 10 \end{array}\right]$$

in Example 2 in Section 4.3. In that example, we found that the final augmented matrix for $\mathbf{A}$ is

$$\left[\begin{array}{cccc|c} ① & 0 & 0 & 0 & 2 \\ 0 & ① & 0 & 0 & -4 \\ 0 & 0 & ① & 0 & -1 \\ 0 & 0 & 0 & ① & 3 \\ 0 & 0 & 0 & 0 & 0 \end{array}\right],$$

which has exactly four circled pivots. Therefore, $\text{rank}(\mathbf{A}) = 4$. ∎

Example 2. The initial augmented matrix

$$\mathbf{B} = \left[\begin{array}{rrrr|r} 1 & -3 & -4 & -7 & -51 \\ -3 & 9 & 14 & 23 & 169 \\ -2 & 6 & 13 & 14 & 117 \\ 2 & -6 & -10 & -13 & -103 \end{array}\right]$$

from Example 1 in Section 4.4 has

$$\left[\begin{array}{cccc|c} ① & -3 & 0 & 0 & -4 \\ 0 & 0 & ① & 0 & 3 \\ 0 & 0 & 0 & ① & 5 \\ 0 & 0 & 0 & 0 & 0 \end{array}\right]$$

as its final augmented matrix. Since we obtained three circled pivots here, $\text{rank}(\mathbf{B}) = 3$. ∎

Markov Chains, Revisited

Recall that in Section 3.1, we considered the following Markov chain matrix, representing the probabilities that in the next week, a carrier would either remain on one of three routes or switch to a new route.

$$
\mathbf{M} = \begin{array}{c} \textbf{Next} \\ \textbf{Week} \end{array} \begin{array}{c} \\ \text{Route A} \\ \text{Route B} \\ \text{Route C} \end{array} \overset{\displaystyle \begin{array}{c} \textbf{Current Week} \\ \text{Route A} \quad \text{Route B} \quad \text{Route C} \end{array}}{\begin{bmatrix} .60 & .20 & .05 \\ .30 & .50 & .15 \\ .10 & .30 & .80 \end{bmatrix}}.
$$

In Section 4.1, we learned that a steady-state vector for $\mathbf{M}$ would have to be a solution of the following linear system:

$$
\begin{cases}
-.40x & + & .20y & + & .05z & = & 0 & \text{(Route A)} \\
.30x & - & .50y & + & .15z & = & 0 & \text{(Route B)} \\
.10x & + & .30y & - & .20z & = & 0 & \text{(Route C)} \\
x & + & y & + & z & = & 1 & \text{(Sum Condition)}
\end{cases}
$$

When solving for a steady-state vector for a Markov chain, the calculations involved would often be cumbersome if performed by hand. However, the result can generally be found relatively quickly using "rref" on your calculator. Applying "rref" to the augmented matrix for this system, we obtain

$$
\left[\begin{array}{ccc|c}
1 & 0 & 0 & 0.2037 \\
0 & 1 & 0 & 0.2778 \\
0 & 0 & 1 & 0.5185 \\
0 & 0 & 0 & 0
\end{array} \right],
$$

and so the unique steady-state vector for $\mathbf{M}$ is

$$
\begin{bmatrix} x \\ y \\ z \end{bmatrix} = \begin{bmatrix} 20.37\% \\ 27.78\% \\ 51.85\% \end{bmatrix},
$$

thus confirming the solution for this system that was given in Section 4.1. We also noted there that since all entries of $\mathbf{M}$ are nonzero, this steady-state vector is actually $\mathbf{p}_{\lim}$ for *any* initial probability vector $\mathbf{p}$ for this Markov chain.

Glossary

- **Rank of a matrix:** The rank of a (possibly augmented) matrix is the number of circled pivots in its unique corresponding reduced row echelon form matrix.

- Reduced row echelon form of a matrix: A (possibly augmented) matrix is in reduced row echelon form if and only if the following four conditions hold:

 (1) The first nonzero entry in each row is a "1" (the unique circled pivot entry for that row).

 (2) Each successive row has its circled pivot in a later column.

 (3) All entries above and below each circled pivot are " 0."

 (4) Any rows consisting entirely of zeros are the final rows.

Every (augmented) matrix corresponds to a unique reduced row echelon form matrix when the Gauss-Jordan Method is carried out on all possible columns.

- Staircase pattern of pivots: The circled pivot entries in the final augmented matrix after the Gauss-Jordan Method occur in a descending pattern. Each successive circled pivot entry is in the row following the previous circled pivot entry, but one or more columns may possibly be skipped over between the circled pivot entries.

Exercises for Section 4.5

1. Determine whether each of the following augmented matrices is in reduced row echelon form. (Any pivots that should be circled are left uncircled here.) If not, what is the next row operation that should be applied in the Gauss-Jordan Method?

(a) $\begin{bmatrix} 1 & 0 & -2 & -3 \\ 0 & 1 & 7 & 4 \\ 0 & 0 & 0 & 0 \end{bmatrix}$

(b) $\begin{bmatrix} 2 & 0 & 6 & -2 \\ 0 & -3 & 5 & 4 \\ 0 & 0 & -1 & 3 \end{bmatrix}$

(c) $\begin{bmatrix} 1 & -3 & -1 & 2 & -1 \\ 0 & 1 & 2 & -3 & 5 \\ 0 & -4 & -5 & 8 & -2 \\ -2 & 5 & 6 & 4 & 9 \end{bmatrix}$

(d) $\begin{bmatrix} 1 & 0 & 3 & -5 & 12 \\ 0 & 1 & -5 & 6 & -9 \\ 0 & 0 & 0 & -1 & 13 \\ 0 & 0 & 1 & -2 & 11 \end{bmatrix}$

(e) $\begin{bmatrix} 1 & 0 & -8 & 0 & 2 \\ 0 & 1 & 3 & 0 & -5 \\ 0 & 0 & 0 & 1 & 4 \\ 0 & 0 & 0 & 0 & 0 \end{bmatrix}$

(f) $\begin{bmatrix} 1 & 0 & -4 & -1 & 2 & -8 \\ 0 & 1 & 6 & -5 & 0 & 3 \\ 0 & 0 & 0 & 0 & 1 & -11 \\ 0 & 0 & 0 & 0 & 0 & 0 \end{bmatrix}$

(g) $\begin{bmatrix} 1 & 0 & -3 & 4 & 2 & 5 \\ 0 & 1 & 5 & -2 & 0 & -1 \\ 0 & 0 & 0 & 3 & 6 & -3 \\ 0 & 0 & 0 & 0 & -6 & 7 \end{bmatrix}$

(h) $\begin{bmatrix} 1 & 0 & -8 & 0 & -1 & 5 & 4 \\ 0 & 1 & 3 & 0 & 2 & -3 & -2 \\ 0 & 0 & 0 & 1 & -4 & -1 & -6 \\ 0 & 0 & 0 & 0 & 0 & 8 & -3 \\ 0 & 0 & 0 & 0 & -5 & -2 & -1 \end{bmatrix}$

(i) $\begin{bmatrix} 1 & 0 & -5 & 7 & 0 & 15 \\ 0 & 1 & 2 & 6 & 0 & 32 \\ 0 & 0 & 0 & 0 & 1 & 3 \\ 0 & 0 & 0 & 3 & -9 & 12 \end{bmatrix}$

2. Determine whether each of the following augmented matrices is in re-

duced row echelon form. If not, what is the next row operation that should be applied in the Gauss-Jordan Method?

(a) $\begin{bmatrix} 1 & -5 & -2 & | & 8 \\ 0 & 3 & 7 & | & -2 \\ -5 & 4 & -1 & | & 5 \end{bmatrix}$

(b) $\begin{bmatrix} 1 & 0 & -2 & | & 3 \\ 0 & 0 & -3 & | & -1 \\ 0 & -6 & 9 & | & 10 \end{bmatrix}$

(c) $\begin{bmatrix} 1 & -4 & 3 & 0 & | & 2 \\ 0 & 0 & 1 & -5 & | & -8 \\ 0 & 0 & 0 & 4 & | & -6 \\ 0 & 0 & 0 & 0 & | & 0 \end{bmatrix}$

(d) $\begin{bmatrix} 1 & 4 & -5 & 6 & 8 & | & 18 \\ 0 & 0 & 0 & 1 & 2 & | & 5 \\ 0 & 0 & 0 & 3 & 9 & | & 27 \\ 0 & 0 & 2 & 4 & -8 & | & 16 \end{bmatrix}$

(i) $\begin{bmatrix} 1 & 0 & -8 & 7 & -1 & 5 & | & 3 \\ 0 & 1 & 3 & -5 & -9 & -3 & | & 6 \\ 0 & 0 & 0 & -2 & -1 & -1 & | & -4 \\ 0 & 0 & 0 & 0 & 3 & 8 & | & 2 \\ 0 & 0 & 0 & 0 & 0 & -5 & | & -1 \end{bmatrix}$

(e) $\begin{bmatrix} 1 & -6 & 8 & 0 & | & -3 \\ 0 & 1 & -4 & 0 & | & 5 \\ 0 & 0 & 0 & 1 & | & -2 \\ 0 & 0 & 0 & 0 & | & 0 \end{bmatrix}$

(f) $\begin{bmatrix} 1 & 0 & -8 & -7 & | & -5 \\ 0 & 1 & 0 & 3 & | & 2 \\ 0 & 0 & 1 & -2 & | & 11 \\ 0 & 0 & 0 & 4 & | & -12 \end{bmatrix}$

(g) $\begin{bmatrix} 1 & -4 & 6 & 9 & 2 & | & 4 \\ 0 & 0 & 0 & -5 & 0 & | & 3 \\ 0 & 0 & -3 & 1 & 1 & | & -7 \\ 0 & 0 & 0 & 0 & 0 & | & 0 \end{bmatrix}$

(h) $\begin{bmatrix} 1 & 0 & 0 & -3 & 0 & -6 & | & 7 \\ 0 & 1 & 0 & -5 & 0 & -1 & | & -2 \\ 0 & 0 & 1 & 4 & 0 & 8 & | & -3 \\ 0 & 0 & 0 & 0 & 1 & -2 & | & 5 \end{bmatrix}$

3. Use a calculator to find the rank of each of the following matrices.

(a) $\begin{bmatrix} 3 & -9 & -4 & | & 30 \\ -5 & 15 & 3 & | & -39 \\ -1 & 3 & 1 & | & -9 \end{bmatrix}$

(b) $\begin{bmatrix} -2 & -8 & 1 & | & -35 \\ 4 & 19 & -3 & | & 88 \\ -3 & -11 & 1 & | & -46 \\ 2 & 9 & -1 & | & 40 \end{bmatrix}$

(c) $\begin{bmatrix} -5 & -1 & 15 & -1 & | & 30 \\ -5 & -3 & 5 & 1 & | & 52 \\ 3 & 2 & -2 & -1 & | & -35 \\ 4 & 1 & -11 & 1 & | & -23 \end{bmatrix}$

(d) $\begin{bmatrix} -5 & 16 & 127 & 13 & 9 & | & -63 \\ 14 & 4 & -14 & 6 & 3 & | & 50 \\ -2 & 4 & 34 & 3 & 2 & | & -20 \\ 11 & 10 & 37 & 11 & 7 & | & 25 \end{bmatrix}$

(e) $\begin{bmatrix} -9 & -2 & 37 & 3 & -3 & | & -18 \\ 6 & -3 & -42 & 2 & -36 & | & 53 \\ -4 & -5 & 0 & 5 & -37 & | & 30 \\ 3 & -4 & -31 & 2 & -36 & | & 41 \\ 1 & -1 & -9 & -2 & -2 & | & -6 \end{bmatrix}$

4. Use a calculator to find the rank of each of the following matrices.

(a) $\begin{bmatrix} -2 & 2 & 15 & -61 \\ -2 & 1 & 10 & -44 \\ -3 & 1 & 13 & -59 \end{bmatrix}$

(b) $\begin{bmatrix} -7 & -4 & -8 & 11 \\ 3 & 3 & 6 & -15 \\ 2 & 1 & 2 & -2 \\ -4 & -3 & -6 & 12 \end{bmatrix}$

(c) $\begin{bmatrix} -8 & 11 & 5 & 4 & 75 \\ 13 & -7 & -2 & -3 & -101 \\ 23 & -2 & 5 & 2 & -147 \\ 7 & 0 & 2 & 1 & -43 \end{bmatrix}$

(d) $\begin{bmatrix} -5 & 3 & -36 & 1 & 3 & 36 \\ 4 & -1 & 19 & -5 & -2 & -24 \\ -2 & 1 & -13 & 1 & 1 & 14 \\ 5 & -2 & 29 & -4 & -1 & -36 \end{bmatrix}$

(e) $\begin{bmatrix} -11 & 55 & -7 & -4 & 3 & 5 & 2 \\ -14 & 70 & 7 & 1 & -50 & -1 & 26 \\ 12 & -60 & 8 & 4 & -4 & -7 & 5 \\ 13 & -65 & 60 & 24 & -178 & -32 & 83 \\ -12 & 60 & 1 & 0 & -27 & 4 & 2 \end{bmatrix}$

5. Use a calculator to find the complete solution set for each of the following linear systems.

(a) $\begin{cases} 46x_1 + x_2 + 40x_3 - 6x_4 = -53 \\ -53x_1 - 6x_2 - 44x_3 + 7x_4 = 25 \\ 38x_1 + 5x_2 + 31x_3 - 5x_4 = -19 \\ 63x_1 + x_2 + 61x_3 - 9x_4 = 79 \end{cases}$

(b) $\begin{cases} -5x_1 + 15x_2 - 2x_3 + 114x_4 = -410 \\ 7x_1 - 10x_2 + 8x_3 + 125x_4 = 131 \\ 14x_1 - 2x_2 + 13x_3 + 451x_4 = -103 \\ 21x_1 - 17x_2 + 19x_3 + 455x_4 = 218 \end{cases}$

(c) $\begin{cases} -3x_1 + 5x_2 - 47x_3 - 3x_4 - 40x_5 = 31 \\ -3x_1 + x_2 - 31x_3 \qquad\quad - 14x_5 = 44 \\ -2x_1 + 3x_2 - 30x_3 - 4x_4 - 29x_5 = -12 \\ 3x_1 - 2x_2 + 35x_3 + 5x_4 + 29x_5 = 23 \end{cases}$

(d) $\begin{cases} -5x_1 + 20x_2 + 18x_3 + 123x_4 - 61x_5 = -94 \\ -3x_1 + 12x_2 + 3x_3 + 27x_4 - 21x_5 = -33 \\ 4x_1 - 16x_2 - 6x_3 - 48x_4 + 32x_5 = 50 \\ -7x_1 + 28x_2 + 8x_3 + 69x_4 - 51x_5 = -80 \\ 10x_1 - 40x_2 - 24x_3 - 174x_4 + 98x_5 = 152 \end{cases}$

6. Use a calculator to find the complete solution set for each of the following linear systems.

(a) $\begin{cases} 7x_1 - 7x_2 + 7x_3 = 86 \\ 7x_1 - 5x_2 + 6x_3 = 66 \\ 3x_1 - 3x_2 + x_3 = 99 \\ 8x_1 - 5x_2 + 6x_3 = 84 \end{cases}$

(b)
$$\begin{cases} 3x_1 - x_2 - 23x_3 + 13x_4 = 29 \\ -7x_1 + 6x_2 + 72x_3 - 45x_4 = -97 \\ 6x_1 - 5x_2 - 61x_3 + 38x_4 = 82 \\ 2x_1 - x_2 - 17x_3 + 10x_4 = 22 \end{cases}$$

(c)
$$\begin{cases} 27x_1 - x_2 - 25x_3 + 17x_4 + 16x_5 = -22 \\ -29x_1 + x_2 + 27x_3 - 18x_4 - 17x_5 = 24 \\ 26x_1 - x_2 - 25x_3 + 17x_4 + 16x_5 = -38 \\ -57x_1 + x_2 + 48x_3 - 32x_4 - 30x_5 = -35 \\ 32x_1 - x_2 - 30x_3 + 20x_4 + 19x_5 = -30 \end{cases}$$

(d)
$$\begin{cases} 4x_1 + 7x_2 - 18x_4 + 147x_5 - 89x_6 = 102 \\ 5x_1 + 12x_2 + 13x_3 - 33x_4 + 246x_5 - 167x_6 = 189 \\ 3x_1 + 4x_2 - 5x_3 - 9x_4 + 84x_5 - 43x_6 = 51 \\ 8x_1 + 28x_2 + 56x_3 - 82x_4 + 566x_5 - 422x_6 = 472 \end{cases}$$

7. Justin is the owner of Blues Moody, Inc. (located at Pinder Lodge near the edge of Lake Thomas), which sells vacuum cleaners. The *Lite Delight* model weighs 15 pounds and comes in an 18 cubic foot box. The *Deep Sweep* model weighs 25 pounds and comes in a 12 cubic foot box. The *Clean Machine* model weighs 20 pounds and comes in a 15 cubic foot box. The company's delivery van holds a maximum of 1500 pounds and has 1044 cubic feet of space. Suppose also that the total number of vacuum cleaners in the van is 72, and that the van is fully loaded, both by weight and by volume.

   (a) Create a system of linear equations whose solution indicates the number of each type of vacuum cleaner involved. Indicate what your variables represent.

   (b) For the linear system in part (a), there is more than one possible solution. Use a calculator to determine the complete solution set for this system by row reduction, where the unknowns represent the number of each type of vacuum cleaner in the van. Indicate at least two different particular solutions.

   (c) Suppose we are given the additional information that the total number of *Deep Sweep* and *Clean Machine* models together equals three times the number of *Lite Delight* models. Find the solution set using row reduction for the associated linear system.

8. A manufacturer of stuffed animals produces teddy bears, koalas, and tigers. Each teddy bear requires 3 square feet of cloth, 2 units of stuffing, and 1 packet of dye. Each koala requires 4 square feet of cloth, 3 units of stuffing, and 2 packets of dye. Each tiger requires 2 square feet of cloth, 4 units of stuffing, and 6 packets of dye. The manufacturer has available 2740 square feet of cloth, 2480 units of stuffing, and 2220 packets of dye. Assume that the manufacturer wants to use up all of the materials at hand.

(a) Create a system of linear equations whose solution indicates the number of each type of stuffed animal involved. Indicate what your variables represent.

(b) For the linear system in part (a), there is more than one possible solution. Use a calculator to determine the complete solution set for this system by row reduction, where the unknowns represent the number of each type of stuffed animals manufactured. Indicate at least two different particular solutions.

(c) Suppose we are given the additional information that the total number of stuffed animals manufactured is 860. Find the solution set using row reduction for the associated linear system.

9. An appliance repair shop has three different types of employees: highly skilled workers, regular workers, and unskilled workers. The highly skilled workers repair an average of 10 items per day, earn $14 per hour, and require 500 square feet of work space. The regular workers can repair an average of 6 items per day, earn $10 per hour, and require 350 square feet of work space. The unskilled workers repair an average of 2 items per day, earn $8 per hour, and require 200 square feet of work space. The repair shop owner wants to repair an average of 480 items each day, wants to spend a total of $790 per hour on salaries, and expects to use 26750 square feet of work space, and wants the total number of highly skilled and regular workers together to equal three times the number of unskilled workers.

(a) Create a system of linear equations whose solution indicates the number of each type of worker involved. Indicate what your variables represent.

(b) Use your calculator to determine the complete solution set for the linear system in part (a) using row reduction.

(c) Remove the condition that the total number of highly skilled and regular workers together must equal three times the number of unskilled workers, and then use your calculator to determine the complete solution set for the revised linear system.

10. A nutritionist needs to design a new supplement containing exactly 1000 mg of calcium, 260 mg of Vitamin C, 270 mg of Vitamin E, and 200 mg of iron. She has three powdered supplements available. Supplement A contains 200 mg of calcium, 20 mg of Vitamin C, 10 mg of Vitamin E, and 22 mg of iron per gram of powder. Supplement B contains 100 mg of calcium, 40 mg of Vitamin C, 30 mg of Vitamin E, and 16 mg of iron per gram of powder. Supplement C contains 50 mg of calcium, 10 mg of Vitamin C, 50 mg of Vitamin E, and 28 mg of iron per gram of powder.

(a) Create a system of linear equations whose solution indicates the number of grams of each supplement to be combined. Indicate what your variables represent.

(b) Use your calculator to determine the complete solution set for the linear system in part (a) using row reduction.

(c) Suppose the nutritionist decides instead to have 180 mg of iron in the new supplement. Use your calculator to determine the complete solution set for the revised linear system.

11. Prove that the rank of a matrix is less than or equal to the number of rows in the matrix.

12. Prove that the rank of a matrix is less than or equal to the number of columns in the matrix.

13. Suppose the matrix for a given Markov chain is

$$\mathbf{M} = \begin{bmatrix} .36 & .43 & .24 \\ .26 & .17 & .25 \\ .38 & .40 & .51 \end{bmatrix}.$$

Set up an appropriate system of linear equations (with all variables placed on the left-hand side of each equation, and with like terms combined) whose solution is a steady-state vector for $\mathbf{M}$ (as in Exercises 11 and 12 of Section 4.1). Then use the "rref" function on your calculator to find the unique steady-state vector for this linear system. (Since all entries of $\mathbf{M}$ are nonzero, this steady-state vector is actually $\mathbf{p}_{\lim}$ for *any* initial probability vector $\mathbf{p}$ for this Markov chain.)

14. Suppose the matrix for a given Markov chain is

$$\mathbf{M} = \begin{bmatrix} .17 & .31 & .24 & .32 \\ .48 & .12 & .25 & .23 \\ .26 & .22 & .18 & .41 \\ .09 & .35 & .33 & .04 \end{bmatrix}.$$

Set up an appropriate system of linear equations (with all variables placed on the left-hand side of each equation, and with like terms combined) whose solution is a steady-state vector for $\mathbf{M}$ (as in Exercises 11 and 12 of Section 4.1). Then use the "rref" function on your calculator to find the unique steady-state vector for this linear system. (Since all entries of $\mathbf{M}$ are nonzero, this steady-state vector is actually $\mathbf{p}_{\lim}$ for *any* initial probability vector $\mathbf{p}$ for this Markov chain.)

15. For the Markov chain matrix

$$\mathbf{M} = \begin{bmatrix} 0 & 0 & 1 \\ 1 & 0 & 0 \\ 0 & 1 & 0 \end{bmatrix},$$

with the initial probability vector $\mathbf{p}_0 = [.40, .25, .35]$, calculate $\mathbf{p}_1$, $\mathbf{p}_2$, and $\mathbf{p}_3$. From these results, explain why $\mathbf{p}_0$ has no limit vector $\mathbf{p}_{\lim}$.

Then, use the "rref" function on your calculator to show that there is
a unique steady-state vector for $\mathbf{M}$. (However, repeated multiplications
by $\mathbf{M}$ do not cause the initial probability vector $\mathbf{p}_0 = [.40, .25, .35]$ to
converge to this steady-state vector. Notice this does not contradict the
comments in Section 4.1 about limit vectors, since we only guaranteed
there that a limit vector would occur as long as some positive power
of $\mathbf{M}$ had no zero entries. Here, all positive powers of $\mathbf{M}$ are equal to
either $\mathbf{M}$, $\mathbf{M}^2$, or $\mathbf{M}^3$, but all three of these matrices do contain zero
entries.)

16. For the Markov chain matrix

$$\mathbf{M} = \begin{bmatrix} 0 & .50 & 0 & .50 \\ .50 & 0 & .50 & 0 \\ 0 & .50 & 0 & .50 \\ .50 & 0 & .50 & 0 \end{bmatrix},$$

with the initial probability vector $\mathbf{p}_0 = [.50, 0, .50, 0]$, calculate $\mathbf{p}_1$, $\mathbf{p}_2$,
and $\mathbf{p}_3$. From these results, explain why $\mathbf{p}_0$ has no limit vector $\mathbf{p}_{\lim}$.
Then, use the "rref" function on your calculator to show that there is
a unique steady-state vector for $\mathbf{M}$. (However, repeated multiplications
by $\mathbf{M}$ do not cause the initial probability vector $\mathbf{p}_0 = [.50, 0, .50, 0]$ to
converge to this steady-state vector. Notice this does not contradict the
comments in Section 4.1 about limit vectors, since we only guaranteed
there that a limit vector would occur as long as some positive power
of $\mathbf{M}$ had no zero entries. Here, all positive powers of $\mathbf{M}$ are equal to
either $\mathbf{M}$ or $\mathbf{M}^2$, but both of these matrices do contain zero entries.)

4.6   Homogeneous Systems and Mutually Orthogonal Sets

In this section, we introduce a special type of system of linear equations.

Homogeneous Systems of Linear Equations

A system of linear equations of the form $\mathbf{AX} = \mathbf{0}$ is called a **homogeneous
system of linear equations**. For example, the linear system

$$\begin{cases} 2x + 10y - 14z = 0 \\ 3x + 11y - 13z = 0 \end{cases}$$

is homogeneous because the constants on the right side of the equation are all
zero.

The major advantage in working with a homogeneous linear system over
any other linear system is that a homogeneous system of linear equations
must be consistent; that is, it must have a solution! The reason for this is

simple. Setting every variable in a homogeneous system equal to zero always satisfies every equation in the system. No matter what coefficients are used in the system, if every variable equals zero, the left side of every equation will equal zero. Hence, it will equal the zero on the right side of the equation. Therefore, we never have to worry about the pesky "no solutions" case when working with homogeneous systems. The special solution for a homogeneous system in which every variable equals zero is called the **trivial solution** for the system. Any other solution having at least one variable with a nonzero value, if such a solution exists, is called a **nontrivial solution**. Of course, if even one nontrivial solution exists for a homogeneous system, then, since we also have the trivial solution, the system will have at least two solutions, implying that it actually has an infinite number of solutions!

Example 1. Consider the homogeneous linear system

$$\begin{cases} 2x + 10y - 14z = 0 \\ 3x + 11y - 13z = 0 \end{cases}.$$

Of course, the system has the trivial solution $(0, 0, 0)$. However, we want to find the complete solution set. We do this by using Gauss-Jordan row reduction, as before. Now, the augmented matrix

$$\begin{bmatrix} 2 & 10 & -14 & | & 0 \\ 3 & 11 & -13 & | & 0 \end{bmatrix} \text{ row reduces to } \begin{bmatrix} ① & 0 & 3 & | & 0 \\ 0 & ① & -2 & | & 0 \end{bmatrix},$$

giving us the complete solution set $\{(-3z, 2z, z)\}$. This system has an infinite number of nontrivial solutions, which can be found by plugging in any nonzero number for z. ∎

In Example 1, we saw that every solution for the homogeneous system was of the form $(-3z, 2z, z)$. We could also express this in the form $z(-3, 2, 1)$; that is, a scalar multiple of the nontrivial solution $(-3, 2, 1)$, which is obtained when $z = 1$.

Example 2. Consider the following homogeneous linear system:

$$\begin{cases} 4x_1 + 8x_2 + 6x_3 + 2x_4 = 0 \\ 3x_1 + 6x_2 + 4x_3 + 3x_4 = 0 \\ 5x_1 + 10x_2 + 14x_3 - 17x_4 = 0 \end{cases}.$$

Using the Gauss-Jordan Method, we row reduce

$$\begin{bmatrix} 4 & 8 & 6 & 2 & | & 0 \\ 3 & 6 & 4 & 3 & | & 0 \\ 5 & 10 & 14 & -17 & | & 0 \end{bmatrix} \text{ to obtain } \begin{bmatrix} ① & 2 & 0 & 5 & | & 0 \\ 0 & 0 & ① & -3 & | & 0 \\ 0 & 0 & 0 & 0 & | & 0 \end{bmatrix}.$$

Therefore, the complete solution set for the system is

$$\{(-2x_2 - 5x_4, x_2, 3x_4, x_4)\}.$$

This system has the trivial solution, plus an infinite number of nontrivial solutions. ∎

In Example 2, we can find some particular nontrivial solutions by substituting in numbers for x_2 and x_4 in the general form $(-2x_2 - 5x_4, x_2, 3x_4, x_4)$ for all solutions. For example, letting $x_2 = 1$ and $x_4 = 0$, we get the particular solution $(-2, 1, 0, 0)$. Using $x_2 = 0$ and $x_4 = 1$ produces the solution $(-5, 0, 3, 1)$. Notice that

$$(-2x_2 - 5x_4, \ x_2, \ 3x_4, \ x_4) = x_2\,(-2, 1, 0, 0) + x_4\,(-5, 0, 3, 1).$$

That is, every solution can be expressed as a linear combination of the two particular solutions $(-2, 1, 0, 0)$ and $(-5, 0, 3, 1)$ that we computed.

The homogeneous linear systems in Examples 1 and 2 both had nontrivial solutions. In both cases, after the example, we saw that the solutions in the complete solution set could be expressed as a linear combination of a few particular solutions. We can always do this for any homogeneous linear system that has nontrivial solutions. The process is simple. We look at the general description of the complete solution set. If there are nontrivial solutions, then there are undetermined variables in the general solution – variables that can equal any number. We take each undetermined variable in turn and set it equal to 1, while setting all of the other undetermined variables (if there are any) equal to zero. This process gives us one particular solution for each undetermined variable in the description of the complete solution set. These particular solutions are called **fundamental solutions** for the homogeneous system of linear equations. In Example 1, we had one fundamental solution, $(-3, 2, 1)$, for the system. In Example 2, we found two fundamental solutions, $(-2, 1, 0, 0)$ and $(-5, 0, 3, 1)$. In general, if a homogeneous linear system has nontrivial solutions, then every solution can be expressed as a linear combination of its fundamental solutions. But, of course, if a homogeneous linear system has only the trivial solution, then it does not have any fundamental solutions. In that case, the complete solution set is just $\{\mathbf{0}\}$.

Example 3. The complete solution set for the homogeneous linear system

$$\begin{cases} x_1 + 2x_2 - 2x_3 - x_4 - 6x_5 = 0 \\ 3x_1 + 6x_2 - 5x_3 - x_4 - 13x_5 = 0 \\ -2x_1 - 4x_2 + 7x_3 + 8x_4 + 27x_5 = 0 \\ 5x_1 + 10x_2 - 8x_3 - x_4 - 20x_5 = 0 \end{cases}$$

is

$$\{(-2x_2 - 3x_4 - 4x_5, \ x_2, \ -2x_4 - 5x_5, \ x_4, \ x_5)\}.$$

There are three undetermined variables that can take on any value, namely x_2, x_4, and x_5. Hence, there are three fundamental solutions for this system. These are found by setting each undetermined variable in turn equal to 1, while we set each of the others equal to zero. Setting $x_2 = 1$, $x_4 = 0$, and $x_5 = 0$ yields the fundamental solution $(-2, 1, 0, 0, 0)$. Substituting $x_2 = 0$, $x_4 = 1$, and $x_5 = 0$ produces the fundamental solution $(-3, 0, -2, 1, 0)$. Finally, using $x_2 = 0$, $x_4 = 0$, and $x_5 = 1$ gives us the fundamental solution $(-4, 0, -5, 0, 1)$.

Every solution for the homogeneous linear system can be expressed as a linear combination of these fundamental solutions. In particular, every solution is of the form

$$x_2\,(-2, 1, 0, 0, 0) + x_4\,(-3, 0, -2, 1, 0) + x_5\,(-4, 0, -5, 0, 1).\ \blacksquare$$

Example 4. Consider the following homogeneous system of linear equations:

$$\begin{cases} x - 2y - z = 0 \\ 2x - 3y + z = 0 \\ -3x + 8y + 7z = 0 \\ x - 3y = 0 \end{cases}.$$

The corresponding augmented matrix

$$\left[\begin{array}{ccc|c} 1 & -2 & -1 & 0 \\ 2 & -3 & 1 & 0 \\ -3 & 8 & 7 & 0 \\ 1 & -3 & 0 & 0 \end{array}\right] \quad \text{row reduces to} \quad \left[\begin{array}{ccc|c} ① & 0 & 0 & 0 \\ 0 & ① & 0 & 0 \\ 0 & 0 & ① & 0 \\ 0 & 0 & 0 & 0 \end{array}\right].$$

Therefore, this linear system has $(0, 0, 0)$, the trivial solution, as its only solution. That is, there are no nontrivial solutions. Hence, the system does not have any fundamental solutions, but it does have the trivial solution. ∎

Homogeneous Systems and Rank

As we know, every homogeneous system of linear equations is consistent. The trivial solution is always in the solution set. So, we get *only* the trivial solution in the case when the system's solution is *unique*. In Section 4.4 we learned that we get a unique solution for a consistent linear system when, in the Gauss-Jordan Method, every column to the left of the bar in the associated reduced row echelon form augmented matrix has a circled pivot in every column. Clearly, this happens when the rank of the matrix (the number of circled pivots) equals the number of variables in the system (the number of columns to the left of the bar).[2] If the rank of the matrix is less than the number of variables, then there will be a column to the left of the bar having no circled pivot. Therefore, according to Section 4.4, there will be an infinite number of nontrivial solutions in addition to the trivial solution. Hence, we have the following result:

Theorem 2. *Suppose* $\mathbf{A}$ *is an* $m \times n$ *matrix, and* $\operatorname{rank}(\mathbf{A}) = k$. *Then:*

If $k = n$, *then the homogeneous linear system* $\mathbf{AX} = \mathbf{0}$ *has only the trivial solution;*

If $k < n$, *then the homogeneous linear system* $\mathbf{AX} = \mathbf{0}$ *has the trivial solution plus an infinite number of nontrivial solutions. In this case, the system has* $(n - k)$ *fundamental solutions, and every solution in the complete solution set for the system is a linear combination of these fundamental solutions.*

---

[2] A homogeneous system will never have a circled pivot to the right of the bar, since those entries are all zeros.

For example, the linear system in Example 1 has three variables (so $n = 3$). The coefficient matrix **A** has rank 2. (Check out the reduced row echelon form in Example 1.) So, Theorem 2 tells us that there are nontrivial solutions. Also, every solution is a linear combination of the one ($= 3 - 2$) fundamental solution (which means, just a scalar multiple of it). This is consistent with what we saw in and directly after Example 1.

In Example 2, the coefficient matrix again has rank 2. In that example, there are four variables (so $n = 4$). Since the rank is less than the number of variables, Theorem 2 accurately shows that there are nontrivial solutions. After Example 2, we saw that the system has 2 ($= 4 - 2$) fundamental solutions.

Although we did not show the work to compute the rank of the coefficient matrix in Example 3, we can figure out what it must be. In the example there are five variables, and we found three fundamental solutions. Hence, the rank of the coefficient matrix must be 2.

In Example 4, the rank of the coefficient matrix is 3 and there are three variables. Thus, Theorem 2 accurately tells us that there is only the trivial solution for that system.

Now every circled pivot in a reduced row echelon form matrix must be in a different row, and so the rank of any matrix cannot exceed the number of rows in the matrix. (See Exercise 11 in Section 4.5.) Combining this fact with Theorem 2, and the fact that rows in the coefficient matrix for a linear system correspond to equations in the linear system, we get the following:

Corollary 3. *If a homogeneous system of linear equations has more variables than equations, then the system has an infinite number of nontrivial solutions, in addition to the trivial solution.*

Example 5. Because the homogeneous linear system

$$\begin{cases} 2x_1 - 4x_2 - 5x_3 + x_4 = 0 \\ 6x_1 - 8x_2 - 3x_3 - 2x_4 = 0 \end{cases}$$

has more variables than equations, Corollary 3 tells us that the system has an infinite number of nontrivial solutions.[3] ∎

Mutually Orthogonal Sets of Vectors

The set $\{[2, 1, -2], [1, 2, 2]\}$ is a mutually orthogonal set of vectors. We would like to add a third nonzero vector to this set and still have a mutually orthogonal set. Thus, we need to find a third nonzero vector that is orthogonal to both $[2, 1, -2]$ and $[1, 2, 2]$. Such a vector $[x, y, z]$ would have to be nonzero and have

$$[2, 1, -2] \cdot [x, y, z] = 0$$
$$\text{and } [1, 2, 2] \cdot [x, y, z] = 0.$$

---

[3]In Exercise 1(b), you are asked to solve this linear system.

Computing the dot products shows that $[x, y, z]$ would be a nontrivial solution for the homogeneous linear system

$$\begin{cases} 2x + y - 2z = 0 \\ x + 2y + 2z = 0 \end{cases}.$$

Now, the number of equations in this system is less than the number of variables. Hence, Corollary 3 assures us that there exist nontrivial solutions! If you solve the system,[4] you will find the one fundamental solution $(2, -2, 1)$. By Theorem 2, all solutions are scalar multiples of this fundamental solution. But, we now know we can expand the original set $\{[2, 1, -2], [1, 2, 2]\}$ by adding the vector $[2, -2, 1]$ to obtain the mutually orthogonal set

$$\{[2, 1, -2], [1, 2, 2], [2, -2, 1]\}.$$

We have illustrated a general principle:

Theorem 4. *If S is a set of k distinct n-dimensional mutually orthogonal vectors and $k < n$, then there is a nonzero vector* **v** *that is orthogonal to all of the vectors in S. That is, adding* **v** *to the set S will form a larger mutually orthogonal set of vectors.*

The truth of Theorem 4 follows directly from Corollary 3. Solving for the vector **v** in Theorem 4 amounts to solving a homogeneous linear system having k equations (each equation claiming that **v** is orthogonal to one of the vectors in S) and n variables (one for each of the coordinates of **v**).

Example 6. Consider the mutually orthogonal set $\{[1, 1, 1, 3], [1, 1, 1, -1]\}$ of 4-dimensional vectors. Since there are only two vectors in the set and the dimension is 4, Theorem 4 says we can find a nonzero vector **v** that is orthogonal to all of the vectors in the set. Let us solve for $\mathbf{v} = [v_1, v_2, v_3, v_4]$. We do this by solving the equations

$$[1, 1, 1, 3] \cdot [v_1, v_2, v_3, v_4] = 0$$
$$[1, 1, 1, -1] \cdot [v_1, v_2, v_3, v_4] = 0.$$

Solving these equations is equivalent to solving the homogeneous linear system

$$\begin{cases} v_1 + v_2 + v_3 + 3v_4 = 0 \\ v_1 + v_2 + v_3 - v_4 = 0 \end{cases}.$$

Using the Gauss-Jordan Method, we row reduce

$$\left[\begin{array}{cccc|c} 1 & 1 & 1 & 3 & 0 \\ 1 & 1 & 1 & -1 & 0 \end{array} \right] \quad \text{to obtain} \quad \left[\begin{array}{cccc|c} ① & 1 & 1 & 0 & 0 \\ 0 & 0 & 0 & ① & 0 \end{array} \right].$$

The general solution set is $\{(-v_2 - v_3, v_2, v_3, 0)\}$. There are two fundamental solutions: $[-1, 1, 0, 0]$ and $[-1, 0, 1, 0]$, both of which are orthogonal to the two

---

[4]In Exercise 1(a), you are asked to solve this linear system.

vectors we already have, but not to each other. We can use either of them for
v. Thus, the set

$$\{[1,1,1,3],[1,1,1,-1],[-1,1,0,0]\}$$

is an expanded mutually orthogonal set.

Now, our new set still only has three vectors. But the dimension in this
example is 4. So Theorem 4 says that we can find yet another vector orthogonal to all three of these, and extend the set even further. We could proceed
in the same way we found the third vector, setting up a system of three equations with four variables. However, we already have another vector that is
orthogonal to the first two vectors in our expanded set, namely the second
fundamental solution we found. Our problem is that it is not orthogonal to the
third vector. So, an easier approach is to take the two fundamental solutions
and apply the Gram-Schmidt Process to adjust the second vector to make it
orthogonal to the first. (See Section 3.5.) Using the Gram-Schmidt Process
on the set $\{[-1,1,0,0],[-1,0,1,0]\}$ of fundamental solutions, the new second
vector is

$$[-1,0,1,0] - \mathbf{proj}_{[-1,1,0,0]}[-1,0,1,0]$$

$$= [-1,0,1,0] - \left(\frac{[-1,0,1,0]\cdot[-1,1,0,0]}{\|[-1,1,0,0]\|^2}\right)[-1,1,0,0]$$

$$= [-1,0,1,0] - \frac{1}{2}[-1,1,0,0]$$

$$= \left[-\frac{1}{2},-\frac{1}{2},1,0\right].$$

We can multiply this vector by 2 to eliminate the fractions, as we did in
Section 3.5. Doing so and adding the resulting vector to the set of three
vectors we already have produces the larger mutually orthogonal set

$$\{[1,1,1,3],[1,1,1,-1],[-1,1,0,0],[-1,-1,2,0]\}. \quad \blacksquare$$

Theorem 4 does not assure us that we can expand the final set in Example
6 any further, because the number of vectors in that final set equals the
dimension of the vectors. In Chapter 6 we will prove that the number of
vectors in a mutually orthogonal set of nonzero vectors cannot be higher than
the dimension of the vectors in the set. Using that fact, we know for certain
that we cannot expand the set in Example 6 any further.

Glossary

- Fundamental solution: A particular nontrivial solution for a homogeneous system of linear equations found by setting one of the undetermined variables in the description of the general solution equal to 1,
  and all other undetermined variables (if there are any) in the general

solution equal to 0. If a homogeneous linear system has only the trivial solution, then it has no fundamental solutions. If a homogeneous system has any nontrivial solutions at all, then every solution for the homogeneous system can be expressed as a linear combination of its fundamental solutions.

- Homogeneous system of linear equations: A linear system of the form $\mathbf{AX} = \mathbf{0}$; that is a system of linear equations in which the constant on the right side of every equation is zero.

- Nonhomogeneous system of linear equations: A linear system that is not homogeneous; that is, it is of the form $\mathbf{AX} = \mathbf{B}$, where $\mathbf{B} \neq \mathbf{0}$. (See the exercises.)

- Nontrivial solution: A solution for a homogeneous system of linear equations in which at least one variable has a nonzero value. If at least one nontrivial solution exists for a given homogeneous linear system, then there are an infinite number of nontrivial solutions in the complete solution set for the system.

- Trivial solution: The solution for a homogeneous system of linear equations in which every variable has the value zero. Every homogeneous linear system has the trivial solution in its complete solution set.

Exercises for Section 4.6

1. In each part, find the complete solution set for the given homogeneous system of linear equations. If the linear system has an infinite number of solutions, express the solution set in terms of a linear combination of fundamental solutions. Use the rref function on your calculator.

(a) $\begin{cases} 2x + y - 2z = 0 \\ x + 2y + 2z = 0 \end{cases}$

(b) $\begin{cases} 2x_1 - 4x_2 - 5x_3 + x_4 = 0 \\ 6x_1 - 8x_2 - 3x_3 - 2x_4 = 0 \end{cases}$

(c) $\begin{cases} 2x - y + 8z = 0 \\ -x - 7y + 3z = 0 \\ x + 2y + 3z = 0 \end{cases}$

(d) $\begin{cases} 4x_1 + 8x_2 - 13x_3 + x_4 + 60x_5 = 0 \\ 3x_1 + 6x_2 - x_3 - 8x_4 + 10x_5 = 0 \\ x_1 + 2x_2 + x_3 - 4x_4 - 2x_5 = 0 \\ -2x_1 - 4x_2 + 6x_4 - 4x_5 = 0 \end{cases}$

(e) $\begin{cases} 3v_1 + 11v_3 = 0 \\ 5v_1 + 2v_2 + 21v_3 = 0 \\ 2v_1 - v_2 + 5v_3 = 0 \\ -3v_1 - v_2 - 13v_3 = 0 \end{cases}$

2. In each part, find the complete solution set for the given homogeneous system of linear equations. If the linear system has an infinite number of solutions, express the solution set in terms of a linear combination of fundamental solutions. Use the rref function on your calculator.

(a) $\begin{cases} 5x - 8y + 4z = 0 \\ -3x + 7y + 9z = 0 \\ 7x - 15y - 17z = 0 \end{cases}$

(b) $\begin{cases} 2x + y + 8z = 0 \\ 3x + 10y - 5z = 0 \\ 2x + 7y - 4z = 0 \end{cases}$

(c) $\begin{cases} 2x_1 - 3x_2 - 5x_4 = 0 \\ -6x_1 + 9x_2 - 2x_3 + 9x_4 = 0 \\ 8x_1 - 12x_2 + 4x_3 - 8x_4 = 0 \\ 4x_1 - 6x_2 + 3x_3 - x_4 = 0 \end{cases}$

(d) $\begin{cases} 10x_1 - 3x_2 + x_3 + 2x_4 = 0 \\ 2x_1 - 10x_2 + 6x_3 - x_4 = 0 \\ -5x_1 - 4x_2 + 3x_3 - 2x_4 = 0 \\ 3x_1 + x_2 - x_3 + x_4 = 0 \end{cases}$

(e) $\begin{cases} 5v_1 - 2v_2 + 6v_3 + 9v_4 - 4v_5 = 0 \\ 10v_1 - 3v_2 + 8v_3 + 12v_4 - v_5 = 0 \\ -5v_1 + 5v_2 - 18v_3 - 27v_4 + 25v_5 = 0 \end{cases}$

3. In each part, find a nonzero vector **v** that is orthogonal to every vector in the given mutually orthogonal set. Use the rref function on your calculator.

(a) $\{[1, 1, 3], [4, 2, -2]\}$

(b) $\{[2, 4, -1], [10, -1, 16]\}$

(c) $\{[0, 1, 1, 0], [-6, -7, 7, 16], [3, 1, -1, 2]\}$

(d) $\{[1, -1, 1, 0, -1], [1, 0, -1, 0, 0], [1, -3, 1, -2, 5], [1, 2, 1, -2, 0]\}$

4. In each part, find a nonzero vector **v** that is orthogonal to every vector in the given mutually orthogonal set. Use the rref function on your calculator.

(a) $\{[2, 1, -1], [1, -1, 1]\}$

(b) $\{[11, 1, 16], [3, -1, -2]\}$

(c) $\{[4, 3, -11, 8], [4, 1, 1, -1], [-2, 9, 9, 10]\}$

(d) $\{[1, -2, 2, 1, -4], [125, 49, -23, 99, 20], [1, 0, 2, -1, 1], [-2, -9, 1, 6, 6]\}$

5. Expand the mutually orthogonal set $\{[3, -9, -14, 7], [-28, 84, 19, 158]\}$ of 4-dimensional vectors to a mutually orthogonal set containing four nonzero vectors by first solving a homogeneous system of linear equations, and then applying the Gram-Schmidt Process, as was done in Example 6. Use your calculator for computations.

6. Expand the mutually orthogonal set $\{[3, 1, 4, 3, -5], [-2, 1, -1, 3, 0]\}$ of 5-dimensional vectors to a mutually orthogonal set containing five nonzero vectors by first solving a homogeneous system of linear equations, and then applying the Gram-Schmidt Process in a manner analogous to that used in Example 6. Use your calculator for computations.

7. Explain why $\mathbf{X} = \mathbf{0}$ is *never* a solution for a nonhomogeneous system of linear equations; that is, to a system of the form $\mathbf{AX} = \mathbf{B}$ for which $\mathbf{B} \neq \mathbf{0}$.

8. If $\mathbf{A}$ is an $m \times n$ matrix, explain why $\text{rank}(\mathbf{A}) = \text{rank}([\mathbf{A} \,|\, \mathbf{0}])$. That is, explain why augmenting $\mathbf{A}$ with a column of zeros does not change its rank.

9. Suppose that $\mathbf{u}$ and $\mathbf{v}$ are both solutions to the nonhomogeneous linear system $\mathbf{AX} = \mathbf{B}$. Prove that $\mathbf{w} = \mathbf{u} - \mathbf{v}$ is a solution to the homogeneous linear system $\mathbf{AX} = \mathbf{0}$.

10. Use Exercise 9 to show that if the nonhomogeneous linear system $\mathbf{AX} = \mathbf{B}$ has an infinite number of solutions, then the homogeneous linear system $\mathbf{AX} = \mathbf{0}$ has a nontrivial solution.

11. Suppose that $\{\mathbf{v}_1, \ldots, \mathbf{v}_k\}$ is an orthonormal set of n-dimensional vectors. Let $\mathbf{A}$ be the matrix whose rows are the vectors $\mathbf{v}_1, \ldots, \mathbf{v}_k$ and let $\mathbf{B}$ be the matrix whose columns are the vectors $\mathbf{v}_1, \ldots, \mathbf{v}_k$. What is the matrix $\mathbf{AB}$?

12. Suppose that $\{\mathbf{v}_1, \ldots, \mathbf{v}_k\}$ is a mutually orthogonal set of n-dimensional vectors. Let $\mathbf{A}$ be the matrix whose rows are the vectors $\mathbf{v}_1, \ldots, \mathbf{v}_k$ and let $\mathbf{B}$ be the matrix whose columns are the vectors $\mathbf{v}_1, \ldots, \mathbf{v}_k$. What is the matrix $\mathbf{AB}$?

4.7 Chapter 4 Test

1. Express the given system of linear equations in the form $\mathbf{AX} = \mathbf{B}$ and in the form $[\mathbf{A} \,|\, \mathbf{B}]$:

$$\begin{cases} x_1 + 2x_2 + 6x_3 + 5x_4 + 18x_5 = 85 \\ + 5x_2 + 20x_3 + 6x_4 + 20x_5 = 92 \\ -x_1 + x_2 + 6x_3 - x_4 - 4x_5 = -23 \end{cases}.$$

2. Determine whether each of the following augmented matrices is in reduced row echelon form. If not, what is the next row operation that should be applied in the Gauss-Jordan Method? Then (if appropriate) perform that one row operation to obtain the next matrix.

(a) $\begin{bmatrix} 1 & 2 & -3 & 5 & 16 \\ 0 & 0 & 9 & 18 & 29 \\ -2 & 0 & 1 & 8 & 15 \end{bmatrix}$

(d) $\begin{bmatrix} 1 & -5 & 2 & -3 & 8 \\ 0 & 0 & 0 & 2 & 10 \\ 0 & 0 & 3 & 9 & 54 \end{bmatrix}$

(b) $\begin{bmatrix} 1 & 0 & 3 & 0 & 7 \\ 0 & 1 & -2 & 0 & 4 \\ 0 & 0 & 0 & 1 & 1 \end{bmatrix}$

(e) $\begin{bmatrix} 1 & 0 & 3 & 7 \\ 0 & 1 & -5 & 3 \\ 0 & 0 & 0 & 0 \\ 0 & 0 & 0 & 0 \end{bmatrix}$

(c) $\begin{bmatrix} -4 & 0 & 8 & -12 & 36 \\ 0 & 1 & 5 & 6 & 19 \\ 0 & 0 & 7 & 21 & 28 \end{bmatrix}$

(f) $\begin{bmatrix} 1 & 3 & 0 & 2 & 20 \\ 0 & 1 & 1 & 5 & 22 \\ 0 & 0 & 0 & 2 & 6 \\ 0 & 0 & -1 & 4 & 28 \end{bmatrix}$

3. Each of the following augmented matrices is in reduced row echelon form. Give the complete solution set for the associated linear system. If the linear system has an infinite number of solutions, also provide three different particular solutions. Use $x_1, x_2, \dots$ as the variables.

(a) $\begin{bmatrix} 1 & -2 & 0 & 4 & 7 & 18 \\ 0 & 0 & 1 & -3 & 5 & 34 \end{bmatrix}$

(c) $\begin{bmatrix} 1 & 0 & 5 & 0 & 0 \\ 0 & 1 & -2 & 0 & 0 \\ 0 & 0 & 0 & 1 & 0 \\ 0 & 0 & 0 & 0 & 1 \end{bmatrix}$

(b) $\begin{bmatrix} 1 & 0 & 0 & 5 \\ 0 & 1 & 0 & 7 \\ 0 & 0 & 1 & 2 \\ 0 & 0 & 0 & 0 \end{bmatrix}$

(d) $\begin{bmatrix} 1 & 0 & 3 & 0 & 0 \\ 0 & 1 & -6 & 0 & 0 \\ 0 & 0 & 0 & 1 & 0 \end{bmatrix}$

4. Write a system of linear equations to represent the given situation. Be sure to explicitly explain the meaning of each of the variables you use. You are not required to solve the system of linear equations.

Madeline is at the candy shop. She wants to purchase five pounds of candy. The vanilla fudge costs \$2 per pound, the chocolate fudge costs \$2.50 per pound, and the salt water taffy costs \$1.50 per pound. She would like to have twice as much chocolate fudge as vanilla fudge. Madeline plans to spend exactly \$10. How many pounds of each type of candy should Madeline buy to meet all of these conditions?

5. Determine the rank of each of the following matrices. Use the rref function on your calculator.

(a) $\begin{bmatrix} 6 & 3 & 9 & | & 24 \\ 5 & 3 & 7 & | & 21 \\ 2 & 1 & 3 & | & 8 \end{bmatrix}$

(c) $\begin{bmatrix} 8 & -5 & 1 & 35 \\ -2 & 6 & -3 & -34 \\ 40 & -15 & 1 & 129 \\ 11 & -4 & 0 & 34 \\ 7 & -3 & 0 & 23 \end{bmatrix}$

(b) $\begin{bmatrix} 16 & 4 & 32 & 9 & | & 2 \\ 5 & 2 & 7 & 1 & | & 0 \\ 1 & 2 & -5 & -1 & | & -1 \\ 1 & -1 & 7 & 2 & | & 1 \end{bmatrix}$

6. In each part, find the complete solution set for the given homogeneous system of linear equations. If the linear system has an infinite number of solutions, express the solution set in terms of a linear combination of fundamental solutions. Use the rref function on your calculator.

(a) $\begin{cases} 4x + 3y & = 0 \\ -2y - z & = 0 \\ 22x + 29y + 7z & = 0 \\ 3x + 4y + z & = 0 \end{cases}$

(b) $\begin{cases} 2x_1 + x_2 - 6x_3 + 19x_4 - 25x_5 = 0 \\ -3x_1 + 7x_2 - 5x_3 + 19x_4 - 7x_5 = 0 \\ x_1 - x_2 - x_3 + 3x_4 - 7x_5 = 0 \end{cases}$

7. Expand the mutually orthogonal set $\{[2, -1, 1, -3], [1, 1, -1, 0]\}$ of 4-dimensional vectors to a mutually orthogonal set containing 4 nonzero vectors by first solving a homogeneous system of linear equations, and then applying the Gram-Schmidt Process. Use your calculator for computations. Be sure to multiply the last vector by an appropriate scalar to eliminate fractions.

8. In each part, indicate whether the given statement is true or whether it is false.

   (a) A system of linear equations cannot have more equations than unknowns.

   (b) If a system of linear equations has five equations, then a solution to the system, if one exists, is an ordered 5-tuple.

   (c) A system of two linear equations with two unknowns for which the equations are the formulas for two parallel lines has no solution.

   (d) The equation $4w - 3x + 2y - 18z - 54 = 10$ is a linear equation.

   (e) The ordered triple $(3, 8, -4)$ is a solution for the linear system

$$\begin{cases} 4x + 2y + 7z = 0 \\ 3x - y + 5z = -19 \end{cases}.$$

   (f) Squaring every entry in a row of an augmented matrix is one of the allowable row operations in the Gauss-Jordan Method.

(g) Multiplying an entire row by zero is one of the allowable row operations in the Gauss-Jordan Method.

(h) Subtracting the first row from the third row is one of the allowable row operations in the Gauss-Jordan Method.

(i) When solving a linear system using the Gauss-Jordan Method, if you encounter a row of all zeroes at any step, then the system has no solution.

(j) The standard row operation used to target a specific entry when using the Gauss-Jordan Method is

$$- \text{(target entry)} < \text{pivot row} > + < \text{target row} >$$
$$\implies < \text{target row} > .$$

(k) A type (III) row operation is used when the intended pivot entry is 0, and there is a nonzero number in the pivot column below the intended pivot.

(l) In the Gauss-Jordan Method, you should never perform a type (III) operation to swap the current pivot row with a row above it.

(m) In the Gauss-Jordan Method, if the final matrix does not have a circled pivot in every column to the left of the bar, then the original linear system has an infinite number of solutions.

(n) In the Gauss-Jordan Method, if the final matrix has a circled pivot in every column to the left of the bar, then the original linear system has a unique solution.

(o) If the final reduced row echelon form matrix for a specific linear system is

$$\left[\begin{array}{cccc|c} 1 & 0 & 3 & 0 & 10 \\ 0 & 1 & -7 & 0 & 3 \\ 0 & 0 & 0 & 1 & 0 \end{array} \right],$$

then the linear system has no solution.

(p) If the final reduced row echelon form matrix for a specific linear system is as given in part (o), then the linear system has an infinite number of solutions.

(q) If the final reduced row echelon form matrix for a specific linear system is as given in part (o), then, in every solution, at least one of the variables has the value 0.

(r) It is possible to have two circled pivots in the same row in a reduced row echelon form matrix.

(s) The rank of a matrix **A** cannot exceed the number of rows in **A**.

(t) It is possible to have two circled pivots in the same column in a reduced row echelon form matrix.

(**u**) The rank of a matrix **A** cannot exceed the number of columns in **A**.

(**v**) The final reduced row echelon form obtained when solving a linear system depends upon the order in which the equations are listed in a linear system.

(**w**) It is possible to get an incorrect solution for a linear system when row reducing on a calculator because of error induced by rounding.

(**x**) The column to the right of the bar is irrelevant when computing the rank of an augmented matrix.

(**y**) A system of linear equations that is not homogeneous is said to be nonhomogeneous.

(**z**) If **A** is a 5×4 matrix, then the system of linear equations $\mathbf{AX} = \mathbf{B}$, where $\mathbf{B} \neq \mathbf{0}$ has no solutions.

(**aa**) If **A** is a 5×4 matrix, then the system of linear equations $\mathbf{AX} = \mathbf{0}$ has no solutions.

(**bb**) Every homogeneous system of linear equations has the trivial solution.

(**cc**) If a homogeneous system of linear equations has at least one nontrivial solution, then it has an infinite number of nontrivial solutions.

(**dd**) If a homogeneous system of linear equations has a nontrivial solution, then every solution for the system can be expressed as a linear combination of fundamental solutions.

(**ee**) Every homogeneous system of linear equations has at least one fundamental solution.

(**ff**) When solving a homogeneous linear system using the Gauss-Jordan Method, if the final matrix does not have a circled pivot in every column to the left of the bar, then the original linear system has an infinite number of solutions.

(**gg**) When solving a homogeneous linear system using the Gauss-Jordan Method, if the final matrix has a circled pivot in every column to the left of the bar, then the original linear system has a unique solution.

(**hh**) If the rank of a matrix **A** equals the number of rows in **A**, then the homogeneous linear system $\mathbf{AX} = \mathbf{0}$ has only the trivial solution.

(**ii**) If the rank of a matrix **A** is less than the number of columns in **A**, then the homogeneous linear system $\mathbf{AX} = \mathbf{0}$ has only the trivial solution.

(**jj**) If rank $(\mathbf{A}) = 5$ for a 6×7 matrix **A**, then the homogeneous linear system $\mathbf{AX} = \mathbf{0}$ has exactly one fundamental solution.

(**kk**) If rank $(\mathbf{A}) = 5$ for a 6×7 matrix $\mathbf{A}$, then the homogeneous linear system $\mathbf{AX} = \mathbf{0}$ has exactly two fundamental solutions.

(**ll**) If a homogeneous linear system $\mathbf{AX} = \mathbf{0}$ with six variables has exactly two fundamental solutions, then the rank of $\mathbf{A}$ equals 4.

(**mm**) If a homogeneous system of linear equations has more variables than equations, then the system has an infinite number of nontrivial solutions, in addition to the trivial solution.

(**nn**) If $\mathbf{A}$ is a 3×7 matrix, then the homogeneous linear system $\mathbf{AX} = \mathbf{0}$ has exactly four fundamental solutions.

(**oo**) Every set of mutually orthogonal vectors can be expanded by adding one more vector to obtain a larger mutually orthogonal set.

(**pp**) It is impossible to have a mutually orthogonal set of n-dimensional vectors that contains more than n elements.

(**qq**) It is impossible to have a mutually orthogonal set of n-dimensional nonzero vectors that contains more than n elements.

(**rr**) A Markov chain matrix always converges to a final limit vector for any given initial probability vector.

Chapter 5

Application: Least–Squares

In certain real-world problems, we have data indicating that two given variables are related to each other, but we might not actually know a particular equation that can predict the value of one variable from the other. In such a case, our goal is to find a formula that provides a very good model of the relationship between the variables. Such a formula is often the first step toward using other mathematical techniques to analyze the real-world situation. In this chapter we consider the problem of finding a function or formula that fits a given data set as closely as possible.

5.1 Approximation by a Linear Function

Frequently, the relationship between two variables x and y is almost linear; that is, there is an equation of a line in the form $y = mx + b$ that comes close to describing how the two quantities are linked.

The Goal: A Least-Squares Line

Let us consider a simple example.

Example 1. Maxwell Mir, manager of the Mir Resort (see Section 2.1), has collected the following data on gross revenues (in thousands of $) during the first seven years of operation:

Year of Operation	1	2	3	4	5	6	7
Gross Revenue	702	1351	2874	3768	4245	5404	5755

For planning purposes, Max would like to establish a formula, or function, that, given the year, will give him an estimated gross revenue for that year. Such a function would allow Max to project his revenues in future years. To begin, Max uses the data to create a set of points in a 2-dimensional coordinate system:

$$\Big\{ (1, 702), \, (2, 1351), \, (3, 2874), \, (4, 3768), \, (5, 4245), \, (6, 5404), \, (7, 5755) \Big\}.$$

Figure 5.1 shows these points plotted in a coordinate plane. Although these points clearly do not lie along a common straight line, the points do appear to be *almost* linear. Perhaps a linear function of the form $y = mx + b$, in which y

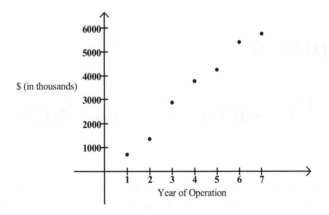

Figure 5.1: Gross Revenue at the Mir Resort

represents revenue in thousands of $, and x represents the year (time), could give a reasonable estimate for the connection between x and y. The question, then, is "Which line should we choose?" That is, which values of m and b in the formula $y = mx + b$ gives the best fit for the data?

Notice that no single line can fit the data *exactly* since the points do not form a perfectly linear pattern. Figure 5.2 illustrates two different lines, both of which come very close to the data points, and there are many other possible choices of lines that lie near to all of the data points as well. To determine which of the infinite number of possible lines is best, we need a mathematical definition of precisely what we mean by "best." ■

Least-Squares: Smallest Error

Consider Figure 5.3, in which a set of points is given along with a line that is meant to approximate the data. Think of this line[1] as a function that is intended to predict the relationship between x and y. We would like to measure how closely this particular line models the data. To do this, we draw vertical line segments from the data points to the approximating line, and measure the lengths of these segments. (See Figure 5.4.) At each x-value, the vertical distance in Figure 5.4 represents the error between the actual y-value from the given data and its related y-value on the approximating line. To calculate the overall error involved in using this particular line to

---

[1]In some applications, a more general curve is used (rather than a line) to approximate the data. In Section 5.2, we learn how to approximate data using a quadratic function instead.

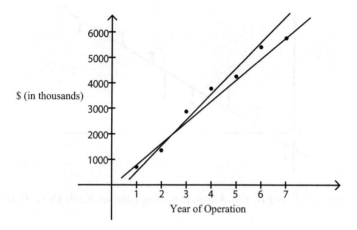

Figure 5.2: Two Different Lines Approximating the Data in Example 1

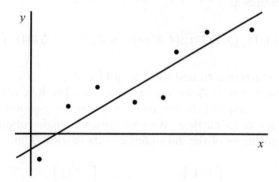

Figure 5.3: A Line Approximating a Set of Points

estimate the data, we add up the *squares* of the lengths of these line segments.[2] Mathematicians take the line having the smallest overall error (that is, having the smallest sum of the squares of the individual errors) as the one that best fits the data. For this reason, this method is often referred to as the "least-squares method" for finding a line of best fit.

Linear Regression

Another name for the process of applying the least-squares method to find the line that best fits a given set of data is **linear regression**.[3] We will illustrate how linear regression is performed using the data in Example 1.

---

[2]It is natural to wonder why we take the *squares* of the individual errors before adding them together. There is a good reason for this, which will be discussed at the end of the chapter, after we have a more thorough understanding of how the method works.

[3]In the next section, we consider quadratic regression, where we approximate the data with a quadratic, rather than a linear, function.

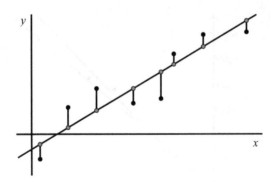

Figure 5.4: Error of the Approximating Line at Each Data Point

Example 2. Consider the gross revenue data points from the Mir Resort presented in Example 1:

$$\left\{ (1, 702), (2, 1351), (3, 2874), (4, 3768), (5, 4245), (6, 5404), (7, 5755) \right\}.$$

We apply linear regression to find the line of best fit.

First, we create a matrix $\mathbf{A}$ having two columns. The first column has the number 1 in every entry, and the second column has the x-coordinates of each of the data points as its entries. We also create a single-column matrix $\mathbf{B}$ having the y-coordinates of the data points as its entries. Thus, we get

$$\mathbf{A} = \begin{bmatrix} 1 & 1 \\ 1 & 2 \\ 1 & 3 \\ 1 & 4 \\ 1 & 5 \\ 1 & 6 \\ 1 & 7 \end{bmatrix} \quad \text{and} \quad \mathbf{B} = \begin{bmatrix} 702 \\ 1351 \\ 2874 \\ 3768 \\ 4245 \\ 5404 \\ 5755 \end{bmatrix}.$$

We want to find the values of m and b for the line $y = mx + b$ that best fits the data. Consider the column matrix $\mathbf{X} = \begin{bmatrix} b \\ m \end{bmatrix}$. Then, $\mathbf{AX} = \mathbf{B}$ amounts to

$$\begin{bmatrix} b + 1m \\ b + 2m \\ b + 3m \\ b + 4m \\ b + 5m \\ b + 6m \\ b + 7m \end{bmatrix} = \begin{bmatrix} 702 \\ 1351 \\ 2874 \\ 3768 \\ 4245 \\ 5404 \\ 5755 \end{bmatrix}, \quad \text{or,} \quad \begin{bmatrix} m(1) + b \\ m(2) + b \\ m(3) + b \\ m(4) + b \\ m(5) + b \\ m(6) + b \\ m(7) + b \end{bmatrix} = \begin{bmatrix} 702 \\ 1351 \\ 2874 \\ 3768 \\ 4245 \\ 5404 \\ 5755 \end{bmatrix}.$$

Notice in each equation that the column on the left side represents the y-values that we get if we plug the actual x-values of the data points into the equation $y = mx + b$, while the column on the right side represents the y-values of the actual data points.

Now, in either equation, both columns can only be equal if the line $y = mx + b$ actually goes through all seven of the data points, in which case $\mathbf{X}$ would be the solution to the system of equations $\mathbf{AX} = \mathbf{B}$. But, in fact, these seven points do not all line on a common line. Hence, there are no values for m and b in either equation for which the columns are equal, and so the system of equations $\mathbf{AX} = \mathbf{B}$ has no solution. However, we are not trying to find an *exact* solution, but just trying to get as close as we can to a solution.

To accomplish this, we modify the equation $\mathbf{AX} = \mathbf{B}$ by multiplying both sides by $\mathbf{A}^T$. This gives a related system of linear equations $\left(\mathbf{A}^T\mathbf{A}\right)\mathbf{X} = \mathbf{A}^T\mathbf{B}$ that does have a solution. Computing $\mathbf{A}^T\mathbf{A}$ and $\mathbf{A}^T\mathbf{B}$, we see that the system $\left(\mathbf{A}^T\mathbf{A}\right)\mathbf{X} = \mathbf{A}^T\mathbf{B}$ is

$$\begin{bmatrix} 7 & 28 \\ 28 & 140 \end{bmatrix}\begin{bmatrix} b \\ m \end{bmatrix} = \begin{bmatrix} 24099 \\ 121032 \end{bmatrix}.$$

The augmented matrix for this linear system is

$$\left[\begin{array}{cc|c} 7 & 28 & 24099 \\ 28 & 140 & 121032 \end{array}\right], \text{ which row reduces to } \left[\begin{array}{cc|c} 1 & 0 & -76.71 \\ 0 & 1 & 879.86 \end{array}\right],$$

where we have rounded to two places after the decimal point. That is, $m = 879.86$ and $b = -76.71$. It can be shown mathematically that the line $y = 879.86x - 76.71$ is the one line, out of all possible lines, that gives the smallest sum of squares of the individual errors from each data point. This line is illustrated along with the original data points in Figure 5.5.

Now Max Mir can use the equation of this least-squares line to predict future gross revenue, under the assumption that current revenue trends will continue. For example, to estimate his revenue in year 8, Max plugs $x = 8$ into the equation $y = 879.86x - 76.71$, resulting in $y = (879.86)(8) - 76.71 = 6962.17$. So his estimated revenue for year 8 is $\$6,962,170$. ∎

We now summarize the general method of linear regression, as illustrated in Example 2:

The Five-Step Method for Linear Regression

Given a set of points $\left\{(x_1, y_1), \ldots, (x_n, y_n)\right\}$, use the following steps to find values of m and b such that the line $y = mx + b$ best fits the data:

Step 1: Construct an $n \times 2$ matrix $\mathbf{A}$ in which the entries in the first column all equal 1, and the entries in the second column are, in order, $x_1, \ldots, x_n$.

Step 2: Create an $n \times 1$ matrix $\mathbf{B}$ whose entries are, in order, $y_1, \ldots, y_n$.

Step 3: Compute the 2×2 matrix $\mathbf{A}^T\mathbf{A}$ and the 2×1 matrix $\mathbf{A}^T\mathbf{B}$.

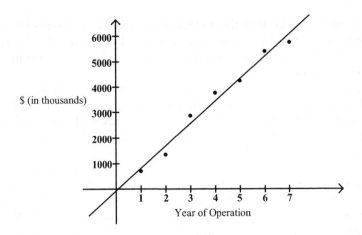

Figure 5.5: The Line $y = 879.86x - 76.71$ of Best Fit for Example 2

Step 4: Solve the linear system $\left(\mathbf{A}^T\mathbf{A}\right)\mathbf{X} = \mathbf{A}^T\mathbf{B}$, where $\mathbf{X} = \begin{bmatrix} b \\ m \end{bmatrix}$, by using the Gauss-Jordan Method on the augmented matrix $\left[\mathbf{A}^T\mathbf{A} \,\middle|\, \mathbf{A}^T\mathbf{B}\right]$.

Step 5: The line $y = mx+b$ is the line that fits the data as closely as possible, where m and b are determined from the solution to the linear system obtained in Step 4.

We have not yet explained why multiplying both sides of the equation $\mathbf{A}\mathbf{X} = \mathbf{B}$ by $\mathbf{A}^T$ produces a linear system that yields the least-squares line. We will give the "flavor" of an explanation for this at the end of the chapter. However, the formal details of why this technique works involve topics in linear algebra that are beyond the level of this course.

Let us consider another example of the linear regression method.

Example 3. In June of 2012, the Smith family signed up for a cell-phone data plan that allowed them to share 6 gigabytes of Internet data on their phones each month for several months. The following chart shows the actual amount of data they transferred each month, in gigabytes:

Month	June	Aug	Sept	Oct	Nov
Amount of data	0.41	1.12	1.87	2.31	3.39

Unfortunately, the Smiths lost the information for their July usage, so that is unavailable. We will use linear regression to find the equation of the line that best fits the information in the chart.

Before we start, we need to express the given information as a set of points having x- and y-coordinates. We have numbers for the y-values, but we need numbers for the x-values; that is, the months. The simplest thing to do is use $x = 6$ for June, $x = 8$ for August, $x = 9$ for September, etc. This produces the 5 points

$$\Big\{ (6, 0.41), \ (8, 1.12), \ (9, 1.87), \ (10, 2.31), \ (11, 3.39) \Big\}.$$

Step 1: We create the 5×2 matrix $\mathbf{A}$ having all 1's in its first column and having the x-values of the given points in its second column:

$$\mathbf{A} = \begin{bmatrix} 1 & 6 \\ 1 & 8 \\ 1 & 9 \\ 1 & 10 \\ 1 & 11 \end{bmatrix}.$$

Step 2: We construct the 5×1 matrix $\mathbf{B}$ having the y-values of the points in its only column:

$$\mathbf{B} = \begin{bmatrix} 0.41 \\ 1.12 \\ 1.87 \\ 2.31 \\ 3.39 \end{bmatrix}.$$

Step 3: We compute $\mathbf{A}^T\mathbf{A}$ and $\mathbf{A}^T\mathbf{B}$:

$$\mathbf{A}^T\mathbf{A} = \begin{bmatrix} 5 & 44 \\ 44 & 402 \end{bmatrix} \quad \text{and} \quad \mathbf{A}^T\mathbf{B} = \begin{bmatrix} 9.10 \\ 88.64 \end{bmatrix}.$$

Step 4: We row reduce the augmented matrix

$$\left[\begin{array}{cc|c} 5 & 44 & 9.10 \\ 44 & 402 & 88.64 \end{array} \right] \quad \text{to obtain} \quad \left[\begin{array}{cc|c} 1 & 0 & -3.270 \\ 0 & 1 & 0.578 \end{array} \right],$$

where we have rounded to three places after the decimal point. Thus, $m = 0.578$ and $b = -3.270$.

Step 5: The least-squares line for this data is $y = 0.578x - 3.270$. We can also express this as a function: $f(x) = 0.578x - 3.270$.

The Smiths want to use this analysis to predict when they will go over the 6 gigabyte limit, assuming that their usage continues to increase at the same rate. That is, they need to know at what x-value does $y = f(x) = 6$? Hence, we must solve $0.578x - 3.270 = 6$. Solving for x produces $x = 16.038$. Thus, the linear regression line predicts that they will hit their data ceiling during month 16. Now since 12 represents December 2012, and 13 represents January 2013, we see that 16 represents April 2013. At that point, they should reevaluate their situation and determine whether they need to enhance their data usage plan. ∎

One change we could have made in Example 3 is to use a different numbering scheme for the months so that the x-values are smaller and easier to manipulate. For example, we could have let $x = 0$ represent June, $x = 2$ represent August, $x = 3$ represent September, etc. (Notice that we skipped $x = 1$ for July.) Whenever we change the meanings of the x-values, it is important that we keep the same proportional spacing between them. As long as we do this, the linear regression function that we obtain still gives a best fit to the data. However, the actual equation that we derive is going to be different than the equation that we obtained in Example 3, because the x-values now have a different meaning. With the new formula, we would discover that the Smiths reach 6 gigabytes of usage at $x = 10$ instead of $x = 16$. But, under the new scheme, $x = 10$ represents April 2013, and so the final result is unchanged. In Exercise 7, you will be asked to redo Example 3 using this different numbering scheme to verify these assertions.

Example 3 illustrates one of the most important uses of linear regression: using the line of best fit to predict y-values for x-values outside the range of the given data. That is, in Example 3, we assumed that the equation for the line of best fit would provide a reasonable estimate of future y-values, and so we used this equation to predict the Smiths' upcoming gigabyte usage.

The process of predicting the y-values corresponding to x-values that lie *outside* the range of the given data is called **extrapolation**. But, we can also use the least-squares formula to estimate the y-values corresponding to x-values that lie *within* the range of the given data. This process is called **interpolation**. For example, in Example 3, the Smiths could make a good estimate of what their data plan usage was in July 2012 by plugging $x = 7$ into the line of best fit, yielding $f(7) = 0.578\,(7) - 3.270 = 0.776$ gigabytes.

Glossary

- Extrapolation: Estimating the corresponding y-value for an x-value that lies outside the range of the given data.

- Interpolation: Estimating the corresponding y-value for an x-value that lies within the range of the given data.

- Least-squares method: Modeling a given set of data points using a function having a minimum overall error relative to the data. The overall error between a function and a set of data is calculated by summing the squares of the differences between the y-values of the given data and the corresponding y-values of the function for each of the x-values in turn.

- Linear regression: The least-squares method in which the desired function is linear; that is, the equation of a line.

- Line of best fit: The line obtained by applying linear regression to a given set of data.

Exercises for Section 5.1

1. In each part, use linear regression to find the equation of the line that best fits the given data. Round your coefficients to three places after the decimal point. Then use that equation to predict the y-values corresponding to $x = 4$ and $x = 10$.

   (a) $\left\{(1,3),\,(2,5),\,(5,8),\,(7,13)\right\}$

   (b) $\left\{(0,2),\,(3,4),\,(6,4),\,(7,4),\,(8,7)\right\}$

2. In each part, use linear regression to find the equation of the line that best fits the given data. Round your coefficients to three places after the decimal point. Then use that equation to predict the y-values corresponding to $x = 5$ and $x = 9$.

   (a) $\left\{(0,-0.5),\,(3,1.5),\,(4,2),\,(7,5),\,(8,6.5)\right\}$

   (b) $\left\{(-1,-21),\,(2,10),\,(3,24),\,(4,42),\,(6,76),\,(7,73)\right\}$

3. The chart gives the annual usage of electricity per capita in kilowatt-hours in the United States. (Based on IEA data © OECD/IEA 2014, www.iea.org/statistics. Licence: www.iea.org/t&c; as modified by David A. Hecker and Stephen F. Andrilli.)

Year	1960	1965	1970	1975	1980
Usage in kwh	4049.79	5234.69	7236.66	8522.39	9862.37

Year	1985	1990	1995	2000
Usage in kwh	10414.21	11713.33	12659.61	13671.05

   (a) Use linear regression to find the equation of a line that approximates this data. Have $x = 0$ represent 1960, $x = 1$ represent 1965, etc. Round your coefficients to two places after the decimal point.

   (b) Use your answer in part (a) to estimate the annual electricity usage per capita in 2005 and in 2009. Round your answers to two places after the decimal point. (Note: Actual usage in 2005 was only 13704.58 kwh, and, surprisingly, in 2009 was 12913.71 kwh.)

   (c) For which answers in part (b) were you extrapolating? For which were you interpolating?

4. The chart gives the annual CO_2 emissions per capita in metric tons in the United States. (Data is from The World Bank.)

Year	1990	1992	1994	1996	1998
CO_2 Emissions	19.55	19.01	19.87	19.84	19.75

Year	2000	2001	2002	2004	2006
CO_2 Emissions	19.54	18.91	18.91	19.00	18.48

(a) Use linear regression to find the equation of a line that approximates this data. Have $x = 0$ represent 1990, $x = 1$ represent 1991, etc. Round your coefficients to three places after the decimal point.

(b) Use your answer in part (a) to estimate the annual CO_2 emissions per capita in 1980, 1985, 1995, and in 2008. Round your answers to two places after the decimal point. (Note: Actual emissions in metric tons: 1980: 20.78; 1985: 18.86; 1995: 19.67; 2008: 17.96.)

(c) For which answers in part (b) were you extrapolating? For which were you interpolating?

5. The chart gives the annual yield of cereal grains in the United States in kilograms per hectare planted. (Data is from The World Bank, and does not include animal feed.)

Year	1965	1970	1975	1980	1985
Cereal Yield	3040.8	3154.9	3461.3	3771.9	4763.3

Year	1990	1995	2000	2005	2010
Cereal Yield	4755.1	4644.8	5854.3	6451.0	6987.6

(a) Use linear regression to find the equation of a line that approximates this data. Have $x = 0$ represent 1965, $x = 1$ represent 1970, etc. Round your coefficients to two places after the decimal point.

(b) Use your answer in part (a) to estimate the annual cereal yield in 1961 and in 1998. Round your answers to one place after the decimal point. (Note: Actual yield in kg per hectare: 1961: 2522.3; 1998: 5676.1.)

(c) For which answers in part (b) did you use extrapolation? For which did you use interpolation?

(d) Use linear regression to find the equation of a line that approximates just the data in the second line of the chart. Have $x = 0$ represent 1990, $x = 1$ represent 1995, etc. Round your coefficients to two places after the decimal point.

(e) Use your answer to part (d) to obtain new estimates for the annual cereal yield in 1961 and in 1998. Round your answers to one place after the decimal point.

(f) Considering the actual data given in part (b), compare the predictions from parts (b) and (e). Which gave a more accurate estimate in each year? Can you explain this?

6. The chart gives the GDP (Gross Domestic Product) per unit of energy consumed in the United States in 2012 dollars per kg of oil equivalent. (The term "kg of oil equivalent" refers to the amount of energy obtained from one kilogram of oil. However, the data takes into account all energy

consumed, not just oil. The GDP per unit of energy is a measure of how energy efficient the economy is.) (Based on IEA data © OECD/IEA 2014, www.iea.org/statistics. Licence: www.iea.org/t&c; as modified by David A. Hecker and Stephen F. Andrilli.)

Year	1980	1983	1986	1990	1991	1993
GDP per unit energy	1.53	2.08	2.50	3.00	3.07	3.28

Year	1996	2000	2001	2003	2005	2008
GDP per unit energy	3.67	4.35	4.59	4.90	5.42	6.24

(a) Use linear regression to find the equation of a line that approximates this data. Have $x = 0$ represent 1980, $x = 1$ represent 1981, etc. Round your coefficients to three places after the decimal point.

(b) Use your answer in part (a) to estimate the GDP per unit energy in 1992, in 2006, in 2009, and in 2010. Round your answers to two places after the decimal point. (Note: Actual GDP per unit energy: 1992: 3.18; 2006: 5.80; 2009: 6.41; 2010: 6.46.)

(c) For which answers in part (b) did you use extrapolation? For which did you use interpolation?

(d) Use linear regression to find the equation of a line that approximates just the data in the second line of the chart. Have $x = 0$ represent 1990, $x = 1$ represent 1991, etc. Round your coefficients to three places after the decimal point.

(e) Use your answer to part (d) to obtain new estimates for the GDP per unit energy in 1992, in 2006, in 2009, and in 2010. Round your answers to two places after the decimal point.

(f) Considering the actual data given in part (b), compare the predictions from parts (b) and (e). Which gave a more accurate estimate in each year? Why do you think that happened?

7. Using the data from Example 3, renumber the months as described in the text, letting $x = 0$ represent June 2012, etc. Use linear regression to find the equation of the line that best fits the data using this renumbering. Then interpolate to estimate the Smiths' data usage in July 2012, and extrapolate to predict when the Smiths' data usage will hit 6 gigabytes.

8. The Mir Resort's first year of operation was 2004. Replace the x-values used in Examples 1 and 2 with the actual year; that is, replace $x = 1$ with $x = 2004$, $x = 2$ with $x = 2005$, etc. Use linear regression to find the equation of the line that best fits the data using this renumbering. Round your coefficients to three places after the decimal point. Then use this equation to extrapolate to predict the Mir Resort's gross revenue in 2011 (in thousands of $), rounded to the nearest whole number.

9. Consider the following set of data:

$$\Big\{(2, -0.6),\ (3, 1.5),\ (4, 4.0),\ (5, 4.4),\ (6, 6.3)\Big\}.$$

(a) The line $y = 2x - 5$ appears to align well with these points. Compute the predicted y-values at $x = 2$, $x = 3$, $x = 4$, $x = 5$, and $x = 6$ from the line $y = 2x - 5$. Then compute the sum of the squares of the individual differences between each of these predictions and the corresponding y-values from the actual data points. This gives the sum of the squares of the errors involved when using this line to estimate the data.

(b) Suppose we were to replace the line $y = 2x - 5$ with the linear regression line for the given data and then re-calculate the error involved. Without actually doing any computations, how would you expect the new error to compare with the error obtained in part (a)?

(c) The actual least-squares line for the given data is $y = 1.67x - 3.56$. Compute the sum of the squares of the individual errors involved using this line as you did for the line $y = 2x - 5$ in part (a). Was your prediction in part (b) correct?

10. Consider the following set of data:

$$\Big\{(0, 3.0),\ (1, 2.9),\ (2, 2.2),\ (4, 1.0),\ (5, 0.9),\ (6, 0.4)\Big\}.$$

(a) The line $y = -0.5x + 3.3$ appears to align well with these points. Compute the predicted y-values at $x = 0$, $x = 1$, $x = 2$, $x = 4$, $x = 5$, and $x = 6$ from the line $y = -0.5x + 3.3$. Then compute the sum of the squares of the individual differences between each of these predictions and the corresponding y-values from the actual data points. This gives the sum of the squares of the errors involved when using this line to estimate the data.

(b) Suppose we were to replace the line $y = -0.5x + 3.3$ with the linear regression line for the given data and then re-calculate the error involved. Without actually doing any computations, how would you expect the new error to compare with the error obtained in part (a)?

(c) The actual least-squares line for the given data is $y = -0.46x + 3.12$, rounded to two places after the decimal point. (Thus, the line $y = -0.5x + 3.3$ is very close to the least-squares line.) Compute the sum of the squares of the individual errors involved using this line as you did for the line $y = -0.5x + 3.3$ in part (a). Was your prediction in part (b) correct?

11. Consider the general set of points $\left\{(x_1, y_1), \ldots, (x_n, y_n)\right\}$. Let $\mathbf{A}$ be the $n \times 2$ matrix in which the entries in the first column all equal 1, and the entries in the second column are, in order, $x_1, \ldots, x_n$, as described in the general linear regression method.

   (a) Explain why the $(1, 1)$ entry of $\mathbf{A}^T \mathbf{A}$ equals n, the number of points in the data set.

   (b) Explain why the $(2, 1)$ entry of $\mathbf{A}^T \mathbf{A}$ equals $x_1 + \cdots + x_n$, the sum of the x-coordinates of the data points.

   (c) The $(1, 2)$ entry of $\mathbf{A}^T \mathbf{A}$ also equals $x_1 + \cdots + x_n$. Use part (b) and a theorem from an earlier chapter to explain why.

   (d) Explain why the $(2, 2)$ entry of $\mathbf{A}^T \mathbf{A}$ equals $x_1^2 + \cdots + x_n^2$, the sum of the squares of the x-coordinates of the data points.

12. Consider the general set of points $\left\{(x_1, y_1), \ldots, (x_n, y_n)\right\}$. Let $\mathbf{A}$ be the $n \times 2$ matrix in which the entries in the first column all equal 1, and the entries in the second column are, in order, $x_1, \ldots, x_n$, and let $\mathbf{B}$ be the $n \times 1$ column matrix in which the entries are, in order, $y_1, \ldots, y_n$, as described in the general linear regression method.

   (a) Explain why the $(1, 1)$ entry of $\mathbf{A}^T \mathbf{B}$ equals $y_1 + \cdots + y_n$, the sum of the y-coordinates of the data points.

   (b) Explain why the $(2, 1)$ entry of $\mathbf{A}^T \mathbf{B}$ equals $x_1 y_1 + \cdots + x_n y_n$, the sum of the products of the x-coordinates and corresponding y-coordinates of the data points.

5.2 Higher Degree Polynomials

We have seen how linear regression can find the equation of the line that comes closest to fitting given data. However, not all data is linear. The least-squares method generalizes to a method to find a polynomial of any desired degree – the polynomial of that degree (or lower) that best fits the data.

Quadratic Polynomials

Our next goal is to fit data to a quadratic polynomial; that is, a function of the form $y = f(x) = ax^2 + bx + c$, whose graph is a parabola. We only need to make minor modifications to the method of linear regression in order to find the least-squares quadratic polynomial. Here is the method:

The Five-Step Method for Quadratic Regression

Given a set of points $\left\{(x_1, y_1), \ldots, (x_n, y_n)\right\}$, take the following steps to find values of a, b, and c such that the parabola $y = ax^2 + bx + c$ best fits the data:

Step 1: Construct an $n \times 3$ matrix $\mathbf{A}$ in which the entries in the first column all equal 1, the entries in the second column are, in order, $x_1, \ldots, x_n$, and the entries in the third column are, in order, $x_1^2, \ldots, x_n^2$.

Step 2: Create an $n \times 1$ matrix $\mathbf{B}$ whose entries are, in order, $y_1, \ldots, y_n$.

Step 3: Compute the 3×3 matrix $\mathbf{A}^T\mathbf{A}$ and the 3×1 matrix $\mathbf{A}^T\mathbf{B}$.

Step 4: Solve the linear system $\left(\mathbf{A}^T\mathbf{A}\right)\mathbf{X} = \mathbf{A}^T\mathbf{B}$, where $\mathbf{X} = \begin{bmatrix} c \\ b \\ a \end{bmatrix}$, by

using the Gauss-Jordan Method on the augmented matrix $\left[\mathbf{A}^T\mathbf{A} \,\middle|\, \mathbf{A}^T\mathbf{B}\right]$.

Step 5: The parabola $y = f(x) = ax^2 + bx + c$ is the quadratic that fits the data as closely as possible, where a, b, and c are determined by the solution to the linear system obtained in Step 4. (It is possible that the solution could have $a = 0$, in which case the least-squares quadratic polynomial is actually linear. In that case, the linear function is a better fit than all possible quadratic functions. This rarely occurs.)

Just as with linear regression, when dealing with quadratic polynomials, a solution to the system of equations $\mathbf{AX} = \mathbf{B}$ would be a *perfect* fit to the data, not just a close fit. However, the linear system $\mathbf{AX} = \mathbf{B}$ has no solution unless the data points all happen to fall on the same parabola. That almost never happens. It is by multiplying both sides of the equation $\mathbf{AX} = \mathbf{B}$ by $\mathbf{A}^T$ that we, instead, get a linear system that gets us as close as possible to a solution in a situation in which no perfect solution exists. We will discuss this in more detail at the end of the chapter.

Also, what we mean by "best fit" is the same in the quadratic case as it is in the linear case. That is, the best quadratic equation is the one for which the sum of the squares of the error at each data point is as small as possible.

Let us consider an example.

Example 1. Recall the data from Example 3 in Section 5.1 involving the Smith family's data usage:

$$\Big\{(6, 0.41), \ (8, 1.12), \ (9, 1.87), \ (10, 2.31), \ (11, 3.39)\Big\}.$$

We will find the least-squares quadratic polynomial for this data set.

Step 1: We construct a 5×3 matrix $\mathbf{A}$ in which the entries in the first column all equal 1, the entries in the second column are, in order, $x_1, \ldots, x_5$, and the entries in the third column are, in order, $x_1^2, \ldots, x_5^2$:

$$\mathbf{A} = \begin{bmatrix} 1 & 6 & 36 \\ 1 & 8 & 64 \\ 1 & 9 & 81 \\ 1 & 10 & 100 \\ 1 & 11 & 121 \end{bmatrix}.$$

Step 2: We create a 5×1 matrix $\mathbf{B}$ whose entries are, in order, $y_1, \ldots, y_5$:

$$\mathbf{B} = \begin{bmatrix} 0.41 \\ 1.12 \\ 1.87 \\ 2.31 \\ 3.39 \end{bmatrix}.$$

Step 3: We compute $\mathbf{A}^T\mathbf{A}$ and $\mathbf{A}^T\mathbf{B}$:

$$\mathbf{A}^T\mathbf{A} = \begin{bmatrix} 5 & 44 & 402 \\ 44 & 402 & 3788 \\ 402 & 3788 & 36594 \end{bmatrix} \quad \text{and} \quad \mathbf{A}^T\mathbf{B} = \begin{bmatrix} 9.10 \\ 88.64 \\ 879.10 \end{bmatrix}.$$

Step 4: We solve the linear system $(\mathbf{A}^T\mathbf{A})\mathbf{X} = \mathbf{A}^T\mathbf{B}$ by row reducing

$$\begin{bmatrix} 5 & 44 & 402 & | & 9.10 \\ 44 & 402 & 3788 & | & 88.64 \\ 402 & 3788 & 36594 & | & 879.10 \end{bmatrix} \quad \text{to obtain} \quad \begin{bmatrix} 1 & 0 & 0 & | & 1.645 \\ 0 & 1 & 0 & | & -0.636 \\ 0 & 0 & 1 & | & 0.072 \end{bmatrix},$$

where we have rounded to three places after the decimal point.

Step 5: The function $y = g(x) = 0.072x^2 - 0.636x + 1.645$ is the least-squares quadratic polynomial for this situation.

Let us compare the linear function we obtained in Example 3 in Section 5.1 with this second-degree polynomial. Figure 5.6 shows the graphs of both functions, as well as the five data points, all on the same coordinate graph.

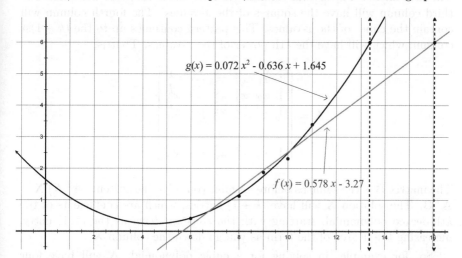

Figure 5.6: Linear and Quadratic Approximations for the Smiths' Data Usage

The figure illustrates that linear regression predicts that the Smiths will be using 6 gigabytes of data per month near $x = 16$, which corresponds to April 2013. The quadratic model predicts that the Smiths will hit the 6 gigabyte rate between $x = 13$ and $x = 14$, which is in mid-January 2013.

(You can solve for this x-value more precisely by solving for x in the equation $0.072x^2 - 0.636x + 1.645 = 6$ using the quadratic formula, which you should remember from high school algebra. Doing so gives the solution $x \approx 13.36$.)

So which least-squares function should we use? To decide you need to determine whether you expect your data, in the future, to be more linear or act more like a quadratic function? Do you expect the data to "go straight" or curve up more? In this case, the Smiths could look at their usage in December 2012 and see which function comes closest to having predicted it. In any case, to play it safe, the Smiths should probably increase their data limit sooner rather than later. ■

Cubic Polynomials and Beyond

The least-squares method generalizes to polynomials of any degree. You can solve for the cubic (third degree) polynomial that best fits the data. Or you can solve for a fourth or fifth degree polynomial. You should choose the degree that you expect will best model the behavior of the given real-world situation in the long run – not necessarily the one that best fits just the data you have.

To adapt the general methods we presented to produce a higher-degree polynomial, you only need to add more columns to the matrix $\mathbf{A}$. Specifically, to get a kth degree polynomial, $\mathbf{A}$ will need $(k + 1)$ columns. The first column will have all 1's. The second column will have the x-values of the data. The third column will have the squares of the x-values. The fourth column will contain the cubes of the x-values. This pattern continues up to the $(k + 1)$st column, which will have the kth power of the x-values. That is,

$$\mathbf{A} = \begin{bmatrix} 1 & x_1 & x_1^2 & x_1^3 & \cdots & x_1^k \\ 1 & x_2 & x_2^2 & x_2^3 & \cdots & x_2^k \\ 1 & x_3 & x_3^2 & x_3^3 & \cdots & x_3^k \\ \vdots & \vdots & \vdots & \vdots & \ddots & \vdots \\ 1 & x_n & x_n^2 & x_n^3 & \cdots & x_n^k \end{bmatrix}.$$

The matrix $\mathbf{B}$ is the same as before. Again, you solve the system $\left(\mathbf{A}^T\mathbf{A}\right)\mathbf{X} = \mathbf{A}^T\mathbf{B}$. The solution $\mathbf{X}$ will have $(k + 1)$ entries, which are coefficients of the kth-degree polynomial, starting with the *constant term* as the $(1, 1)$ entry, increasing the degree of the term as you go down the column $\mathbf{X}$.

So, for example, in solving for a cubic polynomial, $\mathbf{A}$ will have four columns, with the fourth column containing the cubes of the x-values. The least-squares cubic polynomial will be of the form $y = ax^3 + bx^2 + cx + d$, where the values of a, b, c, and d are the solution to the system $\left(\mathbf{A}^T\mathbf{A}\right)\mathbf{X} = \mathbf{A}^T\mathbf{B}$, with d being the *first* entry in the solution, c being the *second* entry, b being the *third* entry, and a being the last entry. However, we will not solve for any polynomials having a degree higher than two in this textbook.

Glossary

- Least-squares polynomial: A polynomial function found by using the least-squares method. This section discussed least-squares quadratic polynomials in detail, and also mentioned cubic and higher-degree polynomials.

- Quadratic regression: The least-squares method in which the desired function is a quadratic polynomial; that is, the equation of a parabola.

Exercises for Section 5.2

1. In each part, use the least-squares method to find the quadratic polynomial that best fits the given data. Round your coefficients to four places after the decimal point. Then use that function to predict the y-values corresponding to $x = 4$ and $x = 10$.

   (a) $\left\{ (0, 9.5), (1, 8.2), (2, 7.8), (5, 5.0), (8, 4.7) \right\}$

   (b) $\left\{ (-1, 2.8), (1, 4.2), (2, 5.3), (5, 5.1), (7, 5.2), (8, 4.4) \right\}$

2. In each part, use the least-squares method to find the quadratic polynomial that best fits the given data. Round your coefficients to four places after the decimal point. Then use that function to predict the y-values corresponding to $x = 3$ and $x = 9$.

   (a) $\left\{ (-1, 13.6), (0, 12.7), (1, 12.8), (2, 12.2), (4, 10.8), \right.$
   $\left. (6, 8.4), (8, 6.0) \right\}$

   (b) $\left\{ (-2, 7.46), (0, 5.42), (2, 2.46), (4, 0.26), (5, 3.57), (6, 8.02), \right.$
   $\left. (7, 8.36), (8, 14.39) \right\}$

3. A certain set contains 18 points. How many rows and columns would there be in the matrix $\mathbf{A}$ used to find the least-squares polynomial of degree 12 using this data?

4. A certain data set contains 18 points, with the x-coordinates equaling $-1, 0, 1, 2, \ldots, 16$, in that order. Consider the matrix $\mathbf{A}$ used to find the least squares polynomial of degree 12 for this data. What are the first four rows of $\mathbf{A}$?

5. The chart gives the expenditures on the military in the United States as a percentage of the federal budget. (Data is from The World Bank.)

Year	2001	2003	2005	2007	2009
% Military Spending	15.63	17.85	18.86	18.58	18.04

(a) Find the least-squares quadratic polynomial that fits the given data. Have $x = 0$ represent 2000, $x = 1$ represent 2001, etc. Round your coefficients to four places after the decimal point.

(b) Use your answer to part (a) to estimate the percentage of the federal budget spent on the military in 2006 and in 2010. Round your answers to two places after the decimal point. (The actual expenditures were 18.76% in 2006 and 17.89% in 2010.)

6. The chart gives the Wholesale Price Index (WPI) in the United States, normalized so that 2005 has a score of 100. (Data is from The World Bank.)

Year	1980	1985	1990	1995	2000	2005	2010
WPI	57.06	65.53	73.85	79.26	84.32	100.00	117.37

(a) Find the least-squares quadratic polynomial that fits the given data. Have $x = 0$ represent 1980, $x = 1$ represent 1981, etc. Round your coefficients to four places after the decimal point.

(b) Use your answer to part (a) to estimate the Wholesale Price Index in 1970, in 2003 and in 2011. Round your answers to two places after the decimal point. (The actual WPI was 23.42 in 1970, 87.75 in 2003 and 127.73 in 2011.)

(c) Find the least-squares quadratic polynomial that fits the three points at 2000, 2005, and 2010. Have $x = 0$ represent 2000, $x = 1$ represent 2001, etc. Round your coefficients to four places after the decimal point.

(d) Show that your answer to part (c) is actually an exact fit for the three given points by using your answer to compute values for y at $x = 0$, $x = 5$, and $x = 10$ to verify that the answers agree with the given data. (Note: This happens because any three points having different x-coordinates determine a unique parabola, just as two distinct points determine a unique line.)

(e) Use your answer to part (c) to estimate the Wholesale Price Index in 1970, in 2003 and in 2011. Round your answers to two places after the decimal point.

5.3   How Does Least-Squares Work?

In this section, we attempt to explain what is going on mathematically behind the scenes to enable the least-squares method to find the closest polynomial function fitting a given set of data. The discussion will involve several of the more theoretical results from Chapters 1 and 3. But still, the hard-core technical details are beyond the level of this course. Our goal is just to give you the "flavor" of what is going on.

The Geometry of Linear Regression

Suppose we have a set of n data points and have created the matrices $\mathbf{A}$ and $\mathbf{B}$ needed in linear regression. We explained in Example 2 of Section 5.1 that if the n points happen to appear all on the same line, the system of linear equations $\mathbf{AX} = \mathbf{B}$ has $\mathbf{X} = \begin{bmatrix} b \\ m \end{bmatrix}$ as a solution, where m and b are such that $y = mx + b$ is the equation of the line. However, as we have seen, the linear system $\mathbf{AX} = \mathbf{B}$ typically has no solution, requiring us to use linear regression to find the line that *best* fits the data, rather than the line that *exactly* fits the data.

Let us take a closer look at the matrix equation $\mathbf{AX} = \mathbf{B}$. Suppose that $\mathbf{v}_1$ and $\mathbf{v}_2$ are the two columns of $\mathbf{A}$, thought of as vectors. Then, by MSC2 (Section 1.8), $\mathbf{AX} = b\mathbf{v}_1 + m\mathbf{v}_2$. Thus, we are looking for a linear combination of $\mathbf{v}_1$ and $\mathbf{v}_2$ that equals $\mathbf{B}$.

Recall the Vector Etch-a-Sketch that we introduced in Section 3.6. Suppose that we have an n-dimensional Vector Etch-a-Sketch with two knobs programmed for the vectors $\mathbf{v}_1$ and $\mathbf{v}_2$. Let I be the starting point of the stylus. Trying to solve the system $\mathbf{AX} = \mathbf{B}$ is saying we would like to find out how to turn the knobs so that the total movement corresponds to the vector $\mathbf{B}$. If T is the terminal point of $\mathbf{B}$, then we are trying to move the stylus from I to T. The values of m and b tell us how to turn the knobs. Unfortunately, T probably does not lie on the plane of points (in n-dimensional space) that we can reach by turning the two dials that we have available. There is no solution. So, the next question is, if we cannot turn the knobs to get exactly to the terminal point T of $\mathbf{B}$, how close can we get to it? We would like to get as close as possible.

To help you understand this, try imagining the problem taking place in three dimensions. The terminal point T of $\mathbf{B}$ is somewhere off of a given plane in space. The initial point I of $\mathbf{B}$ is on the plane. We would like to find the closest point on the plane to T. To do this geometrically, we "drop a perpendicular" from T down to the plane. That is, we find the one point P on the plane so that the vector from P to T is perpendicular to the plane. P is then the closest point on the plane to T. The vector $\mathbf{C}$ from I to P is called the projection of $\mathbf{B}$ down onto the plane. See Figure 5.7.

In n-dimensions, the situation is completely analogous – just more difficult to draw. We want to find the point P on the plane determined by $\mathbf{v}_1$ and $\mathbf{v}_2$ that is closest to the terminal point T of $\mathbf{B}$. Since P is on the plane, P can be reached by turning the two dials. That is, there exist values of m and b such that $b\mathbf{v}_1 + m\mathbf{v}_2 = \mathbf{C}$, where $\mathbf{C}$ is the vector from I to P. In other words, turning the two dials takes us through the movement given by $\mathbf{C}$ to the point P.

This means that the system of linear equations $\mathbf{AX} = \mathbf{C}$ has a solution! The solution $\mathbf{X} = \begin{bmatrix} b \\ m \end{bmatrix}$ for this system tells us how to get $\mathbf{AX}$ as close to $\mathbf{B}$ as possible, since it gives us $\mathbf{C}$, whose terminal point is as close as possible

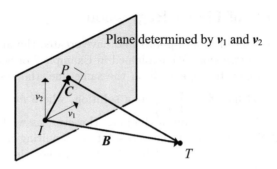

Figure 5.7: Projecting **B** Onto the Plane Determined by $\mathbf{v}_1$ and $\mathbf{v}_2$

to the terminal point of **B** while staying on the plane. Geometrically, this is what linear regression does. It solves for the value of **X** such that **AX** takes us as close as we can get to the terminal point of **B**.

This explains why the least-squares method minimizes the sum of the *squares* of the individual errors. Geometrically, we are finding a point P on a plane closest in distance to the terminal point T of **B**. But, the distance in n-dimensional space is the square root of the sum of the squares of the differences in the coordinates of the two points – that is, the length of the error vector, which is the vector from P to T. So, to make that distance as small as possible, we must make the sum of the squares of the errors as small as possible. The sum of the squares of the errors is just the square of the length of the vector from P to T in n-dimensional space!

The same basic geometric ideas explain what is going on behind the scenes in using the least-squares method to find quadratic and higher-degree polynomial approximations for a data set. The difference is that, since **A** has more than two columns, there are more than two knobs on the Vector Etch-a-Sketch – one knob for each column of **A**. This makes the situation more complex because the set of points you can reach using this dials is no longer a plane, but, instead, some larger, higher-dimensional space. What it means to "drop a perpendicular" onto this higher-dimensional space is difficult to imagine and impossible to draw in a figure. However, the concepts are the same. They are just more difficult to picture in our minds.

The Algebra of Linear Regression

In Exercise 1, we give you a simple data set and ask you to solve the linear system $\mathbf{A}^T\mathbf{A}\mathbf{X} = \mathbf{A}^T\mathbf{B}$ as you would using linear regression. The exercise then guides you step-by-step to compute **C**, the projection of **B** onto the plane determined by the two columns of **A**. Finally, you are asked to solve the linear system $\mathbf{A}\mathbf{X} = \mathbf{C}$ and show that you get the same solution for **X** that you obtained using linear regression. This illustrates (but does not prove) that the two methods are doing the same thing.

Unfortunately, the theory to justify that the solution of $\mathbf{A}^T\mathbf{AX} = \mathbf{A}^T\mathbf{B}$ drops a perpendicular down onto a plane and then solves the corresponding linear system obtained, uses topics in linear algebra that are a bit beyond the level of this textbook. First, we would need to formally generalize Theorem 5 in Section 3.5 (The Projection Theorem) to handle these projections into higher-dimensional spaces. We would also have to be much more technically precise about the meanings of the movements on the Vector Etch-a-Sketch. We have really only used that device as a motivational illustration, but we need more technical detail to use such concepts in a proof. However, because solving $\mathbf{A}^T\mathbf{AX} = \mathbf{A}^T\mathbf{B}$ does, indeed, perform the geometric operations we have described, this system of linear equations will always have a solution.

Another interesting theoretical aspect of the least-squares method is based on Theorem 2 in Section 1.6, which tells us that the matrix $\mathbf{A}^T\mathbf{A}$ is symmetric. (You may have noticed this about $\mathbf{A}^T\mathbf{A}$ in all of the examples and while doing exercises.) In a more advanced course in linear algebra, we would discover important properties of symmetric matrices involving mutually orthogonal sets of vectors. These properties of $\mathbf{A}^T\mathbf{A}$ can be used to analyze geometric properties of the matrix $\mathbf{A}$ in interesting ways that also shed some light on what is happening in least-squares. However, these topics are also theoretically far beyond the level of this textbook.

Exercises for Section 5.3

1. Consider the following set of data points:

$$\Big\{(1,8),\ (3,13),\ (5,20),\ (7,25)\Big\}.$$

The matrices $\mathbf{A}$ and $\mathbf{B}$ for using linear regression on this data are

$$\mathbf{A} = \begin{bmatrix} 1 & 1 \\ 1 & 3 \\ 1 & 5 \\ 1 & 7 \end{bmatrix} \text{ and } \mathbf{B} = \begin{bmatrix} 8 \\ 13 \\ 20 \\ 25 \end{bmatrix}.$$

(a) Use the Gauss-Jordan Method to solve the linear system $\mathbf{A}^T\mathbf{AX} = \mathbf{A}^T\mathbf{B}$ as you would do when applying linear regression to this data.

(b) Use the Gram-Schmidt Process on the two columns $\mathbf{v}_1$ and $\mathbf{v}_2$ of $\mathbf{A}$ to find vectors $\mathbf{w}_1$ and $\mathbf{w}_2$ so that $\{\mathbf{w}_1, \mathbf{w}_2\}$ is an orthogonal set. (The vectors $\mathbf{w}_1$ and $\mathbf{w}_2$ determine the same plane as the two columns of $\mathbf{A}$ and have the advantage of being perpendicular to each other.)

(c) Compute $\mathbf{proj}_{\mathbf{w}_1}\mathbf{B}$ and $\mathbf{proj}_{\mathbf{w}_2}\mathbf{B}$.

(d) Calculate $\mathbf{C} = \mathbf{proj}_{\mathbf{w}_1}\mathbf{B} + \mathbf{proj}_{\mathbf{w}_2}\mathbf{B}$. (The vector $\mathbf{C}$ is the result of "dropping a perpendicular" from the vector $\mathbf{B}$ onto the plane determined by the two columns of **A**. If you use the x-values from the given data with the corresponding y-values in **C**, the four points will all lie on the least-squares line.)

(e) Use the Gauss-Jordan Method to solve the linear system $\mathbf{AX} = \mathbf{C}$.

(f) Compare the solution in part (e) to the solution of the linear system $\mathbf{A}^T\mathbf{AX} = \mathbf{A}^T\mathbf{B}$ from part (a).

2. Consider the following set of data points:

$$\left\{(0, 2.5),\ (1, 5.5),\ (2, 7.0),\ (3, 9.5),\ (4, 10.5)\right\}.$$

The matrices **A** and **B** for using linear regression on this data are

$$\mathbf{A} = \begin{bmatrix} 1 & 0 \\ 1 & 1 \\ 1 & 2 \\ 1 & 3 \\ 1 & 4 \end{bmatrix} \text{ and } \mathbf{B} = \begin{bmatrix} 2.5 \\ 5.5 \\ 7.0 \\ 9.5 \\ 10.5 \end{bmatrix}.$$

(a) Use the Gauss-Jordan Method to solve the linear system $\mathbf{A}^T\mathbf{AX} = \mathbf{A}^T\mathbf{B}$ as you would do when applying linear regression to this data.

(b) Use the Gram-Schmidt Process on the two columns $\mathbf{v}_1$ and $\mathbf{v}_2$ of **A** to find vectors $\mathbf{w}_1$ and $\mathbf{w}_2$ so that $\{\mathbf{w}_1, \mathbf{w}_2\}$ is an orthogonal set. (The vectors $\mathbf{w}_1$ and $\mathbf{w}_2$ determine the same plane as the two columns of **A** and have the advantage of being perpendicular to each other.)

(c) Compute $\mathbf{proj}_{\mathbf{w}_1}\mathbf{B}$ and $\mathbf{proj}_{\mathbf{w}_2}\mathbf{B}$.

(d) Calculate $\mathbf{C} = \mathbf{proj}_{\mathbf{w}_1}\mathbf{B} + \mathbf{proj}_{\mathbf{w}_2}\mathbf{B}$. (The vector **C** is the result of "dropping a perpendicular" from the vector **B** onto the plane determined by the two columns of **A**. If you use the x-values from the given data with the corresponding y-values in **C**, the five points will all lie on the least-squares line.)

(e) Use the Gauss-Jordan Method to solve the linear system $\mathbf{AX} = \mathbf{C}$.

(f) Compare the solution in part (e) to the solution of the linear system $\mathbf{A}^T\mathbf{AX} = \mathbf{A}^T\mathbf{B}$ from part (a).

5.4   Chapter 5 Test

1. Use linear regression to find the equation of the line that best fits the following data:

$$\left\{(0, 12.62),\ (1, 11.83),\ (3, 10.87),\ (4, 9.91),\ (5, 8.78)\right\}.$$

Round your coefficients to three places after the decimal point. Then use that equation to predict the y-values corresponding to $x = 2$ and $x = 7$.

2. Use the least-squares method to find the quadratic polynomial that best fits the following data:

$$\Big\{ (-2, 14.12), \, (0, 22.92), \, (1, 22.41), \, (3, 29.5), \, (6, 26.53), \, (8, 25.51),$$

$$(10, 17.28) \Big\}.$$

Round your coefficients to four places after the decimal point. Then use that function to predict the y-values corresponding to $x = 2$ and $x = 11$.

3. The chart gives the number of motor vehicles per 1000 people in the United States in various years. (Data is from The World Bank.)

Year	2003	2004	2006	2007	2008
Motor Vehicles per 1000 people	796	810	818	820	815

  (a) Use linear regression to find the equation of a line that approximates this data. Have $x = 0$ represent 2000, $x = 1$ represent 2001, etc. Round your coefficients to two places after the decimal point.

  (b) Use your answer in part (a) to estimate the number of motor vehicles per 1000 people in 2004 and in 2009. Round your answers to the nearest whole number. (Note: Actual number of motor vehicles per 1000 people: 2005: 816; 2009: 802.)

4. Consider the data in Problem 3 regarding the number of motor vehicles per 1000 people.

  (a) Find the least-squares quadratic polynomial that fits the given data. Have $x = 0$ represent 2000, $x = 1$ represent 2001, etc. Round your coefficients to four places after the decimal point.

  (b) Use your answer to part (a) to estimate the number of motor vehicles per 1000 people in 2005 and in 2009. Round your answers to the nearest whole number.

5. In each part, indicate whether the given statement is true or whether it is false.

  (a) Linear Regression refers to the least-squares method applied to find the least-squares polynomial of degree 1.

  (b) Least-squares is a method in which, given a set of points in a plane, one computes a polynomial function having a desired degree (or less) such that the sum of the vertical distances from each given point to the graph of the polynomial function is as small as possible.

  (c) A quadratic polynomial has degree 2, and a cubic polynomial has degree 3.

(**d**) Given a set of points, the least-squares method will find the polynomial of the desired degree that exactly passes through the largest number of the points possible.

(**e**) In the least-squares method, the number of rows in the matrix **A** is one more than the degree of the desired polynomial.

(**f**) In the least-squares method, the number of columns in the matrix **A** is one more than the degree of the desired polynomial.

(**g**) In the least-squares method, the number of rows in the matrix **A** equals the number of data points.

(**h**) In the least-squares method, none of the rows of the matrix **A** can have all of its entries equal to zero.

(**i**) The matrix $\mathbf{A}^T\mathbf{A}$ used in the least-squares method is symmetric.

(**j**) The matrix $\mathbf{A}^T\mathbf{B}$ used in the least-squares method is symmetric.

(**k**) If $(2, 47)$ and $(6, 192)$ are two of the several given points in a least-squares problem, then, after computing the least-squares polynomial of the desired degree, one would use extrapolation to find the predicted value of the function when $x = 5$.

(**l**) Interpolation is the process of predicting a y-value corresponding to an x-value that is within the range of the x-values in the original data set.

(**m**) Given the matrices **A** and **B** determined as in a least-squares problem, the system of linear equations given by $\mathbf{A}^T\mathbf{A}\mathbf{X} = \mathbf{A}^T\mathbf{B}$ must have a solution.

(**n**) Given the matrices **A** and **B** determined as in a least-squares problem, the system of linear equations given by $\mathbf{A}\mathbf{X} = \mathbf{B}$ typically does not have a solution, although it is possible that it does.

(**o**) In the least-squares method, it is possible to end up with the equation of a line as the answer, even if you are solving for the equation of a quadratic polynomial, although this is unlikely.

(**p**) The theory behind linear regression involves "dropping a perpendicular" in n-dimensional space from the vector **B** onto the plane determined by the columns of **A**.

(**q**) A line obtained using linear regression on a given set of points must pass through at least one of the points.

Chapter 6

Inverses for Matrices

In this chapter, we discuss multiplicative inverses for matrices – their applications and how to calculate them.

6.1 Multiplicative Inverses

Solving Matrix Equations

In your algebra course in high school, you learned how to solve an equation such as $5x = 15$. Although you may have skipped some of the steps below, you essentially proceeded as follows:

$$
\begin{aligned}
5x &= 15 \\
\frac{1}{5}(5x) &= \frac{1}{5}(15) \\
\left(\frac{1}{5} \cdot 5\right)x &= 3 \\
1x &= 3 \\
x &= 3.
\end{aligned}
$$

You found the reciprocal of the coefficient 5, which is its *multiplicative inverse*, and multiplied both sides of the equation by that number. On the left side of the equation, after using the associative law for multiplication, you computed $\left(\frac{1}{5} \cdot 5\right)$, resulting in 1, which is the identity element for multiplication of numbers. Of course, $1x = x$, no matter what the value is for the variable x. This completes the simplification of the left side of the equation, making the solution for x obvious – it equals the number computed on the right side of the equation.

Our goal is to use a similar method to solve systems of linear equations that have been written as a matrix equation of the form $\mathbf{Ax} = \mathbf{b}$. First, we see that we will need an identity element for *matrix* multiplication to play the role of the number 1 we used in solving our equation involving numbers. We know that an appropriately-sized identity matrix $\mathbf{I}$ has the desired property that $\mathbf{Ix} = \mathbf{x}$ for any vector $\mathbf{x}$, and in Chapter 1 we called $\mathbf{I}$ the identity element for matrix multiplication.

Next, we need to find a multiplicative inverse for the matrix $\mathbf{A}$, just as we found a multiplicative inverse for the number 5. We want a matrix so that when we multiply it times $\mathbf{A}$ we get the multiplicative identity. More simply, we need a matrix $\mathbf{C}$ such that $\mathbf{CA} = \mathbf{I}$.

Example 1. We would like to solve the linear system

$$\begin{cases} 5x + 7y = 22 \\ 2x + 3y = 9 \end{cases}.$$

First, we express this system using matrix multiplication as follows:

$$\begin{bmatrix} 5 & 7 \\ 2 & 3 \end{bmatrix} \begin{bmatrix} x \\ y \end{bmatrix} = \begin{bmatrix} 22 \\ 9 \end{bmatrix}.$$

Next, we multiply both sides of this equation by

$$\begin{bmatrix} 3 & -7 \\ -2 & 5 \end{bmatrix},$$

and simplify:

$$\begin{bmatrix} 3 & -7 \\ -2 & 5 \end{bmatrix} \left(\begin{bmatrix} 5 & 7 \\ 2 & 3 \end{bmatrix} \begin{bmatrix} x \\ y \end{bmatrix} \right) = \begin{bmatrix} 3 & -7 \\ -2 & 5 \end{bmatrix} \begin{bmatrix} 22 \\ 9 \end{bmatrix}$$

$$\left(\begin{bmatrix} 3 & -7 \\ -2 & 5 \end{bmatrix} \begin{bmatrix} 5 & 7 \\ 2 & 3 \end{bmatrix} \right) \begin{bmatrix} x \\ y \end{bmatrix} = \begin{bmatrix} 3 \\ 1 \end{bmatrix}$$

$$\begin{bmatrix} 1 & 0 \\ 0 & 1 \end{bmatrix} \begin{bmatrix} x \\ y \end{bmatrix} = \begin{bmatrix} 3 \\ 1 \end{bmatrix}$$

$$\begin{bmatrix} x \\ y \end{bmatrix} = \begin{bmatrix} 3 \\ 1 \end{bmatrix}.$$

Hence, we get the unique solution $x = 3$, $y = 1$ for the linear system. You can verify that this is correct by plugging these values into the original equations.
■

You should compare the steps used to solve the matrix equation in Example 1 with the steps we took to solve the equation $5x = 15$ at the beginning of the chapter. The only two differences between the two solution methods are, first, that we were using matrices in Example 1 instead of numbers, and second, that it is unclear how we computed the multiplicative inverse for the 2×2 matrix involved in Example 1. We will learn how to compute multiplicative inverses for square matrices later in this section.

What is an Inverse for a Matrix?

Throughout this chapter, we will be mostly interested in multiplicative inverses for *square* matrices. So, first, we must explicitly state what this means:

If A is an $n \times n$ matrix, then a **multiplicative inverse** for A is an $n \times n$ matrix B such that $AB = I_n$ and $BA = I_n$.

We usually call a multiplicative inverse for a square matrix just an *inverse* for the matrix, dropping the word "multiplicative."

Notice that we require that an inverse for a square matrix to be two-sided; that is, multiplying a matrix A by an inverse B *in either order* produces the identity matrix. That is, a square matrix and an inverse for it must commute. However, we have the following theorem, which makes this a less stringent requirement:

Theorem 1. *If* A *and* B *are* $n \times n$ *matrices such that* $AB = I_n$*, then* $BA = I_n$.

So, by Theorem 1, we see that if the product of two square matrices in one order is I_n, then the product in the reverse order is also I_n. The matrices *must* commute. Hence, although the definition of an inverse matrix requires the product in both orders equalling I_n, Theorem 1 tells us that we need only check the product in one order to see that it equals I_n, since the other order then follows automatically.

It is important to note that Theorem 1 only applies to square matrices. The analogous statement for non-square matrices is false.

A second important fact about inverses for square matrices is that, when they exist, they are *unique*. That is, we have the following:

Theorem 2. *If* A*,* B*, and* C *are* $n \times n$ *matrices such that both* B *and* C *are inverses for* A*, then* $B = C$.

Theorem 2 is actually easy to prove.
Proof of Theorem 2:
If both B and C are inverses for A, then $BA = I_n$ and $AC = I_n$. Using these equations, we get

$$B = BI_n = B(AC) = (BA)C = I_nC = C. \; \blacklozenge$$

Therefore, by Theorem 2, a square matrix can have at most one inverse, since any other inverse would have to equal the first. Hence, we can speak about *the* inverse for a square matrix, if one exists, rather than *an* inverse for a square matrix. So, if a square matrix A has an inverse, we will use the symbol A^{-1} to represent the unique inverse for A.

Some Square Matrices Do Not Have an Inverse

We have been careful in our discussion about inverses for matrices to include the phrase "if one exists" when needed. This is because not every square matrix has an inverse. For example, a square matrix having all zero entries cannot have an inverse because $\mathbf{O}_{nn}\mathbf{A} = \mathbf{O}_{nn}$ for any $n \times n$ matrix $\mathbf{A}$. You cannot multiply $\mathbf{O}_{nn}$ by another matrix and end with $\mathbf{I}_n$. However, many other square matrices also do not have inverses – even matrices that do not have any zero entries! In particular, if a square matrix $\mathbf{A}$ is a zero divisor, then it cannot have an inverse.[1] Recall that we defined a nonzero matrix $\mathbf{A}$ to be a zero divisor if there is a nonzero matrix $\mathbf{B}$ such that either $\mathbf{AB}$ or $\mathbf{BA}$ is a zero matrix. We can prove that a square zero divisor cannot have an inverse using a proof by contradiction.

Well, suppose that both $\mathbf{AB} = \mathbf{O}$ and $\mathbf{A}$ *does* have an inverse. Then

$$
\begin{aligned}
\mathbf{A}^{-1}\left(\mathbf{AB}\right) &= \mathbf{A}^{-1}\mathbf{O} \\
\left(\mathbf{A}^{-1}\mathbf{A}\right)\mathbf{B} &= \mathbf{O} \\
\mathbf{IB} &= \mathbf{O} \\
\mathbf{B} &= \mathbf{O}.
\end{aligned}
$$

Therefore, $\mathbf{B}$ would have to be a zero matrix. But that contradicts the assumption that $\mathbf{B}$ is not a zero matrix from the definition of a zero divisor. Hence, we could not have $\mathbf{AB} = \mathbf{O}$ with $\mathbf{A}$ having an inverse. (An analogous proof works if you assume $\mathbf{BA} = \mathbf{O}$.)

Example 2. In Section 1.7, we saw that the matrix

$$
\mathbf{A} = \begin{bmatrix} 4 & 2 \\ 6 & 3 \end{bmatrix}
$$

is a zero divisor. We will show why $\mathbf{A}$ cannot have an inverse by trying to solve for an inverse for $\mathbf{A}$.

Consider the equation

$$
\begin{bmatrix} 4 & 2 \\ 6 & 3 \end{bmatrix}\begin{bmatrix} w & x \\ y & z \end{bmatrix} = \begin{bmatrix} 1 & 0 \\ 0 & 1 \end{bmatrix},
$$

which would have to be true for some values of w, x, y, and z in order for $\mathbf{A}$ to have an inverse. Perform the matrix product on the left. The resulting matrix must equal $\mathbf{I}_2$ (on the right). Setting corresponding entries equal produces the following system of linear equations:

$$
\begin{cases}
4w & + \; 2y & & = 1 \\
& 4x & + \; 2z & = 0 \\
6w & + \; 3y & & = 0 \\
& 6x & + \; 3z & = 1.
\end{cases}
$$

---

[1]The converse of this statement is also true; that is, if a nonzero square matrix does not have an inverse, then it is a zero divisor. Exercise 5, below, explores this idea a little further.

Using rref on our calculator on the associated augmented matrix for this system yields

$$\begin{bmatrix} 1 & 0 & 0.5 & 0 & | & 0 \\ 0 & 1 & 0 & 0.5 & | & 0 \\ 0 & 0 & 0 & 0 & | & 1 \\ 0 & 0 & 0 & 0 & | & 0 \end{bmatrix}.$$

(The calculator has created a pivot in the last column.) Because of the form of the third row of the matrix, this system clearly has no solution. Hence, the matrix **A** cannot have an inverse. ∎

A square matrix that does *not* have an inverse is said to be **singular**. And so, a square matrix that *does* have an inverse is said to be **nonsingular**.

How to Calculate the Inverse of a Square Matrix

Next, we describe how to determine whether a particular square matrix is nonsingular (has an inverse) or singular (does not have an inverse), and, if it is nonsingular, to calculate the unique inverse for the matrix.

Given an $n \times n$ matrix **A**, take the following steps to compute its inverse, or to determine that it is singular:

Step 1: Form the augmented $n \times (2n)$ matrix $[\mathbf{A} \,|\, \mathbf{I}_n]$, having the n columns of the identity matrix *after* the vertical bar.

Step 2: Put the matrix $[\mathbf{A} \,|\, \mathbf{I}_n]$ from Step 1 into reduced row echelon form.

Step 3: If the n columns to the left of the bar in the reduced row echelon form for $[\mathbf{A} \,|\, \mathbf{I}_n]$ form the identity matrix, then **A** is nonsingular. Otherwise, **A** is singular.

Step 4: If **A** has been determined to be nonsingular in Step 3, then the reduced row echelon form matrix will be in the form $\left[\mathbf{I}_n \,|\, \mathbf{A}^{-1}\right]$. That is, the n columns to the right of the bar in the reduced row echelon form matrix make up the inverse for the matrix **A**.

Let us consider some examples.

Example 3. Suppose $\mathbf{A} = \begin{bmatrix} 5 & 7 \\ 2 & 3 \end{bmatrix}$, the matrix from Example 1. Let us compute its inverse.

Step 1: We form the augmented matrix

$$\begin{bmatrix} 5 & 7 & | & 1 & 0 \\ 2 & 3 & | & 0 & 1 \end{bmatrix}.$$

Step 2: Using our calculator, we find that the reduced row echelon form for this augmented matrix is

$$\begin{bmatrix} 1 & 0 & | & 3 & -7 \\ 0 & 1 & | & -2 & 5 \end{bmatrix}.$$

Step 3: Because the matrix to the left of the bar in the matrix from Step 2 is I_2, the matrix $\mathbf{A}$ is nonsingular.

Step 4: The inverse for $\mathbf{A}$ is the matrix to the right of the bar, namely

$$\begin{bmatrix} 3 & -7 \\ -2 & 5 \end{bmatrix}. \quad \blacksquare$$

Example 4. Suppose $\mathbf{A} = \begin{bmatrix} 4 & 2 \\ 6 & 3 \end{bmatrix}$, the matrix from Example 2. We will follow the method for computing the inverse of $\mathbf{A}$.

Step 1: We form the augmented matrix

$$\begin{bmatrix} 4 & 2 & | & 1 & 0 \\ 6 & 3 & | & 0 & 1 \end{bmatrix}.$$

Step 2: Using our calculator, we find that the reduced row echelon form for this augmented matrix is

$$\begin{bmatrix} 1 & \frac{1}{2} & | & 0 & \frac{1}{6} \\ 0 & 0 & | & 1 & -\frac{2}{3} \end{bmatrix}.$$

Step 3: Because the matrix to the left of the bar in the matrix from Step 2 is not I_2, the matrix $\mathbf{A}$ is singular. It does not have an inverse. $\blacksquare$

Example 5. Suppose $\mathbf{B} = \begin{bmatrix} 2 & 6 & 1 & -1 \\ 2 & 3 & 1 & 0 \\ 5 & 1 & 3 & 2 \\ 2 & 1 & 2 & 1 \end{bmatrix}$. We compute the inverse of $\mathbf{B}$.

Step 1: We form the augmented matrix

$$\begin{bmatrix} 2 & 6 & 1 & -1 & | & 1 & 0 & 0 & 0 \\ 2 & 3 & 1 & 0 & | & 0 & 1 & 0 & 0 \\ 5 & 1 & 3 & 2 & | & 0 & 0 & 1 & 0 \\ 2 & 1 & 2 & 1 & | & 0 & 0 & 0 & 1 \end{bmatrix}.$$

Step 2: Using our calculator, we find that the reduced row echelon form for this augmented matrix is

$$\begin{bmatrix} 1 & 0 & 0 & 0 & | & 1 & -2 & 1 & -1 \\ 0 & 1 & 0 & 0 & | & -\frac{3}{2} & \frac{7}{2} & -1 & \frac{1}{2} \\ 0 & 0 & 1 & 0 & | & \frac{5}{2} & -\frac{11}{2} & 1 & \frac{1}{2} \\ 0 & 0 & 0 & 1 & | & -\frac{11}{2} & \frac{23}{2} & -3 & \frac{3}{2} \end{bmatrix}.$$

Step 3: Because the matrix to the left of the bar in the matrix from Step 2 is I_4, the matrix $\mathbf{B}$ is nonsingular.

Step 4: The inverse for $\mathbf{B}$ is the matrix to the right of the bar, namely

$$\mathbf{B}^{-1} = \begin{bmatrix} 1 & -2 & 1 & -1 \\ -\frac{3}{2} & \frac{7}{2} & -1 & \frac{1}{2} \\ \frac{5}{2} & -\frac{11}{2} & 1 & \frac{1}{2} \\ -\frac{11}{2} & \frac{23}{2} & -3 & \frac{3}{2} \end{bmatrix}.$$

We can verify our answer by using matrix multiplication to check that $\mathbf{B}^{-1}\mathbf{B} = \mathbf{I}_4$, or that $\mathbf{B}\mathbf{B}^{-1} = \mathbf{I}_4$. (Theorem 1 shows that if one of these two equations is true, then so is the other.) You should try these computations on your calculator. ■

Solving Systems of Linear Equations, Revisited

If a matrix $\mathbf{A}$ is nonsingular, then the linear system represented by the matrix equation $\mathbf{A}\mathbf{x} = \mathbf{b}$ can be solved by multiplying both sides of the equation $\mathbf{A}\mathbf{x} = \mathbf{b}$ by $\mathbf{A}^{-1}$ and simplifying, leading to the solution $\mathbf{x} = \mathbf{A}^{-1}\mathbf{b}$. Because this computation leaves no other possibilities for the solution to $\mathbf{A}\mathbf{x} = \mathbf{b}$, we know that this is the *unique* solution for the linear system – that is, there cannot be an infinite number of solutions. Therefore, we have the following theorem:

Theorem 3. *If* $\mathbf{A}$ *is a nonsingular matrix, then the system of linear equations represented by* $\mathbf{A}\mathbf{x} = \mathbf{b}$ *has the* unique *solution* $\mathbf{x} = \mathbf{A}^{-1}\mathbf{b}$.

Example 6. Consider the following system of linear equations:

$$\begin{cases} 2w + 6x + y - z = 37 \\ 2w + 3x + y = 24 \\ 5w + x + 3y + 2z = 30 \\ 2w + x + 2y + z = 15 \end{cases}$$

Let us find the complete solution set for this linear system.

The given linear system is equivalent to the matrix equation

$$\begin{bmatrix} 2 & 6 & 1 & -1 \\ 2 & 3 & 1 & 0 \\ 5 & 1 & 3 & 2 \\ 2 & 1 & 2 & 1 \end{bmatrix} \begin{bmatrix} w \\ x \\ y \\ z \end{bmatrix} = \begin{bmatrix} 37 \\ 24 \\ 30 \\ 15 \end{bmatrix}.$$

Now the 4×4 coefficient matrix is just the matrix $\mathbf{B}$ from Example 5. Therefore, by Theorem 3, the unique solution to the system is

$$\begin{bmatrix} w \\ x \\ y \\ z \end{bmatrix} = \mathbf{B}^{-1} \begin{bmatrix} 37 \\ 24 \\ 30 \\ 15 \end{bmatrix} = \begin{bmatrix} 1 & -2 & 1 & -1 \\ -\frac{3}{2} & \frac{7}{2} & -1 & \frac{1}{2} \\ \frac{5}{2} & -\frac{11}{2} & 1 & \frac{1}{2} \\ -\frac{11}{2} & \frac{23}{2} & -3 & \frac{3}{2} \end{bmatrix} \begin{bmatrix} 37 \\ 24 \\ 30 \\ 15 \end{bmatrix} = \begin{bmatrix} 4 \\ 6 \\ -2 \\ 5 \end{bmatrix}. ■$$

Glossary

- Determinant of a 2×2 matrix: The determinant δ of the matrix $\begin{bmatrix} a & b \\ c & d \end{bmatrix}$ is the *number* $\delta = ad - bc$. See Exercise 7, below.

- Multiplicative inverse for a square matrix: If $\mathbf{A}$ is an $n \times n$ matrix, then a multiplicative inverse for $\mathbf{A}$ is an $n \times n$ matrix $\mathbf{B}$ such that $\mathbf{AB} = \mathbf{I}_n$ and $\mathbf{BA} = \mathbf{I}_n$.

- Nonsingular matrix: A square matrix that has an inverse.

- Singular matrix: A square matrix that does not have an inverse.

Exercises for Section 6.1

1. In each part, determine whether the given matrix is singular or nonsingular. If it is nonsingular, calculate the inverse for the matrix.

   (a) $\mathbf{I}_3$

   (b) $\begin{bmatrix} -4 & 3 \\ 8 & -5 \end{bmatrix}$

   (c) $\begin{bmatrix} 6 & -21 \\ 4 & -14 \end{bmatrix}$

   (d) $\begin{bmatrix} 1 & 2 & 3 \\ 4 & 5 & 6 \\ 7 & 8 & 9 \end{bmatrix}$

   (e) $\begin{bmatrix} -5 & 1 & 1 \\ 5 & 11 & 4 \\ 11 & 39 & 15 \end{bmatrix}$

   (f) $\begin{bmatrix} 1 & 2 & 1 & -2 \\ 3 & 1 & 4 & 0 \\ 1 & 1 & 1 & -1 \\ 4 & 3 & 5 & -2 \end{bmatrix}$

2. In each part, determine whether the given matrix is singular or nonsingular. If it is nonsingular, calculate the inverse for the matrix.

   (a) $\begin{bmatrix} 25 & 15 \\ 10 & 6 \end{bmatrix}$

   (b) $\begin{bmatrix} 2 & 3 \\ 9 & 11 \end{bmatrix}$

   (c) $\begin{bmatrix} 6 & 0 & 0 \\ 0 & -4 & 0 \\ 0 & 0 & 0 \end{bmatrix}$

   (d) $\begin{bmatrix} 8 & 0 & 0 \\ 0 & -2 & 0 \\ 0 & 0 & 5 \end{bmatrix}$

   (e) $\begin{bmatrix} 4 & 2 & 1 \\ 3 & 3 & 1 \\ 9 & 10 & 3 \end{bmatrix}$

   (f) $\begin{bmatrix} 2 & 10 & 8 \\ 8 & 26 & 18 \\ 7 & 18 & 11 \end{bmatrix}$

   (g) $\begin{bmatrix} 2 & -10 & -4 & 0 \\ -6 & 10 & 5 & 4 \\ 3 & -10 & -6 & -2 \\ 2 & -6 & -3 & -1 \end{bmatrix}$

3. In each part, use the inverse of the matrix of coefficients to find the unique solution for the given system of linear equations.

   (a) $\begin{cases} 8x - 7y = 26 \\ -3x + 2y = -11 \end{cases}$

   (b) $\begin{cases} y - z = 7 \\ x + z = -2 \\ x + 2y = 6 \end{cases}$

   (c) $\begin{cases} 14x + 6y + z = 100 \\ 3x + 4y + z = 29 \\ 19x + 14y + 3z = 152 \end{cases}$

$$(d) \quad \begin{cases} 2x_1 - x_2 + x_3 + 5x_4 = 10 \\ 3x_1 - 4x_2 + 4x_3 - 5x_4 = 40 \\ x_1 - 2x_2 - x_4 = 10 \\ 4x_1 - 6x_2 + 5x_3 - 6x_4 = 53 \end{cases}$$

4. In each part, use the inverse of the matrix of coefficients to find the unique solution for the given system of linear equations.

$$(a) \quad \begin{cases} -6x + 4y = 46 \\ 5x - 2y = -29 \end{cases}$$

$$(b) \quad \begin{cases} 2x = 18 \\ 3x - 5y = 12 \\ 4x + 3y - 4z = 5 \end{cases}$$

$$(c) \quad \begin{cases} 4x - y + 2z = 37 \\ -15x + 4y - 6z = -133 \\ 3x + y + 2z = 33 \end{cases}$$

$$(d) \quad \begin{cases} x_1 + x_2 + 4x_3 = 67 \\ 2x_1 + 3x_2 + 10x_3 + x_4 = 221 \\ -x_1 + x_2 + 2x_3 - 2x_4 = 12 \\ 3x_1 + 9x_3 - 4x_4 = 85 \end{cases}$$

5. Suppose A is a nonzero $n \times n$ matrix whose rank is less than n.

   (a) What result would you obtain when using the method from this section to find the inverse for A? Why?

   (b) Explain why the homogeneous linear system $Ax = 0$ has a non-trivial solution s.

   (c) Suppose B is the $n \times n$ matrix for which every column is the vector s from part (b). Explain why $AB = O_{nn}$, making A a zero divisor. (Hint: Consider MSC1 from Section 1.8.)

6. Suppose A is an $n \times n$ matrix whose rank equals n.

   (a) What result would you obtain when using the method from this section to find the inverse for A? Why?

   (b) Explain why the ith column of A^{-1} is the unique solution for the linear system $Ax = e_i$. (Hint: Consider Theorem 3 and MSC3 from Section 1.8.)

7. Suppose $A = \begin{bmatrix} a & b \\ c & d \end{bmatrix}$ and $\delta = ad - bc$. (The number δ is called the determinant of A.) Let B be the matrix $\begin{bmatrix} d & -b \\ -c & a \end{bmatrix}$. Compute both AB and BA, writing your results in terms of the number δ – not using a, b, c, or d.

8. Let A, B, and δ be as described in Exercise 7.

(a) If $\delta \neq 0$, use your answer to Exercise 7 to prove that $\mathbf{A}$ is nonsingular, and that $\mathbf{A}^{-1} = \frac{1}{\delta}\mathbf{B}$.

(b) If $\delta = 0$, use your answer to Exercise 7 to prove that either $\mathbf{A} = \mathbf{O}_{22}$, or that $\mathbf{A}$ is a zero divisor. In either case, explain why $\mathbf{A}$ is singular.

9. Suppose $\mathbf{A} = \begin{bmatrix} 8 & 7 & 3 \\ 6 & 5 & 2 \end{bmatrix}$ and $\mathbf{B} = \begin{bmatrix} 1 & -1 \\ -4 & 5 \\ 7 & -9 \end{bmatrix}$.

(a) Compute $\mathbf{AB}$ and $\mathbf{BA}$.

(b) Explain why your results to part (a) do not contradict Theorem 1.

(c) Calculate the rank of $\mathbf{BA}$.

10. Suppose that $\mathbf{A}$ is an $m \times n$ matrix and $\mathbf{B}$ is an $n \times m$ matrix, and that $m < n$.

(a) Explain why rank $(\mathbf{A}) \leq m$ and rank $(\mathbf{B}) \leq m$.

(b) What are the sizes of $\mathbf{AB}$ and $\mathbf{BA}$?

(c) Although we will not prove it in this text, the rank of the product of two matrices cannot exceed the rank of either of the two matrices. (That is, rank $(\mathbf{CD}) \leq$ rank $(\mathbf{C})$ and rank $(\mathbf{CD}) \leq$ rank $(\mathbf{D})$.) Use this fact and part (a) to explain why rank $(\mathbf{AB}) \leq m$ and rank $(\mathbf{BA}) < n$.

(d) Use parts (b) and (c) to explain why it is possible that $\mathbf{AB} = \mathbf{I}_m$, but it is not possible that $\mathbf{BA} = \mathbf{I}_n$. (Exercise 9 gives a specific example of this situation.)

6.2   Properties of Inverse Matrices

In this section, we discuss some important properties of nonsingular matrices and their inverses.

Algebraic Properties of Inverse Matrices

There are several algebraic formulas involving nonsingular matrices and their inverses that can be useful when performing computations with, or simplifications of, algebraic expressions involving inverses. Four basic *inverse matrix properties* are given here:

Suppose $\mathbf{A}$ and $\mathbf{B}$ are nonsingular $n \times n$ matrices. Then,

IMP1: $\mathbf{A}^{-1}$ is nonsingular, and $\left(\mathbf{A}^{-1}\right)^{-1} = \mathbf{A}$.

IMP2: $(\mathbf{AB})$ is nonsingular and $(\mathbf{AB})^{-1} = \mathbf{B}^{-1}\mathbf{A}^{-1}$.

IMP3: $\mathbf{A}^T$ is nonsingular and $\left(\mathbf{A}^T\right)^{-1} = \left(\mathbf{A}^{-1}\right)^T$.

IMP4: If c is a nonzero scalar, then $c\mathbf{A}$ is nonsingular and $(c\mathbf{A})^{-1} = \frac{1}{c}\mathbf{A}^{-1}$.

The first property, IMP1, merely tells us that finding the inverse of an inverse takes us back to where we started. So, because

$$\begin{bmatrix} 8 & 11 \\ 3 & 4 \end{bmatrix}^{-1} = \begin{bmatrix} -4 & 11 \\ 3 & -8 \end{bmatrix}, \text{ we know that } \begin{bmatrix} -4 & 11 \\ 3 & -8 \end{bmatrix}^{-1} = \begin{bmatrix} 8 & 11 \\ 3 & 4 \end{bmatrix}.$$

This should be obvious. The fact that

$$\begin{bmatrix} 8 & 11 \\ 3 & 4 \end{bmatrix}^{-1} = \begin{bmatrix} -4 & 11 \\ 3 & -8 \end{bmatrix}$$

means that

$$\begin{bmatrix} 8 & 11 \\ 3 & 4 \end{bmatrix}\begin{bmatrix} -4 & 11 \\ 3 & -8 \end{bmatrix} = \mathbf{I}_2 \text{ and } \begin{bmatrix} -4 & 11 \\ 3 & -8 \end{bmatrix}\begin{bmatrix} 8 & 11 \\ 3 & 4 \end{bmatrix} = \mathbf{I}_2.$$

But saying that

$$\begin{bmatrix} -4 & 11 \\ 3 & -8 \end{bmatrix}^{-1} = \begin{bmatrix} 8 & 11 \\ 3 & 4 \end{bmatrix}$$

means the exact same thing! (Think about that!)

IMP2 shows us how to distribute the inverse operation over matrix multiplication. Be careful! This is similar to how the transpose operation distributes over matrix multiplication – it changes the order of the product. For example, if

$$\mathbf{A} = \begin{bmatrix} 1 & -2 & -3 \\ -4 & 9 & 13 \\ -1 & 1 & 3 \end{bmatrix} \text{ and } \mathbf{B} = \begin{bmatrix} 1 & -5 & 4 \\ 0 & 1 & -1 \\ 1 & -7 & 7 \end{bmatrix},$$

then using the row reduction method for finding inverses on our calculator (try it!), we obtain

$$\mathbf{A}^{-1} = \begin{bmatrix} 14 & 3 & 1 \\ -1 & 0 & -1 \\ 5 & 1 & 1 \end{bmatrix} \text{ and } \mathbf{B}^{-1} = \begin{bmatrix} 0 & 7 & 1 \\ -1 & 3 & 1 \\ -1 & 2 & 1 \end{bmatrix}.$$

Then, performing matrix multiplication, we get

$$\mathbf{AB} = \begin{bmatrix} -2 & 14 & -15 \\ 9 & -62 & 66 \\ 2 & -15 & 16 \end{bmatrix} \text{ and } \mathbf{B}^{-1}\mathbf{A}^{-1} = \begin{bmatrix} -2 & 1 & 6 \\ -12 & -2 & -3 \\ -11 & -2 & -2 \end{bmatrix}.$$

We can verify that $(\mathbf{AB})^{-1} = \mathbf{B}^{-1}\mathbf{A}^{-1}$ by simply multiplying these last two matrices together and checking that we get $\mathbf{I}_3$ (which we do). Or, you could row reduce

$$\left[\begin{array}{ccc|ccc} -2 & 14 & -15 & 1 & 0 & 0 \\ 9 & -62 & 66 & 0 & 1 & 0 \\ 2 & -15 & 16 & 0 & 0 & 1 \end{array}\right]$$

and see that the result is, indeed,

$$\left[\begin{array}{ccc|ccc} 1 & 0 & 0 & -2 & 1 & 6 \\ 0 & 1 & 0 & -12 & -2 & -3 \\ 0 & 0 & 1 & -11 & -2 & -2 \end{array}\right] \quad \text{(try it!)}.$$

IMP2 is easy to prove.

Proof of IMP2: To show that $(\mathbf{AB})^{-1} = \mathbf{B}^{-1}\mathbf{A}^{-1}$, we only need to multiply $(\mathbf{AB})$ by $(\mathbf{B}^{-1}\mathbf{A}^{-1})$. If the result of this matrix product is $\mathbf{I}_n$, then $(\mathbf{B}^{-1}\mathbf{A}^{-1})$ must be the inverse of $(\mathbf{AB})$, and $(\mathbf{AB})$ is nonsingular. So,

$$\begin{aligned} (\mathbf{AB})\left(\mathbf{B}^{-1}\mathbf{A}^{-1}\right) &= \mathbf{A}\left(\mathbf{B}\left(\mathbf{B}^{-1}\mathbf{A}^{-1}\right)\right) && \text{by MM2 (Section 1.7)} \\ &= \mathbf{A}\left(\left(\mathbf{BB}^{-1}\right)\mathbf{A}^{-1}\right) && \text{by MM2} \\ &= \mathbf{A}\left(\left(\mathbf{I}_n\right)\mathbf{A}^{-1}\right) && \text{because } \mathbf{BB}^{-1} = \mathbf{I}_n \\ &= \mathbf{AA}^{-1} = \mathbf{I}_n. \end{aligned}$$

We do *not* need to check that $\left(\mathbf{B}^{-1}\mathbf{A}^{-1}\right)(\mathbf{AB}) = \mathbf{I}_n$, because Theorem 1 assures us that it works out. ◆

IMP3 tells us that the inverse of the transpose is the transpose of the inverse. You will be asked to prove this in Exercise 15. For an example, consider

$$\mathbf{A} = \left[\begin{array}{cc} 2 & 1 \\ -4 & 3 \end{array}\right], \text{ whose transpose is } \mathbf{A}^T = \left[\begin{array}{cc} 2 & -4 \\ 1 & 3 \end{array}\right].$$

Using our row reduction method on a calculator shows us that

$$\mathbf{A}^{-1} = \left[\begin{array}{cc} \frac{3}{10} & -\frac{1}{10} \\ \frac{2}{5} & \frac{1}{5} \end{array}\right] \text{ and } \left(\mathbf{A}^T\right)^{-1} = \left[\begin{array}{cc} \frac{3}{10} & \frac{2}{5} \\ -\frac{1}{10} & \frac{1}{5} \end{array}\right].$$

You can easily see that the transpose of $\mathbf{A}^{-1}$ equals the computed inverse of $\mathbf{A}^T$.

Finally, IMP4 tells us that a nonzero scalar multiple of a nonsingular matrix is nonsingular. If you already know the inverse of $\mathbf{A}$, then the inverse of $c\mathbf{A}$ is just $\frac{1}{c}\mathbf{A}^{-1}$. So, if we use the matrix

$$\mathbf{A} = \left[\begin{array}{ccc} 1 & -2 & -3 \\ -4 & 9 & 13 \\ -1 & 1 & 3 \end{array}\right],$$

whose inverse we computed earlier, we see that the inverse of

$$3\mathbf{A} = \left[\begin{array}{ccc} 3 & -6 & -9 \\ -12 & 27 & 39 \\ -3 & 3 & 9 \end{array}\right] \text{ is } \frac{1}{3}\mathbf{A}^{-1} = \left[\begin{array}{ccc} \frac{14}{3} & 1 & \frac{1}{3} \\ -\frac{1}{3} & 0 & -\frac{1}{3} \\ \frac{5}{3} & \frac{1}{3} & \frac{1}{3} \end{array}\right],$$

where we have used the inverse for $\mathbf{A}$ that we previously calculated.

We ask you to prove IMP4 in Exercise 16.

Inverses of Special Matrices

In this section, we present a few theorems regarding the inverses of some special types of matrices – without proof.

Theorem 4. *If* $\mathbf{D}$ *is a diagonal* $n \times n$ *matrix having no zero entries on its main diagonal. Then* $\mathbf{D}$ *is nonsingular, and*

$$\mathbf{D}^{-1} = \begin{bmatrix} \frac{1}{d_{11}} & 0 & \cdots & 0 \\ 0 & \frac{1}{d_{22}} & \cdots & 0 \\ \vdots & \vdots & \ddots & \vdots \\ 0 & 0 & \cdots & \frac{1}{d_{nn}} \end{bmatrix}.$$

For example, if

$$\mathbf{D} = \begin{bmatrix} 4 & 0 \\ 0 & -5 \end{bmatrix}, \text{ then } \mathbf{D}^{-1} = \begin{bmatrix} \frac{1}{4} & 0 \\ 0 & -\frac{1}{5} \end{bmatrix}.$$

Theorem 5. *An upper triangular* $n \times n$ *matrix having no zero entries on its main diagonal is nonsingular and its inverse is also upper triangular. If an upper triangular matrix has at least one zero entry on its main diagonal, then it is singular.*

For example, if

$$\mathbf{U} = \begin{bmatrix} 2 & 1 & 3 \\ 0 & -1 & 7 \\ 0 & 0 & 5 \end{bmatrix}, \text{ then, by row reduction, } \mathbf{U}^{-1} = \begin{bmatrix} \frac{1}{2} & \frac{1}{2} & -1 \\ 0 & -1 & \frac{7}{5} \\ 0 & 0 & \frac{1}{5} \end{bmatrix},$$

which is upper triangular.

A result analogous to Theorem 5 is true for lower triangular matrices.

Theorem 6. *If* $\mathbf{S}$ *is a nonsingular symmetric matrix, then* $\mathbf{S}^{-1}$ *is also symmetric.*

For example, if

$$\mathbf{S} = \begin{bmatrix} 11 & -4 & -15 \\ -4 & 2 & 5 \\ -15 & 5 & 21 \end{bmatrix}, \text{ then, by row reduction, } \mathbf{S}^{-1} = \begin{bmatrix} 17 & 9 & 10 \\ 9 & 6 & 5 \\ 10 & 5 & 6 \end{bmatrix},$$

which is symmetric.

Be careful! A symmetric matrix *could* be singular! Try to think of an example of a nonzero singular 3×3 symmetric matrix. (Try a matrix having all zeros in the last row.)

Theorem 7. *Let* $\mathbf{K}$ *be an* $n \times n$ *skew-symmetric matrix. If* n *is odd, then* $\mathbf{K}$ *is singular. If* n *is even and* $\mathbf{K}$ *is nonsingular, then* $\mathbf{K}^{-1}$ *is also skew-symmetric.*

For $n = 3$, consider

$$\mathbf{K} = \begin{bmatrix} 0 & 1 & -2 \\ -1 & 0 & 3 \\ 2 & -3 & 0 \end{bmatrix}.$$

Now

$$\left[\begin{array}{ccc|ccc} 0 & 1 & -2 & 1 & 0 & 0 \\ -1 & 0 & 3 & 0 & 1 & 0 \\ 2 & -3 & 0 & 0 & 0 & 1 \end{array}\right] \text{ row reduces to } \left[\begin{array}{ccc|ccc} 1 & 0 & -3 & 0 & -1 & 0 \\ 0 & 1 & -2 & 0 & -\frac{2}{3} & -\frac{1}{3} \\ 0 & 0 & 0 & 1 & \frac{2}{3} & \frac{1}{3} \end{array}\right],$$

which implies that $\mathbf{K}$ is singular, as expected.

For $n = 2$, if

$$\mathbf{K} = \begin{bmatrix} 0 & 4 \\ -4 & 0 \end{bmatrix}, \text{ then, by row reduction, } \mathbf{K}^{-1} = \begin{bmatrix} 0 & -\frac{1}{4} \\ \frac{1}{4} & 0 \end{bmatrix},$$

which is skew-symmetric.

Be careful! An $n \times n$ skew-symmetric matrix with n even *could* be singular! Can you think of an example of a nonzero 4×4 skew-symmetric matrix that is singular?

Negative Powers of a Nonsingular Matrix

If $\mathbf{A}$ is a nonsingular matrix, since it is square, we learned in Section 1.7 that we can raise $\mathbf{A}$ to nonnegative integer powers. However, using the inverse of $\mathbf{A}$, we can also raise $\mathbf{A}$ to *negative* integer powers. In particular,

If $\mathbf{A}$ is a nonsingular matrix, and k is a positive integer, then $\mathbf{A}^{-k}$ (a negative power of $\mathbf{A}$) is defined to equal $\left(\mathbf{A}^{-1}\right)^{k}$; that is, the inverse of $\mathbf{A}$ raised to the kth power.

This means that the symbol "$\mathbf{A}^{-1}$" can be thought of two different ways – as meaning "the inverse of $\mathbf{A}$," or as "$\mathbf{A}$ raised to the -1 power." With either interpretation, it means the same thing.

So, for example, if

$$\mathbf{A} = \begin{bmatrix} 6 & 9 & 1 \\ 3 & 5 & 1 \\ 2 & 3 & 1 \end{bmatrix}, \text{ then, by row reduction, } \mathbf{A}^{-1} = \begin{bmatrix} 1 & -3 & 2 \\ -\frac{1}{2} & 2 & -\frac{3}{2} \\ -\frac{1}{2} & 0 & \frac{3}{2} \end{bmatrix}.$$

So, $\mathbf{A}^{-1}$ is both the inverse of $\mathbf{A}$ and $\mathbf{A}$ raised to the -1 power. To find "higher" negative powers of $\mathbf{A}$, we just raise $\mathbf{A}^{-1}$ to the appropriate positive power. For example,

$$\mathbf{A}^{-3} = \left(\mathbf{A}^{-1}\right)^3 = \begin{bmatrix} 1 & -3 & 2 \\ -\frac{1}{2} & 2 & -\frac{3}{2} \\ -\frac{1}{2} & 0 & \frac{3}{2} \end{bmatrix}^3 = \begin{bmatrix} \frac{5}{4} & -\frac{45}{2} & \frac{123}{4} \\ -\frac{3}{8} & \frac{53}{4} & -\frac{153}{8} \\ -\frac{21}{8} & \frac{27}{4} & -\frac{23}{8} \end{bmatrix}.$$

The rules for exponents that we learned in Section 1.7 for positive powers of a square matrix are also true for negative powers of a nonsingular matrix. That is:

If $\mathbf{A}$ is a nonsingular matrix and k and l are any integers, and c is a nonzero scalar, then

RFE1: $\mathbf{A}^{k+l} = \mathbf{A}^k \mathbf{A}^l$

RFE2: $\left(\mathbf{A}^k\right)^l = \mathbf{A}^{kl}$

RFE3: $\left(\mathbf{A}^T\right)^k = \left(\mathbf{A}^k\right)^T$

RFE4: $(c\mathbf{A})^k = c^k \mathbf{A}^k$.

We did not change the names of these rules from those used in Section 1.7, because the rules are essentially the same as before, except that, when $\mathbf{A}$ is nonsingular, you may use negative values for k and l. Also notice that the scalar c in RFE4 must be nonzero when negative powers are involved, because $(c\mathbf{A})$ is singular when $c = 0$.

In Section 1.7, we mentioned that RFE1 implies that a square matrix $\mathbf{A}$ commutes with every positive integer power of $\mathbf{A}$. With the revised RFE1 for nonsingular matrices, we can see that every integer power of a nonsingular matrix $\mathbf{A}$ commutes with every other integer power of $\mathbf{A}$. In Exercises 13 through 16 of Section 1.8 we extended this result regarding commuting matrices to polynomials of $\mathbf{A}$. Now, with our revised version of RFE1, we can generalize those results to using all integer powers for a nonsingular matrix. In particular,

CPM: If $\mathbf{A}$ is a nonsingular matrix, and if $\mathbf{B}$ and $\mathbf{C}$ are two linear combinations of integer powers of $\mathbf{A}$, then $\mathbf{BC} = \mathbf{CB}$.

Example 1. Suppose

$$\mathbf{A} = \begin{bmatrix} 7 & -9 \\ -6 & 8 \end{bmatrix}, \text{ then } \mathbf{A}^{-1} = \begin{bmatrix} 4 & \frac{9}{2} \\ 3 & \frac{7}{2} \end{bmatrix}.$$

Also, let

$$\mathbf{B} = 4\mathbf{A}^{-2} - 6\mathbf{A}^{-1} + 3\mathbf{A}^0 - 5\mathbf{A} + 2\mathbf{A}^3$$

$$= 4\begin{bmatrix} 4 & \frac{9}{2} \\ 3 & \frac{7}{2} \end{bmatrix}^2 - 6\begin{bmatrix} 4 & \frac{9}{2} \\ 3 & \frac{7}{2} \end{bmatrix} + 3\begin{bmatrix} 1 & 0 \\ 0 & 1 \end{bmatrix} - 5\begin{bmatrix} 7 & -9 \\ -6 & 8 \end{bmatrix} + 2\begin{bmatrix} 7 & -9 \\ -6 & 8 \end{bmatrix}^3$$

$$= \begin{bmatrix} 3124 & -3861 \\ -2574 & 3553 \end{bmatrix},$$

and let

$$\mathbf{C} = 8\mathbf{A}^{-3} - 10\mathbf{A}^{-1} + 3\mathbf{A}^2 = \begin{bmatrix} 2023 & 1557 \\ 1038 & 1850 \end{bmatrix}.$$

So $\mathbf{B}$ and $\mathbf{C}$ are both linear combinations of integer powers of $\mathbf{A}$. Therefore, $\mathbf{B}$ must commute with $\mathbf{C}$. We can easily check that both $\mathbf{BC}$ and $\mathbf{CB}$ equal

$$\begin{bmatrix} 2312134 & -2278782 \\ -1519188 & 2565332 \end{bmatrix}. \quad \blacksquare$$

Glossary

- Negative power of a nonsingular matrix: If $\mathbf{A}$ is a nonsingular matrix, and k is a positive integer, then $\mathbf{A}^{-k}$ (a negative power of $\mathbf{A}$) is defined to equal $\left(\mathbf{A}^{-1}\right)^k$; that is, the inverse of $\mathbf{A}$ raised to the kth power.

Exercises for Section 6.2

1. Suppose $\mathbf{A} = \begin{bmatrix} 1 & 3 \\ -3 & -5 \end{bmatrix}$ and $(\mathbf{AB})^{-1} = \begin{bmatrix} 9 & 4 \\ -\frac{11}{4} & -\frac{5}{4} \end{bmatrix}$. Calculate $\mathbf{B}^{-1}$.
   (Hint: Use IMP2.)

2. Suppose $\mathbf{B} = \begin{bmatrix} -1 & 5 & 1 \\ -2 & 2 & 1 \\ -2 & 1 & 1 \end{bmatrix}$ and $(\mathbf{AB})^{-1} = \begin{bmatrix} 4 & -18 & 35 \\ -1 & 4 & -8 \\ 10 & -43 & 84 \end{bmatrix}$. Calculate $\mathbf{A}^{-1}$. Then compute $\mathbf{A}$.

3. The inverse of $\mathbf{A} = \begin{bmatrix} 13 & 10 & 3 \\ 8 & 2 & 1 \\ -2 & -4 & -1 \end{bmatrix}$ is $\mathbf{B} = \begin{bmatrix} 1 & -1 & 2 \\ 3 & -\frac{7}{2} & \frac{11}{2} \\ -14 & 16 & -27 \end{bmatrix}$.

   (a) Compute $\left(\mathbf{A}^T\right)^{-1}$ without using a calculator.
   (b) Calculate the inverse of the matrix $(2\mathbf{A})$ without using a calculator.

4. The inverse of $\mathbf{A} = \begin{bmatrix} 3 & 9 & -2 \\ 1 & 5 & -1 \\ 1 & 2 & 0 \end{bmatrix}$ is the matrix $\mathbf{B} = \begin{bmatrix} \frac{2}{3} & -\frac{4}{3} & \frac{1}{3} \\ -\frac{1}{3} & \frac{2}{3} & \frac{1}{3} \\ -1 & 1 & 2 \end{bmatrix}$.

   (a) Compute $\left(\mathbf{A}^T\right)^{-1}$ without using a calculator.
   (b) Calculate the inverse of the matrix $(-\frac{1}{3}\mathbf{B})$ without using a calculator.

5. (a) Find the inverse for the 2×2 skew-symmetric matrix $\begin{bmatrix} 0 & 1 \\ -1 & 0 \end{bmatrix}$.
   (b) Use your answer to part (a) and IMP4 to find a formula for the inverse to the 2×2 skew-symmetric matrix $\begin{bmatrix} 0 & a \\ -a & 0 \end{bmatrix}$ when $a \neq 0$.

6. (a) Find the inverse for the 2×2 symmetric matrix $\begin{bmatrix} 0 & 1 \\ 1 & 0 \end{bmatrix}$.

(b) Use your answer to part (a) and IMP4 to find a formula for the inverse to the 2×2 symmetric matrix $\begin{bmatrix} 0 & a \\ a & 0 \end{bmatrix}$ when $a \neq 0$.

7. State the theorem for lower triangular matrices that is analogous to Theorem 5.

8. Without doing any computation, explain why the matrix product

$$\left(\begin{bmatrix} 1 & 2 & 3 \\ 0 & 4 & 5 \\ 0 & 0 & 6 \end{bmatrix} \begin{bmatrix} 7 & 0 & 0 \\ 8 & 9 & 0 \\ 10 & 11 & 12 \end{bmatrix} \right)$$

is nonsingular. (Hint: Use Theorem 5, Exercise 7, and IMP2.)

9. Do not use a calculator or row reduction in this problem. In each part, determine whether the given matrix is singular or whether it is nonsingular. Give a reason for your answer. If the matrix is nonsingular, give its inverse. You might want to consider Exercises 5 and 6 for some parts of this exercise.

(a) $\begin{bmatrix} 0 & -3 \\ 3 & 0 \end{bmatrix}$

(b) $\begin{bmatrix} 4 & 0 \\ 0 & -5 \end{bmatrix}$

(c) $\begin{bmatrix} 0 & 6 & 7 \\ -6 & 0 & -5 \\ -7 & 5 & 0 \end{bmatrix}$

(d) $\begin{bmatrix} 4 & 3 & 2 & 5 \\ 0 & 0 & 6 & 9 \\ 0 & 0 & 5 & 8 \\ 0 & 0 & 0 & 7 \end{bmatrix}$

10. Do not use a calculator or row reduction in this problem. In each part, determine whether the given matrix is singular or whether it is nonsingular. Give a reason for your answer. If the matrix is nonsingular, give its inverse. You might want to consider Exercises 5 and 6 for some parts of this exercise.

(a) $\begin{bmatrix} -2 & 0 \\ 0 & 9 \end{bmatrix}$

(c) $\begin{bmatrix} 0 & 7 \\ 7 & 0 \end{bmatrix}$

(b) $\begin{bmatrix} 3 & 0 & 0 & 0 \\ 6 & 2 & 0 & 0 \\ 7 & 9 & 0 & 0 \\ 4 & 5 & 1 & 8 \end{bmatrix}$

(d) $\begin{bmatrix} 0 & 3 & -4 & 2 & -1 \\ -3 & 0 & 9 & -6 & 8 \\ 4 & -9 & 0 & -7 & -2 \\ -2 & 6 & 7 & 0 & -5 \\ 1 & -8 & 2 & 5 & 0 \end{bmatrix}$

11. Give an example of a nonzero 3×3 symmetric matrix that is singular.

12. Give an example of a nonzero 4×4 skew-symmetric matrix that is singular.

13. Suppose $\mathbf{A} = \begin{bmatrix} 2 & 4 \\ 1 & 3 \end{bmatrix}$ and $\mathbf{B} = \begin{bmatrix} 0 & 1 & 1 \\ -4 & 1 & 2 \\ -3 & 0 & 1 \end{bmatrix}$.

(a) Compute $\mathbf{A}^3$, $\mathbf{A}^{-2}$, $\mathbf{B}^{-3}$, and $\mathbf{B}^2$.

(b) Using only matrix multiplication and your answers to part (a), recalculate $\mathbf{A}^{-1}$ and $\mathbf{B}^{-1}$.

(c) Use your answers to part (a) to determine $\left(\frac{1}{4}\mathbf{A}\right)^{-4}$ and $\left(2\mathbf{B}^T\right)^{-4}$.

(d) Using your answers to previous parts, calculate $\mathbf{C} = 5\mathbf{A}^3 - \left(\frac{1}{4}\mathbf{A}\right)^{-4}$ and $\mathbf{D} = 4\mathbf{A}^{-2} - 10\mathbf{A}^{-1} + 5\mathbf{A}$. Then compute both $\mathbf{CD}$ and $\mathbf{DC}$ to show that $\mathbf{C}$ and $\mathbf{D}$ commute, as promised by CPM.

14. Suppose $\mathbf{A}$ and $\mathbf{B}$ are 3×3 nonsingular matrices such that
$$\mathbf{A}^5 = \begin{bmatrix} 17 & 75 & 49 \\ 26 & 115 & 75 \\ 23 & 101 & 66 \end{bmatrix}, \mathbf{A}^{-3} = \begin{bmatrix} -4 & 0 & 3 \\ -3 & 2 & 0 \\ 6 & -3 & -1 \end{bmatrix}, \text{ and }$$
$$\mathbf{A}^T\mathbf{B}^T = \begin{bmatrix} 0 & 1 & 1 \\ 3 & 5 & 3 \\ 2 & 4 & 3 \end{bmatrix}.$$

(a) Find $\mathbf{A}$, $\mathbf{A}^{-1}$ and $\mathbf{B}$, using only matrix multiplication and the transpose operation.

(b) Compute $\left(\frac{1}{3}\mathbf{A}\right)^{-2}$.

(c) Calculate $\mathbf{C} = \mathbf{A}^5 - 25\mathbf{A} + 12\mathbf{A}^{-3}$ and $\mathbf{D} = \left(\frac{1}{3}\mathbf{A}\right)^{-2} - 9\mathbf{I}_3 + 4\mathbf{A}^{-1}$. Then compute both $\mathbf{CD}$ and $\mathbf{DC}$ to show that $\mathbf{C}$ and $\mathbf{D}$ commute, as promised by CPM.

15. Prove IMP3. You can do this by showing that performing the multiplication $\left(\mathbf{A}^T\right)\left(\left(\mathbf{A}^{-1}\right)^T\right)$ results in $\mathbf{I}_n$, which implies that $\left(\left(\mathbf{A}^{-1}\right)^T\right)$ is, indeed, the inverse for $\left(\mathbf{A}^T\right)$.

16. Prove IMP4. You can do this by showing that performing the multiplication $(c\mathbf{A})\left(\frac{1}{c}\mathbf{A}^{-1}\right)$ results in $\mathbf{I}_n$, which implies that $\left(\frac{1}{c}\mathbf{A}^{-1}\right)$ is, indeed, the inverse for $(c\mathbf{A})$.

6.3 Orthogonal Matrices and Mutually Orthogonal Sets

In this section, we will use properties of matrix inverses to further explore sets of mutually orthogonal vectors.

Orthogonal Matrices

An $n \times n$ matrix $\mathbf{A}$ is called an **orthogonal matrix** if the rows of $\mathbf{A}$ form an *orthonormal* set of vectors. Recall that an orthonormal set of vectors is a mutually orthogonal set of *unit* vectors. Because the rows of an orthogonal matrix form an orthonormal set of vectors, not just a mutually orthogonal set of vectors, you might think it it is more reasonable to call such a matrix an orthonormal matrix. But, no. Unfortunately, that is not the standard mathematical term used, and so we are stuck with calling such a matrix an orthogonal matrix. The following matrices are all orthogonal matrices:

$$\mathbf{I}_n, \quad \begin{bmatrix} \frac{3}{5} & \frac{4}{5} \\ -\frac{4}{5} & \frac{3}{5} \end{bmatrix}, \quad \begin{bmatrix} \frac{3}{7} & -\frac{2}{7} & \frac{6}{7} \\ \frac{6}{7} & \frac{3}{7} & -\frac{2}{7} \\ -\frac{2}{7} & \frac{6}{7} & \frac{3}{7} \end{bmatrix}, \quad \begin{bmatrix} \frac{1}{2} & -\frac{1}{2} & \frac{1}{2} & \frac{1}{2} \\ \frac{1}{2} & \frac{1}{2} & -\frac{1}{2} & \frac{1}{2} \\ \frac{1}{2} & \frac{1}{2} & \frac{1}{2} & -\frac{1}{2} \\ -\frac{1}{2} & \frac{1}{2} & \frac{1}{2} & \frac{1}{2} \end{bmatrix}.$$

Of course, there are many more examples of orthogonal matrices. In fact, to find an orthogonal matrix, you can take any nonsingular matrix, apply the Gram-Schmidt Process to its rows, normalize each of the resulting vectors to form an orthonormal set, and then use those vectors as the rows of a matrix.[2] This will give you an orthogonal matrix. However, the entries will typically be very "ugly" numbers, as you will see in the next example.

Example 1. Let us start with the nonsingular matrix

$$\mathbf{A} = \begin{bmatrix} 3 & 5 & 1 \\ 14 & 8 & 7 \\ 2 & 1 & 1 \end{bmatrix}.$$

We will perform the Gram-Schmidt Process using the rows of $\mathbf{A}$ as our starting vectors. So $\mathbf{v}_1 = [3, 5, 1]$, $\mathbf{v}_2 = [14, 8, 7]$, and $\mathbf{v}_3 = [2, 1, 1]$.

For our mutually orthogonal set, we start with $\mathbf{x}_1 = \mathbf{v}_1 = [3, 5, 1]$. Then,

$$\mathbf{x}_2 = \mathbf{v}_2 - \mathbf{proj}_{\mathbf{x}_1} \mathbf{v}_2 = \left[\frac{223}{35}, -\frac{165}{35}, \frac{156}{35} \right].$$

We adjust $\mathbf{x}_2$ by multiplying by 35 to get the new $\mathbf{x}_2 = [223, -165, 156]$. (Ugly – and it gets worse!)

Next,

$$\mathbf{x}_3 = \mathbf{v}_3 - \mathbf{proj}_{\mathbf{x}_1} \mathbf{v}_3 - \mathbf{proj}_{\mathbf{x}_2} \mathbf{v}_3 = \left[\frac{27}{2894}, -\frac{7}{2894}, -\frac{46}{2894} \right].$$

We adjust $\mathbf{x}_3$ by multiplying by 2894 to get the new $\mathbf{x}_3 = [27, -7, -46]$.

This gives us the mutually orthogonal set

$$\{[3, 5, 1], \ [223, -165, 156], \ [27, -7, -46]\}.$$

Finally, we normalize each of these vectors and use them as rows to form the orthogonal matrix

$$\begin{bmatrix} \frac{3}{\sqrt{35}} & \frac{5}{\sqrt{35}} & \frac{1}{\sqrt{35}} \\ \frac{223}{\sqrt{101290}} & -\frac{165}{\sqrt{101290}} & \frac{156}{\sqrt{101290}} \\ \frac{27}{\sqrt{2894}} & -\frac{7}{\sqrt{2894}} & \frac{46}{\sqrt{2894}} \end{bmatrix} \approx \begin{bmatrix} .5071 & .8452 & .1690 \\ .7007 & -.5184 & .4902 \\ .5019 & -.1301 & -.8551 \end{bmatrix}.$$

[2]Starting with a nonsingular matrix assures us that the Gram-Schmidt process will produce n *nonzero* vectors. If we start with a singular matrix instead, some of the resulting vectors will be zero vectors. Hence, we will not have enough nonzero vectors to create an orthogonal matrix. We will not prove this fact in this textbook.

As you can see, the resulting orthogonal matrix is very ugly, as we had antic-
ipated. ∎

The most important result we need regarding orthogonal matrices is given
in our next theorem:

Theorem 8. *If* $\mathbf{A}$ *is an orthogonal* $n \times n$ *matrix, then* $\mathbf{A}$ *is nonsingular and*
$\mathbf{A}^{-1} = \mathbf{A}^T$.

We present a simple proof of this theorem:

Proof of Theorem 8:

We will show that $\mathbf{A}\mathbf{A}^T = \mathbf{I}_n$. Doing so will prove that $\mathbf{A}^T$ is the inverse
for $\mathbf{A}$.

Now, since $\mathbf{A}$ is an $n \times n$ matrix, then $\mathbf{A}^T$ is also $n \times n$, and so $\mathbf{A}\mathbf{A}^T$ is
$n \times n$. Hence, the product $\mathbf{A}\mathbf{A}^T$ is the correct size to equal $\mathbf{I}_n$.

Let $\mathbf{v}_i$ be the ith row of $\mathbf{A}$, thought of as a vector. Then, by the definition
of an orthogonal matrix, $\{\mathbf{v}_1, \ldots, \mathbf{v}_n\}$ is an orthonormal set of vectors.

Now, the (i, j) entry of $\mathbf{A}\mathbf{A}^T$ equals the dot product of the ith row of $\mathbf{A}$
and the jth column of $\mathbf{A}^T$. (See Section 3.3.) But this is just $\mathbf{v}_i \cdot \mathbf{v}_j$. Thus,
the (i, j) entry of $\mathbf{A}\mathbf{A}^T$ equals $\mathbf{v}_i \cdot \mathbf{v}_j$.

Because $\{\mathbf{v}_1, \ldots, \mathbf{v}_n\}$ is an orthonormal set of vectors, $\mathbf{v}_i \cdot \mathbf{v}_j = 0$ whenever
$i \neq j$. Therefore, the (i, j) entry of $\mathbf{A}\mathbf{A}^T$ is zero whenever $i \neq j$, and so
all of the entries off the main diagonal of $\mathbf{A}\mathbf{A}^T$ are zero. Thus, $\mathbf{A}\mathbf{A}^T$ is a
diagonal matrix. But, because $\{\mathbf{v}_1, \ldots, \mathbf{v}_n\}$ is an orthonormal set of vectors,
$\mathbf{v}_i \cdot \mathbf{v}_i = \|\mathbf{v}_i\|^2 = 1$. Hence, the (i, i) entry of $\mathbf{A}\mathbf{A}^T$ is 1, and so all of the entries
of $\mathbf{A}\mathbf{A}^T$ *on* the main diagonal equal 1. We conclude, then, that $\mathbf{A}\mathbf{A}^T = \mathbf{I}_n$.
(Note: This proof is similar to the answer for Exercise 11 in Section 4.6.) ♦

It is relatively easy to follow the proof of Theorem 8 backwards to see that
its converse is also true; that is, *if* $\mathbf{A}^{-1} = \mathbf{A}^T$, *then* $\mathbf{A}$ *is an orthogonal matrix*
– its rows form an orthonormal set of vectors. Using this, we can combine the
proof of Theorem 8 with Theorem 1 and properties of the transpose operation
to prove that $\mathbf{A}^T\mathbf{A} = \mathbf{I}_n$ as well, and show that $\mathbf{A}^T$ is also an orthogonal
matrix. This means that the *columns* of an orthogonal matrix also form an
orthonormal set of vectors! (See Exercises 9 through 12.)

You should test each of the orthogonal matrices we have seen as examples
and verify for yourself that $\mathbf{A}\mathbf{A}^T = \mathbf{I}_n$, $\mathbf{A}^T\mathbf{A} = \mathbf{I}_n$, and that the columns of
those matrices form orthonormal sets of vectors.

Mutually Orthogonal Sets of Vectors

Our next goal is to use Theorem 8 to prove our final result of the chapter – that
it is impossible to have a mutually orthogonal set of nonzero n-dimensional
vectors that contains more than n vectors.

Theorem 9. *(Size of a Mutually Orthogonal Set of Vectors)*
(SMOSV) A set of mutually orthogonal n*-dimensional nonzero vectors cannot contain more than* n *vectors.*

Proof: We will use a proof by contradiction to prove the SMOSV Theorem. That is, we will start by assuming that this theorem is false and show that this assumption leads to a contradiction.

Since we are assuming the theorem is false, we can assume that *there is* a set of mutually orthogonal n-dimensional nonzero vectors that contains more than n vectors. So, suppose that

$$S = \{\mathbf{v}_1, \mathbf{v}_2, \ldots, \mathbf{v}_m\}$$

is a set of mutually orthogonal n-dimensional nonzero vectors and that $m > n$.

Now, since all of the vectors in the set S are nonzero, they can all be normalized to create a set of unit vectors $T = \{\mathbf{u}_1, \mathbf{u}_2, \ldots, \mathbf{u}_m\}$, where $\mathbf{u}_i = \left(\frac{1}{\|\mathbf{v}_i\|}\right)\mathbf{v}_i$ for each i. The set T is actually an orthonormal set containing m vectors, because all of the vectors in the set are unit vectors, and, when $i \neq j$,

$$\mathbf{u}_i \cdot \mathbf{u}_j = \left(\left(\frac{1}{\|\mathbf{v}_i\|}\right)\mathbf{v}_i\right) \cdot \left(\left(\frac{1}{\|\mathbf{v}_j\|}\right)\mathbf{v}_j\right)$$

$$= \left(\left(\frac{1}{\|\mathbf{v}_i\|}\right)\left(\frac{1}{\|\mathbf{v}_j\|}\right)\right)(\mathbf{v}_i \cdot \mathbf{v}_j)$$

$$= \left(\left(\frac{1}{\|\mathbf{v}_i\|}\right)\left(\frac{1}{\|\mathbf{v}_j\|}\right)\right)(0) = 0.$$

Next, consider the matrix $\mathbf{A}$ whose rows are the vectors $\mathbf{u}_1, \ldots, \mathbf{u}_n$, the first n vectors in T. Because T is an orthonormal set of vectors, so is $\{\mathbf{u}_1, \ldots, \mathbf{u}_n\}$, which just contains some, but not all, of the vectors in T. Therefore, $\mathbf{A}$ is an orthogonal matrix.

Now, in Section 4.6, we saw that the solutions to a homogeneous linear system $\mathbf{Ax} = \mathbf{0}$ are the vectors that are orthogonal to all of the rows of $\mathbf{A}$. So, we consider the homogeneous linear system $\mathbf{Ax} = \mathbf{0}$ using the orthogonal matrix $\mathbf{A}$ that we have created. Any vector that is orthogonal to all of the rows of $\mathbf{A}$ is a solution to this homogeneous linear system. Therefore, the vectors $\mathbf{u}_{n+1}, \ldots, \mathbf{u}_m$ are all solutions to this system, because they are all orthogonal to $\mathbf{u}_1, \ldots, \mathbf{u}_n$.

Now, by Theorem 8, because $\mathbf{A}$ is an orthogonal matrix, $\mathbf{A}$ is nonsingular, with inverse $\mathbf{A}^T$. So, by Theorem 3 in Section 6.1, the linear system $\mathbf{Ax} = \mathbf{0}$ has a *unique* solution, namely $\mathbf{A}^{-1}\mathbf{0} = \mathbf{A}^T\mathbf{0} = \mathbf{0}$. Thus, the trivial solution is the only solution to the linear system $\mathbf{Ax} = \mathbf{0}$. However, we have already established that all of the vectors $\mathbf{u}_{n+1}, \ldots, \mathbf{u}_m$ are solutions to this system. They must be nontrivial solutions, because they are all unit vectors, and so they do not equal the zero vector. This is a contradiction, since we also proved that the trivial solution is the *only* solution. This contradiction means that we cannot have $m > n$; that is, we cannot have more than n vectors in a set of mutually orthogonal n-dimensional nonzero vectors. ♦

When the SMOSV Theorem is taken together with Theorem 4 from Section 4.6, we see that, in n-dimensional space, we can expand any set of mutually orthogonal nonzero vectors up to the point where we have n of them, but not beyond that point.[3] These results also combine nicely with the LCMOV Theorem from Section 3.6, which tells us that, for a set of mutually orthogonal nonzero vectors, no vector in the set can be expressed as a linear combination of the other vectors in the set. Back in Section 3.6, we described this as saying that, if we program the dials on a Vector Etch-a-Sketch with the nonzero vectors in a mutually orthogonal set, then each knob takes the stylus in a genuinely new direction. Theorem 4 in Section 4.6 says that if we have fewer than n dials on an n-dimensional Vector Etch-a-Sketch (using mutually orthogonal nonzero vectors), we can always add more dials, orthogonal to those we already have, and thus take the stylus off into new directions. But, by the SMOSV Theorem, once we have n dials, we cannot add any more. Essentially, there are no more new directions left to explore! Hence, what we have called n-dimensional space all along really does have exactly n dimensions, no matter how we try to set up mutually orthogonal axes!

Assignment 1: Go back and reread Sections 3.6 and 4.6 to refresh your memory and help you understand how they mesh with the SMOSV Theorem in this section.

Glossary

- **Orthogonal matrix**: A square matrix in which the rows of the matrix form an orthonormal set of vectors. Equivalently, a square matrix $\mathbf{A}$ is orthogonal if and only if $\mathbf{A}$ is nonsingular and $\mathbf{A}^{-1} = \mathbf{A}^T$.

Exercises for Section 6.3

1. Determine which of the following matrices are orthogonal matrices. For those that are, also verify that the columns of the matrix form an orthonormal set of vectors. For those that are not, explain why not.

   (a) $\mathbf{I}_3$

   (b) $\begin{bmatrix} 3 & -4 \\ 4 & 3 \end{bmatrix}$

   (c) $\begin{bmatrix} \frac{5}{13} & \frac{12}{13} \\ \frac{12}{13} & -\frac{5}{13} \end{bmatrix}$

   (d) $\begin{bmatrix} \frac{\sqrt{3}}{3} & \frac{\sqrt{3}}{3} & \frac{\sqrt{3}}{3} \\ \frac{\sqrt{14}}{14} & \frac{\sqrt{56}}{14} & -\frac{\sqrt{126}}{14} \\ -\frac{5}{\sqrt{42}} & \frac{4}{\sqrt{42}} & \frac{1}{\sqrt{42}} \end{bmatrix}$

   (e) $\begin{bmatrix} \frac{4}{\sqrt{69}} & \frac{2}{\sqrt{69}} & \frac{7}{\sqrt{69}} \\ \frac{3}{\sqrt{14}} & \frac{1}{\sqrt{14}} & -\frac{2}{\sqrt{14}} \\ \frac{5}{\sqrt{57}} & \frac{4}{\sqrt{57}} & -\frac{4}{\sqrt{57}} \end{bmatrix}$

2. Determine which of the following matrices are orthogonal matrices. For those that are, also verify that the columns of the matrix form an orthonormal set of vectors. For those that are not, explain why not.

---

[3] Although the zero vector is orthogonal to every vector, we can not include it in the set because we are considering only nonzero vectors here.

(a) $\begin{bmatrix} 0 & 1 \\ 1 & 0 \end{bmatrix}$

(b) $\begin{bmatrix} \frac{4}{\sqrt{41}} & \frac{5}{\sqrt{41}} \\ -\frac{5}{\sqrt{41}} & \frac{4}{\sqrt{41}} \end{bmatrix}$

(c) $\begin{bmatrix} \frac{2}{15} & \frac{11}{15} & \frac{2}{3} \\ \frac{14}{15} & \frac{2}{15} & -\frac{1}{3} \\ \frac{1}{3} & -\frac{2}{3} & \frac{2}{3} \end{bmatrix}$

(d) $\begin{bmatrix} \frac{4}{\sqrt{26}} & \frac{1}{\sqrt{26}} & -\frac{3}{\sqrt{26}} \\ \frac{5}{\sqrt{65}} & \frac{-2}{\sqrt{65}} & \frac{6}{\sqrt{65}} \\ \frac{2}{\sqrt{69}} & \frac{8}{\sqrt{69}} & \frac{1}{\sqrt{69}} \end{bmatrix}$

(e) $\begin{bmatrix} \frac{4\sqrt{18}}{18} & \frac{\sqrt{18}}{18} & -\frac{\sqrt{18}}{18} \\ \frac{\sqrt{11}}{11} & -\frac{3\sqrt{11}}{11} & \frac{\sqrt{11}}{11} \\ \frac{1}{5\sqrt{2}} & \frac{1}{2\sqrt{2}} & \frac{13}{10\sqrt{2}} \end{bmatrix}$

3. In each part, start with the given nonsingular matrix and perform the Gram-Schmidt Process on the rows of the matrix to find an orthonormal set of vectors. Use that orthonormal set to create an orthogonal matrix.

(a) $\begin{bmatrix} 4 & 7 \\ 3 & 9 \end{bmatrix}$

(b) $\begin{bmatrix} 10 & -45 & 30 \\ 24 & -31 & 6 \\ 6 & -49 & -15 \end{bmatrix}$

4. In each part, start with the given nonsingular matrix and perform the Gram-Schmidt Process on the rows of the matrix to find an orthonormal set of vectors. Use that orthonormal set to create an orthogonal matrix.

(a) $\begin{bmatrix} 3 & 7 \\ -4 & 2 \end{bmatrix}$

(b) $\begin{bmatrix} 36 & -27 & 24 \\ 36 & -61 & 58 \\ 87 & 7 & 58 \end{bmatrix}$

5. Explain why performing a type (III) row operation on an orthogonal matrix results in another orthogonal matrix.

6. In general, does performing either a type (I) or type (II) row operation on an orthogonal matrix result in another orthogonal matrix? Explain the reasoning behind your answer.

7. Suppose $\mathbf{A}$ is a matrix whose rows are the vectors in the mutually orthogonal set

$$\{[-28, 84, 19, 158], [7, 0, 2, 1], [15, 54, -42, -21], [3, -9, -14, 7]\}.$$

Compute $\mathbf{AA}^T$, and show that the result is a diagonal matrix for which the ith diagonal entry is the square of the norm of the ith vector in the set. Then compute $\mathbf{A}^T\mathbf{A}$ and show that the result is *not* a diagonal matrix.

8. Let $\mathbf{A}$ be the matrix from Exercise 7. Show that the columns of $\mathbf{A}$ do *not* form a mutually orthogonal set of vectors.

9. Suppose $\mathbf{A}$ is an orthogonal matrix. Use Theorem 1 and the proof of Theorem 8 to prove that $\mathbf{A}^T\mathbf{A} = \mathbf{I}_n$.

10. Suppose $\mathbf{A}$ is an orthogonal matrix. Prove that $\mathbf{A}^T$ is nonsingular and that $\left(\mathbf{A}^T\right)^{-1} = \mathbf{A}$. (Hint: See IMP3 in Section 6.2.)

11. Suppose $\mathbf{A}$ is an orthogonal matrix. Explain why the equation $\mathbf{A}^T\mathbf{A} = \mathbf{I}_n$ proved in Exercise 9 shows that the columns of $\mathbf{A}$ form an orthonormal set of vectors.

12. Suppose $\mathbf{A}$ is an orthogonal matrix. Explain why Exercise 11 shows that the rows of $\mathbf{A}^T$ form an orthonormal set of vectors, making $\mathbf{A}^T$ an orthogonal matrix.

6.4 Chapter 6 Test

1. In each part, determine whether the matrix is singular or nonsingular. If it is nonsingular, find its inverse. (Use your calculator, if needed.)

(a) $\begin{bmatrix} \frac{1}{3} & \frac{2}{3} & -\frac{2}{3} \\ -\frac{2}{3} & \frac{2}{3} & \frac{1}{3} \\ \frac{2}{3} & \frac{1}{3} & \frac{2}{3} \end{bmatrix}$ (b) $\begin{bmatrix} 0 & 9 & 2 \\ 14 & 17 & 5 \\ 4 & 3 & 1 \end{bmatrix}$ (c) $\begin{bmatrix} -2 & 4 & 2 \\ 3 & 4 & 7 \\ 1 & 3 & 4 \end{bmatrix}$

2. Use an inverse matrix to solve the following system of linear equations. (Use your calculator, if needed.)

$$\begin{cases} 4x_1 + 2x_2 + 9x_3 & = 47 \\ 3x_1 + 3x_2 + 5x_3 - 3x_4 = 18 \\ -3x_1 - 3x_2 \qquad\ + 7x_4 = 17 \\ 2x_1 + 2x_2 + 2x_3 - 3x_4 = 3 \end{cases}$$

3. Determine which of the following matrices are orthogonal matrices. For those that are, find the inverse of the matrix.

(a) $\begin{bmatrix} 3 & 4 & 12 \\ 4 & -12 & 3 \\ 12 & 3 & -4 \end{bmatrix}$

(b) $\begin{bmatrix} \frac{3}{13} & \frac{4}{13} & -\frac{12}{13} \\ \frac{8}{\sqrt{82}} & \frac{3}{\sqrt{82}} & \frac{3}{\sqrt{82}} \\ \frac{6}{\sqrt{182}} & -\frac{5}{\sqrt{182}} & -\frac{11}{\sqrt{182}} \end{bmatrix}$

(c) $\begin{bmatrix} \frac{1}{3} & 0 & \frac{2}{3} & -\frac{2}{3} \\ \frac{2}{3} & -\frac{2}{3} & 0 & \frac{1}{3} \\ 0 & \frac{1}{3} & \frac{2}{3} & \frac{2}{3} \\ \frac{2}{3} & \frac{2}{3} & -\frac{1}{3} & 0 \end{bmatrix}$

4. The inverse of $\mathbf{A} = \begin{bmatrix} 9 & 7 & 1 \\ 9 & 7 & 2 \\ 4 & 3 & 1 \end{bmatrix}$ is the matrix $\mathbf{B} = \begin{bmatrix} 1 & -4 & 7 \\ -1 & 5 & -9 \\ -1 & 1 & 0 \end{bmatrix}$.

(a) Compute $\left(\mathbf{A}^T\right)^{-1}$ without using a calculator.

(b) Calculate the inverse of the matrix $(-2\mathbf{B})$ without using a calculator.

5. Suppose $\mathbf{B} = \begin{bmatrix} 10 & 3 & 2 \\ 5 & 2 & 1 \\ 4 & 1 & 1 \end{bmatrix}$ and $(\mathbf{AB})^{-1} = \begin{bmatrix} -23 & -13 & -3 \\ 44 & 18 & 3 \\ 50 & 37 & 10 \end{bmatrix}$. Calculate $\mathbf{A}^{-1}$. Then compute $\mathbf{A}$.

6. Suppose $\mathbf{A}$ and $\mathbf{B}$ are 3×3 nonsingular matrices such that
$$\mathbf{A}^4 = \begin{bmatrix} 153 & 74 & 22 \\ 362 & 175 & 52 \\ 244 & 118 & 35 \end{bmatrix}, \mathbf{A}^{-3} = \begin{bmatrix} -4 & -1 & 4 \\ 8 & 0 & -5 \\ 1 & 7 & -11 \end{bmatrix},$$
and $\mathbf{B}^T\mathbf{A}^T = \begin{bmatrix} 8 & 19 & 13 \\ -1 & -1 & -3 \\ 0 & 6 & 7 \end{bmatrix}$.

   (a) Find $\mathbf{A}$, $\mathbf{A}^{-1}$ and $\mathbf{B}$, using only matrix multiplication and the transpose operation.

   (b) Compute $(2\mathbf{A})^{-2}$.

   (c) Calculate $\mathbf{C} = \mathbf{A}^4 - 5\mathbf{A} + 3\mathbf{I}_3 + 2\mathbf{A}^{-3}$ and $\mathbf{D} = 12(2\mathbf{A})^{-2} - 4\mathbf{I}_3 + 3\mathbf{A}^{-1}$. Then compute both $\mathbf{CD}$ and $\mathbf{DC}$ to show that $\mathbf{C}$ and $\mathbf{D}$ commute, as promised by CPM.

7. Suppose that $\mathbf{A}$ is a nonsingular symmetric matrix. Prove that $\mathbf{A}^{-1}$ is symmetric. (You are being asked to prove Theorem 6. Hint: You need to prove that $(\mathbf{A}^{-1})^T = \mathbf{A}^{-1}$. Use IMP3.)

8. Starting with the nonsingular matrix $\mathbf{A} = \begin{bmatrix} 1 & 3 & 0 \\ -4 & 13 & 5 \\ 5 & 10 & 13 \end{bmatrix}$, perform the Gram-Schmidt Process on the rows of $\mathbf{A}$ to find an orthonormal set of vectors. Use that orthonormal set to create an orthogonal matrix.

9. In each part, indicate whether the given statement is true or whether it is false.

   (a) If $\mathbf{A}$ is a nonsingular matrix, then the system of linear equations represented by $\mathbf{Ax} = \mathbf{b}$ has a unique solution.

   (b) If $\mathbf{A}$ is a nonsingular matrix, then the system of linear equations represented by $\mathbf{Ax} = \mathbf{b}$ has $\mathbf{x} = \mathbf{Ab}$ as a solution.

   (c) If $\mathbf{A}$ is a nonsingular matrix, then the homogeneous system of linear equations $\mathbf{Ax} = \mathbf{0}$ has a nontrivial solution.

   (d) If $\mathbf{A}$ and $\mathbf{B}$ are square matrices such that $\mathbf{AB} = \mathbf{I}$, then $\mathbf{AB} = \mathbf{BA}$.

   (e) If $\mathbf{A}$ and $\mathbf{B}$ are non-square matrices such that $\mathbf{AB} = \mathbf{I}$, then $\mathbf{BA} = \mathbf{I}$.

(f) If $\mathbf{A}$, $\mathbf{B}$ and $\mathbf{C}$ are matrices such that $\mathbf{AB} = \mathbf{I}$ and $\mathbf{CA} = \mathbf{I}$, then $\mathbf{B} = \mathbf{C}$.

(g) If $\mathbf{v}$ is the 3rd column of the inverse of a nonsingular matrix $\mathbf{A}$, then $\mathbf{Av} = \mathbf{e}_3$.

(h) If $\mathbf{A}$ and $\mathbf{B}$ are nonsingular matrices of the same size, then $\mathbf{AB}$ is also nonsingular and $(\mathbf{AB})^{-1} = \mathbf{A}^{-1}\mathbf{B}^{-1}$.

(i) If a square matrix $\mathbf{A}$ is a zero divisor, then $\mathbf{A}$ is a singular matrix.

(j) The determinant of the matrix $\begin{bmatrix} 4 & 5 \\ 3 & 6 \end{bmatrix}$ is the number 9.

(k) The rank of a nonsingular $n \times n$ matrix must equal n.

(l) If $\mathbf{A}$ and $\mathbf{B}$ are nonsingular matrices, then
$$\left((\mathbf{AB})^T\right)^{-1} = \left(\mathbf{B}^{-1}\right)^T \left(\mathbf{A}^{-1}\right)^T.$$

(m) The inverse of $\begin{bmatrix} 4 & 0 \\ 0 & -\frac{1}{4} \end{bmatrix}$ is $\begin{bmatrix} \frac{1}{4} & 0 \\ 0 & -4 \end{bmatrix}$.

(n) The inverse of $\begin{bmatrix} 0 & 4 \\ -\frac{1}{4} & 0 \end{bmatrix}$ is $\begin{bmatrix} 0 & -4 \\ \frac{1}{4} & 0 \end{bmatrix}$.

(o) The matrix $\begin{bmatrix} 0 & 5 & -2 \\ -5 & 0 & 8 \\ 2 & -8 & 0 \end{bmatrix}$ is nonsingular.

(p) The matrix product $\left(\begin{bmatrix} 2 & 8 & -7 \\ 0 & 4 & 6 \\ 0 & 0 & 17 \end{bmatrix} \begin{bmatrix} 19 & 0 & 0 \\ 4 & -5 & 0 \\ 3 & -1 & 14 \end{bmatrix} \right)$ is nonsingular.

(q) If $\mathbf{A}$ is a nonsingular matrix, then $\mathbf{A}^3\mathbf{A}^{-7}$ is the inverse of $\mathbf{A}^4$.

(r) If $\mathbf{A}$ is a nonsingular matrix, then $\mathbf{A}^4\mathbf{A}^{-3} = \mathbf{A}^6\mathbf{A}^{-2}$.

(s) The inverse of a nonsingular upper triangular matrix is lower triangular.

(t) A singular upper triangular matrix must have a zero on its main diagonal.

(u) A singular symmetric matrix must have a zero on its main diagonal.

(v) A singular skew-symmetric matrix must have a zero on its main diagonal.

(w) If the rows of a square matrix form a set of nonzero mutually orthogonal vectors, then the matrix is an orthogonal matrix.

(x) If $\mathbf{A}$ is a nonsingular matrix such that $\mathbf{A}^T = \mathbf{A}^{-1}$, then the columns of $\mathbf{A}$ form an orthonormal set of vectors.

(y) Performing a type (III) row operation on an orthogonal matrix results in another orthogonal matrix.

(**z**) If $\mathbf{A}$ and $\mathbf{B}$ are orthogonal $n \times n$ matrices, then $\mathbf{AB}$ is also an orthogonal matrix.

(**aa**) If $\mathbf{A}$ and $\mathbf{B}$ are orthogonal $n \times n$ matrices, then $\mathbf{A} + \mathbf{B}$ is also an orthogonal matrix.

(**bb**) Every orthonormal set of n-dimensional vectors contains exactly n vectors.

(**cc**) No orthonormal set of n-dimensional vectors contains exactly $n+1$ vectors.

(**dd**) If $\mathbf{A}$ is a nonsingular matrix, then $(2\mathbf{A})^{-1}\mathbf{A} = \frac{1}{2}$.

(**ee**) If $\mathbf{A}$ is a nonsingular matrix, then $\left(4\mathbf{A}^{-1}\right)^{T} = \left(\frac{1}{4}\mathbf{A}^{T}\right)^{-1}$.

Chapter 7

Geometric Transformations with Matrices

In this chapter, we examine how matrix multiplication can be used to produce rotations, reflections, and other geometric transformations of the plane.

7.1 Matrix Transformations of the Plane

In this section, we assume that all points and vectors are in a two-dimensional plane. To simplify matters, we also assume that all vectors in this section have their initial points at the origin. Therefore, the coordinates of the terminal points of the vectors will be the same as the coordinates of the vectors themselves. In so doing, we merge the concepts of vectors and points, sometimes thinking of vectors as points, and at other times, imagining points as vectors.

Images of Points and Figures Under Matrix Multiplication

If we multiply a 2×2 matrix by a 2-dimensional point (thought of as a vector), the result is another 2-dimensional point. For example,

$$\begin{bmatrix} 3 & -4 \\ 4 & 3 \end{bmatrix} \begin{bmatrix} -1 \\ 2 \end{bmatrix} = \begin{bmatrix} -11 \\ 2 \end{bmatrix}.$$

Thus, we can think of the matrix $\mathbf{A} = \begin{bmatrix} 3 & -4 \\ 4 & 3 \end{bmatrix}$ as moving the point $(-1, 2)$ to the point $(-11, 2)$ through matrix multiplication. We call the point $(-11, 2)$ the **image** of the point $(-1, 2)$ under multiplication by the matrix $\mathbf{A}$.

We can discover exactly how multiplication by the same matrix $\mathbf{A}$ moves several points by calculating their images.

298 Chapter 7: Geometric Transformations with Matrices

Example 1. Under multiplication by $\mathbf{A} = \begin{bmatrix} 3 & -4 \\ 4 & 3 \end{bmatrix}$, the points $(1,3)$, $(3,1)$, $(2,0)$, and $(1,1)$ have the following images:

$$\begin{bmatrix} 3 & -4 \\ 4 & 3 \end{bmatrix} \begin{bmatrix} 1 \\ 3 \end{bmatrix} = \begin{bmatrix} -9 \\ 13 \end{bmatrix}; \quad \begin{bmatrix} 3 & -4 \\ 4 & 3 \end{bmatrix} \begin{bmatrix} 3 \\ 1 \end{bmatrix} = \begin{bmatrix} 5 \\ 15 \end{bmatrix};$$

$$\begin{bmatrix} 3 & -4 \\ 4 & 3 \end{bmatrix} \begin{bmatrix} 2 \\ 0 \end{bmatrix} = \begin{bmatrix} 6 \\ 8 \end{bmatrix}; \quad \begin{bmatrix} 3 & -4 \\ 4 & 3 \end{bmatrix} \begin{bmatrix} 1 \\ 1 \end{bmatrix} = \begin{bmatrix} -1 \\ 7 \end{bmatrix}. \quad \blacksquare$$

When multiplying a matrix $\mathbf{A}$ by several different points, we can compute all of the multiplications simultaneously by first creating a matrix whose columns contain all of the points, and then multiplying $\mathbf{A}$ by this larger matrix. For example,

$$\begin{bmatrix} 3 & -4 \\ 4 & 3 \end{bmatrix} \begin{bmatrix} -1 & 1 & 3 & 2 & 1 \\ 2 & 3 & 1 & 0 & 1 \end{bmatrix} = \begin{bmatrix} -11 & -9 & 5 & 6 & -1 \\ 2 & 13 & 15 & 8 & 7 \end{bmatrix}$$

calculates all five images that we previously computed using just a single matrix multiplication. (This follows from property MSC1 from Section 1.8, which verifies that the same result is obtained from multiplying $\mathbf{A}$ by each column in turn of the second matrix in the product.)

However, we are not just interested in the images of a few points. Instead, we want to understand exactly how multiplication by $\mathbf{A}$ affects the movement of *all* of the points in the plane. That is, how does multiplying by $\mathbf{A}$ *transform* the plane? A **matrix transformation** of the plane is the movement of all of the points in the plane that results from a matrix multiplication.

We can get a better idea of how a particular matrix moves the points in the plane by considering a small set of points that make up some sort of figure, and then calculating the images of the points to obtain the image of the overall figure.

Example 2. Consider again the five points $(-1,2)$, $(1,3)$, $(3,1)$, $(2,0)$, $(1,1)$, and their respective images $(-11,2)$, $(-9,13)$, $(5,15)$, $(6,8)$, $(-1,7)$ under the matrix $\mathbf{A} = \begin{bmatrix} 3 & -4 \\ 4 & 3 \end{bmatrix}$. Connect the original points with line segments in the given order to form the smaller five-sided polygon shown in Figure 7.1. We then connect the corresponding images of the five points in the same order to construct the image of the polygon. This is the larger polygon in Figure 7.1. $\blacksquare$

The technique shown in the last example works because the image of a line segment under a matrix transformation is always another line segment. Therefore, we can determine the image of an entire line segment under a matrix transformation just by finding the images of the segment's two endpoints, as we did in Figure 7.1. In Exercise 9 you will show that multiplication by the matrix $\mathbf{A}$ has taken the original polygon, rotated it about 53° counterclockwise around the origin, and then enlarged it to 5 times its size by moving all of

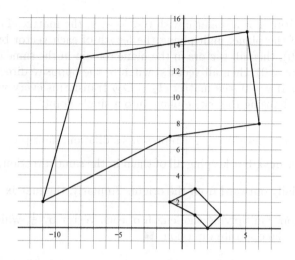

Figure 7.1: Figure Formed by Five Points and Its Image Using **A**

the points in the polygon 5 times as far away from the origin as they originally were. In fact, this is precisely the transformation that multiplication by this matrix **A** accomplishes on the entire plane! That is, each vector is rotated about 53° around the origin, and then multiplied by the scalar 5.

Certain transformations of the plane are particularly important in computer graphics and related fields. Computer software is used in many applications to move figures to new positions on the computer screen. In so doing, there is a need to calculate the new locations for these figures using matrix multiplication and other tools. We next examine particular geometric transformations of the plane that have useful properties, and discover their corresponding matrices.

Scalings: Contractions and Dilations

In Section 3.2, we saw the effect of scalar multiplication on vectors. In particular, when both c and $\mathbf{v}$ are nonzero, $c\mathbf{v}$ is a vector parallel to $\mathbf{v}$. If $|c| > 1$, then $c\mathbf{v}$ is longer than $\mathbf{v}$, representing a *dilation*. Similarly, if $0 < |c| < 1$, then $c\mathbf{v}$ is shorter than $\mathbf{v}$, representing a *contraction*. Both of these transformations of the plane are also called **scalings**.

Now, we can perform scalings on vectors using matrix multiplication instead. Because $(c\mathbf{I}_2)\mathbf{v} = c(\mathbf{I}_2\mathbf{v}) = c\mathbf{v}$, multiplying by the matrix $c\mathbf{I}_2$ has the same effect on $\mathbf{v}$ as multiplying by the scalar c. The number c is called the **scaling factor** of the transformation. Hence we have:

Dilation: For a scalar c with $|c| > 1$, the matrix $c\mathbf{I}_2$ performs a *dilation* of the plane, multiplying the length of each vector by $|c|$. If c is positive $(c > 0)$, then all vectors in the plane remain in the same direction. If c is negative $(c < 0)$, then all vectors switch to the opposite direction.

Contraction: For a scalar c with $0 < |c| < 1$, the matrix $c\mathbf{I}_2$ performs a *contraction* of the plane, multiplying the length of each vector by $|c|$. If c is positive $(c > 0)$, then all vectors in the plane remain in the same direction. If c is negative $(c < 0)$, then all vectors switch to the opposite direction.

Collapse: If $c = 0$, the matrix $c\mathbf{I}_2 = \mathbf{O}_{22}$ transforms every vector to the zero vector, collapsing the entire plane into a single point.

Example 3. The matrix $\begin{bmatrix} 5 & 0 \\ 0 & 5 \end{bmatrix}$ represents a dilation, multiplying the length of every vector by 5, but keeping all of them in the same direction, while moving them radially outward from the origin. Similarly, the matrix $\begin{bmatrix} -\frac{1}{2} & 0 \\ 0 & -\frac{1}{2} \end{bmatrix}$ is a contraction, multiplying the length of each vector by $\frac{1}{2}$, while switching all of them to the opposite direction. ∎

Rotations

Another basic geometric transformation of the plane is a rotation of the entire plane about the origin. In particular, we have:

Rotation: Multiplying by the matrix $\begin{bmatrix} \cos\theta & -\sin\theta \\ \sin\theta & \cos\theta \end{bmatrix}$ performs a *counterclockwise rotation* of the plane about the origin through an angle with measure θ. To perform a *clockwise* rotation of the plane through the origin with measure θ, we can perform a *counterclockwise rotation* of the plane about the origin through an angle with measure $(-\theta)$ instead. Such a transformation involves multiplying by the matrix

$$\begin{bmatrix} \cos(-\theta) & -\sin(-\theta) \\ \sin(-\theta) & \cos(-\theta) \end{bmatrix} = \begin{bmatrix} \cos\theta & \sin\theta \\ -\sin\theta & \cos\theta \end{bmatrix}.$$

Notice that the matrix for a clockwise rotation through angle θ is the *transpose* of the matrix for the corresponding counterclockwise rotation through the angle θ.

Example 4. For $\theta = 30°$, since $\cos 30° = \frac{\sqrt{3}}{2}$ and $\sin 30° = \frac{1}{2}$, the matrix

$$\begin{bmatrix} \frac{\sqrt{3}}{2} & -\frac{1}{2} \\ \frac{1}{2} & \frac{\sqrt{3}}{2} \end{bmatrix}$$

performs a counterclockwise rotation about the origin through an angle of 30°. We illustrate this in Figure 7.2 by showing the image of a four-sided polygon (quadrilateral) using this transformation. The four vertices of the quadrilateral are $(4, 1)$, $(5, 4)$, $(6, 3)$, and $(6, 0)$. Hence the images of these vertices are computed as follows:

$$\begin{bmatrix} \frac{\sqrt{3}}{2} & -\frac{1}{2} \\ \frac{1}{2} & \frac{\sqrt{3}}{2} \end{bmatrix} \begin{bmatrix} 4 & 5 & 6 & 6 \\ 1 & 4 & 3 & 0 \end{bmatrix} = \begin{bmatrix} 3.0 & 2.3 & 3.7 & 5.2 \\ 2.9 & 6.0 & 5.6 & 3.0 \end{bmatrix},$$

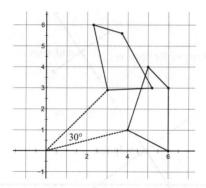

Figure 7.2: Counterclockwise 30° Rotation of a Quadrilateral

rounded to one place after the decimal point. This produces the following four vertices for the image of the quadrilateral: $(3.0, 2.9)$, $(2.3, 6.0)$, $(3.7, 5.6)$, and $(5.2, 3.0)$. ∎

Reflections

Another important basic geometric transformation of the plane is a reflection about a line through the origin. For such a reflection, the image of each point in the plane is on the opposite side of the line of reflection, the segment connecting a point and its image is perpendicular to the line of reflection, and that segment has its midpoint on the line of reflection. Points on the line of reflection are **fixed points**, meaning that they remain unchanged by the transformation. This type of transformation is illustrated in Figure 7.3.

For a simple example, consider the line $y = 0$, which is the x-axis. To find the image of a point reflected about the x-axis, we merely change the sign of the y-coordinate of the point. So, the image of $(2, 3)$ is $(2, -3)$, and the image of $(4, -2)$ is $(4, 2)$. Thus, the image is on the opposing side of the x-axis from the initial point, and the distance from the image to the x-axis is the same as the distance from the x-axis to the initial point.

Now, a reflection about a line that passes through the origin is also a matrix transformation of the plane.

Reflection: The matrix representing a reflection about the line $y = mx$ passing through the origin having slope m is

$$\frac{1}{1+m^2}\begin{bmatrix} 1-m^2 & 2m \\ 2m & m^2-1 \end{bmatrix}.$$

In the special case where the line of reflection is the y-axis, there is no number m for its slope. Instead, the matrix for the reflection about the y-axis

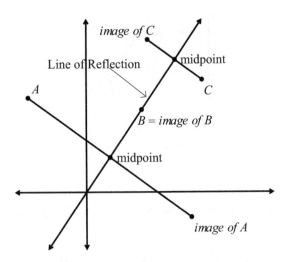

Figure 7.3: A Reflection of 3 Points About a Line Through the Origin

is

$$\begin{bmatrix} -1 & 0 \\ 0 & 1 \end{bmatrix}.$$

Example 5. The matrix for the reflection about the line $y = -2x$ is

$$\frac{1}{5}\begin{bmatrix} -3 & -4 \\ -4 & 3 \end{bmatrix} = \begin{bmatrix} -\frac{3}{5} & -\frac{4}{5} \\ -\frac{4}{5} & \frac{3}{5} \end{bmatrix}.$$

We can use this matrix to calculate the images of the four vertices of the quadrilateral in Figure 7.4:

$$\begin{bmatrix} -\frac{3}{5} & -\frac{4}{5} \\ -\frac{4}{5} & \frac{3}{5} \end{bmatrix}\begin{bmatrix} 4 & -1 & -4 & 0 \\ 1 & 2 & 4 & 7 \end{bmatrix} = \begin{bmatrix} -3.2 & -1.0 & -0.8 & -5.6 \\ -2.6 & 2.0 & 5.6 & 4.2 \end{bmatrix}.$$

Figure 7.4 also illustrates the image of the quadrilateral using the images of the vertices that we have computed. ∎

Translations

Another important type of geometric transformation is a translation. A translation merely involves moving every point in the plane the same distance in the same direction. That is, there is a corresponding translation vector **t** such that the image of any vector **v** is just **v** + **t**. Figure 7.5 illustrates the translation of a quadrilateral by the vector **t** = $[3, -2]$.

Unfortunately, translations are not matrix transformations. There is no 2×2 matrix that can represent a nonzero translation. Because **A0** = **0** for

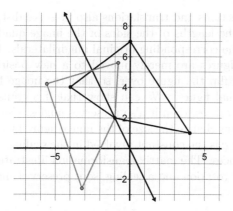

Figure 7.4: Reflection of a Quadrilateral About the Line $y = -2x$

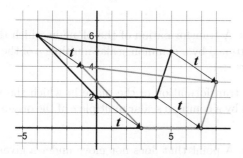

Figure 7.5: Translation of a Quadrilateral by $[3, -2]$

any matrix **A**, the image of the origin under any matrix transformation must be the origin. However, the image of the origin under a translation by the vector **t** is $\mathbf{0} + \mathbf{t} = \mathbf{t}$. So, unless $\mathbf{t} = \mathbf{0}$, a translation cannot be a matrix transformation.

Later in the chapter, we will introduce a more general system of coordinates for the plane that will allow us to perform translations using matrix multiplication with 3×3 matrices.

Preservation of Angles and Lengths

All of the geometric transformations of the plane that we have introduced so far, except for the *collapse*, are **similarities** of the plane. This means that they preserve angles between vectors, so they do not alter the angles within the figures that we draw. Notice in Figures 7.2, 7.4, and 7.5 that the angles in the image quadrilaterals have the same measure as the corresponding angles in the original quadrilaterals.

Rotations, reflections and translations also preserve distances. In Figures 7.2, 7.4, and 7.5, the lengths of the sides of the image quadrilaterals are the same as those of the corresponding original quadrilaterals. For rotations and translations, the figures are merely moved to a new position in the plane. Reflections simply flip the figure over, so that the image figure is a mirror image of the original figure. We can show that if a transformation of the plane preserves distances, then it also preserves angles. (See Exercise 13.)

However, scalings change the size of the image figure by a factor of $|c|$ from the original, where c is the constant involved in the scaling. Thus, scalings do *not* preserve distances. The matrix transformation illustrated in Figure 7.1 is a combination of a rotation and a scaling, so it preserves angles, but does not preserve distances.

Glossary

- **Collapse:** A transformation of the plane that sends all vectors to a single point.

- **Contraction:** A transformation of the plane in which the resultant vectors are shortened by a common scalar multiple c of the original vectors, where $0 < |c| < 1$.

- **Dilation:** A transformation of the plane in which the resultant vectors are enlarged by a common scalar multiple c of the original vectors, where $|c| > 1$.

- **Fixed point:** A point that does not move under a given transformation of the plane; that is, a point that is its own image.

- **Geometric transformation:** A transformation of the plane whose movement of points can be described geometrically. Rotations, reflections, scalings, and translations are examples of geometric transformations.

- **Image:** The resulting point (or vector) obtained from a given point (or vector) from a transformation of the plane.

- **Matrix transformation of the plane:** A movement of the points in the plane obtained by matrix multiplication.

- **Preserving angles:** A transformation of the plane preserves angles if the angle between all pairs of line segments that share a common endpoint is the same as the angle between their images. Similarities preserve angles.

- **Preserving distances:** A transformation of the plane preserves distances if the distance between any pair of points in the plane equals the distance between their images. Rotations, reflections, and translations all preserve distances. Any matrix transformation that preserves distances also preserves angles. (See Exercise 13.)

- Reflection about a line: A transformation of the plane in which the image of each point in the plane is on the opposite side of the line of reflection, the segment connecting a point and its image is perpendicular to the line of reflection, and that segment has its midpoint on the line of reflection. Points on the line of reflection remain fixed.

- Reflection through a point: A scaling with scaling factor -1. (See Exercise 8 in this section, and Exercises 3(a) and 4(d) in Section 7.3.)

- Rotation: A transformation of the plane in which every point is rotated by the same angle about a fixed point. Unless stated otherwise, rotations are assumed to be counterclockwise.

- Scaling: A transformation of the plane that multiplies every vector by the same scalar c.

- Scaling factor: The scalar used in a scaling transformation.

- Similarity: A transformation of the plane that preserves angles. Scalings, rotations, reflections, and translations are all similarities.

- Transformation of the plane: A function which moves each point of the plane to some (image) point in the plane. Examples include contractions, dilations, rotations, reflections, and translations.

- Translation: A transformation of the plane in which every point is moved the same distance in the same direction as a given vector $\mathbf{t}$. The distance moved is the length of $\mathbf{t}$. For any vector $\mathbf{v}$ in the plane, the image of $\mathbf{v}$ under the translation is $\mathbf{v} + \mathbf{t}$.

Exercises for Section 7.1

1. In each part, if possible, find a 2×2 matrix that represents the given geometric transformation. Also, in each case, specify whether the transformation preserves angles and whether it preserves distances.

   (a) A scaling that doubles the lengths of all vectors, but keeps them in the same direction.

   (b) A scaling that multiplies the length of vectors by $\frac{1}{3}$, but switches them to the opposite direction.

   (c) A reflection about the line $y = \frac{1}{2}x$.

   (d) A counterclockwise rotation about the origin through an angle of $75°$. Round the entries of your matrix to three places after the decimal point.

   (e) A clockwise rotation about the origin through an angle of $140°$. Round the entries of your matrix to three places after the decimal point.

(f) A transformation for which every point is moved 4 units to the left and 7 units up.

2. In each part, if possible, find a 2×2 matrix that represents the given geometric transformation. Also, in each case, specify whether the transformation preserves angles and whether it preserves distances.

(a) A scaling that halves the lengths of all vectors, but keeps them in the same direction.

(b) A scaling for which vectors are switched to the opposite direction and their lengths are tripled.

(c) A transformation for which the image of every vector is the zero vector.

(d) A transformation for which every point is moved 6 units to the right.

(e) A reflection about the line $y = 3x$.

(f) A clockwise rotation about the origin through an angle of $200°$. Round the entries of your matrix to three places after the decimal point.

(g) A transformation for which the image of every vector has the same length as the original vector, but is in the opposite direction.

3. For each of the given geometric transformations, find the image of the quadrilateral whose vertices are $(-1, 2)$, $(1, -1)$, $(6, 1)$, $(5, 5)$. Describe the image of the quadrilateral by giving its vertices, rounded to two places after the decimal point. Do not round the entries of any matrix you use.

(a) A reflection about the line $y = 1.5x$.

(b) A scaling using scalar $c = 3.4$.

(c) A counterclockwise rotation about the origin through an angle of $84°$.

4. For each of the given geometric transformations, find the image of the quadrilateral whose vertices are $(-1, 3)$, $(4, 0)$, $(5, 3)$, $(2, 8)$. Describe the image of the quadrilateral by giving its vertices, rounded to two places after the decimal point. Do not round the entries of any matrix you use.

(a) A clockwise rotation about the origin through an angle of $24°$.

(b) A reflection about the line $y = -2.5x$.

(c) A translation by the vector $\mathbf{t} = [3, -5]$.

5. Consider the quadrilateral whose vertices are $(-1, 1)$, $(1, -1)$, $(5, 2)$, $(1, 6)$. Draw both this quadrilateral and its image under a clockwise rotation through an angle of $40°$ about the origin on the same coordinate grid.

6. Consider the quadrilateral whose vertices are $(-1, 0)$, $(3, 1)$, $(3, 6)$, $(-2, 8)$. Draw both this quadrilateral and its image under a reflection about the line $y = -2.2x$ on the same coordinate grid. Also draw the graph of the line on the same grid.

7. What movement is caused by a scaling having scaling factor equal to 1?

8. A scaling centered at the origin having scaling factor equal to -1 is equivalent to what rotation? (This scaling is also called a *reflection through the point* $(0, 0)$.)

9. Consider the matrix transformation using the matrix $\mathbf{A} = \begin{bmatrix} 3 & -4 \\ 4 & 3 \end{bmatrix}$ acting on the five-sided polygon whose vertices are $(1, 1)$, $(-1, 2)$, $(1, 3)$, $(3, 1)$, and $(2, 0)$, as illustrated in Figure 7.1.

   (a) Thinking of each point as a vector from the origin to that point, show that the lengths of the images of each of the vertices of the polygon are five times the lengths of the original corresponding points (ignoring discrepancies due to rounding errors). (Round lengths to two places after the decimal point.)

   (b) Again, thinking of each point as a vector from the origin to that point, show that the cosine of the angle between each of the vertices in the polygon and its corresponding image is $\frac{3}{5}$. (Note: Since $\arccos\left(\frac{3}{5}\right) \approx 53°$, this transformation actually represents a counterclockwise rotation of $\approx 53°$.)

10. Consider the matrix transformation using the matrix $\mathbf{A} = \begin{bmatrix} 3 & -4 \\ 4 & 3 \end{bmatrix}$ acting on the five-sided polygon whose vertices are $(1, 1)$, $(-1, 2)$, $(1, 3)$, $(3, 1)$, and $(2, 0)$, as illustrated in Figure 7.1.

   (a) Compute the vectors $\mathbf{v}_1, \ldots, \mathbf{v}_5$ whose initial and terminal points are adjacent vertices of the polygon, starting with $(1, 1)$ and going around the polygon clockwise. (These vectors correspond to the five sides of the polygon.)

   (b) Calculate the vectors $\mathbf{w}_1, \ldots, \mathbf{w}_5$ whose initial and terminal points are adjacent vertices of the image polygon, starting with the first image point and going around the image clockwise. (These vectors correspond to the five sides of the image polygon.)

(c) Show that this matrix transformation has multiplied the lengths of each of the sides of the original polygon by 5 by verifying that $\|\mathbf{w}_i\| = 5 \|\mathbf{v}_i\|$ for each i (ignoring discrepancies due to rounding errors). (Round your computations to two places after the decimal point.)

(d) Show that the cosine of the angle between each $\mathbf{v}_i$ and corresponding $\mathbf{w}_i$ is $\frac{3}{5}$. (See the Note in part (b) of Exercise 9 above. This gives another verification that the given transformation represents a counterclockwise rotation of $\approx 53°$.)

(e) Compute the angle at each vertex of the original polygon. These equal the (smaller of the) angles between $-\mathbf{v}_1$ and $\mathbf{v}_2$, between $-\mathbf{v}_2$ and $\mathbf{v}_3$, between $-\mathbf{v}_3$ and $\mathbf{v}_4$, between $-\mathbf{v}_4$ and $\mathbf{v}_5$, and between $-\mathbf{v}_5$ and $\mathbf{v}_1$. Round your answers to the nearest degree.

(f) Compute the angle at each vertex of the image polygon. These equal the (smaller of the) angles between $-\mathbf{w}_1$ and $\mathbf{w}_2$, between $-\mathbf{w}_2$ and $\mathbf{w}_3$, between $-\mathbf{w}_3$ and $\mathbf{w}_4$, between $-\mathbf{w}_4$ and $\mathbf{w}_5$, and between $-\mathbf{w}_5$ and $\mathbf{w}_1$. Round your answers to the nearest degree, and show that these are equal to the corresponding angles from part (e). This illustrates that the given transformation preserves angles.

11. Suppose that a 2×2 matrix $\mathbf{A}$ represents a matrix transformation of the plane. Use MSC3 from Section 1.8 to show that the first column of $\mathbf{A}$ is the image of $[1,0]$, and the second column of $\mathbf{A}$ is the image of $[0,1]$.

12. This exercise uses the result proved in Exercise 11 to verify that the formula given in this section for the matrix for a reflection about the line $y = mx$ is correct. We do this by computing the images of $[1,0]$ and $[0,1]$ for such a reflection.

(a) Show that the points $(0,0)$ and $(1,m)$ are on the line $y = mx$. Use this to explain why the vector $[1,m]$ is parallel to the line $y = mx$.

(b) Figure 7.6 illustrates that the image of $[1,0]$ under the reflection about the line $y = mx$ is the vector

$$\mathbf{proj}_{[1,m]} [1,0] + \left(\left(\mathbf{proj}_{[1,m]} [1,0] \right) - [1,0] \right).$$

Compute this vector. Simplify your result and show that it equals the first column of the matrix given in this section for the reflection.

(c) A drawing similar to Figure 7.6 would illustrate that the image of $[0,1]$ under the reflection about the line $y = mx$ is the vector

$$\mathbf{proj}_{[1,m]} [0,1] + \left(\left(\mathbf{proj}_{[1,m]} [0,1] \right) - [0,1] \right).$$

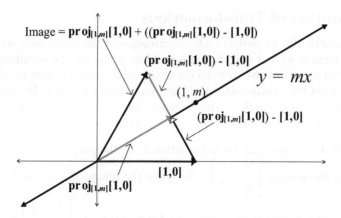

Figure 7.6: Reflection of $[1,0]$ about the line $y = mx$

Compute this vector. Simplify your result and show that it equals the second column of the matrix given in this section for the reflection.

13. Use the Polarization Identity (see Section 3.3) to prove that any matrix transformation of the plane that preserves distances also preserves angles. (That is, if $\mathbf{A}$ is a 2×2 matrix, and $\|\mathbf{Ax}\| = \|\mathbf{x}\|$ for all vectors $\mathbf{x}$ in the plane, show that the angle between any two vectors $\mathbf{v}$ and $\mathbf{w}$ in the plane is the same as the angle between the vectors $\mathbf{Av}$ and $\mathbf{Aw}$.)

14. A square matrix for which the columns form an orthonormal set of vectors is called an orthogonal matrix (see Section 6.3). (It can be shown that if a matrix is an orthogonal matrix, then it preserves both angles and distances. We will not prove this fact here.)

   (a) Prove that for every value of m, the matrix for the reflection in the plane about the line $y = mx$ is an orthogonal matrix.

   (b) Prove that for every value of θ, the matrix for a counterclockwise rotation in the plane about the origin through an angle θ is an orthogonal matrix. (Use the trigonometric identity $\sin^2 \theta + \cos^2 \theta = 1$ in your proof.)

7.2   Compositions and Inverses of Transformations

In this section, we consider several transformations acting on the plane in sequence.

Composition of Transformations

We frequently wish to perform a transformation of the plane on a vector, and then perform a second transformation of the plane on the resulting vector. That is, we carry out two movements of the plane, one followed by the other. This is called the **composition** of the two transformations. Because transformations are just functions, this concept should be familiar from your study of the composition of functions in high school algebra.

Example 1. Suppose the transformation T_1 of the plane represents multiplication by the matrix $\begin{bmatrix} \frac{3}{5} & -\frac{4}{5} \\ \frac{4}{5} & \frac{3}{5} \end{bmatrix}$. We write the image of a vector $\mathbf{v}$ under T_1 using function notation as $T_1(\mathbf{v})$, and hence $T_1(\mathbf{v}) = \begin{bmatrix} \frac{3}{5} & -\frac{4}{5} \\ \frac{4}{5} & \frac{3}{5} \end{bmatrix}\mathbf{v}$. (This is actually a counterclockwise rotation of the plane about the origin through an angle θ, where $\cos\theta = \frac{3}{5}$, $\sin\theta = \frac{4}{5}$. The angle θ is about $53°$.) Next, let T_2 be the dilation of the plane that multiplies every vector by 5. The matrix for T_2 is $5\mathbf{I}_2$. Therefore, for any vector $\mathbf{w}$, $T_2(\mathbf{w}) = (5\mathbf{I}_2)\mathbf{w}$. Now, if we want to first rotate the plane using the transformation T_1, and then perform the dilation T_2, we compute the image of a vector $\mathbf{v}$ as follows:

$$T_2(T_1(\mathbf{v})) = \begin{bmatrix} 5 & 0 \\ 0 & 5 \end{bmatrix}\left(\begin{bmatrix} \frac{3}{5} & -\frac{4}{5} \\ \frac{4}{5} & \frac{3}{5} \end{bmatrix}\mathbf{v} \right)$$

$$= \left(\begin{bmatrix} 5 & 0 \\ 0 & 5 \end{bmatrix}\begin{bmatrix} \frac{3}{5} & -\frac{4}{5} \\ \frac{4}{5} & \frac{3}{5} \end{bmatrix} \right)\mathbf{v}$$

$$= \begin{bmatrix} 3 & -4 \\ 4 & 3 \end{bmatrix}\mathbf{v}.$$

Notice that the matrix for the combined transformation is just the product of the matrices for T_1 and T_2, but *from right to left*. ∎

Notice also that the matrix for the composition in Example 1 is the same as the matrix $\mathbf{A}$ that we introduced at the beginning of the chapter. (The image of a particular polygon under the matrix $\mathbf{A}$ was shown in Figure 7.1.)

In general, we have the following:

CMT1: If T_1 and T_2 are two matrix transformations of the plane represented by the matrices $\mathbf{A}_1$ and $\mathbf{A}_2$, respectively, then the composition transformation T resulting from first applying T_1, and then applying T_2 is represented by the matrix product $(\mathbf{A}_2\mathbf{A}_1)$, where the product is written in the *reverse* order than the order in which the transformations are applied.

The composition transformation T is usually written symbolically as $T_2 \circ T_1$. When T is applied to find the image of a vector $\mathbf{v}$, we write

$T(\mathbf{v}) = (T_2 \circ T_1)(\mathbf{v}) = T_2(T_1(\mathbf{v}))$, where the last expression shows that $T_1(\mathbf{v})$ is calculated first, and then the result of that computation is plugged into the transformation T_2.

A composition involving a transformation that is not represented by a matrix can be trickier, because we do not have two matrices that can be multiplied together as in Example 1.

Example 2. Suppose T_1 is the translation of the plane by the vector $[4, 7]$, and T_2 is the reflection of the plane about the line $y = \frac{2}{3}x$. Let $T = T_2 \circ T_1$. Now, since T_1 is a translation, it is not a matrix transformation. Similarly, the composition T cannot be a matrix transformation because the image of the zero vector under T is not the zero vector. However, T_2 is represented by the matrix $\mathbf{B} = \frac{1}{13}\begin{bmatrix} 5 & 12 \\ 12 & -5 \end{bmatrix}$. Therefore, the image of a general vector $\mathbf{v}$ under T is given by

$$
\begin{aligned}
T(\mathbf{v}) &= (T_2 \circ T_1)(\mathbf{v}) = T_2(T_1(\mathbf{v})) \\
&= T_2(\mathbf{v} + [4, 7]) = \frac{1}{13}\begin{bmatrix} 5 & 12 \\ 12 & -5 \end{bmatrix}(\mathbf{v} + [4, 7]) \\
&= \frac{1}{13}\begin{bmatrix} 5 & 12 \\ 12 & -5 \end{bmatrix}\mathbf{v} + \frac{1}{13}\begin{bmatrix} 5 & 12 \\ 12 & -5 \end{bmatrix}\begin{bmatrix} 4 \\ 7 \end{bmatrix} \\
&= T_2(\mathbf{v}) + T_2\left(\begin{bmatrix} 4 \\ 7 \end{bmatrix}\right) = T_2(\mathbf{v}) + \begin{bmatrix} 8 \\ 1 \end{bmatrix}. \quad \blacksquare
\end{aligned}
$$

Notice in Example 2 that $T(\mathbf{0}) = [8, 1]$. In other words, if we let T_3 be the translation of the plane by the vector $[8, 1]$, then $T = T_3 \circ T_2$; that is, we get the same total movement of the plane by first doing the reflection T_2 and then performing the translation T_3. A similar argument for any translation of the plane gives the following general principle:

CMT2: If M is a matrix transformation of the plane and T represents a translation of the plane by the vector $\mathbf{t}$, then, for every vector $\mathbf{v}$, $(M \circ T)(\mathbf{v}) = M(\mathbf{v} + \mathbf{t}) = M(\mathbf{v}) + M(\mathbf{t})$.

Inverses of Transformations

Almost all of the transformations we have studied so far can be undone, or reversed. That is, they have inverses. The inverse of a transformation T is a transformation T^{-1} such that, for every vector $\mathbf{v}$, $(T^{-1} \circ T)(\mathbf{v}) = \mathbf{v}$ and $(T \circ T^{-1})(\mathbf{v}) = \mathbf{v}$. If T is a matrix transformation represented by a nonsingular matrix $\mathbf{A}$, then T has an inverse, which is the matrix transformation represented by $\mathbf{A}^{-1}$. This is because multiplying by $\mathbf{A}^{-1}$ undoes the effect of multiplying by $\mathbf{A}$. The only non-invertible transformation we have studied so far is the "collapse," in which all vectors are sent to a single point.

We now examine the inverses of several particular types of transformations:

Scaling: If S is a scaling transformation of the plane about the origin, it is represented by a matrix of the form $c\mathbf{I}_2$ for some scalar c. As long as $c \neq 0$, S^{-1} is the matrix transformation represented by the matrix $\left(\frac{1}{c}\right)\mathbf{I}_2$. For example, to undo the multiplication of all vectors by 7, we merely multiply all vectors by $\frac{1}{7}$.

Reflection: If R is the reflection of the plane about a line, then R is its own inverse; that is, $R^{-1} = R$. For, when we perform the reflection of a vector about a line, repeating the same reflection a second time sends the vector "back through the mirror" to its original position. Using a little algebra, you can check that if $\mathbf{R}$ is the matrix for a reflection of the plane about the particular line $y = mx$, then $\mathbf{RR} = \mathbf{I}_2$, verifying again that R is its own inverse. So, for example, to undo a reflection of the plane about the line $y = 6x$, merely repeat the reflection about that same line!

Rotation: To undo a *counterclockwise* rotation of the plane about the origin through an angle θ, we just perform a *clockwise* rotation about the origin through the same angle θ. Similarly, to undo a *clockwise* rotation of the plane about the origin through an angle θ, we perform a *counterclockwise* rotation about the origin through the angle θ. This is because in each case we can undo the first rotation by merely rotating the plane back again in the opposite direction! So, for example, the matrix for a counterclockwise rotation of the plane about the origin through an angle of $100°$ is

$$\begin{bmatrix} \cos 100° & -\sin 100° \\ \sin 100° & \cos 100° \end{bmatrix} \approx \begin{bmatrix} -0.1736 & -0.9848 \\ 0.9848 & -0.1736 \end{bmatrix}.$$

The inverse for this rotation is a counterclockwise rotation about the origin through a $-100°$ angle, or equivalently, a *clockwise* rotation about the origin through an angle of $100°$, as we observed in Section 7.1. The matrix for this inverse transformation is

$$\begin{bmatrix} \cos(-100°) & -\sin(-100°) \\ \sin(-100°) & \cos(-100°) \end{bmatrix} = \begin{bmatrix} \cos 100° & \sin 100° \\ -\sin 100° & \cos 100° \end{bmatrix}$$

$$\approx \begin{bmatrix} -0.1736 & 0.9848 \\ -0.9848 & -0.1736 \end{bmatrix}.$$

You can check that multiplying the matrix for the original transformation by the matrix for its inverse transformation yields $\mathbf{I}_2$; that is, the matrices are inverses of each other, as expected.

We also noted in Section 7.1 that the original matrix and its inverse are *transposes* of each other as well. This is because we know from part (b) of Exercise 14 in Section 7.1 that the matrix for a rotation of the plane about the origin is an *orthogonal* matrix; that is, it is a square matrix whose columns form an orthonormal set of vectors. But, from Theorem 8 in Section 6.3, every orthogonal matrix has the property that its transpose equals its inverse!

Translations: To undo the translation of the plane by a vector $\mathbf{t}$, we simply move the vectors in the plane back to where they originally were by performing the translation by the vector $-\mathbf{t}$.   That is, if we move all the points in the plane a certain distance in one direction, they are restored to their original positions by moving them the same distance in the opposite direction. For example, the inverse of the translation by the vector $[3, -8]$ is the translation by the vector $[-3, 8]$.

Conjugation

If T and M are two transformations of the plane so that T has an inverse, then one frequently executed operation is the composition $T^{-1} \circ M \circ T$. This is called the **conjugation** of M by T. Now, composition of transformations, just like matrix multiplication, does *not* have the commutative property, and so we *cannot* cancel the T and the T^{-1} in the conjugation formula. However, because composition of transformations *does* have the associative property, we can compute this triple composition either as $T^{-1} \circ (M \circ T)$ or $\left(T^{-1} \circ M\right) \circ T$.

We will see shortly that it is frequently useful to conjugate by a translation. If T is the translation by a vector $\mathbf{t}$, then we can just incorporate T^{-1} into CMT2 to obtain:

CMT3: If M is a matrix transformation and T is the translation by the vector $\mathbf{t}$, then, for every vector $\mathbf{v}$, the result of applying the conjugation of M by T to $\mathbf{v}$ is $\left(T^{-1} \circ M \circ T\right)(\mathbf{v}) = M(\mathbf{v}) + M(\mathbf{t}) - \mathbf{t}$.

Example 3. If T is the translation of the plane by the vector $[2, 3]$, M is the reflection of the plane about the line $y = x$, and $\mathbf{v}$ is the vector $[4, -5]$, then

$$\begin{aligned}
\left(T^{-1} \circ M \circ T\right)(\mathbf{v}) &= M(\mathbf{v}) + M(\mathbf{t}) - \mathbf{t} \\
&= \begin{bmatrix} 0 & 1 \\ 1 & 0 \end{bmatrix} \begin{bmatrix} 4 \\ -5 \end{bmatrix} + \begin{bmatrix} 0 & 1 \\ 1 & 0 \end{bmatrix} \begin{bmatrix} 2 \\ 3 \end{bmatrix} - \begin{bmatrix} 2 \\ 3 \end{bmatrix} \\
&= \begin{bmatrix} -5 \\ 4 \end{bmatrix} + \begin{bmatrix} 3 \\ 2 \end{bmatrix} - \begin{bmatrix} 2 \\ 3 \end{bmatrix} = \begin{bmatrix} -4 \\ 3 \end{bmatrix}.\quad \blacksquare
\end{aligned}$$

Transformations Not Centered at the Origin

Except for translations, all of the transformations of the plane that we have considered so far in this chapter have been centered at the origin. In particular, for every matrix transformation of the plane, the origin is a fixed point – that is, a point that is not moved by the transformation. However, in this subsection, we show how conjugation is used to describe transformations of the plane that are not centered at the origin. We consider three cases in turn:

(1) a scaling of the plane centered at a point other than the origin,
(2) a rotation of the plane about a point other than the origin, and
(3) a reflection of the plane about a line that does not pass through the origin.

Scaling: In the next example, we illustrate how to perform a scaling of the plane that is centered at a point other than the origin.

Example 4. Consider a scaling of the plane with factor $c = 3$ centered at $(4, -1)$. This means that $(4, -1)$ stays fixed, but all other points in the plane are moved 3 times as far (in the same direction) from $(4, -1)$ than they were. (That is, the points are moving radially away from $(4, -1)$.) We want to compute the image of a given point (x, y) under this scaling.

Method 1 (without conjugation): Consider the vector $[x - 4, \ y + 1]$ having initial point $(4, -1)$ and terminal point (x, y). Now, because of the scaling factor, we need to multiply this vector by 3, yielding $[3x - 12, \ 3y + 3]$. (This describes the movement of (x, y) from $(4, -1)$ as a result of the scaling.) Since the initial point of this vector is $(4, -1)$, the terminal point of the vector is $(3x - 8, \ 3y + 2)$. Therefore, this point is the image of (x, y) under this scaling. (See Figure 7.7.)

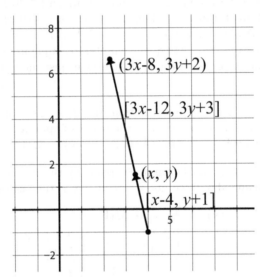

Figure 7.7: A Scaling of the Plane with Factor 3 Centered at $(4, -1)$

Method 2 (using conjugation): Let T be the translation by the vector $\mathbf{t} = [-4, 1]$ (whose coordinates have opposite signs from the fixed point). Also, let M be the scaling with factor 3 *centered at the origin*. Then M is a matrix transformation with matrix $3\mathbf{I}_2$. The conjugation $T^{-1} \circ M \circ T$ then produces the correct scaling with factor 3 centered at $(4, -1)$. (This is because we are first moving the center of the scaling to the origin using the translation T. Next, we are applying the scaling M *centered at the origin*. Finally, we are moving the origin back to where the center of the scaling is supposed to be, using T^{-1}.) Then, according to CMT3, given a point (x, y) (or vector $[x, y]$),

$$
\begin{aligned}
(T^{-1} \circ M \circ T)\,([x, y]) &= M\,([x, y]) + M\,([-4, 1]) - [-4, 1] \\
&= [3x, \ 3y] + [-12, 3] - [-4, 1] \\
&= [3x - 8, \ 3y + 2].
\end{aligned}
$$

This agrees with the image of the point (x, y) that we computed earlier. Equivalently,

$$(T^{-1} \circ M \circ T)\,([x, y]) = [3x,\ 3y] + [-8, 2].\quad\blacksquare$$

We can use the formula we derived in Example 4 to determine the image of any point. For example, for the quadrilateral with vertices $(2, 2)$, $(-1, 4)$, $(-2, 1)$, and $(1, -2)$, the images of the vertices under this scaling are, respectively, $(-2, 8)$, $(-11, 14)$, $(-14, 5)$, and $(-5, -4)$. Figure 7.8 illustrates the movement of the quadrilateral as a result of this scaling.

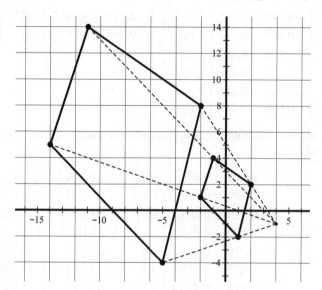

Figure 7.8: A Quadrilateral and its Image Under the Scaling in Example 4

Example 4 illustrated the following important principle:

ConjTrans: A transformation of the plane centered at a point (a, b) can be obtained by conjugating the corresponding transformation *centered at the origin* with the translation by the vector $\mathbf{t} = [-a, -b]$.

Rotation: By using ConjTrans, we can perform a rotation about any point in the plane – not just the origin.

Example 5. Let R be the counterclockwise rotation of the plane through an angle of $136°$ centered at the point $(1, -2)$. We will find the general formula for R and compute the image under R of the quadrilateral with vertices $(4, -1)$, $(10, 2)$, $(6, 4)$, and $(1, 2)$.

First, we let T be the translation of the plane by the vector $\mathbf{t} = [-1, 2]$, which moves the center of the rotation to the origin. Also, suppose M is the counterclockwise rotation of the plane through an angle of 136° *centered at the origin.* The matrix for M is

$$\begin{bmatrix} \cos 136° & -\sin 136° \\ \sin 136° & \cos 136° \end{bmatrix} \approx \begin{bmatrix} -0.719 & -0.695 \\ 0.695 & -0.719 \end{bmatrix}.$$

Then according to ConjTrans, $R = T^{-1} \circ M \circ T$, and using CMT3, the image of a vector $\mathbf{v} = [v_1, v_2]$ under R is

$$R(\mathbf{v}) = M(\mathbf{v}) + M(\mathbf{t}) - \mathbf{t}$$

$$\approx \begin{bmatrix} -0.719 & -0.695 \\ 0.695 & -0.719 \end{bmatrix}\begin{bmatrix} v_1 \\ v_2 \end{bmatrix} + \begin{bmatrix} -0.719 & -0.695 \\ 0.695 & -0.719 \end{bmatrix}\begin{bmatrix} -1 \\ 2 \end{bmatrix} - \begin{bmatrix} -1 \\ 2 \end{bmatrix}$$

$$= \begin{bmatrix} -0.719 & -0.695 \\ 0.695 & -0.719 \end{bmatrix}\begin{bmatrix} v_1 \\ v_2 \end{bmatrix} + \begin{bmatrix} -0.671 \\ -2.133 \end{bmatrix} - \begin{bmatrix} -1 \\ 2 \end{bmatrix}$$

$$= \begin{bmatrix} -0.719 & -0.695 \\ 0.695 & -0.719 \end{bmatrix}\begin{bmatrix} v_1 \\ v_2 \end{bmatrix} + \begin{bmatrix} 0.329 \\ -4.133 \end{bmatrix}.$$

We now use this formula for R is compute the images of the vertices of the quadrilateral:

$$R(4, -1) \approx \begin{bmatrix} -0.719 & -0.695 \\ 0.695 & -0.719 \end{bmatrix}\begin{bmatrix} 4 \\ -1 \end{bmatrix} + \begin{bmatrix} 0.329 \\ -4.133 \end{bmatrix} \approx \begin{bmatrix} -1.852 \\ -0.634 \end{bmatrix};$$

$$R(10, 2) \approx \begin{bmatrix} -0.719 & -0.695 \\ 0.695 & -0.719 \end{bmatrix}\begin{bmatrix} 10 \\ 2 \end{bmatrix} + \begin{bmatrix} 0.329 \\ -4.133 \end{bmatrix} \approx \begin{bmatrix} -8.251 \\ 1.379 \end{bmatrix};$$

$$R(6, 4) \approx \begin{bmatrix} -0.719 & -0.695 \\ 0.695 & -0.719 \end{bmatrix}\begin{bmatrix} 6 \\ 4 \end{bmatrix} + \begin{bmatrix} 0.329 \\ -4.133 \end{bmatrix} \approx \begin{bmatrix} -6.765 \\ -2.839 \end{bmatrix};$$

$$R(1, 2) \approx \begin{bmatrix} -0.719 & -0.695 \\ 0.695 & -0.719 \end{bmatrix}\begin{bmatrix} 1 \\ 2 \end{bmatrix} + \begin{bmatrix} 0.329 \\ -4.133 \end{bmatrix} \approx \begin{bmatrix} -1.780 \\ -4.876 \end{bmatrix}.$$

The image of the quadrilateral is shown in Figure 7.9. ■

Figure 7.9 shows a particular example of the following general principle: If P is the center of a rotation of the plane with angle θ, with a given direction (clockwise or counterclockwise), then the image of a point Q in the plane is found by drawing an arc, centered at P, starting at Q, through an angle measuring θ in the given direction. The end of this arc is the image of Q. (The center, P, remains fixed under this transformation.)

Reflection: To perform a reflection of the plane about a line that does not pass through the origin, we consider a related reflection about the line *through the origin having the same slope* and conjugate that related reflection using a translation that moves any point on the original line to the origin. Typically, the easiest point to select on the line $y = mx + b$ is its y-intercept, $(0, b)$.

Example 6. Let W be the reflection of the plane about the line $y = -\frac{1}{4}x + 3$. We will use ConjTrans to find the formula for W.

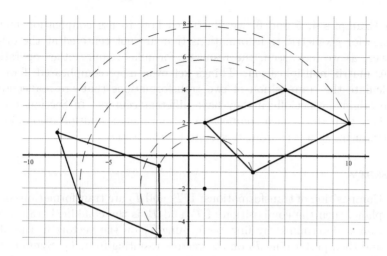

Figure 7.9: A Quadrilateral and its Image under a Rotation of 136° about $(1, -2)$

First, let T be the translation of the plane by the vector $\mathbf{t} = [0, -3]$, which sends the y-intercept $(0, 3)$ of the line $y = -\frac{1}{4}x + 3$ to the origin. Also, let M be the reflection of the plane about the line $y = -\frac{1}{4}x$. The matrix for M is

$$\frac{16}{17} \begin{bmatrix} \frac{15}{16} & -\frac{1}{2} \\ -\frac{1}{2} & -\frac{15}{16} \end{bmatrix} = \begin{bmatrix} \frac{15}{17} & -\frac{8}{17} \\ -\frac{8}{17} & -\frac{15}{17} \end{bmatrix}.$$

Then, by ConjTrans, $W = T^{-1} \circ M \circ T$. Using CMT3, the image of a vector $\mathbf{v}$ under W is

$$
\begin{aligned}
W(\mathbf{v}) &= M(\mathbf{v}) + M(\mathbf{t}) - \mathbf{t} \\
&= \begin{bmatrix} \frac{15}{17} & -\frac{8}{17} \\ -\frac{8}{17} & -\frac{15}{17} \end{bmatrix} \begin{bmatrix} v_1 \\ v_2 \end{bmatrix} + \begin{bmatrix} \frac{15}{17} & -\frac{8}{17} \\ -\frac{8}{17} & -\frac{15}{17} \end{bmatrix} \begin{bmatrix} 0 \\ -3 \end{bmatrix} - \begin{bmatrix} 0 \\ -3 \end{bmatrix} \\
&= \begin{bmatrix} \frac{15}{17} & -\frac{8}{17} \\ -\frac{8}{17} & -\frac{15}{17} \end{bmatrix} \begin{bmatrix} v_1 \\ v_2 \end{bmatrix} + \begin{bmatrix} \frac{24}{17} \\ \frac{45}{17} \end{bmatrix} - \begin{bmatrix} 0 \\ -\frac{51}{17} \end{bmatrix} \\
&= \begin{bmatrix} \frac{15}{17} & -\frac{8}{17} \\ -\frac{8}{17} & -\frac{15}{17} \end{bmatrix} \begin{bmatrix} v_1 \\ v_2 \end{bmatrix} + \begin{bmatrix} \frac{24}{17} \\ \frac{96}{17} \end{bmatrix}. \quad \blacksquare
\end{aligned}
$$

In Example 6, we could have used a different translation to get the formula for W. In fact, we can translate by any vector that moves some point on the line $y = -\frac{1}{4} + 3$ to the origin. For example, the point $(4, 2)$ lies on the line $y = -\frac{1}{4} + 3$. In Exercise 10 you are asked to show that we obtain the same formula if we use a translation by the vector $[-4, -2]$ instead of by the vector $[0, -3]$.

In Examples 4, 5 and 6, we illustrated how to use conjugation to find formulas for geometric transformations that are not centered at the origin. However, as we have seen, the resulting formulas require other operations besides matrix multiplication. Things get even more complicated if we try to calculate the composition of two transformations that are not centered at the origin – especially if they have different centers. In Section 7.3, we show how to perform such compositions using matrix multiplication alone – by introducing a new coordinate system for the plane.

Glossary

- Composition of transformations: If T_1 and T_2 are transformations of the plane, then the composition of T_1 and T_2 is the combined transformation T determined by first performing T_1 on the vectors in the plane and then performing T_2 on the vectors resulting from T_1. Symbolically, $T = T_2 \circ T_1$, and $T(\mathbf{v}) = (T_2 \circ T_1)(\mathbf{v}) = T_2(T_1(\mathbf{v}))$. Composition of transformations is associative; that is, for transformations T_1, T_2, T_3 of the plane, we have $T_3 \circ (T_2 \circ T_1) = (T_3 \circ T_2) \circ T_1$.

- Conjugation: The conjugation of a transformation Q of the plane by an invertible transformation T of the plane is the transformation $T^{-1} \circ Q \circ T$.

- Inverse of a transformation: The inverse of a transformation T of the plane is a transformation T^{-1} of the plane such that, for every vector $\mathbf{v}$, $(T^{-1} \circ T)(\mathbf{v}) = \mathbf{v}$ and $(T \circ T^{-1})(\mathbf{v}) = \mathbf{v}$.

- Invertible transformation: A transformation of the plane is invertible if and only if there is a corresponding inverse transformation for it.

Exercises for Section 7.2

1. In each part, find a formula for the image of a general vector $\mathbf{v}$ under the composition $T_2 \circ T_1$ for the given transformations T_1 and T_2 of the plane. Then compute $(T_2 \circ T_1)((5, -3))$. If indicated, round answers to three places after the decimal point. Otherwise, use fractions.

   (a) T_1: Translation by the vector $[3, 8]$; T_2: Translation by the vector $[5, -4]$.

   (b) T_1: Dilation centered at the origin with scaling factor -4; T_2: Reflection about the line $y = \frac{5}{3}x$.

   (c) T_1: Reflection about the line $y = -1.2x$; T_2: Counterclockwise rotation about the origin through an angle of $65°$. Use decimals.

   (d) T_1: Reflection about the line $y = 2.7x$; T_2: Translation by the vector $[4, 1]$. Use decimals.

   (e) T_1: Translation by the vector $[4, 1]$; T_2: Reflection about the line $y = 2.7x$. Use decimals.

(f) T_1: Counterclockwise rotation about the origin through an angle of 137°; T_2: Clockwise rotation about the origin through an angle of 84°. Use decimals.

(g) T_1: Clockwise rotation about the point $(3,8)$ through an angle of 48°; T_2: Collapse to the point $(7,3)$.

2. In each part, determine a formula for the image of a general vector **v** under the composition $T_2 \circ T_1$ for the given transformations T_1 and T_2 of the plane. Then compute $(T_2 \circ T_1)((-3,7))$. If indicated, round answers to three places after the decimal point. Otherwise, use fractions.

(a) T_1: Translation by the vector $[-2,7]$; T_2: Translation by the vector $[4,-9]$.

(b) T_1: Counterclockwise rotation about the origin through an angle of 200°; T_2: Contraction of the plane centered at the origin with scaling factor $\frac{1}{5}$. Use decimals.

(c) T_1: Clockwise rotation about the origin through an angle of 17°; T_2: Reflection about the line $y = 4.1x$. Use decimals.

(d) T_1: Counterclockwise rotation about the origin through an angle of 137°; T_2: Counterclockwise rotation about the origin through an angle of 84°. Use decimals.

(e) T_1: Collapse to the point $(-7,9)$; T_2: Reflection about the line $y = 8x$.

(f) T_1: Translation by the vector $[6,5]$; T_2: Counterclockwise rotation about the origin through an angle of 128°. Use decimals.

(g) T_1: Counterclockwise rotation about the origin through an angle of 128°; T_2: Translation by the vector $[6,5]$. Use decimals.

3. In each part, if possible, give a formula for the inverse of the given transformation T of the plane.

(a) T: A translation by the vector $[2,9]$.

(b) T: A contraction centered at the origin with scaling factor 0.64.

(c) T: The reflection about the line $y = \frac{4}{3}x$, which is represented by the matrix $\begin{bmatrix} -\frac{7}{25} & \frac{24}{25} \\ \frac{24}{25} & \frac{7}{25} \end{bmatrix}$.

(d) T: The rotation about the origin represented by the matrix $\begin{bmatrix} \frac{7}{25} & \frac{24}{25} \\ -\frac{24}{25} & \frac{7}{25} \end{bmatrix}$. (Note: T is a counterclockwise rotation about the origin through an angle of about 286.26°. This is the only angle θ between 0° and 360° for which $\cos\theta = \frac{7}{25}$ and $\sin\theta = -\frac{24}{25}$.)

(e) T: The collapse to the point $(2,1)$.

4. In each part, if possible, determine a formula for the inverse of the given transformation T of the plane.

   (a) T: A translation by the vector $[-4, 7]$.

   (b) T: A dilation centered at the origin with scaling factor 3.2.

   (c) T: The rotation about the origin represented by the matrix
   $\begin{bmatrix} -\frac{40}{41} & -\frac{9}{41} \\ \frac{9}{41} & -\frac{40}{41} \end{bmatrix}$. (Note: T is a counterclockwise rotation about the origin through an angle of about $167.32°$. This is the only angle θ between $0°$ and $360°$ for which $\cos\theta = -\frac{40}{41}$ and $\sin\theta = \frac{9}{41}$.)

   (d) T: The reflection about the line $y = 9x$, which is represented by the matrix $\begin{bmatrix} -\frac{40}{41} & \frac{9}{41} \\ \frac{9}{41} & \frac{40}{41} \end{bmatrix}$.

5. In each part, determine a formula for $T(\mathbf{v})$, where T is the described geometric transformation of the plane. Use decimals, and round your answers to three places after the decimal point.

   (a) A dilation centered at $(5, 12)$ with scaling factor 6.

   (b) A clockwise rotation through an angle of $15°$ centered at $(16, 3)$.

   (c) A reflection about the line $y = -\frac{1}{9}x + 7$.

6. In each part, determine a formula for $T(\mathbf{v})$, where T is the described geometric transformation of the plane. Use fractions.

   (a) A contraction centered at $(-2, 8)$ with scaling factor $-\frac{1}{2}$.

   (b) A counterclockwise rotation through an angle measuring $\theta = \arccos\left(\frac{20}{29}\right)$ centered at $(9, 3)$. (Hint: $\sin\theta = \frac{21}{29}$.)

   (c) A reflection about the line $y = \frac{3}{7}x - 5$.

   (d) A collapse centered at the point $(-2, 1)$.

7. Suppose T is a reflection of the plane about the line $y = 7x - 3$. Find the images of the vertices of the quadrilateral whose vertices are $(0, 3)$, $(1, -2)$, $(5, 3)$, and $(2, 5)$. Then draw the quadrilateral and its image, as well as the line $y = 7x - 3$, all on the same coordinate plane.

8. Suppose T is a clockwise rotation of the plane centered at the point $(2, 3)$ through an angle of $25°$. Find the images of the vertices of the quadrilateral whose vertices are $(0, 2)$, $(1, -2)$, $(6, -1)$, and $(3, 5)$. Round the images to two places after the decimal point. Then draw the quadrilateral and its image, as well as the point $(2, 3)$, all on the same coordinate plane. Also include $25°$ arcs centered at $(2, 3)$ connecting each vertex and its image in your drawing, as in Figure 7.9.

9. Suppose T is the translation of the plane by the vector $[5, -2]$ and M is the reflection of the plane about the line $y = 7x$. Then $W = T^{-1} \circ M \circ T$ is the reflection about some line that does not pass through the origin. What is the equation of that line?

10. Verify the claim made after Example 6, that if we had used the translation by the vector $[-4, -2]$ instead of the vector $[0, -3]$, we would have arrived at the same formula for the image of a vector $\mathbf{v}$ under the reflection of the plane about the line $y = -\frac{1}{4}x + 3$.

11. The composition of two reflections of the plane about two distinct intersecting lines always results in a rotation of the plane about the point of intersection. This exercise illustrates this general principle with a specific example.

    Let R_1 be the reflection of the plane about the line $y = 2x$, and let R_2 be the reflection of the plane about the line $y = -3x$.

    (a) Find the matrices $\mathbf{A}_1$ and $\mathbf{A}_2$ that represent R_1 and R_2, respectively.

    (b) Compute the matrix for the transformation T_1 obtained by first performing R_1 and then performing R_2.

    (c) Calculate the matrix for a counterclockwise rotation about the origin through an angle of $90°$, and verify that it agrees with the matrix obtained in part (b).

    (d) Compute the matrix for the transformation T_2 obtained by first performing R_2 and then performing R_1.

    (e) Calculate the matrix for a clockwise rotation about the origin through an angle of $90°$, and verify that it agrees with the matrix obtained in part (d).

12. As we claimed in Exercise 11, the composition of two reflections of the plane about two distinct intersecting lines always results in a rotation of the plane about the point of intersection. This exercise illustrates this general principle with another specific example.

    Let R_1 be the reflection of the plane about the line $y = \frac{4}{3}x$, and let R_2 be the reflection of the plane about the line $y = 4x$.

    (a) Find the matrices $\mathbf{A}_1$ and $\mathbf{A}_2$ that represent R_1 and R_2, respectively. (Express the entries of $\mathbf{A}_1$ and $\mathbf{A}_2$ as fractions.)

    (b) Compute the matrix $\mathbf{A}$ for the transformation T obtained by first performing R_1 and then performing R_2. (Express $\mathbf{A}$ in both fractional form and decimal form.)

    (c) Calculate $\theta = \arccos(a_{11})$, where a_{11} is the $(1,1)$ entry of the answer to part (b).

(d) Verify that $\sin\theta$ is the $(2,1)$ entry of the answer to part (b), where θ is the answer to part (c).

(e) Explain why parts (b), (c), and (d) show that T is the counterclockwise rotation of the plane about the origin through the angle θ.

13. Suppose M is the reflection of the plane about the x-axis, and R is the counterclockwise rotation of the plane about the origin through an angle of $30°$. In this exercise, we perform conjugation by the rotation R in a manner analogous to conjugation by a translation.

(a) Compute the matrix for the conjugation $T = R^{-1} \circ M \circ R$. (Hint: Use the earlier observation that the inverse of the matrix for a rotation about the origin is its transpose.)

(b) Notice that the overall effect of the conjugation T in part (a) is to reflect the plane about the line ℓ that passes through the origin at an angle of $-30°$. (This is because the conjugation operation first rotates the plane counterclockwise $30°$ using R. After that, M performs the reflection about the x-axis. Finally, R^{-1} rotates the plane back clockwise $30°$.) Also notice that the slope of ℓ is $m = -\frac{\sqrt{3}}{3}$. (This is because the slope of a nonvertical line is equal to the tangent ("tan" on the calculator) of the angle between the line and the positive x-axis, and because $\tan(-30°) = -\frac{\sqrt{3}}{3}$.) Show that using the formula for a reflection of the plane about $y = mx$ together with this value for m produces the same matrix obtained in part (a). (Thus, the conjugation process gives us an alternate way to compute the matrix for certain reflections.)

(c) State a generalization for the result in part (b) about the conjugation T in the case where R is the counterclockwise rotation of the plane about the origin through an angle θ.

14. Suppose M is the reflection of the plane about the x-axis, and R is the counterclockwise rotation of the plane about the origin through an angle of $58°$. Repeat the following steps from Exercise 13 using M and R:

(a) Compute the matrix for the conjugation $T = R^{-1} \circ M \circ R$. (Hint: Use the earlier observation that the inverse of the matrix for a rotation about the origin is its transpose.) Round your answers to three places after the decimal point.

(b) Notice, in a manner analogous to part (b) in Exercise 13, that the overall effect of the conjugation T in part (a) is to reflect the plane about the line ℓ that passes through the origin at an angle of $-58°$. Show that the slope of ℓ is $m \approx -1.6$. Then, show that using the formula for a reflection of the plane about $y = mx$ together with this value for m produces the same matrix obtained in part (a).

7.3 Homogeneous Coordinates

In the previous sections, we discovered how to represent any scaling, rotation, or reflection of the plane using a single matrix multiplication, but we were unable to do this for a general translation of the plane. To remedy that, in this section we introduce a new way to represent points in the plane which will allow us to express translations using matrix multiplication. This will eliminate the clumsiness that we have encountered up to this point in dealing with translations.

The Homogeneous Coordinate System

Another method to represent points in the plane is to use **homogeneous coordinates**. In this system, points are represented by vectors having 3 coordinates instead of 2, with the stipulation that the third coordinate never equals 0. In homogeneous coordinates, the vector $[a, b, c]$ represents the two-dimensional point $\left(\frac{a}{c}, \frac{b}{c}\right)$. So, for example, the vector $[6, 3, 2]$ represents the point $(3, 1.5)$ in the plane, and the vector $[9, -6, -3]$ represents the point $(-3, 2)$. Unlike the Cartesian coordinate system, the same point in the plane has more than one representation in homogeneous coordinates. For example, the vectors $[2, 3, 1]$, $[4, 6, 2]$, $[6, 9, 3]$, and $[12, 18, 6]$ in homogeneous coordinates, which are all nonzero scalar multiples of each other, all represent the point $(2, 3)$ in Cartesian coordinates. We can easily convert from Cartesian coordinates to homogeneous coordinates: the point (x, y) in Cartesian coordinates is represented by the vector $[x, y, 1]$ in homogeneous coordinates. In fact, throughout this section, we will exclusively use vectors in homogeneous coordinates having the value 1 in the third coordinate to simplify things as much as possible.

Translations in Homogeneous Coordinates

The translation of the plane by a vector $[x, y]$ (Cartesian coordinates) can be implemented in homogeneous coordinates by matrix multiplication using the 3×3 matrix

$$\begin{bmatrix} 1 & 0 & x \\ 0 & 1 & y \\ 0 & 0 & 1 \end{bmatrix}.$$

This works because

$$\begin{bmatrix} 1 & 0 & x \\ 0 & 1 & y \\ 0 & 0 & 1 \end{bmatrix} \begin{bmatrix} a \\ b \\ 1 \end{bmatrix} = \begin{bmatrix} a + x \\ b + y \\ 1 \end{bmatrix},$$

and so the image of the point (a, b) is the point $(a + x, b + y)$, which is just (a, b) translated by $[x, y]$.

Example 1. Consider the translation T of the plane by the vector $[3, -5]$. Then $T((2,7)) = (5,2)$. We can arrive at the same result using homogeneous coordinates as follows: We let $(2,7)$ be represented by $[2,7,1]$, and let $[x,y] = [3,-5]$. Then

$$\begin{bmatrix} 1 & 0 & x \\ 0 & 1 & y \\ 0 & 0 & 1 \end{bmatrix} = \begin{bmatrix} 1 & 0 & 3 \\ 0 & 1 & -5 \\ 0 & 0 & 1 \end{bmatrix}, \text{ so}$$

$$T\left(\begin{bmatrix} 2 \\ 7 \\ 1 \end{bmatrix}\right) = \begin{bmatrix} 1 & 0 & 3 \\ 0 & 1 & -5 \\ 0 & 0 & 1 \end{bmatrix} \begin{bmatrix} 2 \\ 7 \\ 1 \end{bmatrix} = \begin{bmatrix} 5 \\ 2 \\ 1 \end{bmatrix},$$

which represents $(5,2)$ in Cartesian coordinates. ∎

Notice in Example 1 that if we had used a different representation for $(2,7)$ in homogeneous coordinates, such as $[8, 28, 4]$, the same matrix still gives the correct result:

$$T\left(\begin{bmatrix} 8 \\ 28 \\ 4 \end{bmatrix}\right) = \begin{bmatrix} 1 & 0 & 3 \\ 0 & 1 & -5 \\ 0 & 0 & 1 \end{bmatrix} \begin{bmatrix} 8 \\ 28 \\ 4 \end{bmatrix} = \begin{bmatrix} 20 \\ 8 \\ 4 \end{bmatrix},$$

which represents the point $\left(\frac{20}{4}, \frac{8}{4}\right) = (5,2)$ in Cartesian coordinates, giving the same answer as before. However, it is simply easier to use vectors having third coordinate equal to 1.

Other Geometric Transformations

Next, we present the corresponding matrices in homogeneous coordinates for each of the other geometric transformations that we have already studied.

Scaling: A scaling of the plane centered at the origin with scaling factor c can be represented in homogeneous coordinates by the 3×3 matrix

$$\begin{bmatrix} c & 0 & 0 \\ 0 & c & 0 \\ 0 & 0 & 1 \end{bmatrix}.$$

This includes all dilations ($|c| > 1$), contractions ($0 < |c| < 1$), and collapses ($c = 0$).

Rotation: A *counterclockwise* rotation of the plane centered at the origin through an angle θ can be represented in homogeneous coordinates by the 3×3 matrix

$$\begin{bmatrix} \cos\theta & -\sin\theta & 0 \\ \sin\theta & \cos\theta & 0 \\ 0 & 0 & 1 \end{bmatrix}.$$

A *clockwise* rotation of the plane centered at the origin through an angle θ is equivalent to a *counterclockwise* rotation through the angle $-\theta$. The corresponding matrix for this rotation is

$$
\begin{bmatrix} \cos(-\theta) & -\sin(-\theta) & 0 \\ \sin(-\theta) & \cos(-\theta) & 0 \\ 0 & 0 & 1 \end{bmatrix} = \begin{bmatrix} \cos\theta & \sin\theta & 0 \\ -\sin\theta & \cos\theta & 0 \\ 0 & 0 & 1 \end{bmatrix},
$$

which is the *transpose* of the matrix for the counterclockwise rotation through angle θ. Be careful! When computing rotations *not* centered at the origin, the matrix for a clockwise rotation is *not* the transpose of the matrix for the corresponding counterclockwise rotation.

Reflection: A reflection of the plane about the line $y = mx$ can be represented in homogeneous coordinates by the 3×3 matrix

$$
\frac{1}{1+m^2} \begin{bmatrix} 1 - m^2 & 2m & 0 \\ 2m & m^2 - 1 & 0 \\ 0 & 0 & 1 \end{bmatrix} = \begin{bmatrix} \frac{1-m^2}{1+m^2} & \frac{2m}{1+m^2} & 0 \\ \frac{2m}{1+m^2} & \frac{m^2-1}{1+m^2} & 0 \\ 0 & 0 & 1 \end{bmatrix}.
$$

Although the y-axis cannot be expressed in the form $y = mx$, a reflection about the y-axis can be represented in homogeneous coordinates by the 3×3 matrix

$$
\begin{bmatrix} -1 & 0 & 0 \\ 0 & 1 & 0 \\ 0 & 0 & 1 \end{bmatrix}.
$$

Composition of Transformations

Rule CMT1 applies to transformations in homogeneous coordinates, just as it did for transformations using Cartesian coordinates. The advantage of using homogeneous coordinates is that translation is a matrix transformation! This means that we have no need for CMT2 and CMT3 when working with homogeneous coordinates. All of the compositions in which we are interested can now be implemented using matrix multiplication!

Example 2. Suppose T_1 is the translation by the vector $[7, 8]$, and T_2 is the reflection about the line $y = \frac{1}{7}x$. Then, the matrices for these transformations in homogeneous coordinates are

$$
\begin{bmatrix} 1 & 0 & 7 \\ 0 & 1 & 8 \\ 0 & 0 & 1 \end{bmatrix} \text{ for } T_1, \text{ and } \begin{bmatrix} \frac{24}{25} & \frac{7}{25} & 0 \\ \frac{7}{25} & -\frac{24}{25} & 0 \\ 0 & 0 & 1 \end{bmatrix} \text{ for } T_2.
$$

Therefore, the matrix for the composition $T_2 \circ T_1$ is the product of these two matrices in reverse order; namely,

$$
\begin{bmatrix} \frac{24}{25} & \frac{7}{25} & 0 \\ \frac{7}{25} & -\frac{24}{25} & 0 \\ 0 & 0 & 1 \end{bmatrix} \begin{bmatrix} 1 & 0 & 7 \\ 0 & 1 & 8 \\ 0 & 0 & 1 \end{bmatrix} = \begin{bmatrix} \frac{24}{25} & \frac{7}{25} & \frac{224}{25} \\ \frac{7}{25} & -\frac{24}{25} & -\frac{143}{25} \\ 0 & 0 & 1 \end{bmatrix}.
$$

We can use this matrix to compute the images of various points. For example,

$$(T_2 \circ T_1)\left(\begin{bmatrix} 1 \\ -4 \\ 1 \end{bmatrix}\right) = \begin{bmatrix} \frac{24}{25} & \frac{7}{25} & \frac{224}{25} \\ \frac{7}{25} & -\frac{24}{25} & -\frac{143}{25} \\ 0 & 0 & 1 \end{bmatrix} \begin{bmatrix} 1 \\ -4 \\ 1 \end{bmatrix} = \begin{bmatrix} \frac{44}{5} \\ -\frac{8}{5} \\ 1 \end{bmatrix}.$$

Therefore, $(T_2 \circ T_1)\,((1,-4)) = \left(\frac{44}{5}, -\frac{8}{5}\right)$. ∎

Transformations Not Centered at the Origin

Along with CMT1, ConjTrans also works for homogeneous coordinates! Thus, 3×3 matrices can be used to represent various geometric transformations that are centered at any point!

Example 3. Recall the transformation R from Example 5 in Section 7.2, which is a counterclockwise rotation of the plane through an angle of $136°$ centered at the point $(1, -2)$. We will redo the calculation in that example using homogeneous coordinates, showing how we can find the image of the quadrilateral with vertices $(4, -1)$, $(10, 2)$, $(6, 4)$, and $(1, 2)$ by using only matrix multiplication.

Let T be the translation of the plane by the vector $\mathbf{t} = [-1, 2]$, which moves the center of the rotation to the origin, and let M be the counterclockwise rotation of the plane through an angle of $136°$ centered at the origin. Rounding to three places after the decimal point, we get the matrices

$$\begin{bmatrix} -0.719 & -0.695 & 0 \\ 0.695 & -0.719 & 0 \\ 0 & 0 & 1 \end{bmatrix} \text{ for } M,$$

$$\begin{bmatrix} 1 & 0 & -1 \\ 0 & 1 & 2 \\ 0 & 0 & 1 \end{bmatrix} \text{ for } T, \text{ and } \begin{bmatrix} 1 & 0 & 1 \\ 0 & 1 & -2 \\ 0 & 0 & 1 \end{bmatrix} \text{ for } T^{-1}.$$

Then, the matrix $\mathbf{H}$ for $R = T^{-1} \circ M \circ T$ is

$$\mathbf{H} \approx \begin{bmatrix} 1 & 0 & 1 \\ 0 & 1 & -2 \\ 0 & 0 & 1 \end{bmatrix} \begin{bmatrix} -0.719 & -0.695 & 0 \\ 0.695 & -0.719 & 0 \\ 0 & 0 & 1 \end{bmatrix} \begin{bmatrix} 1 & 0 & -1 \\ 0 & 1 & 2 \\ 0 & 0 & 1 \end{bmatrix}$$

$$= \begin{bmatrix} -0.719 & -0.695 & 0.329 \\ 0.695 & -0.719 & -4.133 \\ 0 & 0 & 1 \end{bmatrix}.$$

We now compute the images of the vertices of the quadrilateral. We can calculate them all at once by putting the four vectors together into one matrix, as we did in Cartesian coordinates:

$$\mathbf{H} \begin{bmatrix} 4 & 10 & 6 & 1 \\ -1 & 2 & 4 & 2 \\ 1 & 1 & 1 & 1 \end{bmatrix} \approx \begin{bmatrix} -1.852 & -8.251 & -6.765 & -1.780 \\ -0.634 & 1.379 & -2.839 & -4.876 \\ 1 & 1 & 1 & 1 \end{bmatrix}.$$

Hence, the images of the vertices are:

$$R((4,-1)) \approx (-1.852,-0.634), \qquad R((10,2)) \approx (-8.251,1.379),$$
$$R((6,4)) \approx (-6.765,-2.839), \quad \text{and} \quad R((1,2)) \approx (-1.780,-4.876).$$

These images agree with those we computed in Example 5 in Section 7.2. The image of the quadrilateral is drawn in Figure 7.9 in that section. ∎

The real benefit of using homogeneous coordinates is apparent when we want to compose two transformations that are not centered at the origin. Trying to use ConjTrans with CMT3 twice in succession in Cartesian coordinates is complicated and tedious. But, in homogeneous coordinates, such a composition simply involves an additional matrix multiplication.

Example 4. Suppose R_1 is the reflection of the plane about the line $y = 2x-1$ and R_2 is the reflection of the plane about the line $y = -\frac{1}{3}x + 6$. We will compute the matrix representing $R_2 \circ R_1$ in homogeneous coordinates.

First, we need the matrix for R_1. Let T_1 be the translation by the vector $[0,1]$, which moves the y-intercept of $y = 2x - 1$ to the origin. Suppose M_1 is the reflection about the line $y = 2x$. Then, by ConjTrans, $R_1 = T_1^{-1} \circ M_1 \circ T_1$. So, by CMT1, the matrix for R_1 is

$$\begin{bmatrix} 1 & 0 & 0 \\ 0 & 1 & -1 \\ 0 & 0 & 1 \end{bmatrix} \begin{bmatrix} -\frac{3}{5} & \frac{4}{5} & 0 \\ \frac{4}{5} & \frac{3}{5} & 0 \\ 0 & 0 & 1 \end{bmatrix} \begin{bmatrix} 1 & 0 & 0 \\ 0 & 1 & 1 \\ 0 & 0 & 1 \end{bmatrix} = \begin{bmatrix} -\frac{3}{5} & \frac{4}{5} & \frac{4}{5} \\ \frac{4}{5} & \frac{3}{5} & -\frac{2}{5} \\ 0 & 0 & 1 \end{bmatrix}.$$

Next, we compute the matrix for R_2. Let T_2 be the translation by the vector $[0, -6]$, which moves the y-intercept of $y = -\frac{1}{3}x + 6$ to the origin. Suppose M_2 is the reflection about the line $y = -\frac{1}{3}x$. Then by ConjTrans, $R_2 = T_2^{-1} \circ M_2 \circ T_2$. Hence, the matrix for R_2 in homogeneous coordinates is

$$\begin{bmatrix} 1 & 0 & 0 \\ 0 & 1 & 6 \\ 0 & 0 & 1 \end{bmatrix} \begin{bmatrix} \frac{4}{5} & -\frac{3}{5} & 0 \\ -\frac{3}{5} & -\frac{4}{5} & 0 \\ 0 & 0 & 1 \end{bmatrix} \begin{bmatrix} 1 & 0 & 0 \\ 0 & 1 & -6 \\ 0 & 0 & 1 \end{bmatrix} = \begin{bmatrix} \frac{4}{5} & -\frac{3}{5} & \frac{18}{5} \\ -\frac{3}{5} & -\frac{4}{5} & \frac{54}{5} \\ 0 & 0 & 1 \end{bmatrix}.$$

Therefore, the matrix representing $R_2 \circ R_1$ in homogeneous coordinates is

$$\begin{bmatrix} \frac{4}{5} & -\frac{3}{5} & \frac{18}{5} \\ -\frac{3}{5} & -\frac{4}{5} & \frac{54}{5} \\ 0 & 0 & 1 \end{bmatrix} \begin{bmatrix} -\frac{3}{5} & \frac{4}{5} & \frac{4}{5} \\ \frac{4}{5} & \frac{3}{5} & -\frac{2}{5} \\ 0 & 0 & 1 \end{bmatrix} = \begin{bmatrix} -\frac{24}{25} & \frac{7}{25} & \frac{112}{25} \\ -\frac{7}{25} & -\frac{24}{25} & \frac{266}{25} \\ 0 & 0 & 1 \end{bmatrix}.$$

We will explore this example further in Exercises 3 and 4. ∎

Glossary

- Homogeneous coordinates: A certain method for representing points in the plane using a vector having three coordinates. The point (x, y) in ordinary Cartesian coordinates is represented by any vector of the form

$[kx, ky, k]$, where $k \neq 0$. In particular, (x, y) is represented by the vector $[x, y, 1]$, but also by any nonzero scalar multiple of this vector such as $[2x, 2y, 2]$ or $[-3x, -3y, -3]$. In homogeneous coordinates, the vector $[a, b, c]$ represents the point $\left(\frac{a}{c}, \frac{b}{c}\right)$.

Exercises for Section 7.3

1. Determine the 3×3 matrix that represents the given transformation T in homogeneous coordinates. If decimals are indicated, round your answers to three places after the decimal point. Otherwise, use fractions.

    (a) T is the translation of the plane by the vector $[6, -2]$.

    (b) T is the inverse of the translation of the plane that moves the point $(1, 2)$ to the point $(5, -1)$.

    (c) $T = R_1 \circ T_1$, where R_1 is the reflection of the plane about the line $y = 3.5x$ and T_1 is the translation of the plane by the vector $[7, 1]$. Use decimals.

2. Determine the 3×3 matrix that represents the given transformation T in homogeneous coordinates. If decimals are indicated, round your answers to three places after the decimal point. Otherwise, use fractions.

    (a) T is a translation of the plane that moves the point $(4, 7)$ to the point $(3, 1)$.

    (b) T is the inverse of the translation of the plane whose matrix is
    $$\begin{bmatrix} 1 & 0 & 7 \\ 0 & 1 & 2 \\ 0 & 0 & 1 \end{bmatrix}.$$

    (c) $T = R_1 \circ T_1$, where R_1 is the counterclockwise rotation of the plane about the origin through an angle of $45°$, and T_1 is a translation of the plane by the vector $[4, -6]$. Use decimals.

3. Use conjugation in a manner similar to Examples 3 and 4 to determine the 3×3 matrix that represents the given transformation T in homogeneous coordinates. If decimals are indicated, round your answers to three places after the decimal point. Otherwise, use fractions.

    (a) T is the scaling of the plane with scaling factor -1 centered at $(-3, 8)$. (Note: T takes every point in the plane to the point directly opposite to it through the point $(-3, 8)$. This is called a *reflection through the point* $(-3, 8)$. It is also equivalent to a rotation of $180°$ about the point $(-3, 8)$.)

    (b) T is the reflection of the plane about the line $x = -2$. (Hint: Conjugate the matrix for the reflection of the plane about the y-axis.)

(c) T is the clockwise rotation of the plane about the point $(5, -8)$ through an angle of $75°$. Use decimals.

(d) $T = R_2 \circ R_1$, where R_1 is the reflection of the plane about the line $y = 5x - 8$, and R_2 is the counterclockwise rotation of the plane about the point $(3, 7)$ through an angle of $215°$. Use decimals. (Hint: A shortcut could be used here since the point $(3, 7)$ lies on the line $y = 5x - 8$. After translating $(3, 7)$ to the origin, the line of reflection becomes $y = 5x$, a line through the origin.)

(e) $T = R_4 \circ R_3$, where R_3 is the clockwise rotation of the plane about the point $(2, 1)$ through an angle of $20°$, and R_4 is a counterclockwise rotation of the plane about the point $(4, 7)$ through an angle of $35°$. Use decimals.

4. Use conjugation in a manner similar to Examples 3 and 4 to determine the 3×3 matrix that represents the given transformation T in homogeneous coordinates. If decimals are indicated, round your answers to three places after the decimal point. Otherwise, use fractions.

(a) T is a scaling of the plane with scaling factor 3 centered at the point $(2, 9)$.

(b) T is the reflection of the plane about the line $y = 4x - 3$.

(c) T is the reflection of the plane about the line $x = 3$. (Hint: Conjugate the matrix for the reflection of the plane about the y-axis.)

(d) $T = S_1 \circ R_1$, where R_1 is the clockwise rotation about the point $(4, -9)$ through an angle of $180°$, and S_1 is the scaling centered at $(4, -9)$ with scaling factor -1.

(e) $T = R_3 \circ R_2$, where R_2 is the counterclockwise rotation about the point $(3, 7)$ through an angle of $215°$, and R_3 is the reflection about the line $y = 5x - 8$. Use decimals. (The Hint in part (d) of Exercise 3 also applies here.)

(f) $T = R_5 \circ R_4$, where R_4 is the counterclockwise rotation about the point $(1, -3)$ through an angle of $28°$, and R_5 is a clockwise rotation about the point $(-5, 2)$ through an angle of $132°$. Use decimals.

5. Let $T = R_2 \circ S \circ R_1$, where R_1 is the reflection of the plane about the line $y = 7x + 2$, S is the scaling of the plane with factor $\frac{1}{2}$ centered at the point $(6, 2)$, and R_2 is the clockwise rotation of the plane about the point $(-2, -1)$ through an angle of $81°$. Use conjugation in a manner similar to Example 4 to determine the 3×3 matrix for T in homogeneous coordinates. Use decimals and round your answers to three places after the decimal point.

6. Let $T = S \circ R_2 \circ R_1$, where R_1 is the counterclockwise rotation about the point $(-3, 1)$ through an angle of $52°$, R_2 is the reflection about the line

$y = 11x - 8$, and S is the scaling with factor $-\frac{17}{5}$ centered at the point $(4, -5)$. Use conjugation in a manner similar to Example 4 to determine the 3×3 matrix for T in homogeneous coordinates. Use decimals and round your answers to three places after the decimal point.

7. Consider the transformations R_1 and R_2 from Example 4.

   (a) Verify that $(3, 5)$ is the unique solution of the linear system

   $$\begin{cases} y = & 2x - 1 \\ y = & -\frac{1}{3}x + 6 \end{cases}$$

   consisting of the lines of reflection for R_1 and R_2.

   (b) Because the point $(3, 5)$ lies on both of the lines in part (a), we could have first used a single translation T_3 of the plane by the vector $[-3, -5]$ to move both lines so that they pass through the origin. This means we can express R_1 and R_2 instead as $R_1 = T_3^{-1} \circ M_1 \circ T_3$ and $R_2 = T_3^{-1} \circ M_2 \circ T_3$, where M_1 and M_2 are the matrices for the reflections given in Example 4. Verify that these compositions produce the same matrices for R_1 and R_2 that were calculated in Example 4.

   (c) Note from part (b) that $R_2 \circ R_1 = \left(T_3^{-1} \circ M_2 \circ T_3\right) \circ \left(T_3^{-1} \circ M_1 \circ T_3\right)$ $= T_3^{-1} \circ M_2 \circ \left(T_3 \circ T_3^{-1}\right) \circ M_1 \circ T_3$. But the expression $T_3 \circ T_3^{-1}$ cancels out, because it effectively amounts to translating by the vector $[3, 5]$ and then translating by the vector $[-3, -5]$, which returns every point to its original position. Removing $T_3 \circ T_3^{-1}$, we find that $R_2 \circ R_1 = T_3^{-1} \circ M_2 \circ M_1 \circ T_3$. Use this formula to compute the matrix for $R_2 \circ R_1$ in homogeneous coordinates, and verify that this is the same as the final matrix obtained in Example 4.

8. Consider the transformations R_1 and R_2 from Example 4 and the point $(3, 5)$ from Exercise 7, where the two lines of reflection intersect. Suppose $\theta = \arccos\left(-\frac{24}{25}\right)$. Compute the 3×3 matrix that represents the clockwise rotation of the plane about the point $(3, 5)$ through the angle θ. Use fractions. (Hint: $\sin\theta = \frac{7}{25}$.) Verify that this is the same as the final matrix for $R_2 \circ R_1$ obtained in Example 4. (This exercise illustrates the general principle that when two lines of reflection intersect, the composition of their corresponding reflections of the plane represents a rotation of the plane about the intersection point.)

7.4   Chapter 7 Test

1. In each part, if possible, find a 2×2 matrix that represents the given geometric transformation of the plane in Cartesian coordinates. Then compute the image of the point $(4, 1)$. If decimals are indicated, round answers to three places after the decimal point. Otherwise, use fractions.

(a) A dilation with scaling factor 6 centered at the origin.

(b) A translation by the vector $[5, 8]$.

(c) A reflection about the line $y = -\frac{3}{5}x$.

(d) A counterclockwise rotation about the origin through an angle of $8°$. Use decimals.

(e) Reflection about the line $x = 5$.

2. In each part, determine a formula using Cartesian coordinates for the image of a general vector $\mathbf{v}$ under the composition $T_2 \circ T_1$ for the given transformations T_1 and T_2 of the plane. Then compute the image of the point $(-3, 7)$. If decimals are indicated, round answers to three places after the decimal point. Otherwise, use fractions.

(a) T_1: A reflection about the line $y = x$; T_2: A translation by the vector $[-3, 16]$. (Careful! CMT2 does not apply here.)

(b) T_1: A translation by the vector $[-3, 16]$; T_2: A reflection about the line $y = x$.

(c) T_1: A clockwise rotation about the origin through an angle of $90°$; T_2: A contraction centered at the origin with scaling factor $\frac{2}{3}$.

(d) T_1: A translation by the vector $[2, -1]$; T_2: A counterclockwise rotation about the origin through an angle of $163°$. Use decimals.

(e) T_1: A reflection about the line $y = -12x$; T_2: A clockwise rotation about the origin through an angle of $68°$. Use decimals.

3. In each part, if possible, find a 3×3 matrix that represents the given geometric transformation of the plane in homogeneous coordinates. Then compute the image of the point $(4, 1)$. If decimals are indicated, round answers to three places after the decimal point. Otherwise, use fractions.

(a) A contraction centered at the origin with scaling factor $-\frac{5}{9}$.

(b) A translation by the vector $[-6, 11]$.

(c) A reflection about the line $y = -20x$.

(d) A counterclockwise rotation about the origin through an angle of $190°$. Use decimals.

(e) A dilation centered at $(-7, 3)$ with scaling factor 9.

(f) A reflection about the line $y = -7x - 9$.

(g) A clockwise rotation about the point $(5, -5)$ through an angle of $57°$. Use decimals.

(h) $T = R_2 \circ R_1$, where R_1 is the reflection about the line $x = -5$, and R_2 is the counterclockwise rotation about the point $(-3, 2)$ through an angle of $60°$. Use decimals.

4. In each part, indicate whether the given statement is true or whether it is false.

   (a) A translation of the plane can be represented as a matrix transformation using Cartesian coordinates.

   (b) A translation of the plane can be represented as a matrix transformation using homogeneous coordinates.

   (c) In this chapter, we used homogeneous coordinates with three coordinates to compute transformations in three-dimensional space.

   (d) The vector $[4, 8, -2]$ in homogeneous coordinates represents the point $(-2, -4)$ in Cartesian coordinates.

   (e) The 2×2 matrix for a transformation of the plane that undoes a previous matrix transformation is the inverse of the 2×2 matrix for the previous transformation.

   (f) Multiplying a 2-dimensional vector $\mathbf{v}$ by a scalar c has the same effect as multiplying the matrix $c\mathbf{I}_2$ times $\mathbf{v}$.

   (g) For any vector expressed in homogeneous coordinates, its third coordinate cannot equal zero.

   (h) The 2×2 matrix in Cartesian coordinates for a reflection of the plane about a line is its own inverse.

   (i) The transpose of the 3×3 matrix in homogeneous coordinates for a counterclockwise rotation of the plane about the origin through an angle θ equals the matrix for a clockwise rotation of the plane about the origin through the angle θ.

   (j) The 3×3 matrix in homogeneous coordinates for a reflection of the plane about a line passing through the origin is symmetric.

   (k) The 3×3 matrix in homogeneous coordinates for a scaling of the plane centered at the origin is symmetric.

   (l) The 3×3 matrix in homogeneous coordinates for a scaling of the plane centered at a point other than the origin is symmetric.

   (m) The 3×3 matrix in homogeneous coordinates for a rotation of the plane about the origin is symmetric.

   (n) Every point in the plane that lies on a line ℓ is a fixed point under the reflection of the plane about the line ℓ.

   (o) The center of a rotation of the plane is a fixed point for the rotation.

   (p) A nontrivial translation of the plane has no fixed points.

   (q) Every rotation of the plane preserves angles.

   (r) Every rotation of the plane preserves distances.

   (s) Every reflection of the plane about a line preserves angles.

   (t) Every reflection of the plane about a line preserves distances.

(**u**) Every scaling of the plane with a nonzero scaling factor preserves angles.

(**v**) Every scaling of the plane with a nonzero scaling factor preserves distances.

(**w**) Every translation of the plane preserves angles.

(**x**) Every translation of the plane preserves distances.

(**y**) A dilation of the plane brings points closer together.

(**z**) Under every transformation of the plane represented by a 2×2 matrix in Cartesian coordinates, the image of the origin is the origin.

(**aa**) Conjugation can be used to compute the image of a point under a rotation of the plane that is not centered at the origin.

(**bb**) The composition of two translations of the plane is also a translation of the plane.

(**cc**) The composition of a reflection of the plane about a line and a collapse of the plane is also a collapse of the plane.

(**dd**) The inverse of a reflection of the plane about a line is also a reflection of the plane about a line.

(**ee**) The inverse of a rotation of the plane about a point is also a rotation of the plane about the same point.

(**ff**) The inverse of a contraction of the plane is also a contraction of the plane.

Appendix A

Answers to Odd-Numbered Exercises

Section 1.1

1. Size $= 2 \times 4$; $a_{12} = 5$, $a_{21} = 2$, $a_{24} = 4$; sum $= 9$

3. (a) $\mathbf{C}$ has 35 ($= 5 \times 7$) total entries. Since there are fewer rows of $\mathbf{C}$ than columns, the number of main diagonal elements of $\mathbf{C}$ is equal to the number of rows of $\mathbf{C}$, which is 5. Therefore, there are $35 - 5 = 30$ entries of $\mathbf{C}$ *not* on its main diagonal.

   (b) $\mathbf{D}$ has 48 ($= 8 \times 6$) total entries. Since there are fewer columns of $\mathbf{D}$ than rows, the number of main diagonal elements of $\mathbf{D}$ is equal to the number of columns of $\mathbf{D}$, which is 6. Therefore, there are $48 - 6 = 42$ entries of $\mathbf{D}$ *not* on its main diagonal.

5. Since there are fewer columns of $\mathbf{G}$ than rows, the number of main diagonal elements of $\mathbf{G}$ is equal to the number of columns of $\mathbf{G}$, which is 73. Therefore, the main diagonal elements of $\mathbf{G}$ are $g_{11}, g_{22}, g_{33}, \ldots, g_{73,73}$.

7. $\mathbf{M} =$

	A or B in 1st course	C or D in 1st course	F in (or not taking) 1st course
A or B in 2nd course	0.78	0.36	0
C or D in 2nd course	0.19	0.51	0
F in (or not taking) 2nd course	0.03	0.13	1

Section 1.2

1. (a) $\mathbf{A} + \mathbf{B} = \begin{bmatrix} -3 & -8 \\ 0 & 9 \\ 6 & -8 \end{bmatrix}$; $\mathbf{A} - \mathbf{B} = \begin{bmatrix} 7 & 8 \\ -6 & 5 \\ 6 & 4 \end{bmatrix}$; $\mathbf{B} - \mathbf{A} = \begin{bmatrix} -7 & -8 \\ 6 & -5 \\ -6 & -4 \end{bmatrix}$

   (b) sum of main diagonal elements of $\mathbf{A} + \mathbf{B}$ is 6; sum of main diagonal elements of $\mathbf{A} - \mathbf{B}$ is 12

3. (a) $\mathbf{B} + \mathbf{F} = $

	CD 1	CD 2	CD 3
North Branch	40	30	43
West Branch	36	33	29
South Branch	32	24	45
East Branch	33	29	43

   (b) $\mathbf{F} - \mathbf{B} = $

	CD 1	CD 2	CD 3
North Branch	4	2	-3
West Branch	2	-5	-3
South Branch	2	2	3
East Branch	-5	3	-1

   (Note: The matrix $\mathbf{B} - \mathbf{F}$ would also be an acceptable answer.)

5. (a) $1.25\mathbf{L} = $

	Item 1	Item 2	Item 3
Market 1	7.00	9.25	3.50
Market 2	7.50	10.00	3.75
Market 3	8.00	10.25	4.00

   (b) $0.85\mathbf{L} = $

	Item 1	Item 2	Item 3
Market 1	4.76	6.29	2.38
Market 2	5.10	6.80	2.55
Market 3	5.44	6.97	2.72

Section 1.3

1. $\mathbf{A}_1 + \mathbf{A}_2 + \mathbf{A}_3 = \begin{bmatrix} 10 & 5 \\ 10 & 6 \end{bmatrix}$; $5\mathbf{A}_1 - 3\mathbf{A}_2 = \begin{bmatrix} 13 & 15 \\ 13 & 17 \end{bmatrix}$

3. One way: adding $\mathbf{A}_2$ and $\mathbf{A}_5$ first, to get $3(\mathbf{A}_2 + \mathbf{A}_5) = 3\begin{bmatrix} 7 & -4 \\ 15 & 9 \end{bmatrix}$
   $= \begin{bmatrix} 21 & -12 \\ 45 & 27 \end{bmatrix}$; other way: distributing 3 first, to get $3\mathbf{A}_2 + 3\mathbf{A}_5 = $
   $\begin{bmatrix} -3 & 0 \\ 12 & 18 \end{bmatrix} + \begin{bmatrix} 24 & -12 \\ 33 & 9 \end{bmatrix} = \begin{bmatrix} 21 & -12 \\ 45 & 27 \end{bmatrix}$

5. One way: adding 4 and 7 first, to get $(4 + 7)\mathbf{A}_8 = 11\begin{bmatrix} 3 & 0 & 5 \\ -1 & 7 & -4 \end{bmatrix} = $
   $\begin{bmatrix} 33 & 0 & 55 \\ -11 & 77 & -44 \end{bmatrix}$; other way: distributing $\mathbf{A}_8$ first, to get $4\mathbf{A}_8 + 7\mathbf{A}_8 = $
   $\begin{bmatrix} 12 & 0 & 20 \\ -4 & 28 & -16 \end{bmatrix} + \begin{bmatrix} 21 & 0 & 35 \\ -7 & 49 & -28 \end{bmatrix} = \begin{bmatrix} 33 & 0 & 55 \\ -11 & 77 & -44 \end{bmatrix}$

7. One way: calculating $2\mathbf{A}_5$ first, to get $5(2\mathbf{A}_5) = 5\begin{bmatrix} 16 & -8 \\ 22 & 6 \end{bmatrix}$

$= \begin{bmatrix} 80 & -40 \\ 110 & 30 \end{bmatrix}$; other way: multiplying 5 and 2 first, to get $(5 \times 2)\mathbf{A}_5$

$= 10\mathbf{A}_5 = 10\begin{bmatrix} 8 & -4 \\ 11 & 3 \end{bmatrix} = \begin{bmatrix} 80 & -40 \\ 110 & 30 \end{bmatrix}$

9. $\mathbf{A}_1 + \cdots + \mathbf{A}_4 = \begin{bmatrix} 15 & 13 \\ 8 & 6 \end{bmatrix} = \sum_{i=1}^{4} \mathbf{A}_i$

11. $\sum_{i=2}^{5} \mathbf{A}_i = \mathbf{A}_2 + \ldots + \mathbf{A}_5 = \begin{bmatrix} 21 & 6 \\ 14 & 2 \end{bmatrix}$; $\sum_{j=7}^{7} \mathbf{A}_j = \mathbf{A}_7 = \begin{bmatrix} 6 & 1 & 2 \\ 4 & 5 & 9 \end{bmatrix}$

13. $\sum_{i=7}^{8} b_i \mathbf{A}_i = \begin{bmatrix} 15 & 4 & -7 \\ 19 & -1 & 48 \end{bmatrix}$

15. We cannot compute $\sum_{i=1}^{8} \mathbf{A}_i$ because matrices $\mathbf{A}_1, \ldots, \mathbf{A}_6$ are all 2×2 matrices, but matrices $\mathbf{A}_7$ and $\mathbf{A}_8$ are 2×3 matrices, and we cannot add matrices of different sizes.

17. The addition in the expression $(x + y)\mathbf{A}$ represents an addition of two scalars, while the addition in the expression $x\mathbf{A} + y\mathbf{A}$ represents the addition of two matrices.

19. **ESP2:** If $\mathbf{A}$, $\mathbf{B}$ and $\mathbf{C}$ are matrices of the same size, then $(\mathbf{A} + \mathbf{B}) + \mathbf{C} = \mathbf{A} + (\mathbf{B} + \mathbf{C})$.
    Proof:
    Step (1): (Proving both $(\mathbf{A} + \mathbf{B}) + \mathbf{C}$ and $\mathbf{A} + (\mathbf{B} + \mathbf{C})$ have the same size)
    Suppose $\mathbf{A}$, $\mathbf{B}$ and $\mathbf{C}$ are all $m \times n$ matrices.
    Size of $(\mathbf{A} + \mathbf{B}) + \mathbf{C}$: By the definition of matrix addition, $\mathbf{A} + \mathbf{B}$ is also an $m \times n$ matrix. Then, again, by the definition of matrix addition, $(\mathbf{A} + \mathbf{B}) + \mathbf{C}$ is also an $m \times n$ matrix.
    Size of $\mathbf{A} + (\mathbf{B} + \mathbf{C})$: By the definition of matrix addition, $\mathbf{B} + \mathbf{C}$ is also an $m \times n$ matrix. Then, again, by the definition of matrix addition, $\mathbf{A} + (\mathbf{B} + \mathbf{C})$ is also an $m \times n$ matrix.
    Conclusion: Both $(\mathbf{A} + \mathbf{B}) + \mathbf{C}$ and $\mathbf{A} + (\mathbf{B} + \mathbf{C})$ are $m \times n$ matrices, so they both have the same size.

    Step (2): (Proving the (i, j) entry of $(\mathbf{A} + \mathbf{B}) + \mathbf{C}$ equals the (i, j) entry of $\mathbf{A} + (\mathbf{B} + \mathbf{C})$)
    Recall that the (i, j) entries of $\mathbf{A}$, $\mathbf{B}$ and $\mathbf{C}$, respectively, are a_{ij}, b_{ij}, and c_{ij}.
    (i, j) entry of $(\mathbf{A} + \mathbf{B}) + \mathbf{C}$: By the definition of matrix addition, the (i, j) entry of $(\mathbf{A} + \mathbf{B})$ is $a_{ij} + b_{ij}$. Then, again, by the definition of matrix addition, the (i, j) entry of $(\mathbf{A} + \mathbf{B}) + \mathbf{C}$ is $(a_{ij} + b_{ij}) + c_{ij}$.
    (i, j) entry of $\mathbf{A} + (\mathbf{B} + \mathbf{C})$: By the definition of matrix addition, the (i, j) entry of $(\mathbf{B} + \mathbf{C})$ is $b_{ij} + c_{ij}$. Then, again, by the definition of

matrix addition, the (i, j) entry of $\mathbf{A} + (\mathbf{B} + \mathbf{C})$ is $a_{ij} + (b_{ij} + c_{ij})$. But since a_{ij}, b_{ij} and c_{ij} are numbers, the associative property of addition for *numbers* tells us that $a_{ij} + (b_{ij} + c_{ij}) = (a_{ij} + b_{ij}) + c_{ij}$. Thus, the (i, j) entry of $\mathbf{A} + (\mathbf{B} + \mathbf{C})$ is also equal to $(a_{ij} + b_{ij}) + c_{ij}$.

Conclusion: Both the (i, j) entries of $(\mathbf{A} + \mathbf{B}) + \mathbf{C}$ and $\mathbf{A} + (\mathbf{B} + \mathbf{C})$ equal $(a_{ij} + b_{ij}) + c_{ij}$, so *all* corresponding entries of $(\mathbf{A} + \mathbf{B})$ and $(\mathbf{B} + \mathbf{A})$ are equal.

Finally, since both Steps (1) and (2) are proven, the proof of Property ESP2 is complete.

Section 1.4

1. (a) $[-4]$

   (b) $\begin{bmatrix} 1 & -2 \\ 5 & 3 \end{bmatrix}$

   (c) $\begin{bmatrix} 4 & 0 & 0 \\ 3 & 6 & 0 \\ -8 & 5 & 9 \end{bmatrix}$

   (d) $\begin{bmatrix} 8 & 10 \\ -2 & -3 \\ 7 & -6 \end{bmatrix}$

   (e) $\begin{bmatrix} -2 & 4 & -1 & -6 \\ 6 & -5 & 0 & 8 \\ -3 & 7 & -4 & 9 \end{bmatrix}$

3. $a_{12} = 4$, $a_{32} = 6$, $a_{25} = -7$, $a_{12}^t = 1$, $a_{32}^t = -2$, $a_{53}^t = 5$

5. TR1: $(\mathbf{A}^T)^T = \left(\left(\begin{bmatrix} 3 & -2 & -1 & 4 \\ -6 & -5 & 0 & 2 \\ -4 & -3 & 7 & 1 \end{bmatrix} \right)^T \right)^T = \left(\begin{bmatrix} 3 & -6 & -4 \\ -2 & -5 & -3 \\ -1 & 0 & 7 \\ 4 & 2 & 1 \end{bmatrix} \right)^T$

   $= \begin{bmatrix} 3 & -2 & -1 & 4 \\ -6 & -5 & 0 & 2 \\ -4 & -3 & 7 & 1 \end{bmatrix} = \mathbf{A};$

   TR2: $(-3\mathbf{A})^T = \left(-3 \begin{bmatrix} 3 & -2 & -1 & 4 \\ -6 & -5 & 0 & 2 \\ -4 & -3 & 7 & 1 \end{bmatrix} \right)^T = \begin{bmatrix} -9 & 6 & 3 & -12 \\ 18 & 15 & 0 & -6 \\ 12 & 9 & -21 & -3 \end{bmatrix}^T$

   $= \begin{bmatrix} -9 & 18 & 12 \\ 6 & 15 & 9 \\ 3 & 0 & -21 \\ -12 & -6 & -3 \end{bmatrix} = (-3) \begin{bmatrix} 3 & -6 & -4 \\ -2 & -5 & -3 \\ -1 & 0 & 7 \\ 4 & 2 & 1 \end{bmatrix} = (-3)\mathbf{A}^T;$

   TR3: $(\mathbf{A} + \mathbf{B})^T = \left(\begin{bmatrix} 3 & -2 & -1 & 4 \\ -6 & -5 & 0 & 2 \\ -4 & -3 & 7 & 1 \end{bmatrix} + \begin{bmatrix} -6 & 8 & 3 & -2 \\ 4 & -4 & -1 & -3 \\ 0 & 9 & 1 & -5 \end{bmatrix} \right)^T$

   $= \left(\begin{bmatrix} -3 & 6 & 2 & 2 \\ -2 & -9 & -1 & -1 \\ -4 & 6 & 8 & -4 \end{bmatrix} \right)^T = \begin{bmatrix} -3 & -2 & -4 \\ 6 & -9 & 6 \\ 2 & -1 & 8 \\ 2 & -1 & -4 \end{bmatrix}$

$$= \begin{bmatrix} 3 & -6 & -4 \\ -2 & -5 & -3 \\ -1 & 0 & 7 \\ 4 & 2 & 1 \end{bmatrix} + \begin{bmatrix} -6 & 4 & 0 \\ 8 & -4 & 9 \\ 3 & -1 & 1 \\ -2 & -3 & -5 \end{bmatrix} = \mathbf{A}^T + \mathbf{B}^T;$$

$$(\mathbf{A} - \mathbf{B})^T = \left(\begin{bmatrix} 3 & -2 & -1 & 4 \\ -6 & -5 & 0 & 2 \\ -4 & -3 & 7 & 1 \end{bmatrix} - \begin{bmatrix} -6 & 8 & 3 & -2 \\ 4 & -4 & -1 & -3 \\ 0 & 9 & 1 & -5 \end{bmatrix} \right)^T$$

$$= \left(\begin{bmatrix} 9 & -10 & -4 & 6 \\ -10 & -1 & 1 & 5 \\ -4 & -12 & 6 & 6 \end{bmatrix} \right)^T = \begin{bmatrix} 9 & -10 & -4 \\ -10 & -1 & -12 \\ -4 & 1 & 6 \\ 6 & 5 & 6 \end{bmatrix}$$

$$= \begin{bmatrix} 3 & -6 & -4 \\ -2 & -5 & -3 \\ -1 & 0 & 7 \\ 4 & 2 & 1 \end{bmatrix} - \begin{bmatrix} -6 & 4 & 0 \\ 8 & -4 & 9 \\ 3 & -1 & 1 \\ -2 & -3 & -5 \end{bmatrix} = \mathbf{A}^T - \mathbf{B}^T$$

7. This rule is valid: $(c\mathbf{A} + d\mathbf{B})^T = (c\mathbf{A})^T + (d\mathbf{B})^T$ (by TR3)
   $= c\mathbf{A}^T + d\mathbf{B}^T$ (by applying TR2 to both transposes).

9. **TR1:** Let $\mathbf{A}$ be any matrix. Then, $\left(\mathbf{A}^T \right)^T = \mathbf{A}$.

There are two parts to the proof: (1) we must prove that $\left(\mathbf{A}^T \right)^T$ and $\mathbf{A}$ have the same size, and, (2) we must prove that the corresponding entries of $\left(\mathbf{A}^T \right)^T$ and $\mathbf{A}$ are equal.

Step (1): (Proving $\left(\mathbf{A}^T \right)^T$ and $\mathbf{A}$ have the same size)
Suppose $\mathbf{A}$ is an $m \times n$ matrix.
Size of $\mathbf{A}$: Obviously, we already know $\mathbf{A}$ is an $m \times n$ matrix.
Size of $\left(\mathbf{A}^T \right)^T$: First, we find the size of $\mathbf{A}^T$. By the definition of the transpose operation, $\mathbf{A}^T$ is an $n \times m$ matrix. Then, applying the transpose operation to the matrix $\mathbf{A}^T$, we find that $\left(\mathbf{A}^T \right)^T$ is an $m \times n$ matrix.
Conclusion: Both $\left(\mathbf{A}^T \right)^T$ and $\mathbf{A}$ are $m \times n$ matrices, so both have the same size.

Step (2): (Proving that corresponding entries of $\left(\mathbf{A}^T \right)^T$ and $\mathbf{A}$ are equal)
We will show that the (i, j) entries of both matrices are equal.
(i, j) entry of $\mathbf{A}$: The (i, j) entry of $\mathbf{A}$ is a_{ij}.
(i, j) entry of $\left(\mathbf{A}^T \right)^T$: From the transpose operation, the (i, j) entry of $\left(\mathbf{A}^T \right)^T$ equals the (j, i) entry of $\mathbf{A}^T$, which in turn equals the (i, j) entry of $\mathbf{A} = a_{ij}$.
Conclusion: Hence, the (i, j) entries of $\left(\mathbf{A}^T \right)^T$ and $\mathbf{A}$ are both equal to a_{ij}.
Since both Steps (1) and (2) have been proven, the proof is complete.

Section 1.5

1. (a) square

 (b) none

 (c) zero, square, upper triangular, lower triangular, diagonal, symmetric, skew-symmetric

 (d) none

 (e) square, upper triangular

 (f) square, symmetric

 (g) square, symmetric, stochastic

 (h) square, upper triangular, lower triangular, diagonal, symmetric

 (i) square, lower triangular

3. (a) $\mathbf{B} = \begin{bmatrix} 4 & 4 & 1 \\ 4 & 6 & 3 \\ 1 & 3 & 1 \end{bmatrix}$, $\mathbf{C} = \begin{bmatrix} 0 & -1 & 8 \\ 1 & 0 & -5 \\ -8 & 5 & 0 \end{bmatrix}$

 (b) $\mathbf{B} = \begin{bmatrix} -3 & -2 & 5 \\ -2 & 1 & -4 \\ 5 & -4 & 6 \end{bmatrix}$, $\mathbf{C} = \begin{bmatrix} 0 & -4 & 2 \\ 4 & 0 & 7 \\ -2 & -7 & 0 \end{bmatrix}$

 (c) $\mathbf{B} = \begin{bmatrix} 6 & 3 & 8 & -2 \\ 3 & -4 & 1 & 8 \\ 8 & 1 & 5 & -1 \\ -2 & 8 & -1 & -5 \end{bmatrix}$, $\mathbf{C} = \begin{bmatrix} 0 & 7 & 2 & -3 \\ -7 & 0 & 1 & 0 \\ -2 & -1 & 0 & -5 \\ 3 & 0 & 5 & 0 \end{bmatrix}$

5. Let $\mathbf{A}$ be any square matrix. We want to prove that $(\mathbf{A} - \mathbf{A}^T)$ is skew-symmetric. There are two parts to the proof: (1) we must show that $(\mathbf{A} - \mathbf{A}^T)$ is a square matrix, and, (2) we must show that the transpose of $(\mathbf{A} - \mathbf{A}^T)$ is equal to the negative of $(\mathbf{A} - \mathbf{A}^T)$.

 Step (1) is very easy: since $\mathbf{A}$ is square, $\mathbf{A}^T$ is also square (of the same size), and so their difference $(\mathbf{A} - \mathbf{A}^T)$ is again a square matrix. Hence, to finish, we only need to prove Step (2):

 Step (2): (Showing that $(\mathbf{A} - \mathbf{A}^T)^T = -(\mathbf{A} - \mathbf{A}^T)$)
 Our strategy here is to take the transpose of the matrix $(\mathbf{A} - \mathbf{A}^T)$, and simplify the result using the rules we established earlier, to show that it equals the negative of the original expression $(\mathbf{A} - \mathbf{A}^T)$. Now,

$$\begin{aligned} (\mathbf{A} - \mathbf{A}^T)^T &= \mathbf{A}^T - (\mathbf{A}^T)^T && \text{(by TR3)} \\ &= \mathbf{A}^T - \mathbf{A} && \text{(by TR1)} \\ &= -(-\mathbf{A}^T) - \mathbf{A} && \text{(by ESP7)} \\ &= -\mathbf{A} - (-\mathbf{A}^T) && \text{(by ESP1)} \\ &= -(\mathbf{A} - \mathbf{A}^T) && \text{(by ESP6)}, \end{aligned}$$

and so the transpose of $(\mathbf{A} - \mathbf{A}^T)$ is equal to the negative of $\mathbf{A} - \mathbf{A}^T$.

Since we have verified both Steps (1) and (2), the proof is complete.

7. If $\mathbf{A}$ and $\mathbf{B}$ are both stochastic matrices of the same size, then the entries in the first column of $\mathbf{A}$ add up to 1, and the entries in the first column of $\mathbf{B}$ also add up to 1. Now, the entries in the first column of $\mathbf{A} + \mathbf{B}$ are found by adding corresponding elements of the first column of $\mathbf{A}$ to corresponding elements of the first column of $\mathbf{B}$. But then the sum of the elements in the first column of $\mathbf{A} + \mathbf{B}$ is $1 + 1 = 2 \neq 1$, and so $\mathbf{A} + \mathbf{B}$ cannot be stochastic. (A similar result is true for every other column of $\mathbf{A} + \mathbf{B}$.) An analogous argument shows that the entries in the first column of $\mathbf{A} - \mathbf{B}$ add up to $1 - 1 = 0 \neq 1$, and so $\mathbf{A} - \mathbf{B}$ cannot be stochastic. (A similar result is true for every other column of $\mathbf{A} - \mathbf{B}$.) For a particular example with 3×3 stochastic matrices $\mathbf{A}$ and $\mathbf{B}$, let

$$\mathbf{A} = \begin{bmatrix} \frac{1}{4} & 1 & 0 \\ \frac{1}{2} & 0 & \frac{1}{5} \\ \frac{1}{4} & 0 & \frac{4}{5} \end{bmatrix} \text{ and } \mathbf{B} = \begin{bmatrix} \frac{1}{2} & \frac{2}{3} & 0 \\ 0 & \frac{1}{3} & 0 \\ \frac{1}{2} & 0 & 1 \end{bmatrix}. \text{ Then } \mathbf{A} + \mathbf{B} = \begin{bmatrix} \frac{3}{4} & \frac{5}{3} & 0 \\ \frac{1}{2} & \frac{1}{3} & \frac{1}{5} \\ \frac{3}{4} & 0 & \frac{9}{5} \end{bmatrix},$$

which is not stochastic since its first column (and each of its other two columns as well) adds up to 2, rather than 1.

Section 1.6

1. **AB, AC, AG, AH, AI, BA, BD, BE, BF, CJ, DA, DD, DE, DF, EA, ED, EE, EF, FA, FD, FE, FF, GB, GC, GG, GH, GI, HB, HC, HG, HH, HI, IB, IC, IG, IH, II, JA, JD, JE, JF**

3. $\mathbf{AB} = \begin{bmatrix} 3 & -2 \\ -21 & 14 \end{bmatrix}$, $\mathbf{DA} = \begin{bmatrix} 10 & 7 & 11 \\ 84 & 3 & 49 \end{bmatrix}$,

$\mathbf{GC} = \begin{bmatrix} 18 & 73 & 41 & 77 \\ 11 & 18 & 29 & 39 \\ -6 & -7 & 17 & 7 \end{bmatrix}$, $\mathbf{JA} = \begin{bmatrix} 11 & 5 & 10 \\ 24 & -3 & 11 \\ -46 & 2 & -24 \\ 41 & 8 & 29 \end{bmatrix}$

5. $\mathbf{A}$ and $\mathbf{B}$: No. $\mathbf{AB}$ is a 2×2 matrix and $\mathbf{BA}$ is 3×3;
   $\mathbf{A}$ and $\mathbf{C}$: No. The sizes of these matrices are not compatible for calculating $\mathbf{CA}$;
   $\mathbf{D}$ and $\mathbf{E}$: No. $\mathbf{DE} = \begin{bmatrix} 16 & 24 \\ 60 & 90 \end{bmatrix}$, but $\mathbf{ED} = \begin{bmatrix} 46 & 40 \\ 69 & 60 \end{bmatrix}$;
   $\mathbf{G}$ and $\mathbf{I}$: Yes. Both $\mathbf{GI}$ and $\mathbf{IG}$ equal $\begin{bmatrix} 6 & 2 & -5 \\ 4 & 1 & 0 \\ 9 & -8 & 3 \end{bmatrix}$, which equals $\mathbf{G}$.

7. $\mathbf{J}^T = \begin{bmatrix} 5 & -3 & 2 & 8 \\ -1 & 4 & -6 & 1 \end{bmatrix}$, $\mathbf{C}^T = \begin{bmatrix} 2 & 3 & 0 \\ 4 & 2 & -9 \\ 6 & 5 & 1 \\ 8 & 7 & -3 \end{bmatrix}$,

$$\mathbf{J}^T\mathbf{C}^T = \begin{bmatrix} 74 & 75 & 5 \\ -14 & -18 & -45 \end{bmatrix};$$

$$\mathbf{CJ} = \begin{bmatrix} 74 & -14 \\ 75 & -18 \\ 5 & -45 \end{bmatrix}, \ (\mathbf{CJ})^T = \begin{bmatrix} 74 & 75 & 5 \\ -14 & -18 & -45 \end{bmatrix}$$

9. $\mathbf{IB} = \mathbf{B}$, $\mathbf{IC} = \mathbf{C}$, $\mathbf{IG} = \mathbf{G}$, $\mathbf{AI} = \mathbf{A}$, and $\mathbf{GI} = \mathbf{G}$. Multiplying a matrix by $\mathbf{I}$ on either side leaves the matrix unchanged.

11. $\mathbf{HB} = \begin{bmatrix} -18 & 12 \\ 9 & -6 \\ -12 & 8 \end{bmatrix}$, $\mathbf{HC} = \begin{bmatrix} 4 & 8 & 12 & 16 \\ 9 & 6 & 15 & 21 \\ 0 & 9 & -1 & 3 \end{bmatrix}$, $\mathbf{HG} = \begin{bmatrix} 12 & 4 & -10 \\ 12 & 3 & 0 \\ -9 & 8 & -3 \end{bmatrix}$,

$\mathbf{AH} = \begin{bmatrix} 8 & 3 & -3 \\ 18 & 0 & -5 \end{bmatrix}$, and $\mathbf{GH} = \begin{bmatrix} 12 & 6 & 5 \\ 8 & 3 & 0 \\ 18 & -24 & -3 \end{bmatrix}$. Multiplying by $\mathbf{H}$, when $\mathbf{H}$ is on the left side of another matrix, has the effect of multiplying the first row of that matrix by 2, its second row by 3, and its third row by -1. Multiplying by $\mathbf{H}$, when $\mathbf{H}$ is on the right side of another matrix, has the effect of multiplying the first column of that matrix by 2, its second column by 3, and its third column by -1.

13. (a) Either $\mathbf{AB}$ or $\mathbf{B}^T\mathbf{A}^T$ is a correct answer here. Either choice makes the column headings on the first matrix of the product match the row labels on the second matrix of the product. For example,

		Philadelphia	San Diego	Detroit
	Model X	24.25	29.50	19.00
$\mathbf{AB} =$	Model Y	29.30	36.10	23.00
	Model Z	36.25	44.55	28.25

while

		Model X	Model Y	Model Z
	Philadelphia	24.25	29.30	36.25
$\mathbf{B}^T\mathbf{A}^T =$	San Diego	29.50	36.10	44.55
	Detroit	19.00	23.00	28.25

In either product, the entries give the total cost in dollars needed for assembling, testing, and packaging each model of solar panel. Notice that these two products are transposes of each other.

(b) If the answer chosen in part (a) is $\mathbf{AB}$, then we calculate the $(2,3)$ entry of $\mathbf{AB}$ as follows: $(1.00)(17) + (0.20)(15) + (0.30)(10) = 17 + 3 + 3 = 23$, which is the total cost in dollars for assembling, testing, and packaging one Model Y solar panel in Detroit.

On the other hand, if the answer chosen in part (a) is $\mathbf{B}^T\mathbf{A}^T$, then we calculate the $(2,3)$ entry of $\mathbf{B}^T\mathbf{A}^T$ as follows: $(28)(1.25) + (21)(0.30) + (13)(0.25) = 35.00 + 6.30 + 3.25 = 44.55$, which is the total cost in dollars for assembling, testing, and packaging one Model Z solar panel in San Diego.

(c) If the answer chosen in part (a) is $\mathbf{AB}$, then we calculate the $(3,1)$ entry of $\mathbf{AB}$ as follows: $(1.25)\,(22) + (0.30)\,(20) + (0.25)\,(11)$ $= 27.5{+}6{+}2.75 = 36.25$, which is the total cost in dollars for assembling, testing, and packaging one Model Z solar panel in Philadelphia.

On the other hand, if the answer chosen in part (a) is $\mathbf{B}^T\mathbf{A}^T$, then we calculate the $(3,1)$ entry of $\mathbf{B}^T\mathbf{A}^T$ as follows: $(17)\,(0.75) + (15)\,(0.25) + (10)\,(0.25) = 12.75 + 3.75 + 2.50 = 19.00$, which is the total cost in dollars for assembling, testing, and packaging one Model X solar panel in Detroit.

(d) (See the answer to part (a).) If the answer chosen in part (a) is $\mathbf{AB}$, then $\mathbf{B}^T\mathbf{A}^T$ is the desired answer here. If, instead, the answer chosen in part (a) is $\mathbf{B}^T\mathbf{A}^T$, then $\mathbf{AB}$ is the desired answer here.

(e) If the answer chosen in part (d) is $\mathbf{B}^T\mathbf{A}^T$, then we calculate the $(3,2)$ entry of $\mathbf{B}^T\mathbf{A}^T$ as follows: $(17)\,(1.00){+}(15)\,(0.20){+}(10)\,(0.30)$ $= 17{+}3{+}3 = 23$, which is the total cost in dollars for assembling, testing, and packaging one Model Y solar panel in Detroit. (Notice that this answer does *not* agree with the $(3,2)$ entry of $\mathbf{AB}$, but *does* agree with the $(2,3)$ entry of $\mathbf{AB}$.)

On the other hand, if the answer chosen in part (d) is $\mathbf{AB}$, then we calculate the $(3,2)$ entry of $\mathbf{AB}$ as follows: $(1.25)\,(28){+}(0.30)\,(21){+}$ $(0.25)\,(13) = 35.00 + 6.30 + 3.25 = 44.55$, which is the total cost in dollars for assembling, testing, and packaging one Model Z solar panel in San Diego. (Notice that this answer does *not* agree with the $(3,2)$ entry of $\mathbf{B}^T\mathbf{A}^T$, but *does* agree with the $(2,3)$ entry of $\mathbf{B}^T\mathbf{A}^T$.)

(f) If the answer chosen in part (d) is $\mathbf{B}^T\mathbf{A}^T$, then we calculate the $(2,3)$ entry of $\mathbf{B}^T\mathbf{A}^T$ as follows: $(28)\,(1.25){+}(21)\,(0.30){+}(13)\,(0.25)$ $= 35.00 + 6.30 + 3.25 = 44.55$, which is the total cost in dollars for assembling, testing, and packaging one Model Z solar panel in San Diego. (Notice that this answer does *not* agree with the $(2,3)$ entry of $\mathbf{AB}$, but *does* agree with the $(3,2)$ entry of $\mathbf{AB}$.)

On the other hand, if the answer chosen in part (d) is $\mathbf{AB}$, then we calculate the $(2,3)$ entry of $\mathbf{AB}$ as follows: $(1.00)\,(17){+}(0.20)\,(15){+}$ $(0.30)\,(10) = 17 + 3 + 3 = 23$, which is the total cost in dollars for assembling, testing, and packaging one Model Y solar panel in Detroit. (Notice that this answer does *not* agree with the $(2,3)$ entry of $\mathbf{B}^T\mathbf{A}^T$, but *does* agree with the $(3,2)$ entry of $\mathbf{B}^T\mathbf{A}^T$.)

15. (a) Either $\mathbf{AB}^T$ or $\mathbf{BA}^T$ is a correct answer here. Either choice makes the column headings on the first matrix of the product match the

row labels on the second matrix of the product. For example,

$$\mathbf{AB}^T = \begin{array}{c} \text{Style 1} \\ \text{Style 2} \\ \text{Style 3} \end{array} \begin{array}{ccc} \text{Philadelphia} & \text{San Francisco} & \text{Houston} \\ \left[\begin{array}{ccc} 34.50 & 49.25 & 31.50 \\ 46.45 & 65.80 & 42.40 \\ 54.75 & 77.75 & 50.00 \end{array} \right] \end{array},$$

while

$$\mathbf{BA}^T = \begin{array}{c} \text{Philadelphia} \\ \text{San Francisco} \\ \text{Houston} \end{array} \begin{array}{ccc} \text{Style 1} & \text{Style 2} & \text{Style 3} \\ \left[\begin{array}{ccc} 34.50 & 46.45 & 54.75 \\ 49.25 & 65.80 & 77.75 \\ 31.50 & 42.40 & 50.00 \end{array} \right] \end{array}.$$

In either product, the entries give the total labor costs, in dollars, for fabricating, assembling, and packing one chair of each style in each city. Notice that these two products are transposes of each other.

(b) If the answer chosen in part (a) is $\mathbf{AB}^T$, then we calculate the $(2,3)$ entry of $\mathbf{AB}^T$ as follows: $(2.50)(11) + (1.25)(10) + (0.30)(8)$ $= 27.50 + 12.50 + 2.40 = 42.40$, which is the total cost, in dollars, for producing a single Style 2 chair in Houston.

On the other hand, if the answer chosen in part (a) is $\mathbf{BA}^T$, then we calculate the $(2,3)$ entry of $\mathbf{BA}^T$ as follows: $(18)(3.00) + (14)(1.50) + (11)(0.25) = 54.00 + 21.00 + 2.75 = 77.75$, which is the total cost, in dollars, for producing a single Style 3 chair in San Francisco.

(c) If the answer chosen in part (a) is $\mathbf{AB}^T$, then we calculate the $(1,1)$ entry of $\mathbf{AB}^T$ as follows: $(2.00)(12) + (0.75)(11) + (0.25)(9)$ $= 24.00 + 8.25 + 2.25 = 34.50$, which is the total cost, in dollars, for producing a single Style 1 chair in Philadelphia.

On the other hand, if the answer chosen in part (a) is $\mathbf{BA}^T$, then we calculate the $(1,1)$ entry of $\mathbf{BA}^T$ as follows: $(12)(2.00) + (11)(0.75) + (9)(0.25) = 24.00 + 8.25 + 2.25 = 34.50$, which agrees with the previous answer, since it also represents the total cost, in dollars, for producing a single Style 1 chair in Philadelphia.

(d) (See the answer to part (a).) If the answer chosen in part (a) is $\mathbf{AB}^T$, then $\mathbf{BA}^T$ is the desired answer here. If, instead, the answer chosen in part (a) is $\mathbf{BA}^T$, then $\mathbf{AB}^T$ is the desired answer here.

(e) If the answer chosen in part (d) is $\mathbf{AB}^T$, then we calculate the $(3,2)$ entry of $\mathbf{AB}^T$ as follows: $(3.00)(18) + (1.50)(14) + (0.25)(11)$ $= 54.00 + 21.00 + 2.75 = 77.75$, which is the total cost, in dollars, for producing a single Style 3 chair in San Francisco.

On the other hand, if the answer chosen in part (d) is $\mathbf{BA}^T$, then we calculate the $(3,2)$ entry of $\mathbf{BA}^T$ as follows: $(11)(2.50) +$

$(10)\,(1.25) + (8)(0.30) = 27.50 + 12.50 + 2.40 = 42.40$, which represents the total cost, in dollars, for producing a single Style 2 chair in Houston.

17. (a) Either $\mathbf{A}^T\mathbf{B}$ or $\mathbf{B}^T\mathbf{A}$ is a correct answer here. Either choice makes the column headings on the first matrix of the product match the row labels on the second matrix of the product. For example,

$$
\mathbf{A}^T\mathbf{B} = \begin{array}{c} \text{Blend A} \\ \text{Blend B} \\ \text{Blend C} \\ \text{Blend D} \end{array}
\begin{array}{cccc} \text{Atlanta} & \text{Portland} & \text{Cincinnati} & \text{Dallas} \\ \left[\begin{array}{cccc} 4080 & 4600 & 4280 & 3560 \\ 4062 & 4514 & 4192 & 3490 \\ 3920 & 4280 & 3920 & 3260 \\ 3617 & 4029 & 3702 & 3075 \end{array}\right] \end{array}
$$

while

$$
\mathbf{B}^T\mathbf{A} = \begin{array}{c} \text{Atlanta} \\ \text{Portland} \\ \text{Cincinnati} \\ \text{Dallas} \end{array}
\begin{array}{cccc} \text{Blend A} & \text{Blend B} & \text{Blend C} & \text{Blend D} \\ \left[\begin{array}{cccc} 4080 & 4062 & 3920 & 3617 \\ 4600 & 4514 & 4280 & 4029 \\ 4280 & 4192 & 3920 & 3702 \\ 3560 & 3490 & 3260 & 3075 \end{array}\right] \end{array}.
$$

In either product, the entries give the total cost in dollars of ingredients for one tanker truck-load of each blend in each city. Notice that these two products are transposes of each other.

(b) If the answer chosen in part (a) is $\mathbf{A}^T\mathbf{B}$, then we calculate the $(2,1)$ entry of $\mathbf{A}^T\mathbf{B}$ as follows: $(100)\,(33) + (14)\,(45) + (6)\,(22) = 3300 + 630 + 132 = 4062$, which is the total cost, in dollars, for one tanker truck-load of Blend B in Atlanta.

On the other hand, if the answer chosen in part (a) is $\mathbf{B}^T\mathbf{A}$, then we calculate the $(2,1)$ entry of $\mathbf{B}^T\mathbf{A}$ as follows: $(38)\,(110)+(42)\,(10)+(21)(0) = 4180 + 420 + 0 = 4600$, which is the total cost, in dollars, for one tanker truck-load of Blend A in Portland.

(c) If the answer chosen in part (a) is $\mathbf{A}^T\mathbf{B}$, then we calculate the $(1,3)$ entry of $\mathbf{A}^T\mathbf{B}$ as follows: $(110)\,(35) + (10)\,(43) + (0)\,(15) = 3850 + 430 + 0 = 4280$, which is the total cost, in dollars, for one tanker truck-load of Blend A in Cincinnati.

On the other hand, if the answer chosen in part (a) is $\mathbf{B}^T\mathbf{A}$, then we calculate the $(1,3)$ entry of $\mathbf{B}^T\mathbf{A}$ as follows: $(33)\,(85) + (45)\,(15) + (22)(20) = 2805 + 675 + 440 = 3920$, which is the total cost, in dollars, for one tanker truck-load of Blend C in Atlanta.

(d) (See the answer to part (a).) If the answer chosen in part (a) is $\mathbf{A}^T\mathbf{B}$, then $\mathbf{B}^T\mathbf{A}$ is the desired answer here. If, instead, the answer chosen in part (a) is $\mathbf{B}^T\mathbf{A}$, then $\mathbf{A}^T\mathbf{B}$ is the desired answer here.

(e) If the answer chosen in part (d) is $\mathbf{A}^T\mathbf{B}$, then we calculate the $(3,1)$ entry of $\mathbf{A}^T\mathbf{B}$ as follows: $(85)\,(33) + (15)\,(45) + (20)\,(22) =$

$2805 + 675 + 440 = 3920$, which is the total cost, in dollars, for one tanker truck-load of Blend C in Atlanta.

On the other hand, if the answer chosen in part (d) is $\mathbf{B}^T\mathbf{A}$, then we calculate the $(3, 1)$ entry of $\mathbf{B}^T\mathbf{A}$ as follows: $(35)(110)+(43)(10)+(15)(0) = 3850+430+0 = 4280$, which is the total cost, in dollars, for one tanker truck-load of Blend A in Cincinnati.

19. It is shown in Section 1.6 that for the matrices $\mathbf{A} = \begin{bmatrix} 1 & 2 \\ 3 & 4 \end{bmatrix}$ and $\mathbf{B} = \begin{bmatrix} 5 & 6 \\ 7 & 8 \end{bmatrix}$, we have $\mathbf{AB} \neq \mathbf{BA}$.

21. Let $\mathbf{A}$ be an $m \times n$ matrix. We must show: (1) $\mathbf{AA}^T$ is an $m \times m$ matrix, and (2) the transpose of $\mathbf{AA}^T$ is $\mathbf{AA}^T$ again. Step (1) is easy: Since $\mathbf{A}^T$ is an $n \times m$ matrix, the matrices $\mathbf{A}^T$ and $\mathbf{A}$ are compatible, and then by the definition of matrix multiplication, $\mathbf{AA}^T$ is an $m \times m$ matrix. We will prove Step (2) by taking the transpose of $\mathbf{AA}^T$, simplifying it, and showing that it equals $\mathbf{AA}^T$ itself:

$$\left(\mathbf{AA}^T\right)^T = \left(\mathbf{A}^T\right)^T\left(\mathbf{A}^T\right) \quad \text{by TR4}$$
$$= \mathbf{AA}^T \quad\quad\quad \text{by TR1,}$$

which ends the proof.

Section 1.7

1. (a) $\mathbf{A}^2 = \begin{bmatrix} -5 & 15 \\ -10 & 10 \end{bmatrix}$, $\mathbf{A}^3 = \begin{bmatrix} -35 & 45 \\ -30 & 10 \end{bmatrix}$, $\mathbf{F}^2 = \begin{bmatrix} 8 & 0 & 3 \\ -15 & 5 & -12 \\ 36 & -1 & 33 \end{bmatrix}$,

$\mathbf{F}^3 = \begin{bmatrix} 31 & -5 & 18 \\ -110 & -7 & -87 \\ 241 & -1 & 201 \end{bmatrix}$

(b) $\mathbf{A}^5 = \mathbf{A}^2\mathbf{A}^3 = \begin{bmatrix} -5 & 15 \\ -10 & 10 \end{bmatrix}\begin{bmatrix} -35 & 45 \\ -30 & 10 \end{bmatrix} = \begin{bmatrix} -275 & -75 \\ 50 & -350 \end{bmatrix}$,

$\mathbf{F}^5 = \mathbf{F}^2\mathbf{F}^3 = \begin{bmatrix} 8 & 0 & 3 \\ -15 & 5 & -12 \\ 36 & -1 & 33 \end{bmatrix}\begin{bmatrix} 31 & -5 & 18 \\ -110 & -7 & -87 \\ 241 & -1 & 201 \end{bmatrix}$

$= \begin{bmatrix} 971 & -43 & 747 \\ -3907 & 52 & -3117 \\ 9179 & -206 & 7368 \end{bmatrix}$

(c) $\mathbf{A}^6 = (\mathbf{A}^2)^3 = \left(\begin{bmatrix} -5 & 15 \\ -10 & 10 \end{bmatrix}\right)^3$

$= \begin{bmatrix} -5 & 15 \\ -10 & 10 \end{bmatrix}\begin{bmatrix} -5 & 15 \\ -10 & 10 \end{bmatrix}\begin{bmatrix} -5 & 15 \\ -10 & 10 \end{bmatrix} = \begin{bmatrix} -125 & -1125 \\ 750 & -1250 \end{bmatrix}$, or,

$$\mathbf{A}^6 = (\mathbf{A}^3)^2 = \left(\begin{bmatrix} -35 & 45 \\ -30 & 10 \end{bmatrix} \right)^2 = \begin{bmatrix} -35 & 45 \\ -30 & 10 \end{bmatrix} \begin{bmatrix} -35 & 45 \\ -30 & 10 \end{bmatrix}$$

$$= \begin{bmatrix} -125 & -1125 \\ 750 & -1250 \end{bmatrix};$$

$$\mathbf{F}^6 = (\mathbf{F}^2)^3 = \left(\begin{bmatrix} 8 & 0 & 3 \\ -15 & 5 & -12 \\ 36 & -1 & 33 \end{bmatrix} \right)^3$$

$$= \begin{bmatrix} 8 & 0 & 3 \\ -15 & 5 & -12 \\ 36 & -1 & 33 \end{bmatrix} \begin{bmatrix} 8 & 0 & 3 \\ -15 & 5 & -12 \\ 36 & -1 & 33 \end{bmatrix} \begin{bmatrix} 8 & 0 & 3 \\ -15 & 5 & -12 \\ 36 & -1 & 33 \end{bmatrix}$$

$$= \begin{bmatrix} 5849 & -138 & 4611 \\ -23607 & 686 & -18858 \\ 56022 & -1399 & 44826 \end{bmatrix}, \text{ or,}$$

$$\mathbf{F}^6 = (\mathbf{F}^3)^2 = \left(\begin{bmatrix} 971 & -43 & 747 \\ -3907 & 52 & -3117 \\ 9179 & -206 & 7368 \end{bmatrix} \right)^2$$

$$= \begin{bmatrix} 971 & -43 & 747 \\ -3907 & 52 & -3117 \\ 9179 & -206 & 7368 \end{bmatrix} \begin{bmatrix} 971 & -43 & 747 \\ -3907 & 52 & -3117 \\ 9179 & -206 & 7368 \end{bmatrix}$$

$$= \begin{bmatrix} 5849 & -138 & 4611 \\ -23607 & 686 & -18858 \\ 56022 & -1399 & 44826 \end{bmatrix}$$

(d) $(\mathbf{A}^T)^6 = (\mathbf{A}^6)^T = \begin{bmatrix} -125 & 750 \\ -1125 & -1250 \end{bmatrix}$,

$$(\mathbf{F}^T)^6 = (\mathbf{F}^6)^T = \begin{bmatrix} 5849 & -23607 & 56022 \\ -138 & 686 & -1399 \\ 4611 & -18858 & 44826 \end{bmatrix}$$

(e) $(2\mathbf{A})^3 = \left(2 \begin{bmatrix} 1 & 3 \\ -2 & 4 \end{bmatrix} \right)^3 = \begin{bmatrix} 2 & 6 \\ -4 & 8 \end{bmatrix}^3 = \begin{bmatrix} -280 & 360 \\ -240 & 80 \end{bmatrix}$,

$2^3 \mathbf{A}^3 = 8\mathbf{A}^3 = 8 \begin{bmatrix} -35 & 45 \\ -30 & 10 \end{bmatrix}$ (from part (a)) $= \begin{bmatrix} -280 & 360 \\ -240 & 80 \end{bmatrix}$.

We have verified that RFE4 holds in this particular case.

(f) $-4(\mathbf{AB}) = -4 \left(\begin{bmatrix} 1 & 3 \\ -2 & 4 \end{bmatrix} \begin{bmatrix} 6 & -1 \\ -3 & 2 \end{bmatrix} \right)$

$$= -4 \begin{bmatrix} -3 & 5 \\ -24 & 10 \end{bmatrix} = \begin{bmatrix} 12 & -20 \\ 96 & -40 \end{bmatrix},$$

$(-4\mathbf{A})\mathbf{B} = \left(-4 \begin{bmatrix} 1 & 3 \\ -2 & 4 \end{bmatrix} \right) \begin{bmatrix} 6 & -1 \\ -3 & 2 \end{bmatrix}$

$$= \begin{bmatrix} -4 & -12 \\ 8 & -16 \end{bmatrix} \begin{bmatrix} 6 & -1 \\ -3 & 2 \end{bmatrix} = \begin{bmatrix} 12 & -20 \\ 96 & -40 \end{bmatrix}.$$

We have verified that MM1 holds in this particular case.

(g) $\mathbf{A}(\mathbf{BC}) = \begin{bmatrix} 1 & 3 \\ -2 & 4 \end{bmatrix} \left(\begin{bmatrix} 6 & -1 \\ -3 & 2 \end{bmatrix} \begin{bmatrix} 2 & -5 \\ -4 & 1 \end{bmatrix} \right)$

$$= \begin{bmatrix} 1 & 3 \\ -2 & 4 \end{bmatrix} \begin{bmatrix} 16 & -31 \\ -14 & 17 \end{bmatrix} = \begin{bmatrix} -26 & 20 \\ -88 & 130 \end{bmatrix},$$

$$(\mathbf{AB})\mathbf{C} = \left(\begin{bmatrix} 1 & 3 \\ -2 & 4 \end{bmatrix} \begin{bmatrix} 6 & -1 \\ -3 & 2 \end{bmatrix} \right) \begin{bmatrix} 2 & -5 \\ -4 & 1 \end{bmatrix}$$

$$= \begin{bmatrix} -3 & 5 \\ -24 & 10 \end{bmatrix} \begin{bmatrix} 2 & -5 \\ -4 & 1 \end{bmatrix} = \begin{bmatrix} -26 & 20 \\ -88 & 130 \end{bmatrix}.$$

We have verified that MM2 holds in this particular case.

(h) $(\mathbf{D} + \mathbf{E})\mathbf{F} = \left(\begin{bmatrix} 5 & -3 & 4 \\ -2 & 1 & 3 \end{bmatrix} + \begin{bmatrix} -2 & 4 & -5 \\ -1 & 0 & -2 \end{bmatrix} \right) \begin{bmatrix} 2 & -1 & 0 \\ -4 & -2 & -3 \\ 5 & 1 & 6 \end{bmatrix}$

$$= \begin{bmatrix} 3 & 1 & -1 \\ -3 & 1 & 1 \end{bmatrix} \begin{bmatrix} 2 & -1 & 0 \\ -4 & -2 & -3 \\ 5 & 1 & 6 \end{bmatrix} = \begin{bmatrix} -3 & -6 & -9 \\ -5 & 2 & 3 \end{bmatrix},$$

$$\mathbf{DF} + \mathbf{EF} = \begin{bmatrix} 5 & -3 & 4 \\ -2 & 1 & 3 \end{bmatrix} \begin{bmatrix} 2 & -1 & 0 \\ -4 & -2 & -3 \\ 5 & 1 & 6 \end{bmatrix}$$

$$+ \begin{bmatrix} -2 & 4 & -5 \\ -1 & 0 & -2 \end{bmatrix} \begin{bmatrix} 2 & -1 & 0 \\ -4 & -2 & -3 \\ 5 & 1 & 6 \end{bmatrix}$$

$$= \begin{bmatrix} 42 & 5 & 33 \\ 7 & 3 & 15 \end{bmatrix} + \begin{bmatrix} -45 & -11 & -42 \\ -12 & -1 & -12 \end{bmatrix} = \begin{bmatrix} -3 & -6 & -9 \\ -5 & 2 & 3 \end{bmatrix}.$$

We have verified that MM4 holds in this particular case.

(i) $\mathbf{AB} = \begin{bmatrix} 1 & 3 \\ -2 & 4 \end{bmatrix} \begin{bmatrix} 6 & -1 \\ -3 & 2 \end{bmatrix} = \begin{bmatrix} -3 & 5 \\ -24 & 10 \end{bmatrix}$, but

$\mathbf{BA} = \begin{bmatrix} 6 & -1 \\ -3 & 2 \end{bmatrix} \begin{bmatrix} 1 & 3 \\ -2 & 4 \end{bmatrix} = \begin{bmatrix} 8 & 14 \\ -7 & -1 \end{bmatrix}.$

Therefore $\mathbf{AB} \neq \mathbf{BA}$ in this particular case.

(j) $(\mathbf{A} + \mathbf{B})^2 = \left(\begin{bmatrix} 1 & 3 \\ -2 & 4 \end{bmatrix} + \begin{bmatrix} 6 & -1 \\ -3 & 2 \end{bmatrix} \right)^2$

$$= \begin{bmatrix} 7 & 2 \\ -5 & 6 \end{bmatrix}^2 = \begin{bmatrix} 39 & 26 \\ -65 & 26 \end{bmatrix},$$

$$\mathbf{A}^2 + \mathbf{BA} + \mathbf{AB} + \mathbf{B}^2 = \begin{bmatrix} 1 & 3 \\ -2 & 4 \end{bmatrix}^2 + \begin{bmatrix} 6 & -1 \\ -3 & 2 \end{bmatrix} \begin{bmatrix} 1 & 3 \\ -2 & 4 \end{bmatrix}$$

$$+ \begin{bmatrix} 1 & 3 \\ -2 & 4 \end{bmatrix} \begin{bmatrix} 6 & -1 \\ -3 & 2 \end{bmatrix} + \begin{bmatrix} 6 & -1 \\ -3 & 2 \end{bmatrix}^2$$

$$= \begin{bmatrix} -5 & 15 \\ -10 & 10 \end{bmatrix} + \begin{bmatrix} 8 & 14 \\ -7 & -1 \end{bmatrix} + \begin{bmatrix} -3 & 5 \\ -24 & 10 \end{bmatrix} + \begin{bmatrix} 39 & -8 \\ -24 & 7 \end{bmatrix}$$

$$= \begin{bmatrix} 39 & 26 \\ -65 & 26 \end{bmatrix} \quad \text{(which agrees with } (\mathbf{A} + \mathbf{B})^2 \text{), but}$$

$$\mathbf{A}^2 + 2\mathbf{AB} + \mathbf{B}^2 = \begin{bmatrix} 1 & 3 \\ -2 & 4 \end{bmatrix}^2 + 2 \begin{bmatrix} 1 & 3 \\ -2 & 4 \end{bmatrix} \begin{bmatrix} 6 & -1 \\ -3 & 2 \end{bmatrix} + \begin{bmatrix} 6 & -1 \\ -3 & 2 \end{bmatrix}^2$$

$$= \begin{bmatrix} -5 & 15 \\ -10 & 10 \end{bmatrix} + \begin{bmatrix} -6 & 10 \\ -48 & 20 \end{bmatrix} + \begin{bmatrix} 39 & -8 \\ -24 & 7 \end{bmatrix} = \begin{bmatrix} 28 & 17 \\ -82 & 37 \end{bmatrix},$$

which is *not* equal to $(\mathbf{A} + \mathbf{B})^2$.

From part (i), we know that $\mathbf{BA} \neq \mathbf{AB}$ for these particular matrices. Hence, $\mathbf{BA} + \mathbf{AB} \neq \mathbf{AB} + \mathbf{AB}$ ($= 2\mathbf{AB}$). Therefore, the middle terms of both expressions $\mathbf{A}^2 + \mathbf{BA} + \mathbf{AB} + \mathbf{B}^2$ and $\mathbf{A}^2 + 2\mathbf{AB} + \mathbf{B}^2$ do not agree, and this explains why the matrices we obtained for these expressions above are not equal.

3. (a) $\mathbf{A}^{12} = \left(\mathbf{A}^3\right)^4 = \begin{bmatrix} 2838916 & -6404805 \\ 8539740 & 703981 \end{bmatrix}$

(b) $\left(\mathbf{A}^T\right)^8 = \left(\mathbf{A}^8\right)^T = \left(\mathbf{A}^3\mathbf{A}^5\right)^T = \begin{bmatrix} -37052 & 7068 \\ -5301 & -38819 \end{bmatrix}$

(c) $2\mathbf{A}^{11} = 2\mathbf{A}^3\mathbf{A}^3\mathbf{A}^5 = \begin{bmatrix} 4065448 & -613266 \\ 817688 & 3861026 \end{bmatrix}$

(d) $\left(-\mathbf{A}^T\right)^9 = (-1)^9\left(\mathbf{A}^T\right)^9 = (-1)\left(\mathbf{A}^9\right)^T = -\left(\left(\mathbf{A}^3\right)^3\right)^T$

$= \begin{bmatrix} 95308 & 141140 \\ -105855 & 60023 \end{bmatrix}$

5. Computation shows that $\mathbf{RS} = \mathbf{O}_{22}$. and $\mathbf{SR} = \begin{bmatrix} -18 & -96 & -90 \\ 9 & 48 & 45 \\ -6 & -32 & -30 \end{bmatrix}$.

Just because the product of two matrices in one order produces a zero matrix does not guarantee that the product of the same two matrices in the reverse order is also a zero matrix.

7. $\begin{bmatrix} 3 & 1 \\ 0 & 2 \end{bmatrix}\begin{bmatrix} 4 & 5 \\ 0 & -2 \end{bmatrix} = \begin{bmatrix} 12 & 13 \\ 0 & -4 \end{bmatrix}$,

$\begin{bmatrix} 1 & -3 & -1 \\ 0 & 5 & 3 \\ 0 & 0 & -7 \end{bmatrix}\begin{bmatrix} 2 & 8 & 1 \\ 0 & 4 & 5 \\ 0 & 0 & 3 \end{bmatrix} = \begin{bmatrix} 2 & -4 & -17 \\ 0 & 20 & 34 \\ 0 & 0 & -21 \end{bmatrix}$, and

$\begin{bmatrix} 2 & 1 & -1 & 3 \\ 0 & -4 & 3 & 7 \\ 0 & 0 & 5 & 4 \\ 0 & 0 & 0 & 9 \end{bmatrix}\begin{bmatrix} 8 & -1 & 2 & 4 \\ 0 & 1 & 3 & 5 \\ 0 & 0 & -6 & 7 \\ 0 & 0 & 0 & 5 \end{bmatrix} = \begin{bmatrix} 16 & -1 & 13 & 21 \\ 0 & -4 & -30 & 36 \\ 0 & 0 & -30 & 55 \\ 0 & 0 & 0 & 45 \end{bmatrix}$. We notice that,

in each case, the product of upper triangular matrices is upper triangular. This principle is true in general.

9. The (i, j) entry of $\mathbf{AB}$ is found by multiplying each entry of the ith row of $\mathbf{A}$ by its corresponding entry in the jth column of $\mathbf{B}$. For example, the $(1, 2)$ entry of $\mathbf{AB} = (\frac{1}{4})(\frac{3}{8}) + (\frac{1}{2})(\frac{1}{8}) + (\frac{1}{4})(\frac{1}{2}) = \frac{3}{32} + \frac{1}{16} + \frac{1}{8} = \frac{3}{32} + \frac{2}{32} + \frac{4}{32} = \frac{9}{32}$.

11. $\mathbf{B}^2 = \begin{bmatrix} \frac{4}{9} & \frac{1}{3} & \frac{7}{24} \\ \frac{1}{3} & \frac{3}{8} & \frac{1}{4} \\ \frac{2}{9} & \frac{7}{24} & \frac{11}{24} \end{bmatrix}$ and $\mathbf{B}^3 = \begin{bmatrix} \frac{37}{108} & \frac{101}{288} & \frac{7}{18} \\ \frac{5}{18} & \frac{1}{3} & \frac{17}{48} \\ \frac{41}{108} & \frac{91}{288} & \frac{37}{144} \end{bmatrix}$. Both $\mathbf{B}^2$ and $\mathbf{B}^3$ are

stochastic because all of their entries are positive, and the sum of the entries in each column equals 1.

13. $\begin{bmatrix} 5 & -2 \\ -2 & 6 \end{bmatrix} \begin{bmatrix} 4 & 1 \\ 1 & 3 \end{bmatrix} = \begin{bmatrix} 18 & -1 \\ -2 & 16 \end{bmatrix}$ and

$\begin{bmatrix} 7 & 2 & 3 \\ 2 & -1 & 4 \\ 3 & 4 & 0 \end{bmatrix} \begin{bmatrix} 5 & -1 & 2 \\ -1 & 8 & 7 \\ 2 & 7 & 6 \end{bmatrix} = \begin{bmatrix} 39 & 30 & 46 \\ 19 & 18 & 21 \\ 11 & 29 & 34 \end{bmatrix}$. We conclude that the prod-

uct of two symmetric matrices is not necessarily symmetric.

15. (a) $((3\mathbf{A} - 5\mathbf{B})\mathbf{A}^T)^T = (\mathbf{A}^T)^T(3\mathbf{A} - 5\mathbf{B})^T$ (by TR4)
$= \mathbf{A}(3\mathbf{A} - 5\mathbf{B})^T$ (by TR1)
$= \mathbf{A}((3\mathbf{A})^T - (5\mathbf{B})^T)$ (by TR3)
$= \mathbf{A}(3\mathbf{A}^T - 5\mathbf{B}^T)$ (by TR2)
$= 3\mathbf{A}\mathbf{A}^T - 5\mathbf{A}\mathbf{B}^T$ (by MM3).

(b) Assuming $\mathbf{A}$ and $\mathbf{B}$ are symmetric, then $\mathbf{A}^T = \mathbf{A}$, and $\mathbf{B}^T = \mathbf{B}$. Hence, the final result from part (a) reduces to $3\mathbf{A}\mathbf{A} - 5\mathbf{A}\mathbf{B} = 3\mathbf{A}^2 - 5\mathbf{A}\mathbf{B}$.

17. $\mathbf{A}^4 = \mathbf{O}_{44}$. No positive integer power lower than 4 produces $\mathbf{O}_{44}$.

19. See Exercise 13.

21. Since $\mathbf{A}$ is symmetric, $\mathbf{A}$ is square, so $\mathbf{A}^k$ is also a square matrix, since it has the same size as $\mathbf{A}$. Therefore to finish the proof, it is enough to show that when we take the transpose of $\mathbf{A}^k$, and simplify it, we obtain $\mathbf{A}^k$ again:

$$\left(\mathbf{A}^k\right)^T = \left(\mathbf{A}^T\right)^k \quad \text{by RFE3}$$
$$= \mathbf{A}^k \qquad \text{because } \mathbf{A}^T = \mathbf{A}, \text{ since } \mathbf{A} \text{ is symmetric,}$$

which ends the proof.

23. We are given that s is a scalar, $\mathbf{A}$ is an $m \times n$ matrix, and $\mathbf{B}$ is an $n \times p$ matrix.

Size of $s(\mathbf{A}\mathbf{B})$: Since the number of columns of $\mathbf{A} = n =$ the number of rows in $\mathbf{B}$, the product $\mathbf{A}\mathbf{B}$ is defined. Also, $\mathbf{A}\mathbf{B}$ is an $m \times p$ matrix because $\mathbf{A}$ has m rows and $\mathbf{B}$ has p columns. Since scalar multiplication does not affect the size, $s(\mathbf{A}\mathbf{B})$ also has size $m \times p$.

Size of $(s\mathbf{A})\mathbf{B}$: Since $\mathbf{A}$ is an $m \times n$ matrix, and scalar multiplication does not affect the size, $(s\mathbf{A})$ is also an $m \times n$ matrix. Since the number of columns of $(s\mathbf{A}) = n =$ the number of rows in $\mathbf{B}$, the product $(s\mathbf{A})\mathbf{B}$ is defined. Therefore, $(s\mathbf{A})\mathbf{B}$ is an $m \times p$ matrix because $(s\mathbf{A})$ has m rows and $\mathbf{B}$ has p columns.

Size of $\mathbf{A}(s\mathbf{B})$: Since $\mathbf{B}$ is an $n \times p$ matrix, and scalar multiplication does not affect the size, $(s\mathbf{B})$ is also an $n \times p$ matrix. Since the number of columns of $\mathbf{A} = n =$ the number of rows in $(s\mathbf{B})$, the product $\mathbf{A}(s\mathbf{B})$ is defined. The size of $\mathbf{A}(s\mathbf{B})$ is then $m \times p$ because $\mathbf{A}$ has m rows and $(s\mathbf{B})$ has p columns.

Conclusion: All of $s(\mathbf{A}\mathbf{B})$, $(s\mathbf{A})\mathbf{B}$, and $\mathbf{A}(s\mathbf{B})$ have the same size, namely, $m \times p$.

25. Suppose that **A** and **B** are both 3×3 stochastic matrices. We want to prove that **AB** is also a stochastic matrix by showing that the entries of each column of **AB** are positive or zero, and that the entries in each column of **AB** add up to 1.

We first show that each entry of the 1st column of **AB** is positive or zero. The entries in the 1st column of **AB** are:

$$
\begin{aligned}
(1,1) \text{ entry of } \mathbf{AB} &= a_{11}b_{11} + a_{12}b_{21} + a_{13}b_{31} \\
(2,1) \text{ entry of } \mathbf{AB} &= a_{21}b_{11} + a_{22}b_{21} + a_{23}b_{31} \; . \\
(3,1) \text{ entry of } \mathbf{AB} &= a_{31}b_{11} + a_{32}b_{21} + a_{33}b_{31}
\end{aligned}
$$

In the first sum, all of the factors a_{11}, b_{11}, a_{12}, b_{21}, a_{13}, b_{31} are positive or zero. Therefore, any products or sums involving these numbers must also be positive or zero. Thus, the $(1,1)$ entry of **AB** is positive or zero. A similar argument shows that the $(2,1)$ entry of **AB** and the $(3,1)$ entry of **AB** are both positive or zero.

Next, we show that the sum of the entries in the 1st column of **AB** equals 1. After pulling out common factors, this sum is equal to

$$(a_{11} + a_{21} + a_{31})b_{11} + (a_{12} + a_{22} + a_{32})b_{21} + (a_{13} + a_{23} + a_{33})b_{31}.$$

However, the sums in parentheses are the sums of the 1st, 2nd, and 3rd columns of **A**, respectively. But since **A** is a stochastic matrix, each parenthetical sum equals 1. Therefore, this formula reduces to $b_{11} + b_{21} + b_{31}$, which equals 1 because this is the sum of the entries in the 1st column of **B**, and **B** is a stochastic matrix. Thus, the sum of the entries in the 1st column of **AB** equals 1.

Similarly, we show that the sum of the entries in the 2nd column of **AB** equals 1. This sum is

$$
\begin{aligned}
& (1,2) \text{ entry of } \mathbf{AB} + (2,2) \text{ entry of } \mathbf{AB} + (3,2) \text{ entry of } \mathbf{AB} \\
={} & (a_{11}b_{12} + a_{12}b_{22} + a_{13}b_{32}) + (a_{21}b_{12} + a_{22}b_{22} + a_{23}b_{32}) \\
& \quad + (a_{31}b_{12} + a_{32}b_{22} + a_{33}b_{32}) \\
={} & (a_{11} + a_{21} + a_{31})b_{12} + (a_{12} + a_{22} + a_{32})b_{22} + (a_{13} + a_{23} + a_{33})b_{32} \\
={} & (1)b_{12} + (1)b_{22} + (1)b_{32} \text{ (sum of 2nd column of } \mathbf{B}, \\
& \quad \text{which is stochastic)} = 1.
\end{aligned}
$$

Finally, we show that the sum of the entries in the 3rd column of **AB** equals 1. This sum is

$$
\begin{aligned}
& (1,3) \text{ entry of } \mathbf{AB} + (2,3) \text{ entry of } \mathbf{AB} + (3,3) \text{ entry of } \mathbf{AB} \\
={} & (a_{11}b_{13} + a_{12}b_{23} + a_{13}b_{33}) + (a_{21}b_{13} + a_{22}b_{23} + a_{23}b_{33}) \\
& \quad + (a_{31}b_{13} + a_{32}b_{23} + a_{33}b_{33}) \\
={} & (a_{11} + a_{21} + a_{31})b_{13} + (a_{12} + a_{22} + a_{32})b_{23} + (a_{13} + a_{23} + a_{33})b_{33} \\
={} & (1)b_{13} + (1)b_{23} + (1)b_{33} \text{ (sum of 3rd column of } \mathbf{B}, \\
& \quad \text{which is stochastic)} = 1.
\end{aligned}
$$

Section 1.8

1. The second row of the product, by MSR1, is $\begin{bmatrix} -1 & 3 & 6 \end{bmatrix} \begin{bmatrix} 5 & 2 \\ -4 & 8 \\ 9 & 0 \end{bmatrix}$. Calculating this yields $\begin{bmatrix} 37 & 22 \end{bmatrix}$.

3. The desired linear combination is
$$3 \begin{bmatrix} 5 & -6 & 3 & -2 \end{bmatrix} + (-5) \begin{bmatrix} -8 & 2 & -1 & 7 \end{bmatrix} + 6 \begin{bmatrix} -3 & 0 & 9 & -4 \end{bmatrix}.$$
By MSR2, this can be expressed as
$$\begin{bmatrix} 3 & -5 & 6 \end{bmatrix} \begin{bmatrix} 5 & -6 & 3 & -2 \\ -8 & 2 & -1 & 7 \\ -3 & 0 & 9 & -4 \end{bmatrix} = \begin{bmatrix} 37 & -28 & 68 & -65 \end{bmatrix}.$$

5. $\begin{bmatrix} 4 & 7 & -2 \end{bmatrix} \begin{bmatrix} 8 & 3 & -1 & 0 \\ 5 & 6 & 0 & 2 \\ 3 & 1 & -4 & 9 \end{bmatrix}$
$= 4 \begin{bmatrix} 8 & 3 & -1 & 0 \end{bmatrix} + 7 \begin{bmatrix} 5 & 6 & 0 & 2 \end{bmatrix} - 2 \begin{bmatrix} 3 & 1 & -4 & 9 \end{bmatrix}$
$= \begin{bmatrix} 32 & 12 & -4 & 0 \end{bmatrix} + \begin{bmatrix} 35 & 42 & 0 & 14 \end{bmatrix} - \begin{bmatrix} 6 & 2 & -8 & 18 \end{bmatrix}$
$= \begin{bmatrix} 61 & 52 & 4 & -4 \end{bmatrix}.$

7. Using MSR3, $\mathbf{e}_2^T \begin{bmatrix} 6 & 2 & 3 \\ -4 & 5 & 8 \\ 0 & 2 & 1 \end{bmatrix} = \begin{bmatrix} -4 & 5 & 8 \end{bmatrix}$; $\mathbf{e}_3^T \begin{bmatrix} 4 & 2 \\ 5 & 9 \\ 4 & -3 \end{bmatrix} = \begin{bmatrix} 4 & -3 \end{bmatrix}.$

   Using MSC3, $\begin{bmatrix} 0 & 6 \\ 5 & -7 \\ 2 & 4 \end{bmatrix} \mathbf{e}_1 = \begin{bmatrix} 0 \\ 5 \\ 2 \end{bmatrix}$; $\begin{bmatrix} 1 & 2 & 3 & 4 \\ 5 & 6 & 7 & 8 \\ 6 & 5 & 4 & 3 \end{bmatrix} \mathbf{e}_4 = \begin{bmatrix} 4 \\ 8 \\ 3 \end{bmatrix}.$

9. (a) The matrix needed is $\mathbf{e}_3^T$; the product $\mathbf{e}_3^T \mathbf{A}$ gives the third row of $\mathbf{A}$

   (b) The matrix needed is $\mathbf{e}_4$; the product $\mathbf{A}\mathbf{e}_4$ gives the fourth column of $\mathbf{A}$

11. When $k = 0$, RFE1 becomes $\mathbf{A}^0 \mathbf{A}^l = \mathbf{A}^{0+l}$. The left-hand side equals $\mathbf{I}_n \mathbf{A}^l = \mathbf{A}^l$, which is equal to the right-hand side.
   When $k = 0$, RFE2 becomes $\left(\mathbf{A}^0\right)^l = \mathbf{A}^{0l}$. The left-hand side equals $(\mathbf{I}_n)^l = \mathbf{I}_n$, which is equal to the right-hand side since $\mathbf{A}^{0l} = \mathbf{A}^0 = \mathbf{I}_n$.

13. $\mathbf{B} = \begin{bmatrix} -18 & -202 \\ 101 & 184 \end{bmatrix}$; $\mathbf{AB} = \begin{bmatrix} -256 & -974 \\ 487 & 718 \end{bmatrix}$; $\mathbf{BA} = \mathbf{AB}$; $\mathbf{A}$ and $\mathbf{B}$ commute with each other.

15. $\mathbf{AC} = \mathbf{A} (s_1 \mathbf{B}_1 + \cdots + s_k \mathbf{B}_k) = \mathbf{A} (s_1 \mathbf{B}_1) + \cdots + \mathbf{A} (s_k \mathbf{B}_k)$ (by MM3)
   $= s_1 (\mathbf{AB}_1) + \cdots + s_k (\mathbf{AB}_k)$ (by MM1) $= s_1 (\mathbf{B}_1 \mathbf{A}) + \cdots + s_k (\mathbf{B}_k \mathbf{A})$ (because $\mathbf{A}$ commutes with each $\mathbf{B}_i$) $= (s_1 \mathbf{B}_1) \mathbf{A} + \cdots + (s_k \mathbf{B}_k) \mathbf{A}$ (by MM1)
   $= (s_1 \mathbf{B}_1 + \cdots + s_k \mathbf{B}_k) \mathbf{A}$ (by MM4) $= \mathbf{CA}$, completing the proof.

Chapter 1 Test

14. (a) T   (f) F   (k) F   (p) F   (u) T   (z) T   (ee) F
    (b) F   (g) T   (l) T   (q) F   (v) F   (aa) F
    (c) T   (h) F   (m) T   (r) T   (w) T   (bb) T
    (d) T   (i) F   (n) T   (s) F   (x) T   (cc) T
    (e) F   (j) F   (o) F   (t) F   (y) T   (dd) F

Section 2.1

1.

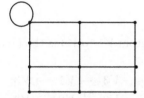

Streets in Isolation, Alaska

5.

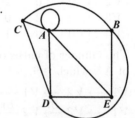

Graph of Figure 2.10
Without Crossing Edges

3.

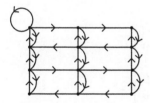

One-Way Streets in Isolation

Section 2.2

1.
$$\begin{array}{c c} & \begin{matrix} A & B & C & D & E \end{matrix} \\ \begin{matrix} A \\ B \\ C \\ D \\ E \end{matrix} & \begin{bmatrix} 1 & 1 & 0 & 2 & 1 \\ 1 & 0 & 1 & 1 & 0 \\ 0 & 1 & 1 & 1 & 0 \\ 2 & 1 & 1 & 0 & 1 \\ 1 & 0 & 0 & 1 & 0 \end{bmatrix} \end{array}$$

5.
$$\begin{array}{c c} & \begin{matrix} A & B & C & D & E \end{matrix} \\ \begin{matrix} A \\ B \\ C \\ D \\ E \end{matrix} & \begin{bmatrix} 0 & 2 & 1 & 0 & 0 \\ 1 & 0 & 0 & 1 & 2 \\ 1 & 0 & 1 & 0 & 0 \\ 0 & 1 & 0 & 0 & 0 \\ 0 & 0 & 0 & 1 & 0 \end{bmatrix} \end{array}$$

3.

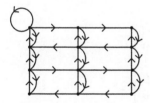

7.

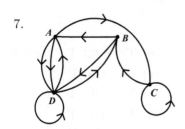

9. $\begin{bmatrix} 0 & 1 & 1 & 1 \\ 1 & 0 & 1 & 1 \\ 1 & 1 & 0 & 1 \\ 1 & 1 & 1 & 0 \end{bmatrix}$

Section 2.3

1. (a)

	$V1$	$V2$	$V3$	$V4$
$V1$	0	1	1	1
$V2$	1	0	2	0
$V3$	1	2	0	1
$V4$	1	0	1	1

(b) The number of paths of length 3 from $V1$ to $V3$ is the $(1, 3)$ entry of $\mathbf{A}^3$, which is 9.

(c) $V1 \longrightarrow V2 \longrightarrow V1 \longrightarrow V3$; $V1 \longrightarrow V3 \longrightarrow V1 \longrightarrow V3$;
$V1 \longrightarrow V3 \xrightarrow{E1} V2 \xrightarrow{E1} V3$; $V1 \longrightarrow V3 \xrightarrow{E1} V2 \xrightarrow{E2} V3$;
$V1 \longrightarrow V3 \xrightarrow{E2} V2 \xrightarrow{E1} V3$; $V1 \longrightarrow V3 \xrightarrow{E2} V2 \xrightarrow{E2} V3$;
$V1 \longrightarrow V3 \longrightarrow V4 \longrightarrow V3$; $V1 \longrightarrow V4 \longrightarrow V1 \longrightarrow V3$;
$V1 \longrightarrow V4 \longrightarrow V4 \longrightarrow V3$

3. (a)

	$V1$	$V2$	$V3$	$V4$	$V5$
$V1$	1	1	1	0	0
$V2$	0	0	0	0	1
$V3$	1	1	0	0	0
$V4$	0	0	1	0	0
$V5$	1	0	0	1	1

(b) The number of paths of length 4 from $V2$ to $V3$ is the $(2, 3)$ entry of $\mathbf{A}^4$, which is 3.

(c) $V2 \longrightarrow V5 \longrightarrow V1 \longrightarrow V1 \longrightarrow V3$;
$V2 \longrightarrow V5 \longrightarrow V5 \longrightarrow V1 \longrightarrow V3$;
$V2 \longrightarrow V5 \longrightarrow V5 \longrightarrow V4 \longrightarrow V3$

5. The adjacency matrix for the graph is

$\mathbf{A} =$

	$V1$	$V2$	$V3$	$V4$	$V5$
$V1$	0	1	0	1	0
$V2$	1	0	2	0	0
$V3$	0	2	0	1	0
$V4$	1	0	1	1	2
$V5$	0	0	0	2	0

(a) The number of paths of length 7 from $V1$ to $V3$ is the $(1, 3)$ entry of $\mathbf{A}^7$, which is 331.

(b) The number of paths of length 7 or less from $V1$ to $V3$ equals
(the $(1,3)$ entry of $\mathbf{A}$) + (the $(1,3)$ entry of $\mathbf{A}^2$)
+ (the $(1,3)$ entry of $\mathbf{A}^3$) + (the $(1,3)$ entry of $\mathbf{A}^4$)
+ (the $(1,3)$ entry of $\mathbf{A}^5$) + (the $(1,3)$ entry of $\mathbf{A}^6$)
+ (the $(1,3)$ entry of $\mathbf{A}^7$) = $0+3+1+26+22+240+331 = 623$.

(c) The length of the shortest path from $V2$ to $V5$ is determined by the lowest power of $\mathbf{A}$ whose $(2,5)$ entry is nonzero. Now,
(the $(2,5)$ entry of $\mathbf{A}$) $= 0$; (the $(2,5)$ entry of $\mathbf{A}^2$) $= 0$;
(the $(2,5)$ entry of $\mathbf{A}^3$) $= 6$.
Therefore, the shortest path from $V2$ to $V5$ has length 3. There are six different shortest paths from $V2$ to $V5$.

7. The adjacency matrix for the digraph is

$$
\mathbf{A} = \begin{array}{c} \\ V1 \\ V2 \\ V3 \\ V4 \\ V5 \\ V6 \end{array}
\begin{array}{c} \begin{array}{cccccc} V1 & V2 & V3 & V4 & V5 & V6 \end{array} \\
\left[\begin{array}{cccccc}
1 & 1 & 1 & 0 & 0 & 0 \\
0 & 0 & 0 & 0 & 0 & 0 \\
0 & 1 & 0 & 1 & 1 & 0 \\
0 & 0 & 0 & 0 & 1 & 0 \\
1 & 0 & 1 & 0 & 0 & 1 \\
0 & 0 & 0 & 0 & 1 & 0
\end{array}\right]. \end{array}
$$

(a) The number of paths of length 7 from $V1$ to $V4$ is the $(1,4)$ entry of $\mathbf{A}^7$, which is 16.

(b) The number of paths of length 7 from $V4$ to $V1$ is the $(4,1)$ entry of $\mathbf{A}^7$, which is 21.

(c) The number of paths of length 4 from $V2$ to $V5$ is the $(2,5)$ entry of $\mathbf{A}^4$, which is 0. This is not surprising, since there are no edges exiting $V2$, and so there are no paths of any length starting at $V2$! (Note, however, that there are 6 paths of length 4 from $V5$ to $V2$.)

(d) The number of paths of length 7 or less from $V1$ to $V3$ equals
(the $(1,3)$ entry of $\mathbf{A}$) + (the $(1,3)$ entry of $\mathbf{A}^2$)
+ (the $(1,3)$ entry of $\mathbf{A}^3$) + (the $(1,3)$ entry of $\mathbf{A}^4$)
+ (the $(1,3)$ entry of $\mathbf{A}^5$) + (the $(1,3)$ entry of $\mathbf{A}^6$)
+ (the $(1,3)$ entry of $\mathbf{A}^7$) = $1+1+2+4+8+16+32 = 64$.

(e) The length of the shortest path from $V4$ to $V2$ is determined by the lowest power of $\mathbf{A}$ whose $(4,2)$ entry is nonzero. Now,
(the $(4,2)$ entry of $\mathbf{A}$) $= 0$; (the $(4,2)$ entry of $\mathbf{A}^2$) $= 0$;
(the $(4,2)$ entry of $\mathbf{A}^3$) $= 2$.
Therefore, the shortest path from $V4$ to $V2$ has length 3. There are two different shortest paths from $V4$ to $V2$.

(f) There is no loop at $V5$, so there is no path of length 1. However, the $(5,5)$ entry of $\mathbf{A}^2$ is nonzero, so the length of the shortest path from $V5$ to $V5$ is 2. Since the $(5,5)$ entry of $\mathbf{A}^2 = 2$, there are actually

2 paths of length 2 from $V5$ to $V5$. These are: $V5 \longrightarrow V3 \longrightarrow V5$ and $V5 \longrightarrow V6 \longrightarrow V5$.

9. (a) Since there are 4 vertices, we raise $(\mathbf{I}_4 + \mathbf{A})$ to the $4 - 1 = 3$ power to obtain: $(\mathbf{I}_4 + \mathbf{A})^3 = \begin{bmatrix} 15 & 0 & 21 & 7 \\ 0 & 8 & 0 & 0 \\ 21 & 0 & 43 & 21 \\ 7 & 0 & 21 & 15 \end{bmatrix}$. Because $(\mathbf{I}_4 + \mathbf{A})^3$ has zero entries, the associated graph is disconnected. $V1$ and $V2$ represent a pair of vertices that are not connected. (In fact, $V2$ is not connected to any of the other vertices.)

   (b) The graph associated with $\mathbf{B}$ is disconnected because $(\mathbf{I}_5 + \mathbf{B})^4$ contains zero entries. $V1$ and $V2$ represent a pair of vertices that are not connected. (In fact, none of the vertices in $\{V1, V3, V5\}$ are connected to any of the vertices in $\{V2,\ V4\}$.)

   (c) The graph associated with $\mathbf{C}$ is connected because $(\mathbf{I}_6 + \mathbf{C})^5$ contains no zero entries.

11. In Exercise 9, only the graph associated with $\mathbf{C}$ is connected. In this case, $(\mathbf{I}_6 + \mathbf{C})^4$ is the highest power of $(\mathbf{I}_6 + \mathbf{C})$ that still contains zero entries. In particular, the $(3, 4)$ and $(4, 3)$ entries of $(\mathbf{I}_6 + \mathbf{C})^4$ equal zero, and are the only zero entries in $(\mathbf{I}_6 + \mathbf{C})^4$. Thus, there are no paths of length 4 between $V3$ and $V4$. However, there are paths of length 4 or less between any two other distinct vertices. Now the $(3, 4)$ and $(4, 3)$ entries of $(\mathbf{I}_6 + \mathbf{C})^5$ are nonzero, so there does exist at least one path of length 5 between $V3$ and $V4$. So, $V3$ and $V4$ are the two vertices that are farthest apart.

13. Because there are theoretically at most six degrees of separation between any two people, there should be a path of length six or less connecting any two vertices of the graph. Therefore $(\mathbf{I} + \mathbf{A})^6$ should not have any zero entries. However, we should expect that $(\mathbf{I} + \mathbf{A})^5$ has some entries that equal zero. Otherwise, the theory would say that there are only 5 degrees of separation between any two people!

Chapter 2 Test

9. (a) T (e) T (i) F (m) F (q) T (u) F (y) T
   (b) F (f) F (j) F (n) T (r) F (v) T (z) F
   (c) F (g) T (k) T (o) F (s) T (w) F
   (d) T (h) T (l) F (p) F (t) F (x) F

Section 3.1

1. (a) 3   (c) 4
   (b) 5   (d) Not possible out of context

3. (a) $(3, 0, -2)$ (c) $(3, 1, 2, -5)$
   (b) $(1, -1, -1, 1)$ (d) $(3, -1, 8)$

5. $[2, -3, 5]$

7. $(2, -5, -1)$

9. (a)   (c)

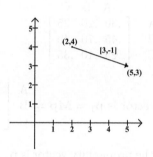

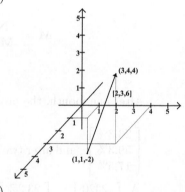

(d)

(b)

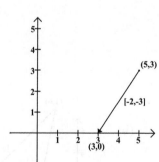

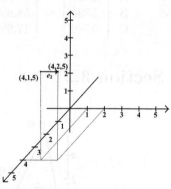

11. The current probability vector is $\mathbf{p} = \begin{matrix} \text{GW} \\ \text{SV} \end{matrix} \begin{bmatrix} .70 \\ .30 \end{bmatrix}$, and the matrix for the Markov chain is

$$\mathbf{M} = \begin{matrix} \textbf{Next} \\ \textbf{Day} \end{matrix} \begin{matrix} \text{GW} \\ \text{SV} \end{matrix} \begin{bmatrix} .60 & .20 \\ .30 & .50 \end{bmatrix} .$$

Current Day
GW SV

After one day, the probability vector is $\mathbf{p}_1 = \mathbf{Mp} = \begin{matrix} GW \\ SV \end{matrix} \begin{bmatrix} .295 \\ .705 \end{bmatrix} =$
$\begin{bmatrix} 29.5\% \\ 70.5\% \end{bmatrix}$, and after two days, the probability vector is $\mathbf{p}_2 = \mathbf{Mp}_1 \approx$
$\begin{matrix} GW \\ SV \end{matrix} \begin{bmatrix} .356 \\ .644 \end{bmatrix} = \begin{bmatrix} 35.6\% \\ 64.4\% \end{bmatrix}$.

13. The current probability vector is $\mathbf{p} = \begin{matrix} A \\ B \\ C \end{matrix} \begin{bmatrix} .25 \\ .60 \\ .15 \end{bmatrix}$, and the matrix for

the Markov chain is

$$\mathbf{M} = \begin{matrix} \textbf{Next} \\ \textbf{Month} \end{matrix} \begin{matrix} \\ A \\ B \\ C \end{matrix} \overset{\begin{matrix} \textbf{Current Month} \\ A \quad B \quad C \end{matrix}}{\begin{bmatrix} .30 & .20 & .25 \\ .45 & .70 & .40 \\ .25 & .10 & .35 \end{bmatrix}}.$$

After one month, the probability vector is $\mathbf{p}_1 = \mathbf{Mp} = \begin{matrix} A \\ B \\ C \end{matrix} \begin{bmatrix} .2325 \\ .5925 \\ .1750 \end{bmatrix} =$
$\begin{bmatrix} 23.3\% \\ 59.3\% \\ 17.5\% \end{bmatrix}$, and after two months, the probability vector is $\mathbf{p}_2 = \mathbf{Mp}_1 \approx$
$\begin{matrix} A \\ B \\ C \end{matrix} \begin{bmatrix} .2320 \\ .5894 \\ .1786 \end{bmatrix} = \begin{bmatrix} 23.2\% \\ 58.9\% \\ 17.9\% \end{bmatrix}$.

Section 3.2

1. (a)

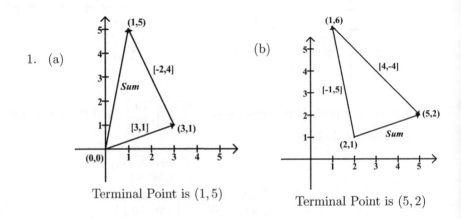

Terminal Point is $(1,5)$

(b)

Terminal Point is $(5,2)$

(c)

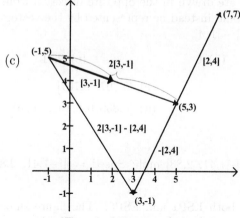

Terminal Point is $(3, -1)$

(d)

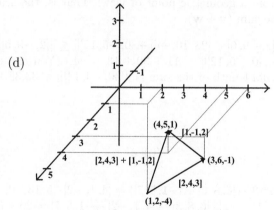

Terminal Point is $(4, 5, 1)$

3.

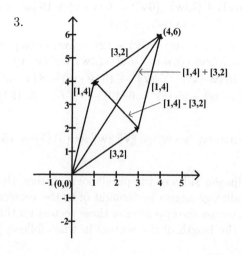

If the vector diagonals are drawn in the opposite direction from those in the diagram, they would instead be represented by the vectors $-[1,4] - [3,2]$ and $[3,2] - [1,4]$.

5. (a) 7   (b) 6   (c) 9.11

7. (a) $\left[\frac{11}{15}, -\frac{2}{15}, \frac{2}{3}\right]$ (c) [0.707, 0.707]

   (b) $\left[-\frac{2}{3}, 0, \frac{4}{9}, \frac{5}{9}, -\frac{2}{9}\right]$ (d) $[0.839, 0, 0.105, -0.105, 0.524]$

9. 72

11. Resultant Velocity $= [-1.41, 2.83]$ mph; Speed $= \|[-1.41, 2.83]\| = 3.16$ mph.

13. Figure 3.14 illustrates both ESP1 and ESP5. The figure shows that $\mathbf{v} + \mathbf{w} = \mathbf{w} + \mathbf{v}$ and $3\mathbf{v} + 3\mathbf{w} = 3\mathbf{w} + 3\mathbf{v}$, which are both examples of ESP1. Also, ESP5 is demonstrated because $3(\mathbf{v} + \mathbf{w})$ and $3\mathbf{v} + 3\mathbf{w}$ are the same vector from a geometric point of view. That is, the scalar 3 distributes over the sum $(\mathbf{v} + \mathbf{w})$.

15. By Corollary 2, $\|[2, -9, 6] + [28, 10, -4] + [0, -5, 12]\| \leq \|[2, -9, 6]\| + \|[28, 10, -4]\| + \|[0, -5, 12]\| = 11 + 30 + 13 = 54$. (You can easily verify that the actual length of the sum is $\|[30, -4, 14]\| \approx 33.35$.)

Section 3.3

1. (a) -24 (b) 0

3. $\mathbf{v} \cdot (\mathbf{w} + \mathbf{z}) = [4, 1, -2] \cdot ([6, 8, 3] + [-1, 5, 7]) = [4, 1, -2] \cdot [5, 13, 10] = 13;$
   $(\mathbf{v} \cdot \mathbf{w}) + (\mathbf{v} \cdot \mathbf{z}) = [4, 1, -2] \cdot [6, 8, 3] + [4, 1, -2] \cdot [-1, 5, 7] = 26 + (-13)$
   $= 13.$

5. (a) $(\mathbf{v} + 3\mathbf{w}) \cdot (6\mathbf{v}) = (\mathbf{v} \cdot (6\mathbf{v})) + ((3\mathbf{w}) \cdot (6\mathbf{v})) = 6(\mathbf{v} \cdot \mathbf{v}) + 18(\mathbf{w} \cdot \mathbf{v})$
     $= 6\|\mathbf{v}\|^2 + 18(\mathbf{v} \cdot \mathbf{w}) = 6(7)^2 + 18(-5) = 204.$

   (b) $(2\mathbf{v} - 5\mathbf{w}) \cdot (3\mathbf{v} + 4\mathbf{w}) = (2\mathbf{v} - 5\mathbf{w}) \cdot (3\mathbf{v}) + (2\mathbf{v} - 5\mathbf{w}) \cdot (4\mathbf{w}) =$
     $(2\mathbf{v}) \cdot (3\mathbf{v}) - (5\mathbf{w}) \cdot (3\mathbf{v}) + (2\mathbf{v}) \cdot (4\mathbf{w}) - (5\mathbf{w}) \cdot (4\mathbf{w}) = 6(\mathbf{v} \cdot \mathbf{v}) -$
     $15(\mathbf{w} \cdot \mathbf{v}) + 8(\mathbf{v} \cdot \mathbf{w}) - 20(\mathbf{w} \cdot \mathbf{w}) = 6\|\mathbf{v}\|^2 - 15(\mathbf{v} \cdot \mathbf{w}) + 8(\mathbf{v} \cdot \mathbf{w})$
     $- 20\|\mathbf{w}\|^2 = 6(7)^2 - 7(\mathbf{v} \cdot \mathbf{w}) - 20(8)^2 = 6(49) - 7(-5) - 20(64)$
     $= -951.$

7. By the Cauchy-Schwarz Inequality, $|\mathbf{v} \cdot \mathbf{w}| \leq \|\mathbf{v}\| \|\mathbf{w}\| = (3)(5) = 15$. Hence, $-15 \leq \mathbf{v} \cdot \mathbf{w} \leq 15$.

9. If $\mathbf{v}$ and $\mathbf{w}$ represent two adjacent sides of the parallelogram, then the distinct diagonals of the parallelogram can be thought of as the vectors $\mathbf{v} + \mathbf{w}$ and $\mathbf{v} - \mathbf{w}$. (If, instead, we use the opposites of these vectors for the diagonals, it does not affect the length of the vectors in what follows.)

Now, since we do not know which diagonal is which, we either have $\|(\mathbf{v}+\mathbf{w})\| = 7$ and $\|(\mathbf{v}-\mathbf{w})\| = 5$, or, $\|(\mathbf{v}+\mathbf{w})\| = 5$ and $\|(\mathbf{v}-\mathbf{w})\| = 7$. Using the Polarization Identity in the first case produces $\mathbf{v}\cdot\mathbf{w} = \frac{1}{4}\left(7^2 - 5^2\right) = 6$. In the second case we get $\mathbf{v}\cdot\mathbf{w} = \frac{1}{4}\left(5^2 - 7^2\right) = -6$. Therefore, $\mathbf{v}\cdot\mathbf{w}$ equals either 6 or -6, and so $|\mathbf{v}\cdot\mathbf{w}| = 6$. We would need to know which diagonal has which length to determine the sign of $\mathbf{v}\cdot\mathbf{w}$.

11. (a) $|\mathbf{v}\cdot\mathbf{w}| = |[1,2,-5]\cdot[1,2,-5]| = 30$;
$\|\mathbf{v}\|\,\|\mathbf{w}\| = \|[1,2,-5]\|\,\|[1,2,-5]\| = \sqrt{30}\sqrt{30} = 30$.

(b) $|\mathbf{v}\cdot\mathbf{w}| = |[2,-1,8]\cdot[-2,1,-8]| = |-69| = 69$;
$\|\mathbf{v}\|\,\|\mathbf{w}\| = \|[2,-1,8]\|\,\|[-2,1,-8]\|$
$= \sqrt{2^2+(-1)^2+8^2}\sqrt{(-2)^2+1^2+(-8)^2} = \sqrt{69}\sqrt{69} = 69$.

(c) $|\mathbf{v}\cdot\mathbf{w}| = |[-4,6,7]\cdot[-12,18,21]| = 303$;
$\|\mathbf{v}\|\,\|\mathbf{w}\| = \|[-4,6,7]\|\,\|[-12,18,21]\|$
$= \sqrt{(-4)^2+6^2+7^2}\sqrt{(-12)^2+18^2+21^2}$
$= \sqrt{101}\sqrt{909} = \sqrt{91809} = 303$.

(d) $|\mathbf{v}\cdot\mathbf{w}| = |[0,0,0]\cdot[8,11,-13]| = 0$;
$\|\mathbf{v}\|\,\|\mathbf{w}\| = \|[0,0,0]\|\,\|[8,11,-13]\| = (0)\sqrt{354} = 0$.

13. $|\mathbf{v}\cdot\mathbf{w}| \leq \|\mathbf{v}\|\,\|\mathbf{w}\|$. Now, $\|\mathbf{v}\|\,\|\mathbf{w}\| = \|[7,-4,4]\|\,\|[2,3,-6]\| = (9)(7) = 63$. Therefore, $-63 \leq \mathbf{v}\cdot\mathbf{w} \leq 63$. That is, -63 is a lower estimate for $\mathbf{v}\cdot\mathbf{w}$, and 63 is an upper estimate for $\mathbf{v}\cdot\mathbf{w}$.

15. The problem is that the expression $\mathbf{x}\cdot(\mathbf{y}\cdot\mathbf{z})$ does not make sense. Computing $(\mathbf{y}\cdot\mathbf{z})$ results in a *scalar*, not a vector. Hence, it is impossible to calculate $\mathbf{x}\cdot(\mathbf{y}\cdot\mathbf{z})$, because this would involve taking the dot product of a vector with a scalar. The dot product requires two vectors. Similarly, $(\mathbf{x}\cdot\mathbf{y})\cdot\mathbf{z}$ would involve the dot product of a scalar with a vector, which does not fit the definition of the dot product.

17. If $\mathbf{v}$ and $\mathbf{w}$ represent two adjacent sides of the parallelogram, then the distinct diagonals of the parallelogram can be thought of as the vectors $\mathbf{v}+\mathbf{w}$ and $\mathbf{v}-\mathbf{w}$. Therefore, the Parallelogram Law states that the sum of the squares of the lengths of two adjacent sides of a parallelogram equals $\frac{1}{2}$ the sum of the squares of the lengths of the (two) distinct diagonals of the parallelogram.

19. Suppose that $\mathbf{v}$ and $\mathbf{w}$ are two vectors having the same dimension.

(a) $\|\mathbf{v}+\mathbf{w}\|^2 = (\mathbf{v}+\mathbf{w})\cdot(\mathbf{v}+\mathbf{w}) = (\mathbf{v}\cdot\mathbf{v}) + (\mathbf{v}\cdot\mathbf{w}) + (\mathbf{w}\cdot\mathbf{v}) + (\mathbf{w}\cdot\mathbf{w})$
$= (\mathbf{v}\cdot\mathbf{v}) + (\mathbf{v}\cdot\mathbf{w}) + (\mathbf{v}\cdot\mathbf{w}) + (\mathbf{w}\cdot\mathbf{w}) = \|\mathbf{v}\|^2 + 2(\mathbf{v}\cdot\mathbf{w}) + \|\mathbf{w}\|^2$.

(b) $(\|\mathbf{v}\| + \|\mathbf{w}\|)^2 = (\|\mathbf{v}\| + \|\mathbf{w}\|)(\|\mathbf{v}\| + \|\mathbf{w}\|) = \|\mathbf{v}\|^2 + \|\mathbf{v}\|\,\|\mathbf{w}\| + \|\mathbf{w}\|\,\|\mathbf{v}\| + \|\mathbf{w}\|^2 = \|\mathbf{v}\|^2 + 2\|\mathbf{v}\|\,\|\mathbf{w}\| + \|\mathbf{w}\|^2$.

(c) Squaring both sides of the Triangle Inequality yields

$$\|\mathbf{v} + \mathbf{w}\|^2 \leq (\|\mathbf{v}\| + \|\mathbf{w}\|)^2.$$

Next, substitute in the results from parts (a) and (b) of this exercise to produce

$$\|\mathbf{v}\|^2 + 2(\mathbf{v} \cdot \mathbf{w}) + \|\mathbf{w}\|^2 \leq \|\mathbf{v}\|^2 + 2\|\mathbf{v}\|\,\|\mathbf{w}\| + \|\mathbf{w}\|^2.$$

Now subtract $\|\mathbf{v}\|^2 + \|\mathbf{w}\|^2$ from both sides of the inequality. This gives $2(\mathbf{v} \cdot \mathbf{w}) \leq 2\|\mathbf{v}\|\,\|\mathbf{w}\|$. Finally, dividing both sides by 2 yields $\mathbf{v} \cdot \mathbf{w} \leq \|\mathbf{v}\|\,\|\mathbf{w}\|$, which is what we were asked to prove.

Section 3.4

1. (a) Angle is a right angle   (c) Angle is obtuse

   (b) Angle is acute   (d) Angle is acute

3. (a) 95.565°   (b) 48.190°   (c) 53.105°   (d) 90°

5. In each part, we must verify that the dot product of each pair of distinct vectors equals zero. (Note that because the dot product is commutative (DP2), we only have to compute each dot product in one of the two possible orders.)

   (a) $[6, 1, -4] \cdot [5, 2, 8] = (6)(5) + (1)(2) + (-4)(8) = 30 + 2 - 32 = 0$
   $[6, 1, -4] \cdot [0, 0, 0] = (6)(0) + (1)(0) + (-4)(0) = 0 + 0 + 0 = 0$
   $[6, 1, -4] \cdot [16, -68, 7] = (6)(16) + (1)(-68) + (-4)(7) = 96 - 68 - 28 = 0$
   $[5, 2, 8] \cdot [0, 0, 0] = (5)(0) + (2)(0) + (8)(0) = 0 + 0 + 0 = 0$
   $[5, 2, 8] \cdot [16, -68, 7] = (5)(16) + (2)(-68) + (8)(7) = 80 - 136 + 56 = 0$
   $[16, -68, 7] \cdot [0, 0, 0] = (16)(0) + (-68)(0) + (7)(0) = 0 + 0 + 0 = 0$

   (b) $[4, 1, 2, -3] \cdot [2, 19, 0, 9] = (4)(2) + (1)(19) + (2)(0) + (-3)(9) = 8 + 19 + 0 - 27 = 0$
   $[4, 1, 2, -3] \cdot [1, -2, 5, 4] = (4)(1) + (1)(-2) + (2)(5) + (-3)(4) = 4 - 2 + 10 - 12 = 0$
   $[2, 19, 0, 9] \cdot [1, -2, 5, 4] = (2)(1) + (19)(-2) + (0)(5) + (9)(4) = 2 - 38 + 0 + 36 = 0$

7. We normalize each of the vectors in the set by dividing each vector by its norm. Now, $\|[2, 10, 11]\| = 15$, $\|[1, 2, -2]\| = 3$, and $\|[14, -5, 2]\| = 15$. These values produce the orthonormal set

$$\left\{ \left[\frac{2}{15}, \frac{2}{3}, \frac{11}{15} \right], \left[\frac{1}{3}, \frac{2}{3}, -\frac{2}{3} \right], \left[\frac{14}{15}, -\frac{1}{3}, \frac{2}{15} \right] \right\}.$$

9. Recall the Polarization Identity: $\mathbf{v} \cdot \mathbf{w} = \frac{1}{4}\left(\|\mathbf{v} + \mathbf{w}\|^2 - \|\mathbf{v} - \mathbf{w}\|^2\right)$. In each part, we think of the given adjacent sides of the parallelogram as the vectors $\mathbf{v}$ and $\mathbf{w}$, with these vectors starting at the point common to both adjacent sides. Then, the vectors $\mathbf{v} + \mathbf{w}$ and $\mathbf{v} - \mathbf{w}$ lie along the diagonals of the parallelogram.

(a) Choose $\mathbf{v} = \overrightarrow{AD}$ and $\mathbf{w} = \overrightarrow{AB}$. From Figure 3.18, $\|\mathbf{v}\| = 5.19$, $\|\mathbf{w}\| = 3.18$, $\|\mathbf{v} + \mathbf{w}\| = 5.00$, and $\|\mathbf{v} - \mathbf{w}\| = 7.01$. Hence, $\mathbf{v} \cdot \mathbf{w} = \frac{1}{4}\left(\|\mathbf{v} + \mathbf{w}\|^2 - \|\mathbf{v} - \mathbf{w}\|^2\right) = \frac{1}{4}\left((5)^2 - (7.01)^2\right) = -6.035025$. Therefore, the measure of the angle between $\mathbf{v}$ and $\mathbf{w}$ is $m\angle BAD = \arccos\left(\frac{-6.035025}{(5.19)(3.18)}\right) \approx \arccos\left(-0.365666\right) \approx 111°$. Then, since adjacent angles of a parallelogram are supplementary (that is, they add up to $180°$), we see that $m\angle ABC \approx 69°$. Also, opposite angles of a parallelogram are equal, so $m\angle BCD \approx 111°$ and $m\angle ADC \approx 69°$.

(b) Choose $\mathbf{v} = \overrightarrow{FG}$ and $\mathbf{w} = \overrightarrow{FE}$. From Figure 3.18, $\|\mathbf{v}\| = 3.29$, $\|\mathbf{w}\| = 3.15$, $\|\mathbf{v} + \mathbf{w}\| = 4.55$, and $\|\mathbf{v} - \mathbf{w}\| = 4.55$. Hence, $\mathbf{v} \cdot \mathbf{w} = \frac{1}{4}\left(\|\mathbf{v} + \mathbf{w}\|^2 - \|\mathbf{v} - \mathbf{w}\|^2\right) = \frac{1}{4}\left((4.55)^2 - (4.55)^2\right) = 0$. Therefore, the angle between $\mathbf{v}$ and $\mathbf{w}$ is a right angle! Then, since adjacent angles of a parallelogram are supplementary, we see that $\angle FEH$ is also a right angle. Also, opposite angles of a parallelogram are equal, so all four angles of this parallelogram are right angles. This parallelogram is a rectangle!

In fact, whenever the diagonals of a parallelogram have the same length, the Polarization Identity shows that the dot product of the vectors along two adjacent sides must equal zero. Hence, the angle between them is a right angle, and therefore, the parallelogram is a rectangle.

11. For two nonzero vectors $\mathbf{v}$ and $\mathbf{w}$, $\mathbf{v} \cdot \mathbf{w} = \|\mathbf{v}\|\|\mathbf{w}\|$ if and only if $\frac{\mathbf{v} \cdot \mathbf{w}}{\|\mathbf{v}\|\|\mathbf{w}\|} = 1$. Hence, using the fact given in the exercise, this occurs if and only if the angle between $\mathbf{v}$ and $\mathbf{w}$ has measure $0°$. This means that $\mathbf{v}$ and $\mathbf{w}$ must be in the same direction.

Section 3.5

1. (a) $\mathbf{proj_w v} \approx [2.17, 0.87, -0.43]$

(b) $\mathbf{proj_w v} \approx [2.67, 0, -2, 67, 1.33]$

3. (a) $\mathbf{p} = \mathbf{proj_w v} = [12, -8, 18]$; $\mathbf{r} = \mathbf{v} - \mathbf{proj_w v} = [5, 3, -2]$; $\mathbf{r} \cdot \mathbf{w} = [5, 3, -2] \cdot [6, -4, 9] = 0$.

(b) $\mathbf{p} = \mathbf{proj_w v} = [-2.4, 1.6, -0.8, 3.2]$; $\mathbf{r} = \mathbf{v} - \mathbf{proj_w v} = [9.4, 7.4, -2.2, 2.8]$; $\mathbf{r} \cdot \mathbf{w} = [9.4, 7.4, -2.2, 2.8] \cdot [3, -2, 1, -4] = 0$.

5. (a) $\mathbf{s} = \mathbf{proj_y x} \approx [1.9, -3.7, -4.7]$; $\mathbf{t} = \mathbf{x} - \mathbf{proj_y x} \approx [1.1, 4.7, -3.3]$;
   $\mathbf{t} \cdot \mathbf{y} \approx [1.1, 4.7, -3.3] \cdot [-2, 4, 5] = 0.1 \approx 0$.

   (b) $\mathbf{s} = \mathbf{proj_y x} \approx [-2.1, 0.0, -1.0, 1.6]$; $\mathbf{t} = \mathbf{x} - \mathbf{proj_y x} \approx$
   $[0.1, 6.0, 5.0, 3.4]$; $\mathbf{t} \cdot \mathbf{y} = [0.1, 6.0, 5.0, 3.4] \cdot [-4, 0, -2, 3] = -0.2 \approx 0$.

7. (a) $c = \left(\frac{\mathbf{z} \cdot \mathbf{y}}{\|\mathbf{y}\|^2}\right) = \left(\frac{[9, -2, 3] \cdot [6, 1, 5]}{\|[6, 1, 5]\|^2}\right) = \frac{67}{62} \approx 1.08$; $\mathbf{r} = \mathbf{z} - c\mathbf{y} \approx$
   $[2.52, -3.08, -2.40]$. These are the only possible answers for c and
   $\mathbf{r}$ according to the uniqueness assertion in the Projection Theorem.

   (b) $c = \left(\frac{\mathbf{z} \cdot \mathbf{y}}{\|\mathbf{y}\|^2}\right) = \left(\frac{[-9, 3, -6] \cdot [6, -2, 4]}{\|[6, -2, 4]\|^2}\right) = \frac{-84}{56} = -1.5$; $\mathbf{r} = \mathbf{z} - c\mathbf{y} =$
   $[0, 0, 0]$. The remainder, $\mathbf{r}$, is the zero vector because $\mathbf{z}$ is a scalar
   multiple of $\mathbf{y}$, and hence, parallel to $\mathbf{y}$, which means the component
   of $\mathbf{z}$ orthogonal to $\mathbf{y}$ is the zero vector. These are the only possible
   answers for c and $\mathbf{r}$ according to the uniqueness assertion in the
   Projection Theorem.

   (c) $c = \left(\frac{\mathbf{z} \cdot \mathbf{y}}{\|\mathbf{y}\|^2}\right) = \left(\frac{[6, -5, -2, -2] \cdot [3, 2, -1, 5]}{\|[3, 2, -1, 5]\|^2}\right) = \frac{0}{39} = 0$; $\mathbf{r} = \mathbf{z} - c\mathbf{y} \approx$
   $[6, -5, -2, -2]$. The scalar c is zero because $\mathbf{z}$ and $\mathbf{y}$ are orthogonal.
   The vector $\mathbf{z}$ has no component parallel to $\mathbf{y}$. These are the only
   possible answers for c and $\mathbf{r}$ according to the uniqueness assertion
   in the Projection Theorem.

9. Force along the ramp $\approx [-6.88, -2.06]$. The magnitude of this force
   ≈ 7.18 lbs. The force against the ramp $\approx [6.88, -22.94]$. The magnitude
   of this force ≈ 23.95 lbs.

11. (a) $\{[1, 3, 2], [1, -1, 1], [5, 1, -4]\}$

    (b) $\{[1, 3, 1, 2], [18, -1, -27, 6], [0, 0, 0, 0], [8, -9, 7, 6]\}$ (When calculat-
    ing the fourth "adjusted" vector, do not use the third "adjusted"
    vector $[0, 0, 0, 0]$, since there is no component of $[0, 0, 0, 0]$ that needs
    to be eliminated.)

13. $\{[0.27, 0.80, 0.53], [0.58, -0.58, 0.58], [0.77, 0.15, -0.62]\}$

Section 3.6

1. Let $\mathbf{v}_1 = [2, 1, -1]$ and $\mathbf{v}_2 = [1, 2, 4]$.

   (a) Neither vector is a scalar multiple of the other, so they have different directions. (Also note that $\{\mathbf{v}_1, \mathbf{v}_2\}$ is a mutually orthogonal
   set, and so the LCMOV Theorem confirms that neither is a scalar
   multiple of the other.)

   (b) Let $\mathbf{v}_3 = [5, -2, -16]$. We are given that $\mathbf{v}_3 = 4\mathbf{v}_1 - 3\mathbf{v}_2$. Hence,
   $-2\mathbf{v}_3 = -8\mathbf{v}_1 + 6\mathbf{v}_2$. Therefore, we can replicate turning the third
   dial twice counterclockwise by, instead, turning the first dial 8 times
   counterclockwise and turning the second dial 6 times clockwise.

(c) Neither vector is a scalar multiple of the other, so they have different directions (although in this case, they are not orthogonal).

(d) Take the equation $\mathbf{v}_3 = 4\mathbf{v}_1 - 3\mathbf{v}_2$ and solve for $\mathbf{v}_2$, producing $\mathbf{v}_2 = \frac{4}{3}\mathbf{v}_1 - \frac{1}{3}\mathbf{v}_3$. Therefore, turning the second dial once clockwise is equivalent to turning the first dial $\frac{4}{3}$ of the way around clockwise and turning the third dial $\frac{1}{3}$ of the way around counterclockwise.

(e) Let $\mathbf{v}_4 = [2, -3, 1]$. Notice that $\{\mathbf{v}_1, \mathbf{v}_2, \mathbf{v}_4\}$ is a mutually orthogonal set. Hence, the LCMOV Theorem tells us that none of these vectors is a linear combination of the others. Therefore, no movement of any one of the first, second, or fourth dials can be replicated by a combination of movements of the other two dials.

3. (a) $\mathbf{v}_1 = [1, 2, 3]$; $\mathbf{v}_2 = [4, 1, -2]$

(b) The first dial is unchanged, and $\mathbf{v}_2 = -\frac{16}{3}\mathbf{v}_1 + \frac{7}{3}[4, 5, 6]$. (These coefficients were found while performing the Gram-Schmidt Process.) Therefore, since $\mathbf{v}_2$ is a linear combination of $[1, 2, 3]$ and $[4, 5, 6]$, it will not take us in a new direction. That is, after resetting the dials, any manipulation of these dials keeps us in the same plane containing $[1, 2, 3]$ and $[4, 5, 6]$.

(c) One possible answer is $\mathbf{v}_3 = [1, -2, 1]$.

(d) The set $\{\mathbf{v}_1, \mathbf{v}_2, \mathbf{v}_3\}$ is a mutually orthogonal set. So, by the LCMOV Theorem, none of these vectors is a linear combination of the others. Therefore, $\mathbf{v}_3$ must go in a new direction, out of the plane containing $[1, 2, 3]$ and $[4, 5, 6]$.

5. (a) If $\{\mathbf{v}_1, \ldots, \mathbf{v}_k\}$ is a mutually orthogonal set of nonzero vectors in $\mathbb{R}^n$, then no vector in this set can be expressed as a linear combination of the other vectors in the set.

(b) Negation: There is a mutually orthogonal set $\{\mathbf{v}_1, \ldots, \mathbf{v}_k\}$ of nonzero vectors such that some vector in this set can be expressed as a linear combination of the other vectors in the set. This negation contains no universal quantifiers, but only existential quantifiers.

Chapter 3 Test

12.

(a) F	(h) T	(o) F	(v) T	(cc) T	(jj) T	(qq) T
(b) F	(i) F	(p) F	(w) F	(dd) T	(kk) F	(rr) T
(c) T	(j) F	(q) F	(x) T	(ee) T	(ll) T	
(d) T	(k) T	(r) F	(y) F	(ff) T	(mm) F	(ss) F
(e) T	(l) F	(s) T	(z) T	(gg) T	(nn) T	(tt) T
(f) F	(m) T	(t) T	(aa) T	(hh) T	(oo) T	
(g) T	(n) F	(u) T	(bb) F	(ii) F	(pp) T	(uu) F

Section 4.1

1. (a) Yes

   (b) No. The first two equations are satisfied, but not the third; the right-hand side of the third equation should equal -129 instead.

3. The system of linear equations corresponding to Figure 4.2 has no solution. A solution to the system would be a point that lies on all three lines. However, even though the lines do intersect in pairs, there is no common point that lies on all three lines.

5. $x =$ the price (per gallon) of regular gasoline; $y =$ the price (per gallon) of premium gasoline; linear system:

$$\begin{cases} 12x & + & 9y & = & 67.50 & \text{(First week)} \\ 11x & + & 10y & = & 68.00 & \text{(Second week)} \end{cases}$$

   (Note: Solution (not required) of system is $x = \$3.00$, $y = \$3.50$.)

7. $x =$ the number of Nanny-Cams made last month; $y =$ the number of Shoplift-Stoppers made last month; $z =$ the number of Weather-Alls made last month; linear system:

$$\begin{cases} x & + & y & + & z & = & 1200 & \text{(Cameras)} \\ 38x & + & 62y & + & 85z & = & 68584 & \text{(Cost)} \\ 57x & + & 91y & + & 104z & = & 94224 & \text{(Profit)} \end{cases}$$

   (Note: Solution (not required) of system is $x = 572$, $y = 284$, $z = 344$.)

9. $x =$ the number of HomeSpace models to manufacture next month; $y =$ the number of Hobbyist models to manufacture next month; $z =$ the number of Warehouser models to manufacture next month; linear system:

$$\begin{cases} 10x & + & 13y & + & 17z & = & 12997 & \text{(Materials)} \\ 1.2x & + & 1.4y & + & 1.7z & = & 1363.2 & \text{(Labor)} \\ & & x & = & & & 2y & \text{(HomeSpace Condition)} \end{cases}$$

   (Note: The last equation can be written as $x - 2y = 0$. The solution (not required) to the system is $x = 254$, $y = 127$, $z = 518$.)

11. A steady-state vector $[x, y, z]$ for the given Markov chain is a solution of the following system of linear equations (with all variables placed on the left-hand side of each equation, and with like terms combined):

$$\begin{cases} -.55x & + & .10y & + & .15z & = & 0 \\ .20x & - & .35y & + & .45z & = & 0 \\ .35x & + & .25y & - & .60z & = & 0 \\ x & + & y & + & z & = & 1 \end{cases}$$

    It is easy to check that $x = 17.81\%$, $y = 50.68\%$, and $z = 31.51\%$ is a solution of this system (after rounding to two places after the decimal point).

Section 4.2

1. Matrix multiplication: $\begin{bmatrix} 5 & 3 & -2 & 4 \\ -2 & 6 & -1 & -5 \\ 7 & -1 & 3 & -2 \end{bmatrix} \begin{bmatrix} x_1 \\ x_2 \\ x_3 \\ x_4 \end{bmatrix} \begin{bmatrix} -36 \\ 28 \\ -1 \end{bmatrix}$;

   augmented matrix $= \left[\begin{array}{cccc|c} 5 & 3 & -2 & 4 & -36 \\ -2 & 6 & -1 & -5 & 28 \\ 7 & -1 & 3 & -2 & -1 \end{array}\right]$

3. (a) (I): $(-7) < 3 > \Longrightarrow < 3 >$

   (b) (II): $4 < 2 > + < 1 > \Longrightarrow < 1 >$

   (c) (III): $< 2 > \Longleftrightarrow < 3 >$

   (d) (II): $(-6) < 4 > + < 3 > \Longrightarrow < 3 >$

   (e) (III): $< 1 > \Longleftrightarrow < 4 >$

   (f) (I): $(\frac{1}{3}) < 2 > \Longrightarrow < 2 >$

5. (a) $\left[\begin{array}{cccc|c} 1 & -3 & 4 & 2 & -16 \\ 0 & 1 & -\frac{1}{2} & 2 & -6 \\ 0 & -2 & 3 & -5 & -15 \end{array}\right]$

   (d) $\left[\begin{array}{cccc|c} 1 & 6 & -3 & 4 & -11 \\ 0 & -5 & 12 & -4 & -8 \\ 0 & 0 & 3 & -5 & -13 \\ 0 & 0 & -2 & -8 & 21 \end{array}\right]$

   (b) $\left[\begin{array}{ccc|c} 1 & 2 & -6 & -5 \\ 0 & -3 & 2 & -22 \\ 0 & 0 & 5 & 12 \end{array}\right]$

   (e) $\left[\begin{array}{ccccc|c} 1 & -1 & -2 & 8 & -3 & -18 \\ 0 & 1 & 5 & -4 & 12 & 2 \\ 0 & 0 & 2 & -3 & 12 & 1 \end{array}\right]$

   (c) $\left[\begin{array}{ccc|c} 1 & 0 & -3 & -8 \\ 0 & 1 & 0 & 6 \\ 0 & 0 & 1 & -6 \\ 0 & 0 & -2 & -9 \end{array}\right]$

   (f) $\left[\begin{array}{ccc|c} 1 & -\frac{1}{2} & \frac{3}{2} & -2 \\ -7 & -3 & 9 & -13 \\ 3 & -8 & -2 & -6 \end{array}\right]$

7. Intermediate steps: $\left[\begin{array}{ccc|c} 1 & 5 & 2 & -27 \\ 0 & -3 & 9 & -12 \\ 0 & 0 & -5 & 15 \\ 0 & 0 & -4 & 12 \end{array}\right]$, $\left[\begin{array}{ccc|c} 1 & 5 & 2 & -27 \\ 0 & 1 & -3 & 4 \\ 0 & 0 & -5 & 15 \\ 0 & 0 & -4 & 12 \end{array}\right]$;

   Final answer: $\left[\begin{array}{ccc|c} 1 & 0 & 17 & -47 \\ 0 & 1 & -3 & 4 \\ 0 & 0 & -5 & 15 \\ 0 & 0 & -4 & 12 \end{array}\right]$

9. (II): $(-4) < 1 > + < 3 > \Longrightarrow < 3 >$, resulting in

   $\left[\begin{array}{cccc|c} 1 & -7 & 7 & -3 & 12 \\ 0 & -1 & 5 & -2 & -9 \\ 0 & 26 & -22 & 7 & -49 \end{array}\right]$.

Section 4.3

1. (a) (II): $(-4) < 1 > + < 3 > \Longrightarrow < 3 >;$ $\begin{bmatrix} 1 & -3 & 5 & -16 \\ 0 & 2 & -7 & 13 \\ 0 & 10 & -25 & 55 \end{bmatrix}$

   (b) (III): $< 2 > \Longleftrightarrow < 3 >;$ $\begin{bmatrix} 1 & 2 & -6 & 40 \\ 0 & -5 & -1 & -10 \\ 0 & 0 & 3 & -15 \end{bmatrix}$

   (c) (I): $\left(-\frac{1}{4}\right) < 2 > \Longrightarrow < 2 >;$ $\begin{bmatrix} 1 & -3 & 2 & -5 & 17 \\ 0 & 1 & \frac{1}{2} & \frac{1}{2} & -\frac{3}{2} \\ 0 & 2 & -4 & 1 & -8 \\ 0 & 5 & 3 & 0 & -2 \end{bmatrix}$

3. (a) $\{(-9,5)\}$ (c) $\{(-6,8,-5,2)\}$

   (b) $\{(4,-5,2)\}$ (d) $\{(8,-2,-1)\}$

5. (a) $\{(8,-5,12)\}$ (b) $\{(-8,9,12,10)\}$

7. (a) $\begin{cases} 3x & + & 2.5y & = & 123 & \text{(Sugar)} \\ 2x & + & 3y & = & 114 & \text{(Butter)} \end{cases};$

   where x represents the number of pounds of vanilla fudge to be made, and y represents the number of pounds of chocolate fudge to be made.

   (b) $(21, 24)$; that is, 21 lbs of vanilla fudge, 24 pounds of chocolate fudge

9. (a) $\begin{cases} .10x & + & .20y & + & .05z & = & 4.0 & \text{(Nitrogen)} \\ .05x & + & .10y & + & .10z & = & 3.2 & \text{(Phosphate)} \\ .03x & + & .14y & + & .02z & = & 2.0 & \text{(Potash)} \end{cases};$

   where x represents the number of tons of *SureGrow*, y represents the number of tons of *QuickGrow*, and z represents the number of tons of *EvenGrow*.

   (b) $(14, 9, 16)$; that is, 14 tons of *SureGrow*, 9 tons of *QuickGrow*, 16 tons of *EvenGrow*

Section 4.4

1. (a) $\{(5z + 3, -6z - 2, z)\}$; two particular solutions (answers can vary) are: $(3, -2, 0)$ (when $z = 0$) and $(8, -8, 1)$ (when $z = 1$)

   (b) Solution set is $\{\ \} = \phi$

   (c) $\{(-12x_3 - 15, 9x_3 + 14, x_3, -8)\}$; two particular solutions (answers can vary) are:
   $(-15, 14, 0, -8)$ (when $x_3 = 0$) and $(-27, 23, 1, -8)$ (when $x_3 = 1$)

(d) $\{(7x_3 - 4x_5 + 13, \ -5x_3 + 11x_5 - 3, \ x_3, \ 6x_5 + 19, \ x_5)\}$; two particular solutions (answers can vary) are: $(20, -8, 1, 19, 0)$ (when $x_3 = 1$ and $x_5 = 0$) and $(9, 8, 0, 25, 1)$ (when $x_3 = 0$ and $x_5 = 1$)

(e) Solution set is $\{\ \} = \phi$

3. (a) Solution set is $\{\ \} = \phi$

(b) $\{(3z + 8, \ -7z + 13, \ z, \ -5)\}$

(c) $\{(6x_2 - 4x_4 + 9, \ x_2, \ 5x_4 - 2, \ x_4)\}$

(d) Solution set is $\{\ \} = \phi$

(e) $\{(5x_3 - 7x_4 + 2, \ -3x_3 + 4x_4 + 14, \ x_3, \ x_4, \ -8)\}$

Section 4.5

1. (a) Matrix is in reduced row echelon form

(b) (I): $(\frac{1}{2}) < 1 > \Longrightarrow < 1 >$

(c) (II): $2 < 1 > + < 4 > \Longrightarrow < 4 >$

(d) (III): $< 3 > \Longleftrightarrow < 4 >$

(e) Matrix is in reduced row echelon form

(f) (II): $(-2) < 3 > + < 1 > \Longrightarrow < 1 >$

(g) (I): $(\frac{1}{3}) < 3 > \Longrightarrow < 3 >$

(h) (III): $< 4 > \Longleftrightarrow < 5 >$

(i) (III): $< 3 > \Longleftrightarrow < 4 >$

3. (a) 2   (b) 3   (c) 3   (d) 4   (e) 3

5. (a) $\{(-24, 17, 23, -19)\}$

(b) Solution set is $\{\ \}$

(c) $\{(-9x_3 - 3x_5 - 12, \ 4x_3 + 5x_5 + 8, \ x_3, \ -2x_5 + 15, \ x_5)\}$

(d) $\{(4x_2 + 3x_4 - 5x_5 + 8, \ x_2, \ -6x_4 + 2x_5 - 3, \ x_4, \ x_5)\}$

7. (a) $\begin{cases} 15x + 25y + 20z = 1500 \ \text{(Pounds)} \\ 18x + 12y + 15z = 1044 \ \text{(Cubic Feet)} \ ; \\ x + y + z = 72 \ \text{(Total)} \end{cases}$

where x represents the number of *Lite Delight* vacuums in the van, y represents the number of *Deep Sweep* vacuums in the van, and z represents the number of *Clean Machine* vacuums in the van.

(b) The general form of the complete solution set for (x, y, z) is $\{(-\frac{1}{2}z + 30, -\frac{1}{2}z + 42, z)\}$. (Notice, however, that x, y must be integers, and so z must be an even integer. Also, z cannot be negative or larger than 72, so, $0 \le z \le 72$. Also notice that if z is greater than 60, then the value of x would be negative, so, in fact,

Appendix A: Answers to Odd-Numbered Exercises

$0 \le z \le 60$.) Two particular solutions are $(x, y, z) = (18, 30, 24)$ (where $z = 24$) and $(x, y, z) = (17, 29, 26)$ (where $z = 26$). Two other particular solutions are $(x, y, z) = (30, 42, 0)$ (where $z = 0$) and $(x, y, z) = (0, 12, 60)$ (where $z = 60$).

(c) The new linear system includes all of the equations in the linear system for part (a), along with the additional equation $3x - y - z = 0$. This linear system has a unique solution: $(x, y, z) = (18, 30, 24)$; that is, 18 *Lite Delight* vacuums, 30 *Deep Sweep* vacuums, and 24 *Clean Machine* vacuums.

9. (a)
$$\begin{cases} 10x + 6y + 2z = 480 \text{ (Items Repaired)} \\ 14x + 10y + 8z = 790 \text{ (Salaries)} \\ 500x + 350y + 200z = 26750 \text{ (Work Space)} \\ x + y - 3z = 0 \text{ (Relation of Workers)} \end{cases} ;$$

where x represents the number of highly skilled workers, y represents the number of regular workers, and z represents the number of unskilled workers.

(b) The solution set is empty. There are no solutions that satisfy all of these conditions.

(c) Solution set for (x, y, z) is $\{(30, 25, 15)\}$; that is, 30 highly skilled workers, 25 regular workers, and 15 unskilled workers.

11. By the first property of reduced row echelon form, the unique reduced row echelon form matrix corresponding to the given matrix has a unique circled pivot in each nonzero row. Therefore, the number of circled pivots cannot be greater than the total number of rows. But the rank is the number of such circled pivots. Therefore, the rank cannot be greater than the number of rows in the matrix.

13. The unique steady-state vector $[x, y, z]$ for the given Markov chain is $x = 32.3\%$, $y = 23.4\%$, and $z = 44.2\%$. (Since all entries of $\mathbf{M}$ are nonzero, this steady-state vector is actually $\mathbf{p}_{\lim}$ for *any* initial probability vector $\mathbf{p}$ for this Markov chain.)

15. $\mathbf{p}_1 = [.35, .40, .25]$, $\mathbf{p}_2 = [.25, .35, .40]$, $\mathbf{p}_3 = [.40, .25, .35]$ ($= \mathbf{p}_0$). To find a potential steady-state vector for $\mathbf{M}$, we row reduce the augmented matrix for the linear system

$$\begin{cases} 0x + 0y + 1z = x \\ 1x + 0y + 0z = y \\ 0x + 1y + 0z = z \\ 1x + 1y + 1z = 1 \end{cases}, \text{ or, } \begin{cases} -1x + 0y + 1z = 0 \\ 1x + -1y + 0z = 0 \\ 0x + 1y + -1z = 0 \\ 1x + 1y + 1z = 1 \end{cases}$$

to obtain

$$\left[\begin{array}{ccc|c} 1 & 0 & 0 & 0.333 \\ 0 & 1 & 0 & 0.333 \\ 0 & 0 & 1 & 0.333 \\ 0 & 0 & 0 & 0.000 \end{array}\right],$$

which has $[0.333, 0.333, 0.333]$ as its unique solution.

Section 4.6

1. (a) $\{z\,(2,-2,1)\}$

   (b) $\left\{x_3\left(-\frac{7}{2},-3,1,0\right)\right.$
   $\left.+x_4\left(2,\frac{5}{4},0,1\right)\right\}$

   (c) $\{(0,0,0)\}$

   (d) $\{x_2\,(-2,1,0,0,0)$
   $+ x_4\,(3,0,1,1,0)$
   $+x_5\,(-2,0,4,0,1)\}$

   (e) $\{(0,0,0)\}$

3. (a) $\mathbf{v} = [4,-7,1]$

   (b) $\mathbf{v} = \left[-\frac{3}{2},1,1\right]$

   (c) $\mathbf{v} = [-2,2,-2,1]$

   (d) $\mathbf{v} = [1,1,1,2,1]$

5. $\{[3,-9,-14,7],[-28,84,19,158],[3,1,0,0],[7,-21,20,10]\}$ or
   $\{[3,-9,-14,7],[-28,84,19,158],[7,0,2,1],[15,54,-42,-21]\}$

7. If $\mathbf{X} = \mathbf{0}$, then $\mathbf{AX} = \mathbf{A0} = \mathbf{0} \neq \mathbf{B}$. So, $\mathbf{X} = \mathbf{0}$ does not satisfy $\mathbf{AX} = \mathbf{B}$.

9. If $\mathbf{u}$ and $\mathbf{v}$ are both solutions to the nonhomogeneous linear system $\mathbf{AX} = \mathbf{B}$, then $\mathbf{Au} = \mathbf{B}$ and $\mathbf{Av} = \mathbf{B}$. Therefore, $\mathbf{Aw} = \mathbf{A}\,(\mathbf{u} - \mathbf{v}) = \mathbf{Au} - \mathbf{Av} = \mathbf{B} - \mathbf{B} = \mathbf{0}$. Hence, $\mathbf{Aw} = \mathbf{0}$, and so $\mathbf{w}$ is a solution to the homogeneous linear system $\mathbf{AX} = \mathbf{0}$.

11. First, because the rows of $\mathbf{A}$ are n-dimensional vectors, $\mathbf{A}$ must have n columns. Similarly, because we have k vectors, $\mathbf{A}$ has k rows. Therefore, $\mathbf{A}$ is a $k \times n$ matrix.

    Next, because the columns of $\mathbf{B}$ are n-dimensional vectors, $\mathbf{B}$ must have n rows. Similarly, because we have k vectors, $\mathbf{B}$ has k columns. Therefore, $\mathbf{B}$ is an $n \times k$ matrix.

    Thus, we conclude that $\mathbf{AB}$ is a $k \times k$ matrix.

    Now, the (i,j) entry of $\mathbf{AB}$ equals the dot product of the ith row of $\mathbf{A}$ and the jth column of $\mathbf{B}$. (See Section 3.3.) But this is just $\mathbf{v}_i \cdot \mathbf{v}_j$. Thus, the (i,j) entry of $\mathbf{AB}$ equals $\mathbf{v}_i \cdot \mathbf{v}_j$.

    Because $\{\mathbf{v}_1,\ldots,\mathbf{v}_k\}$ is an orthonormal set of vectors, $\mathbf{v}_i \cdot \mathbf{v}_j = 0$ whenever $i \neq j$. Therefore, the (i,j) entry of $\mathbf{AB}$ is zero when $i \neq j$, and so all of the entries off the main diagonal of $\mathbf{AB}$ are zero. Thus, $\mathbf{AB}$ is a diagonal matrix. But, because $\{\mathbf{v}_1,\ldots,\mathbf{v}_k\}$ is an orthonormal set of vectors, $\mathbf{v}_i \cdot \mathbf{v}_i = 1$. Hence, the (i,i) entry of $\mathbf{AB}$ is 1, and so all of the entries of **AB** *on* the main diagonal equal 1. We conclude, then, that $\mathbf{AB} = \mathbf{I}_k$.

Chapter 4 Test

8. (a) F   (h) T   (o) F   (v) F   (cc) T   (jj) F   (qq) T
    (b) F   (i) F   (p) T   (w) T   (dd) T   (kk) T   (rr) F
    (c) T   (j) T   (q) T   (x) F   (ee) F   (ll) T
    (d) T   (k) T   (r) F   (y) T   (ff) T   (mm) T
    (e) T   (l) T   (s) T   (z) F   (gg) T   (nn) F
    (f) F   (m) F   (t) F   (aa) F   (hh) F   (oo) F
    (g) F   (n) F   (u) T   (bb) T   (ii) F   (pp) F

Section 5.1

1. (a) $y = f(x) = 1.549x + 1.440$; $f(4) = 7.636$; $f(10) = 16.93$
    (b) $y = f(x) = 0.449x + 2.047$; $f(4) = 3.843$; $f(10) = 6.537$

3. (a) $y = f(x) = 1193.42x + 4489.01$
    (b) 2005: $f(9) = 15229.79$ kwh; 2009: $f(9.8) = 16184.53$ kwh
    (c) Extrapolating for both 2005 and 2009

5. (a) $y = f(x) = 443.45x + 2692.97$
    (b) Estimated cereal yield in kg per hectare: 1961: $f(-0.8) = 2338.2$; 1998: $f(6.6) = 5619.7$
    (c) Extrapolating: 1961; Interpolating: 1998
    (d) $y = f(x) = 627.12x + 4484.32$
    (e) Estimated cereal yield in kg per hectare: 1961: $f(-5.8) = 847.0$; 1998: $f(1.6) = 5487.7$
    (f) The estimate in part (b) is closer to the actual data for the year 1961. This makes sense, because the computations in part (b) use data much closer to the year 1961 than the computations in part (e). For the year 1998, we would normally expect the computations in part (e) to be more accurate, because these calculations emphasize data closer to that year. However, the actual data has an anomaly between the years 1985 and 2000. Notice the drops in the given cereal yield from the year 1985 to the year 1995. The computation in part (e) accentuates this anomaly, because there is less additional data to smooth out its effect. In fact, the actual data makes a sharp increase in the year 1996, and levels off somewhat, rising very slowly, between the years 1996 and 2000. The

linear regression in part (e) smooths all of these drops and spikes into a straight line model that is not as helpful as we would like.

7. $y = f(x) = 0.578x + 0.201$. July's data usage $= f(1) = 0.779$ gigabytes. The small discrepancy between this answer for July's data usage and the one in the text is due to error from rounding off the values of m and b to three places after the decimal point. The formula computed here predicts the Smiths will hit the 6 gigabyte usage rate when $x = 10.033$, which, again, corresponds to the beginning of April 2013.

9. (a) The predicted y-values from the line $y = 2x - 5$ are: for $x = 2$, $y = -1$; for $x = 3$, $y = 1$; for $x = 4$, $y = 3$; for $x = 5$, $y = 5$; for $x = 6$, $y = 7$. The sum of the squares of the errors $= ((-1) - (-0.6))^2 + (1 - 1.5)^2 + (3 - 4.0)^2 + (5 - 4.4)^2 + (7 - 6.3)^2 = 2.26$.

(b) The least-squares method produces the particular linear equation for which the sum of the squares of the errors is the smallest. Therefore, we should expect the error for the linear regression line to be less than the error computed in part (a).

(c) Sum of the squares of the errors $= ((-0.22) - (-0.6))^2 + (1.45 - 1.5)^2 + (3.12 - 4.0)^2 + (4.79 - 4.4)^2 + (6.46 - 6.3)^2 = 1.099$. Yes, the prediction in part (b) was correct, since the new error is less than (half of) the error obtained using the line $y = 2x - 5$ in part (a).

11. (a) The $(1,1)$ entry of $\mathbf{A}^T \mathbf{A}$ is found by using the first row of $\mathbf{A}^T$ and the first column of $\mathbf{A}$. But the first row of $\mathbf{A}^T$ is $[1, 1, \ldots, 1]$, and the first column of $\mathbf{A}$ is a column matrix of all 1's. Multiplying corresponding entries and adding yields $(1 \times 1) + (1 \times 1) + \cdots + (1 \times 1) = 1 + 1 + \cdots + 1$. Since there are n such terms, the sum equals n. Hence, the $(1,1)$ entry of $\mathbf{A}^T \mathbf{A}$ equals n.

(b) The $(2,1)$ entry of $\mathbf{A}^T \mathbf{A}$ is found by using the second row of $\mathbf{A}^T$ and the first column of $\mathbf{A}$. But the second row of $\mathbf{A}^T$ is $[x_1, x_2, \ldots, x_n]$, and the first column of $\mathbf{A}$ is a column matrix of all 1's. Multiplying corresponding entries and adding yields $(x_1 \times 1) + (x_2 \times 1) + \cdots + (x_n \times 1) = x_1 + x_2 + \cdots + x_n$.

(c) Theorem 2 in Section 1.6 states that $\mathbf{A}^T \mathbf{A}$ is a symmetric matrix. Therefore, the $(1,2)$ entry of $\mathbf{A}^T \mathbf{A}$ equals the $(2,1)$ entry of $\mathbf{A}^T \mathbf{A}$. But in part (b) we proved that the $(2,1)$ entry of $\mathbf{A}^T \mathbf{A}$ equals $x_1 + \cdots + x_n$. Hence the $(1,2)$ entry of $\mathbf{A}^T \mathbf{A}$ also equals $x_1 + \cdots + x_n$.

(d) The $(2,2)$ entry of $\mathbf{A}^T \mathbf{A}$ is found by using the second row of $\mathbf{A}^T$ and the second column of $\mathbf{A}$. But the second row of $\mathbf{A}^T$ is $[x_1, x_2, \ldots, x_n]$, and the second column of $\mathbf{A}$ is a column matrix whose entries are $x_1, x_2, \ldots, x_n$, respectively. Multiplying corresponding entries and adding yields $(x_1 \times x_1) + (x_2 \times x_2) + \cdots + (x_n \times x_n) = x_1^2 + x_2^2 + \cdots + x_n^2$.

Section 5.2

1. (a) $y = f(x) = 0.0808x^2 - 1.2647x + 9.5681$; $f(4) = 5.8021$;
   $f(10) = 5.0011$
   (b) $y = f(x) = -0.0853x^2 + 0.7791x + 3.6908$; $f(4) = 5.4424$;
   $f(10) = 2.9518$

3. **A** would be an 18×13 matrix.

5. (a) $y = f(x) = -0.1216x^2 + 1.4936x + 14.3372$
   (b) 2006: 18.92%; 2010: 17.11%

Section 5.3

1. (a) Unique solution; $\mathbf{X} = \begin{bmatrix} 4.9 \\ 2.9 \end{bmatrix}$

   (b) $\mathbf{w}_1 = [1, 1, 1, 1]$;
   $\mathbf{w}_2 = [-3, -1, 1, 3]$

   (c) $\mathbf{proj}_{\mathbf{w}_1} \mathbf{B} = [16.5,\ 16.5,\ 16.5,\ 16.5]$;
   $\mathbf{proj}_{\mathbf{w}_2} \mathbf{B} = [-8.7,\ -2.9,\ 2.9,\ 8.7]$

   (d) $\mathbf{C} = [7.8,\ 13.6,\ 19.4,\ 25.2]$

   (e) Unique solution; $\mathbf{X} = \begin{bmatrix} 4.9 \\ 2.9 \end{bmatrix}$

   (f) They are the same.

Chapter 5 Test

5. (a) T   (d) F   (g) T   (j) F   (m) T   (p) T
   (b) F   (e) F   (h) T   (k) F   (n) T
   (c) T   (f) T   (i) T   (l) T   (o) T   (q) F

Section 6.1

1. Nonsingular: Parts (a), (b), (e); Singular: Parts (c), (d), (f). Inverses
   for the nonsingular matrices are shown here:

   (a) $\mathbf{I}_3$

   (b) $\begin{bmatrix} \frac{5}{4} & \frac{3}{4} \\ 2 & 1 \end{bmatrix}$

   (e) $\begin{bmatrix} -\frac{9}{2} & -12 & \frac{7}{2} \\ \frac{31}{2} & 43 & -\frac{25}{2} \\ -37 & -103 & 30 \end{bmatrix}$

3. (a) $(5, 2)$ (c) $\left(6, \frac{5}{2}, 1\right)$ (d) $(5, -2, 3, -1)$
   (b) $(4, 1, -6)$

5. (a) Because $\text{rank}(\mathbf{A}) < n$, there will be fewer than n circled pivots to the left of the bar when row reducing $[\mathbf{A} \,|\, \mathbf{I}_n]$. Hence, the final reduced row echelon form matrix will not have $\mathbf{I}_n$ to the left of the bar, since at least one column to the left of the bar will not have a circled pivot. Therefore, our method says that the matrix $\mathbf{A}$ is singular.

   (b) This follows directly from Theorem 2 in Chapter 4.

   (c) By MSC1 in Section 1.8, the ith column of $\mathbf{AB}$ equals $\mathbf{A}(i$th column of $\mathbf{B}) = \mathbf{As} = \mathbf{0}$. Therefore, every column of $\mathbf{AB}$ is $\mathbf{0}$. Thus, $\mathbf{AB} = \mathbf{O}_{nn}$.

7. $\mathbf{AB} = \mathbf{BA} = \delta \mathbf{I}_2$

9. (a) $\mathbf{AB} = \mathbf{I}_2$; $\mathbf{BA} = \begin{bmatrix} 2 & 2 & 1 \\ -2 & -3 & -2 \\ 2 & 4 & 3 \end{bmatrix}$

   (b) Theorem 1 requires that $\mathbf{A}$ and $\mathbf{B}$ be square matrices, which is not the case in this exercise.

   (c) $\text{rank}(\mathbf{BA}) = 2$

Section 6.2

1. $\mathbf{B}^{-1} = \mathbf{B}^{-1}\mathbf{I}_2 = \mathbf{B}^{-1}\left(\mathbf{A}^{-1}\mathbf{A}\right) = \left(\mathbf{B}^{-1}\mathbf{A}^{-1}\right)\mathbf{A} = (\mathbf{AB})^{-1}\mathbf{A}$

$$= \begin{bmatrix} 9 & 4 \\ -\frac{11}{4} & -\frac{5}{4} \end{bmatrix} \begin{bmatrix} 1 & 3 \\ -3 & -5 \end{bmatrix} = \begin{bmatrix} -3 & 7 \\ 1 & -2 \end{bmatrix}$$

3. (a) $\left(\mathbf{A}^T\right)^{-1} = \left(\mathbf{A}^{-1}\right)^T = \mathbf{B}^T = \begin{bmatrix} 1 & 3 & -14 \\ -1 & -\frac{7}{2} & 16 \\ 2 & \frac{11}{2} & -27 \end{bmatrix}$

   (b) $(2\mathbf{A})^{-1} = \frac{1}{2}\mathbf{A}^{-1} = \frac{1}{2}\mathbf{B} = \begin{bmatrix} \frac{1}{2} & -\frac{1}{2} & 1 \\ \frac{3}{2} & -\frac{7}{4} & \frac{11}{4} \\ -7 & 8 & -\frac{27}{2} \end{bmatrix}$

5. (a) $\begin{bmatrix} 0 & -1 \\ 1 & 0 \end{bmatrix}$ (b) $\begin{bmatrix} 0 & -\frac{1}{a} \\ \frac{1}{a} & 0 \end{bmatrix}$

7. A lower triangular $n \times n$ matrix having no zero entries on its main diagonal is nonsingular and its inverse is also lower triangular. If a lower triangular matrix has at least one zero entry on its main diagonal, then it is singular.

9. (a) Nonsingular. Inverse $= \begin{bmatrix} 0 & \frac{1}{3} \\ -\frac{1}{3} & 0 \end{bmatrix}$ by Exercise 5(b).

   (b) Nonsingular. Inverse $= \begin{bmatrix} \frac{1}{4} & 0 \\ 0 & -\frac{1}{5} \end{bmatrix}$ by Theorem 4.

   (c) Singular by Theorem 7, because the matrix is skew-symmetric and has an odd number of rows and columns.

   (d) Singular by Theorem 5, because the matrix is upper triangular and has a zero on the main diagonal.

11. One possibility: $\begin{bmatrix} 1 & 2 & 0 \\ 2 & 3 & 0 \\ 0 & 0 & 0 \end{bmatrix}$

13. (a) $\mathbf{A}^3 = \begin{bmatrix} 36 & 92 \\ 23 & 59 \end{bmatrix}$; $\mathbf{A}^{-2} = \begin{bmatrix} \frac{13}{4} & -5 \\ -\frac{5}{4} & 2 \end{bmatrix}$;

$$\mathbf{B}^{-3} = \begin{bmatrix} 47 & -54 & 70 \\ -156 & 179 & -232 \\ 162 & -186 & 241 \end{bmatrix}; \mathbf{B}^2 = \begin{bmatrix} -7 & 1 & 3 \\ -10 & -3 & 0 \\ -3 & -3 & -2 \end{bmatrix}$$

   (b) $\mathbf{A}^{-1} = \mathbf{A}^3 \left(\mathbf{A}^{-2} \right)^2 = \begin{bmatrix} \frac{3}{2} & -2 \\ -\frac{1}{2} & 1 \end{bmatrix}$;

$$\mathbf{B}^{-1} = \mathbf{B}^{-3} \mathbf{B}^2 = \begin{bmatrix} 1 & -1 & 1 \\ -2 & 3 & -4 \\ 3 & -3 & 4 \end{bmatrix}$$

   (c) $\left(\frac{1}{4} \mathbf{A} \right)^{-4} = \left(\frac{1}{4} \right)^{-4} \mathbf{A}^{-4} = 256 \left(\mathbf{A}^{-2} \right)^2 = \begin{bmatrix} 4304 & -6720 \\ -1680 & 2624 \end{bmatrix}$;

$$\left(2\mathbf{B}^T \right)^{-4} = (2)^{-4} \left(\mathbf{B}^T \right)^{-4} = \tfrac{1}{16} \left(\mathbf{B}^{-4} \right)^T = \tfrac{1}{16} \left(\left(\mathbf{B}^{-3} \right)^2 \mathbf{B}^2 \right)^T =$$
$$\begin{bmatrix} \frac{365}{16} & -\frac{605}{8} & \frac{1257}{16} \\ -\frac{419}{16} & \frac{1389}{16} & -\frac{1443}{16} \\ \frac{543}{16} & -\frac{225}{2} & \frac{935}{8} \end{bmatrix}$$

   (d) $\mathbf{C} = \begin{bmatrix} -4124 & 7180 \\ 1795 & -2329 \end{bmatrix}$; $\mathbf{D} = \begin{bmatrix} 8 & 20 \\ 5 & 13 \end{bmatrix}$;

$$\mathbf{CD} = \mathbf{DC} = \begin{bmatrix} 2908 & 10860 \\ 2715 & 5623 \end{bmatrix}$$

15. $\left(\mathbf{A}^T \right) \left(\left(\mathbf{A}^{-1} \right)^T \right) = \left(\mathbf{A}^{-1} \mathbf{A} \right)^T$ (by TR4) $= \left(\mathbf{I}_n \right)^T = \mathbf{I}_n$. Therefore, $\left(\left(\mathbf{A}^{-1} \right)^T \right)$ is the inverse for $\left(\mathbf{A}^T \right)$.

Section 6.3

1. (a) Orthogonal matrix; The columns are e_1, e_2, and e_3. Note that $e_1 \cdot e_2 = 0$, $e_1 \cdot e_3 = 0$, $e_2 \cdot e_3 = 0$, $e_1 \cdot e_1 = 1$, $e_2 \cdot e_2 = 1$, and $e_3 \cdot e_3 = 1$. Therefore, the columns of I_3 form an orthonormal set.

   (b) Not an orthogonal matrix; The rows are not unit vectors. For example, $\|[3, -4]\| = 5$.

   (c) Orthogonal matrix; The columns are $v_1 = \left[\frac{5}{13}, \frac{12}{13}\right]$ and $v_2 = \left[\frac{12}{13}, -\frac{5}{13}\right]$. Note that $v_1 \cdot v_2 = 0$, $v_1 \cdot v_1 = 1$, and $v_2 \cdot v_2 = 1$. Therefore, the columns of the matrix form an orthonormal set.

   (d) Orthogonal matrix; The columns are $v_1 = \left[\frac{\sqrt{3}}{3}, \frac{\sqrt{14}}{14}, -\frac{5}{\sqrt{42}}\right]$, $v_2 = \left[\frac{\sqrt{3}}{3}, \frac{\sqrt{56}}{14}, \frac{4}{\sqrt{42}}\right]$ and $v_3 = \left[\frac{\sqrt{3}}{3}, -\frac{\sqrt{126}}{14}, \frac{1}{\sqrt{42}}\right]$. Note that

   $v_1 \cdot v_2 = \left(\frac{\sqrt{3}}{3}\right)\left(\frac{\sqrt{3}}{3}\right) + \left(\frac{\sqrt{14}}{14}\right)\left(\frac{\sqrt{56}}{14}\right) + \left(-\frac{5}{\sqrt{42}}\right)\left(\frac{4}{\sqrt{42}}\right) = \frac{1}{3} + \frac{1}{7} - \frac{10}{21}$
   $= \frac{7}{21} + \frac{3}{21} - \frac{10}{21} = 0$,

   $v_1 \cdot v_3 = \left(\frac{\sqrt{3}}{3}\right)\left(\frac{\sqrt{3}}{3}\right) + \left(\frac{\sqrt{14}}{14}\right)\left(-\frac{\sqrt{126}}{14}\right) + \left(-\frac{5}{\sqrt{42}}\right)\left(\frac{1}{\sqrt{42}}\right) = \frac{1}{3} - \frac{3}{14} - \frac{5}{42} = \frac{14}{42} - \frac{9}{42} - \frac{5}{42} = 0$,

   $v_2 \cdot v_3 = \left(\frac{\sqrt{3}}{3}\right)\left(\frac{\sqrt{3}}{3}\right) + \left(\frac{\sqrt{56}}{14}\right)\left(-\frac{\sqrt{126}}{14}\right) + \left(\frac{4}{\sqrt{42}}\right)\left(\frac{1}{\sqrt{42}}\right) = \frac{1}{3} - \frac{3}{7} + \frac{2}{21}$
   $= \frac{7}{21} - \frac{9}{21} + \frac{2}{21} = 0$,

   $v_1 \cdot v_1 = \left(\frac{\sqrt{3}}{3}\right)\left(\frac{\sqrt{3}}{3}\right) + \left(\frac{\sqrt{14}}{14}\right)\left(\frac{\sqrt{14}}{14}\right) + \left(-\frac{5}{\sqrt{42}}\right)\left(-\frac{5}{\sqrt{42}}\right) = \frac{1}{3} + \frac{1}{14} + \frac{25}{42} = \frac{14}{42} + \frac{3}{42} + \frac{25}{42} = 1$,

   $v_2 \cdot v_2 = \left(\frac{\sqrt{3}}{3}\right)\left(\frac{\sqrt{3}}{3}\right) + \left(\frac{\sqrt{56}}{14}\right)\left(\frac{\sqrt{56}}{14}\right) + \left(\frac{4}{\sqrt{42}}\right)\left(\frac{4}{\sqrt{42}}\right) = \frac{1}{3} + \frac{2}{7} + \frac{8}{21}$
   $= \frac{7}{21} + \frac{6}{21} + \frac{8}{21} = 1$,

   and $v_3 \cdot v_3 = \left(\frac{\sqrt{3}}{3}\right)\left(\frac{\sqrt{3}}{3}\right) + \left(-\frac{\sqrt{126}}{14}\right)\left(-\frac{\sqrt{126}}{14}\right) + \left(\frac{1}{\sqrt{42}}\right)\left(\frac{1}{\sqrt{42}}\right) = \frac{1}{3} + \frac{9}{14} + \frac{1}{42} = \frac{14}{42} + \frac{27}{42} + \frac{1}{42} = 1$. Therefore, the columns of the matrix form an orthonormal set.

   (e) Not an orthogonal matrix; The second and third rows of the matrix are not orthogonal because $\left[\frac{3}{\sqrt{14}}, \frac{1}{\sqrt{14}}, -\frac{2}{\sqrt{14}}\right] \cdot \left[\frac{5}{\sqrt{57}}, \frac{4}{\sqrt{57}}, -\frac{4}{\sqrt{57}}\right]$
   $= \frac{15}{\sqrt{798}} + \frac{4}{\sqrt{798}} + \frac{8}{\sqrt{798}} = \frac{27}{\sqrt{798}} \neq 0$.

3. (a) We start with
   $$A = \begin{bmatrix} 4 & 7 \\ 3 & 9 \end{bmatrix}.$$

   We will perform the Gram-Schmidt Process using the rows of A as our starting vectors. So $v_1 = [4, 7]$ and $v_2 = [3, 9]$.
   For our mutually orthogonal set, we start with $x_1 = v_1 = [4, 7]$. Then,
   $$x_2 = v_2 - \text{proj}_{x_1} v_2 = \left[-\frac{21}{13}, \frac{12}{13}\right].$$

We adjust $\mathbf{x}_2$ by multiplying by 13 to get the new $\mathbf{x}_2 = [-21, 12]$. This gives us the mutually orthogonal set

$$\{[4, 7], [-21, 12]\}.$$

Finally, we normalize each of these vectors and use them as rows to form the orthogonal matrix

$$\begin{bmatrix} \frac{4}{\sqrt{65}} & \frac{7}{\sqrt{65}} \\ -\frac{7}{\sqrt{65}} & \frac{4}{\sqrt{65}} \end{bmatrix} \approx \begin{bmatrix} .4961 & .8682 \\ -.8682 & .4961 \end{bmatrix}.$$

(b) We start with

$$\mathbf{A} = \begin{bmatrix} 10 & -45 & 30 \\ 24 & -31 & 6 \\ 6 & -49 & -15 \end{bmatrix}.$$

We will perform the Gram-Schmidt Process using the rows of $\mathbf{A}$ as our starting vectors. So $\mathbf{v}_1 = [10, -45, 30]$, $\mathbf{v}_2 = [24, -31, 6]$, and $\mathbf{v}_3 = [6, -49, -15]$.

For our mutually orthogonal set, we start with $\mathbf{x}_1 = \mathbf{v}_1 = [10, -45, 30]$. Then,

$$\mathbf{x}_2 = \mathbf{v}_2 - \text{proj}_{\mathbf{x}_1}\mathbf{v}_2 = \left[18, -4, -12\right].$$

We adjust $\mathbf{x}_2$ by dividing by 2 to get the new $\mathbf{x}_2 = [9, -2, -6]$. Next,

$$\mathbf{x}_3 = \mathbf{v}_3 - \text{proj}_{\mathbf{x}_1}\mathbf{v}_3 - \text{proj}_{\mathbf{x}_2}\mathbf{v}_3 = \left[-18, -18, -21\right].$$

We adjust $\mathbf{x}_3$ by dividing by 3 to get the new $\mathbf{x}_3 = [-6, -6, -7]$. This gives us the mutually orthogonal set

$$\{[10, -45, 30], [9, -2, -6], [-6, -6, -7]\}.$$

Finally, we normalize each of these vectors and use them as rows to form the orthogonal matrix

$$\begin{bmatrix} \frac{2}{11} & -\frac{9}{11} & \frac{6}{11} \\ \frac{9}{11} & -\frac{2}{11} & -\frac{6}{11} \\ -\frac{6}{11} & -\frac{6}{11} & -\frac{7}{11} \end{bmatrix} \approx \begin{bmatrix} .1818 & -.8182 & .5455 \\ .8182 & -.1818 & -.5455 \\ -.5455 & -.5455 & -.6364 \end{bmatrix}.$$

5. An orthogonal matrix is defined as a square matrix whose rows form an orthonormal set of vectors. A type (III) row operation merely changes the order of the rows, and so it does not change the set of vectors formed using the rows. Therefore, if the original matrix has rows that form an orthonormal set of vectors, then the transformed matrix, after a type (III) row operation, will also have rows that form an orthonormal set of vectors, because the set of vectors has not changed.

7. Direct computation shows that

$$\mathbf{A}\mathbf{A}^T = \begin{bmatrix} 33165 & 0 & 0 & 0 \\ 0 & 54 & 0 & 0 \\ 0 & 0 & 5346 & 0 \\ 0 & 0 & 0 & 335 \end{bmatrix},$$

which is a diagonal matrix. Direct computation will also show that the square of the norms of the given vectors appear on the main diagonal of $\mathbf{A}\mathbf{A}^T$.

Calculating further shows that

$$\mathbf{A}^T\mathbf{A} = \begin{bmatrix} 1067 & -1569 & -1190 & -4711 \\ -1569 & 10053 & -546 & 12075 \\ -1190 & -546 & 2325 & 3788 \\ -4711 & 12075 & 3788 & 25455 \end{bmatrix},$$

which is clearly *not* a diagonal matrix.

9. Suppose that $\mathbf{A}$ is an orthogonal matrix. The proof of Theorem 8 shows that $\mathbf{A}\mathbf{A}^T = \mathbf{I}_n$. Theorem 1 then implies that $\mathbf{A}^T\mathbf{A} = \mathbf{I}_n$.

11. Suppose that $\mathbf{A}$ is an orthogonal matrix. Exercise 9 shows that $\mathbf{A}^T\mathbf{A} = \mathbf{I}_n$. Now, the (i, j) entry of $\mathbf{A}^T\mathbf{A} = (i$th row of $\mathbf{A}^T) \cdot (j$th column of $\mathbf{A})$ $= (i$th column of $\mathbf{A}) \cdot (j$th column of $\mathbf{A})$. Therefore, since $\mathbf{A}^T\mathbf{A} = \mathbf{I}_n$, when $i \neq j$, $(i$th column of $\mathbf{A}) \cdot (j$th column of $\mathbf{A}) = 0$ (the (i, j) entry of $\mathbf{I}_n$), and so the columns of $\mathbf{A}$ form a mutually orthogonal set of vectors. When $i = j$,

$$\|i\text{th column of } \mathbf{A}\|^2 = (i\text{th column of } \mathbf{A}) \cdot (i\text{th column of } \mathbf{A}),$$

which is the (i, i) entry of $\mathbf{A}^T\mathbf{A} = \mathbf{I}_n$. Therefore, it equals 1, and so the columns of $\mathbf{A}$ are all unit vectors. Hence, the columns of $\mathbf{A}$ form an orthonormal set of vectors.

Chapter 6 Test

9. (a) T   (f) T   (k) T   (p) T   (u) F   (z) T   (ee) T
    (b) F   (g) T   (l) F   (q) T   (v) T   (aa) F
    (c) F   (h) F   (m) T   (r) F   (w) F   (bb) F
    (d) T   (i) T   (n) T   (s) F   (x) T   (cc) T
    (e) F   (j) T   (o) F   (t) T   (y) T   (dd) F

Section 7.1

1. (a) $\begin{bmatrix} 2 & 0 \\ 0 & 2 \end{bmatrix}$; preserves angles, but not distances.

    (b) $\begin{bmatrix} -\frac{1}{3} & 0 \\ 0 & -\frac{1}{3} \end{bmatrix}$; preserves angles, but not distances.

    (c) $\begin{bmatrix} \frac{3}{5} & \frac{4}{5} \\ \frac{4}{5} & -\frac{3}{5} \end{bmatrix}$; preserves both angles and distances.

    (d) $\begin{bmatrix} 0.259 & -0.966 \\ 0.966 & 0.259 \end{bmatrix}$; preserves both angles and distances.

    (e) $\begin{bmatrix} -0.766 & 0.643 \\ -0.643 & -0.766 \end{bmatrix}$; preserves both angles and distances.

    (f) This is a translation by the vector $\mathbf{t} = [-4, 7]$. It is not a matrix transformation. It preserves both angles and distances.

3. (a) $(2.23, -0.15)$, $(-1.31, 0.54)$, $(-1.38, 5.92)$, $(2.69, 6.54)$
    (b) $(-3.4, 6.8)$, $(3.4, -3.4)$, $(20.4, 3.4)$, $(17, 17)$
    (c) $(-2.09, -0.79)$, $(1.10, 0.89)$, $(-0.37, 6.07)$, $(-4.45, 5.50)$

5. Coordinates of the vertices of the image: $(-0.12, 1.41)$, $(0.12, -1.41)$, $(5.12, -1.68)$, $(4.62, 3.95)$;

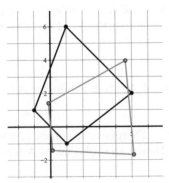

Figure for Exercise 5

7. There is no movement at all! The image of every point is the point itself.

9. (a) $\|[1,1]\| = 1.41$, image $= [-1,7]$, $\|[-1,7]\| = 7.07 \approx 7.05 = 5 \times 1.41$;
$\|[-1,2]\| = 2.24$, image $= [-11,2]$, $\|[-11,2]\| = 11.18$
$\approx 11.20 = 5 \times 2.24$;
$\|[1,3]\| = 3.16$, image $= [-9,13]$, $\|[-9,13]\| = 15.81$
$\approx 15.80 = 5 \times 3.16$;
$\|[3,1]\| = 3.16$, image $= [5,15]$, $\|[5,15]\| = 15.81 \approx 15.80 = 5 \times 3.16$;
$\|[2,0]\| = 2$, image $= [6,8]$, $\|[6,8]\| = 10 = 5 \times 2$.
Small discrepancies are due to the errors introduced by rounding.

(b) Using the formula in Section 3.4, the cosine of the angle between
$[1,1]$ and its image $[-1,7]$ is given by $\frac{[1,1]\cdot[-1,7]}{\|[1,1]\|\,\|[-1,7]\|} \approx \frac{6}{(1.41)(7.07)} \approx$
$\frac{6}{9.97} \approx \frac{3}{5}$. The remaining cosines are calculated in a similar fashion.

11. The image of $[1,0] = \mathbf{A}\mathbf{e}_1 =$ the first column of $\mathbf{A}$, according to MSC3.
The image of $[0,1] = \mathbf{A}\mathbf{e}_2 =$ the second column of $\mathbf{A}$, according to
MSC3.

13. Suppose $\mathbf{A}$ is a matrix for a transformation of the plane, so that for
every vector $\mathbf{x}$, $\|\mathbf{A}\mathbf{x}\| = \|\mathbf{x}\|$. Let $\mathbf{v}$ and $\mathbf{w}$ be vectors in the plane. We
want to prove that the angle between $\mathbf{v}$ and $\mathbf{w}$ equals the angle between
$\mathbf{A}\mathbf{v}$ and $\mathbf{A}\mathbf{w}$.

But, the angle between $\mathbf{A}\mathbf{v}$ and $\mathbf{A}\mathbf{w}$ equals

$$\arccos\left(\frac{(\mathbf{A}\mathbf{v})\cdot(\mathbf{A}\mathbf{w})}{\|\mathbf{A}\mathbf{v}\|\,\|\mathbf{A}\mathbf{w}\|}\right) = \arccos\left(\frac{\frac{1}{4}\left(\|\mathbf{A}\mathbf{v}+\mathbf{A}\mathbf{w}\|^2 - \|\mathbf{A}\mathbf{v}-\mathbf{A}\mathbf{w}\|^2\right)}{\|\mathbf{A}\mathbf{v}\|\,\|\mathbf{A}\mathbf{w}\|}\right)$$

because of the Polarization Identity. Using the distributive law (MM3),
this equals

$$\arccos\left(\frac{\frac{1}{4}\left(\|\mathbf{A}(\mathbf{v}+\mathbf{w})\|^2 - \|\mathbf{A}(\mathbf{v}-\mathbf{w})\|^2\right)}{\|\mathbf{A}\mathbf{v}\|\,\|\mathbf{A}\mathbf{w}\|}\right).$$

Next, we use the fact that the transformation preserves distances. This
allows us to eliminate the factors of $\mathbf{A}$ inside the norms, yielding

$$\arccos\left(\frac{\frac{1}{4}\left(\|(\mathbf{v}+\mathbf{w})\|^2 - \|(\mathbf{v}-\mathbf{w})\|^2\right)}{\|\mathbf{v}\|\,\|\mathbf{w}\|}\right).$$

Applying the Polarization Identity in the numerator gives

$$\arccos\left(\frac{\mathbf{v}\cdot\mathbf{w}}{\|\mathbf{v}\|\,\|\mathbf{w}\|}\right),$$

which, of course, equals the angle between $\mathbf{v}$ and $\mathbf{w}$.

Section 7.2

1. (a) $(T_2 \circ T_1)(\mathbf{v}) = \mathbf{v} + [8,4]$; $(T_2 \circ T_1)((5,-3)) = (13,1)$

   (b) Matrix for T_1: $\mathbf{A}_1 = -4\mathbf{I}_2$; matrix for T_2: $\mathbf{A}_2 = \begin{bmatrix} -\frac{8}{17} & \frac{15}{17} \\ \frac{15}{17} & \frac{8}{17} \end{bmatrix}$;

   $$(T_2 \circ T_1)(\mathbf{v}) = \mathbf{A}_2\mathbf{A}_1\mathbf{v} = \begin{bmatrix} \frac{32}{17} & -\frac{60}{17} \\ -\frac{60}{17} & -\frac{32}{17} \end{bmatrix}\mathbf{v};$$

   $(T_2 \circ T_1)((5,-3)) = (20,-12)$

   (c) Matrix for T_1: $\mathbf{A}_1 = \frac{1}{2.44}\begin{bmatrix} -0.44 & -2.40 \\ -2.40 & 0.44 \end{bmatrix} = \begin{bmatrix} -0.180 & -0.984 \\ -0.984 & 0.180 \end{bmatrix}$;

   matrix for T_2: $\mathbf{A}_2 \approx \begin{bmatrix} 0.423 & -0.906 \\ 0.906 & 0.423 \end{bmatrix}$; $(T_2 \circ T_1)(\mathbf{v}) = \mathbf{A}_2\mathbf{A}_1\mathbf{v} \approx$

   $\begin{bmatrix} 0.815 & -0.579 \\ -0.579 & -0.815 \end{bmatrix}\mathbf{v}$; $(T_2 \circ T_1)((5,-3)) \approx (5.812, -0.450)$

   (d) Matrix for T_1: $\mathbf{A}_1 \approx \begin{bmatrix} -0.759 & 0.651 \\ 0.651 & 0.759 \end{bmatrix}$; $(T_2 \circ T_1)(\mathbf{v}) = T_1(\mathbf{v}) +$

   $[4,1] = \mathbf{A}_1\mathbf{v} + [4,1]$; $(T_2 \circ T_1)((5,-3)) \approx (-1.748, 1.978)$

   (e) Matrix for T_2: $\mathbf{A}_2 \approx \begin{bmatrix} -0.759 & 0.651 \\ 0.651 & 0.759 \end{bmatrix}$; by CMT2, $(T_2 \circ T_1)(\mathbf{v}) =$

   $T_2(\mathbf{v}) + T_2((4,1)) = \mathbf{A}_2\mathbf{v} + \mathbf{A}_2[4,1] \approx \mathbf{A}_2\mathbf{v} + [-2.385, 3.363]$;
   $(T_2 \circ T_1)((5,-3)) \approx (-8.133, 4.341)$

   (f) $(T_2 \circ T_1)$ is a counterclockwise rotation about the origin through
   an angle of 53°. Hence, $(T_2 \circ T_1)(\mathbf{v}) \approx \begin{bmatrix} 0.602 & -0.799 \\ 0.799 & 0.602 \end{bmatrix}\mathbf{v}$;
   $(T_2 \circ T_1)((5,-3)) \approx (5.407, 2.189)$

   (g) Because T_2, the collapse, is performed last, the image of a vector
   under T_1 is irrelevant in computing $(T_2 \circ T_1)(\mathbf{v})$. T_2 sends every
   $T_1(\mathbf{v})$ image to $(7,3)$. Thus, $(T_2 \circ T_1)(\mathbf{v}) = (7,3)$ for all $\mathbf{v}$. Hence,
   $(T_2 \circ T_1)((5,-3)) = (7,3)$.

3. (a) $T^{-1}(\mathbf{v}) = \mathbf{v} + [-2,-9]$ (b) $T^{-1}(\mathbf{v}) = 1.5625\mathbf{I}_2\mathbf{v}$

   (c) A reflection is its own inverse. Hence, $T^{-1}(\mathbf{v}) = \begin{bmatrix} -\frac{7}{25} & \frac{24}{25} \\ \frac{24}{25} & \frac{7}{25} \end{bmatrix}\mathbf{v}$.

   (d) The matrix for the inverse of a rotation about the origin is the
   transpose of the matrix for the original rotation. Hence,
   $$T^{-1}(\mathbf{v}) = \begin{bmatrix} \frac{7}{25} & -\frac{24}{25} \\ \frac{24}{25} & \frac{7}{25} \end{bmatrix}\mathbf{v}.$$

   (e) A collapse does not have an inverse.

5. (a) $T(\mathbf{v}) = 6\mathbf{v} + [-25, -60]$

(b) $T(\mathbf{v}) \approx \begin{bmatrix} 0.966 & 0.259 \\ -0.259 & 0.966 \end{bmatrix} \mathbf{v} + [-0.233,\ 4.246]$

(c) $T(\mathbf{v}) \approx \begin{bmatrix} 0.976 & -0.220 \\ -0.220 & -0.976 \end{bmatrix} \mathbf{v} + [1.540,\ 13.832]$

7. The images of the vertices are: $(1.68, 2.76)$, $(-0.68, -1.76)$, $(-3.12, 4.16)$, and $(0.32, 5.24)$ (see corresponding figure).

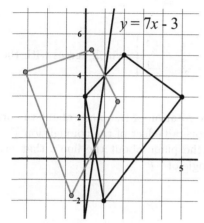

Figure for Exercise 7

9. Notice that the conjugation maps the point $(-5, 2)$ to itself. Therefore, $(-5, 2)$ is on the line of reflection and that line has a slope of 7, since it is parallel to $y = 7x$. Using the Point-Slope Formula from algebra, we find that the line of reflection is $y - 2 = 7(x + 5)$, or more simply, $y = 7x + 37$.

11. (a) $\mathbf{A}_1 = \begin{bmatrix} -\frac{3}{5} & \frac{4}{5} \\ \frac{4}{5} & \frac{3}{5} \end{bmatrix}$, $\mathbf{A}_2 = \begin{bmatrix} -\frac{4}{5} & -\frac{3}{5} \\ -\frac{3}{5} & \frac{4}{5} \end{bmatrix}$

(b) Matrix for $T_1 = \mathbf{A}_2 \mathbf{A}_1 = \begin{bmatrix} 0 & -1 \\ 1 & 0 \end{bmatrix}$

(c) Check that $\cos(90°) = 0$ and that $\sin(90°) = 1$, and so the matrix for T_1 fits the formula for the matrix for a counterclockwise rotation about the origin through an angle of $90°$.

(d) Matrix for $T_2 = \mathbf{A}_1 \mathbf{A}_2 = \begin{bmatrix} 0 & 1 \\ -1 & 0 \end{bmatrix}$

(e) Check that $\cos(-90°) = 0$ and that $\sin(-90°) = -1$, and so the matrix for T_1 fits the formula for the matrix for a clockwise rotation about the origin through an angle of $90°$.

13. (a) $\begin{bmatrix} \frac{1}{2} & -\frac{\sqrt{3}}{2} \\ -\frac{\sqrt{3}}{2} & -\frac{1}{2} \end{bmatrix}$

(b) A line making an angle of $-30°$ with the positive x-axis has a slope equal to $m = \tan(-30°) = -\frac{\sqrt{3}}{3}$. The matrix for the reflection of the plane about $y = -\frac{\sqrt{3}}{3}$ is

$$\frac{1}{1 + \left(-\frac{\sqrt{3}}{3}\right)^2} \begin{bmatrix} 1 - \left(-\frac{\sqrt{3}}{3}\right)^2 & -\frac{2\sqrt{3}}{3} \\ -\frac{2\sqrt{3}}{3} & \left(-\frac{\sqrt{3}}{3}\right)^2 - 1 \end{bmatrix}$$

$$= \frac{3}{4} \begin{bmatrix} \frac{2}{3} & -\frac{2\sqrt{3}}{3} \\ -\frac{2\sqrt{3}}{3} & -\frac{2}{3} \end{bmatrix} = \begin{bmatrix} \frac{1}{2} & -\frac{\sqrt{3}}{2} \\ -\frac{\sqrt{3}}{2} & -\frac{1}{2} \end{bmatrix},$$ which agrees with the result in part (a).

(c) If R is the counterclockwise rotation of the plane about the origin through an angle of θ, the overall effect of the conjugation $T = R^{-1} \circ M \circ R$ is to reflect the plane about the line ℓ that passes through the origin at an angle of $-\theta$.

Section 7.3

1. (a) $\begin{bmatrix} 1 & 0 & 6 \\ 0 & 1 & -2 \\ 0 & 0 & 1 \end{bmatrix}$ (c) $\begin{bmatrix} -0.849 & 0.528 & -5.415 \\ 0.528 & 0.849 & 4.547 \\ 0 & 0 & 1 \end{bmatrix}$

(b) $\begin{bmatrix} 1 & 0 & -4 \\ 0 & 1 & 3 \\ 0 & 0 & 1 \end{bmatrix}$

3. (a) $\begin{bmatrix} -1 & 0 & -6 \\ 0 & -1 & 16 \\ 0 & 0 & 1 \end{bmatrix}$ (d) $\begin{bmatrix} 0.977 & 0.214 & -1.431 \\ 0.214 & -0.977 & 13.194 \\ 0 & 0 & 1 \end{bmatrix}$

(b) $\begin{bmatrix} -1 & 0 & -4 \\ 0 & 1 & 0 \\ 0 & 0 & 1 \end{bmatrix}$ (e) $\begin{bmatrix} 0.966 & -0.259 & 4.130 \\ 0.259 & 0.966 & -0.546 \\ 0 & 0 & 1 \end{bmatrix}$

(c) $\begin{bmatrix} 0.259 & 0.966 & 11.433 \\ -0.966 & 0.259 & -1.100 \\ 0 & 0 & 1 \end{bmatrix}$

5. $\begin{bmatrix} 0.063 & 0.496 & 0.753 \\ 0.496 & -0.063 & -5.343 \\ 0 & 0 & 1 \end{bmatrix}$

Chapter 7 Test

4. (a) F (f) T (k) T (p) T (u) T (z) T (ee) T
   (b) T (g) T (l) F (q) T (v) F (aa) T (ff) F
   (c) F (h) T (m) F (r) T (w) T (bb) T
   (d) T (i) T (n) T (s) T (x) T (cc) T
   (e) T (j) T (o) T (t) T (y) F (dd) T

Appendix B

Properties, Theorems, and Methods

Chapter 1: Matrices

Section 1.3

If $\mathbf{A}$, $\mathbf{B}$, and $\mathbf{C}$ represent matrices that are all the same size, $\mathbf{O}$ is the matrix the same size as $\mathbf{A}$ having all zero entries, and x and y are scalars, then

ESP1: $\mathbf{A} + \mathbf{B} = \mathbf{B} + \mathbf{A}$ (Commutative Property of Addition)

ESP2: $(\mathbf{A} + \mathbf{B}) + \mathbf{C} = \mathbf{A} + (\mathbf{B} + \mathbf{C})$ (Associative Property of Addition)

ESP3: $\mathbf{A} + \mathbf{O} = \mathbf{O} + \mathbf{A} = \mathbf{A}$ (Additive Identity Property of Zero Matrix)

ESP4: $\mathbf{A} + ((-1)\mathbf{A}) = ((-1)\mathbf{A}) + \mathbf{A} = \mathbf{O}$ (Additive Inverse Property)

ESP5: $x(\mathbf{A} \pm \mathbf{B}) = x\mathbf{A} \pm x\mathbf{B}$ (First Distributive Law for Scalar Multiplication over Addition)

ESP6: $(x \pm y)\mathbf{A} = x\mathbf{A} \pm y\mathbf{A}$ (Second Distributive Law for Scalar Multiplication over Addition)

ESP7: $x(y\mathbf{A}) = (xy)\mathbf{A}$ (Associative Property of Scalar Multiplication)

ESP8: $1\mathbf{A} = \mathbf{A}$ (Scalar Multiplicative Identity Property of "1")

Section 1.4

Suppose that $\mathbf{A}$ and $\mathbf{B}$ are both $m \times n$ matrices and that c is a scalar. Then

TR1: $\left(\mathbf{A}^T\right)^T = \mathbf{A}$

TR2: $(c\mathbf{A})^T = c\mathbf{A}^T$

TR3: $(\mathbf{A} \pm \mathbf{B})^T = \mathbf{A}^T \pm \mathbf{B}^T$.

Section 1.5

Theorem 1. *(A Decomposition for Square Matrices) Every square matrix $\mathbf{A}$ can be expressed as the sum of a symmetric matrix $\mathbf{B}$ and a skew-symmetric matrix $\mathbf{C}$ in precisely one way: $\mathbf{B}$ must equal $\frac{1}{2}\left(\mathbf{A} + \mathbf{A}^T\right)$ and $\mathbf{C}$ must equal $\frac{1}{2}\left(\mathbf{A} - \mathbf{A}^T\right)$.*

Section 1.6

TR4: If $\mathbf{A}$ is an $m \times n$ matrix and $\mathbf{B}$ is an $n \times p$ matrix, then $(\mathbf{AB})^T = \mathbf{B}^T\mathbf{A}^T$.

Theorem 2. *If $\mathbf{A}$ is an $m \times n$ matrix, then $\mathbf{A}^T\mathbf{A}$ is a symmetric $n \times n$ matrix.*

Section 1.7

If s is a scalar, $\mathbf{A}$ is an $m \times n$ matrix, $\mathbf{B}$ and $\mathbf{C}$ are $n \times p$ matrices, and $\mathbf{D}$ is a $p \times q$ matrix, then

MM1: $s(\mathbf{AB}) = (s\mathbf{A})\mathbf{B} = \mathbf{A}(s\mathbf{B})$
(Associative Law for Scalar and Matrix Multiplication)

MM2: $\mathbf{A}(\mathbf{BD}) = (\mathbf{AB})\mathbf{D}$
(Associative Law for Matrix Multiplication)

MM3: $\mathbf{A}(\mathbf{B} \pm \mathbf{C}) = \mathbf{AB} \pm \mathbf{AC}$
(First Distributive Law for Matrix Multiplication over Matrix Addition and Subtraction)

MM4: $(\mathbf{B} \pm \mathbf{C})\mathbf{D} = \mathbf{BD} \pm \mathbf{CD}$
(Second Distributive Law for Matrix Multiplication over Matrix Addition and Subtraction)

If $\mathbf{A}$ is a square matrix and k and l are positive integers, and c is a scalar, then

RFE1: $\mathbf{A}^{k+l} = \mathbf{A}^k \mathbf{A}^l$

RFE2: $\left(\mathbf{A}^k\right)^l = \mathbf{A}^{kl}$

RFE3: $\left(\mathbf{A}^T\right)^k = \left(\mathbf{A}^k\right)^T$

RFE4: $(c\mathbf{A})^k = c^k \mathbf{A}^k$.

Theorem 3. *If $\mathbf{A}$ and $\mathbf{B}$ are both $n \times n$ stochastic matrices, then $\mathbf{AB}$ is also an $n \times n$ stochastic matrix.*

STOC: If $\mathbf{A}$ is a stochastic matrix, then for any positive integer n, $\mathbf{A}^n$ is also a stochastic matrix.

Section 1.8

If $\mathbf{A}$ is an $m \times n$ matrix and $\mathbf{B}$ is an $n \times p$ matrix, then

MSR1: The ith row of $\mathbf{AB}$ = (ith row of $\mathbf{A}$)$(\mathbf{B})$

MSC1: The jth column of $\mathbf{AB}$ = $(\mathbf{A})$(jth column of $\mathbf{B}$)

If $\mathbf{A}$ is an $m \times n$ matrix, $\mathbf{R}$ is a $1 \times m$ matrix (single row), and $\mathbf{C}$ is an $m \times 1$ matrix (single column), then

MSR2: The linear combination of the rows of $\mathbf{A}$ which uses the entries in $\mathbf{R}$ as the scalars is equal to $\mathbf{RA}$.

MSC2: The linear combination of the columns of $\mathbf{A}$ which uses the entries in $\mathbf{C}$ as the scalars is equal to $\mathbf{AC}$.

If $\mathbf{A}$ is an $m \times n$ matrix, then

MSR3: $\mathbf{e}_i^T \mathbf{A}$ equals the ith row of $\mathbf{A}$.

MSC3: $\mathbf{Ae}_i$ equals the ith column of $\mathbf{A}$.

IDM: $\mathbf{I}_m \mathbf{A} = \mathbf{A}$ and $\mathbf{AI}_n = \mathbf{A}$.

Chapter 2: Application: Elementary Graph Theory

Section 2.3

Theorem 1. *If $\mathbf{A}$ is the adjacency matrix for a graph or digraph, and if k is a nonnegative whole number, then the (i, j) entry of $\mathbf{A}^k$ equals the number of paths of length k from vertex Vi to vertex Vj.*

Theorem 2. *A graph with n vertices is connected if and only if every entry of $(\mathbf{I}_n + \mathbf{A})^{n-1}$ is a positive integer.*

Chapter 3: Vectors

Section 3.2

LZV: $\|\mathbf{v}\| = 0$ if and only if $\mathbf{v} = \mathbf{0}$.

LSM: If $\mathbf{v}$ is a vector and c is a scalar, then $\|c\mathbf{v}\| = |c|\,\|\mathbf{v}\|$.

Theorem 1. *(Triangle Inequality) Suppose that $\mathbf{v}$ and $\mathbf{w}$ are two n-dimensional vectors. Then*

$$\|\mathbf{v} + \mathbf{w}\| \le \|\mathbf{v}\| + \|\mathbf{w}\|\,.$$

Corollary 2. *Suppose that $\mathbf{v}_1, \ldots, \mathbf{v}_k$ are n-dimensional vectors. Then*

$$\|\mathbf{v}_1 + \mathbf{v}_2 + \cdots + \mathbf{v}_k\| \le \|\mathbf{v}_1\| + \|\mathbf{v}_2\| + \cdots + \|\mathbf{v}_k\|\,.$$

Section 3.3

Suppose that $\mathbf{v}$, $\mathbf{w}$, and $\mathbf{z}$ are n-dimensional vectors and c is a scalar. Then

DP1: $\mathbf{v} \cdot \mathbf{v} = \|\mathbf{v}\|^2$ (**The Dot Product and the Norm**)

DP2: $\mathbf{v} \cdot \mathbf{w} = \mathbf{w} \cdot \mathbf{v}$ (**Commutative Property of the Dot Product**)

DP3: $c(\mathbf{v} \cdot \mathbf{w}) = (c\mathbf{v}) \cdot \mathbf{w} = \mathbf{v} \cdot (c\mathbf{w})$ (**Associative Property of the Dot Product and Scalar Multiplication**)

DP4: $\mathbf{v} \cdot (\mathbf{w} \pm \mathbf{z}) = \mathbf{v} \cdot \mathbf{w} \pm \mathbf{v} \cdot \mathbf{z}$ (**First Distributive Property of the Dot Product**)

DP5: $(\mathbf{v} \pm \mathbf{w}) \cdot \mathbf{z} = \mathbf{v} \cdot \mathbf{z} \pm \mathbf{w} \cdot \mathbf{z}$ (**Second Distributive Property of the Dot Product**)

Let $\mathbf{x}$, $\mathbf{y}$, $\mathbf{w}$, $\mathbf{z}$ be n-dimensional vectors, and let a, b, c, d be scalars. Then

DP-FOIL1: $\begin{aligned}(a\mathbf{x} + b\mathbf{y}) \cdot (c\mathbf{x} + d\mathbf{y}) &= ac\,(\mathbf{x} \cdot \mathbf{x}) + (ad + bc)(\mathbf{x} \cdot \mathbf{y}) + bd(\mathbf{y} \cdot \mathbf{y}) \\ &= ac\,\|\mathbf{x}\|^2 + (ad + bc)(\mathbf{x} \cdot \mathbf{y}) + bd\,\|\mathbf{y}\|^2\end{aligned}$

DP-FOIL2: $(a\mathbf{w} + b\mathbf{x}) \cdot (c\mathbf{y} + d\mathbf{z}) = ac\,(\mathbf{w} \cdot \mathbf{y}) + ad(\mathbf{w} \cdot \mathbf{z}) + bc(\mathbf{x} \cdot \mathbf{y}) + bd(\mathbf{x} \cdot \mathbf{z})$

Theorem 3. *(Polarization Identity) Suppose $\mathbf{v}$ and $\mathbf{w}$ are two vectors having the same dimension. Then*

$$\mathbf{v} \cdot \mathbf{w} = \frac{1}{4}\left(\|\mathbf{v} + \mathbf{w}\|^2 - \|\mathbf{v} - \mathbf{w}\|^2\right).$$

Theorem 4. *(Cauchy-Schwarz Inequality) If $\mathbf{v}$ and $\mathbf{w}$ are two vectors having the same dimension, then*

$$|\mathbf{v} \cdot \mathbf{w}| \le \|\mathbf{v}\|\,\|\mathbf{w}\|\,.$$

Section 3.4

If **v** and **w** are two nonzero vectors having the same dimension, then

AR1: The angle θ between **v** and **w** is acute $(0° \leq \theta < 90°)$ if and only if $\mathbf{v} \cdot \mathbf{w} > 0$.

AR2: The angle θ between **v** and **w** is a right angle $(\theta = 90°)$ if and only if $\mathbf{v} \cdot \mathbf{w} = 0$.

AR3: The angle θ between **v** and **w** is obtuse $(90° < \theta \leq 180°)$ if and only if $\mathbf{v} \cdot \mathbf{w} < 0$.

Section 3.5

Theorem 5. *(Projection Theorem) Suppose **v** and **w** are two vectors having the same dimension with $\mathbf{w} \neq \mathbf{0}$. Then, assuming that $\mathrm{proj}_\mathbf{w}\mathbf{v} \neq \mathbf{0}$, there is one and only one way to express **v** as the sum of two vectors so that one of the vectors in the sum is parallel to **w** and the other is orthogonal to **w**. This unique decomposition of **v** is given by $\mathbf{v} = \mathrm{proj}_\mathbf{w}\mathbf{v} + (\mathbf{v} - \mathrm{proj}_\mathbf{w}\mathbf{v})$.*

Gram-Schmidt Process: Given a set $\{\mathbf{v}_1, \mathbf{v}_2, \ldots, \mathbf{v}_n\}$ of vectors, we create a related set $\{\mathbf{x}_1, \mathbf{x}_2, \ldots, \mathbf{x}_n\}$ of mutually orthogonal vectors as follows:
   Step 1: Let $\mathbf{x}_1 = \mathbf{v}_1$.
   Step 2: Let $\mathbf{x}_2 = \mathbf{v}_2 - (\mathrm{proj}_{\mathbf{x}_1}\mathbf{v}_2)$.
   Step 3: Let $\mathbf{x}_3 = \mathbf{v}_3 - (\mathrm{proj}_{\mathbf{x}_1}\mathbf{v}_3) - (\mathrm{proj}_{\mathbf{x}_2}\mathbf{v}_3)$.
   Step 4: Let $\mathbf{x}_4 = \mathbf{v}_4 - (\mathrm{proj}_{\mathbf{x}_1}\mathbf{v}_4) - (\mathrm{proj}_{\mathbf{x}_2}\mathbf{v}_4) - (\mathrm{proj}_{\mathbf{x}_3}\mathbf{v}_4)$, etc.

Section 3.6

Theorem 6. *(**Linear Combinations of Mutually Orthogonal Vectors**) (**LCMOV**) Suppose that $\{\mathbf{v}_1, \ldots, \mathbf{v}_k\}$ is a mutually orthogonal set of nonzero vectors in $\mathbb{R}^n$. Then, no vector in this set can be expressed as a linear combination of the other vectors in the set.*

Chapter 4: Solving Systems of Linear Equations

Section 4.1

The Five-Step Method for Analyzing Word Problems Involving Linear Systems
   Step 1: Read the problem carefully in its entirety.
   Step 2: Find the question asked by the problem. Use it to determine the values or quantities the problem is asking you to find.
   Step 3: Assign variables to the various values determined in Step 2. Write out the meaning of these variables in detail.
   Step 4: Determine any restrictions or conditions given in the problem.
   Step 5: Write an equation for each restriction found in Step 4.

Section 4.3

In performing the Gauss-Jordan Method, we use the following basic rules:

- We work on one column at a time, proceeding through the columns from left to right, until we reach the bar.

- In each column, we begin by placing the "1" in its proper spot, and then convert the remaining entries of that column into "0."

- To convert a nonzero entry to "1:"

  - We label that entry as the current **pivot entry**, and the row containing that entry as the current **pivot row**. The column containing the pivot entry is the current **pivot column**.

  - We use the type (I) row operation

  $$\left(\frac{1}{\text{pivot entry}}\right) < \text{pivot row} > \implies < \text{pivot row} > .$$

  That is, we multiply the pivot row by the reciprocal of the pivot entry.

- To convert an entry to "0:"

  - We label that entry as the **target entry**, and the row containing that entry as the **target row**.

  - We use the type (II) row operation

  $$- (\text{target entry}) < \text{pivot row} > + < \text{target row} >$$
  $$\implies < \text{target row} > .$$

  That is, we add a multiple of the pivot row to the target row. The number we multiply by is the *negative* of the target entry. We often use **target** as a verb; that is, to "target" an entry means that we use the type (II) row operation above to convert that particular entry into a zero.

- To convert an entry into "1" when that entry is currently "0:"

  - First use a type (III) row operation to switch the row containing that entry with a *later* row so that a nonzero number is placed in the desired position. (We take care not to switch with an earlier row so that we do not destroy the pattern of pivot entries that have already been created.)

  - Then use a type (I) row operation as usual to convert that (nonzero) entry into "1."

Section 4.4

Additional rules for the Gauss-Jordan Method

- If the current pivot entry is "0," and there is no lower row having a nonzero number directly below that "0," then skip over that column. Move instead to the next column to the right, and label that column as the newest pivot column, and label the entry of that column in the *current* pivot row as the new pivot entry. (This ensures that we never skip a row when creating circled pivots.)

- To determine the complete solution set when a linear system has an infinite number of solutions:

  - Write out the linear system corresponding to the final augmented matrix.

  - Assume that the unknowns representing the columns *without* circled pivots have been given any arbitrary values.

  - Use the nonzero rows to solve, where necessary, for the remaining unknowns in terms of those. (That is, solve for the unknowns for columns *with* circled pivots in terms of the unknowns for columns *without* circled pivots.)

Rule for the number of solutions for a given system of linear equations

- If at any point during the Gauss-Jordan Method, the augmented matrix contains a row of the form

$$[0, \ 0, \ 0, \ ..., \ 0 \mid *],$$

where the "$*$" symbol represents a nonzero number, then the original system of linear equations has *no solutions*. It is inconsistent.

- Otherwise, the linear system is consistent, and

  - If there is a circled pivot in every column before the bar, then the original system of linear equations has a *unique solution*. (The number after the bar in each row represents the value of the unknown corresponding to the circled pivot in that row.)

  - If there is at least one column before the bar without a circled pivot, then the original system of linear equations has an *infinite number of solutions*. (The unknowns corresponding to columns without a circled pivot can take on any arbitrary value, and the values of the remaining unknowns can be determined from those values.)

Section 4.5

Theorem 1. *Every (possibly augmented) matrix reduces, by the use of row operations, to a unique corresponding (possibly augmented) matrix in reduced row echelon form.*

Section 4.6

Theorem 2. *Suppose* $\mathbf{A}$ *is an* $m \times n$ *matrix, and* $\operatorname{rank}(\mathbf{A}) = k$. *Then*
If $k = n$, *then the homogeneous linear system* $\mathbf{AX} = \mathbf{0}$ *has only the trivial solution;*
If $k < n$, *then the homogeneous linear system* $\mathbf{AX} = \mathbf{0}$ *has the trivial solution plus an infinite number of nontrivial solutions. In this case, the system has* $(n-k)$ *fundamental solutions, and every solution in the complete solution set for the system is a linear combination of these fundamental solutions.*

Corollary 3. *If a homogeneous system of linear equations has more variables than equations, then the system has an infinite number of nontrivial solutions, in addition to the trivial solution.*

Theorem 4. *If* S *is a set of* k *distinct* n-*dimensional mutually orthogonal vectors and* $k < n$, *then there is a nonzero vector* $\mathbf{v}$ *that is orthogonal to all of the vectors in* S. *That is, adding* $\mathbf{v}$ *to the set* S *will form a larger mutually orthogonal set of vectors.*

Chapter 5: Application: Least Squares

Section 5.1

The Five-Step Method for Linear Regression
Given a set of points $\{(x_1, y_1), \ldots, (x_n, y_n)\}$, take the following steps to find values of m and b such that the line $y = mx + b$ best fits the data:
Step 1: Construct an $n \times 2$ matrix $\mathbf{A}$ in which the entries in the first column all equal 1, and the entries in the second column are, in order, $x_1, \ldots, x_n$.
Step 2: Create an $n \times 1$ matrix $\mathbf{B}$ whose entries are, in order, $y_1, \ldots, y_n$.
Step 3: Compute the 2×2 matrix $\mathbf{A}^T\mathbf{A}$ and the 2×1 matrix $\mathbf{A}^T\mathbf{B}$.
Step 4: Solve the linear system $\left(\mathbf{A}^T\mathbf{A}\right)\mathbf{X} = \mathbf{A}^T\mathbf{B}$, where $\mathbf{X} = \begin{bmatrix} b \\ m \end{bmatrix}$,

by using the Gauss-Jordan Method on the augmented matrix $\left[\, \mathbf{A}^T\mathbf{A} \mid \mathbf{A}^T\mathbf{B} \,\right]$.
Step 5: The line $y = mx + b$ is the line that fits the data as closely as possible, where b and m are determined by the solution to the linear system obtained in Step 4.

Section 5.2

The Five-Step Method for Quadratic Regression

Given a set of points $\{(x_1, y_1), \ldots, (x_n, y_n)\}$, take the following steps to find values of a, b, and c such that the parabola $y = ax^2 + bx + c$ best fits the data:

Step 1: Construct an $n \times 3$ matrix $\mathbf{A}$ in which the entries in the first column all equal 1, the entries in the second column are, in order, $x_1, \ldots, x_n$, and the entries in the third column are, in order, $x_1^2, \ldots, x_n^2$.

Step 2: Create an $n \times 1$ matrix $\mathbf{B}$ whose entries are, in order, $y_1, \ldots, y_n$.

Step 3: Compute the 3×3 matrix $\mathbf{A}^T \mathbf{A}$ and the 3×1 matrix $\mathbf{A}^T \mathbf{B}$.

Step 4: Solve the linear system $\left(\mathbf{A}^T \mathbf{A} \right) \mathbf{X} = \mathbf{A}^T \mathbf{B}$, where $\mathbf{X} = \begin{bmatrix} c \\ b \\ a \end{bmatrix}$,

by using the Gauss-Jordan Method on the augmented matrix $\left[\mathbf{A}^T \mathbf{A} \,\middle|\, \mathbf{A}^T \mathbf{B} \right]$.

Step 5: The parabola $y = f(x) = ax^2 + bx + c$ is the quadratic that fits the data as closely as possible, where a, b, and c are determined by the solution to the linear system obtained in Step 4. (It is possible that the solution could have $a = 0$, in which case the least-squares quadratic polynomial is actually linear. In that case, the linear function is a better fit than all possible quadratic functions. This rarely occurs.)

Chapter 6: Inverses for Matrices

Section 6.1

Theorem 1. *If $\mathbf{A}$ and $\mathbf{B}$ are $n \times n$ matrices such that $\mathbf{AB} = \mathbf{I}_n$, then $\mathbf{BA} = \mathbf{I}_n$.*

Theorem 2. *If $\mathbf{A}$, $\mathbf{B}$, and $\mathbf{C}$ are $n \times n$ matrices such that both $\mathbf{B}$ and $\mathbf{C}$ are inverses for $\mathbf{A}$, then $\mathbf{B} = \mathbf{C}$.*

Given an $n \times n$ matrix $\mathbf{A}$, take the following steps to compute its inverse, or to determine that it is singular:

Step 1: Form the augmented $n \times (2n)$ matrix $[\mathbf{A} \,|\, \mathbf{I}_n]$, having the n columns of the identity matrix *after* the vertical bar.

Step 2: Put the matrix $[\mathbf{A} \,|\, \mathbf{I}_n]$ from Step 1 into reduced row echelon form.

Step 3: If the n columns to the left of the bar in the reduced row echelon form for $[\mathbf{A} \,|\, \mathbf{I}_n]$ form the identity matrix, then $\mathbf{A}$ is nonsingular. Otherwise, $\mathbf{A}$ is singular.

Step 4: If $\mathbf{A}$ has been determined to be nonsingular in Step 3, then the reduced row echelon form matrix will be in the form $\left[\mathbf{I}_n \,\middle|\, \mathbf{A}^{-1} \right]$. That is, the n columns to the right of the bar in the reduced row echelon form matrix make up the inverse for the matrix $\mathbf{A}$.

Theorem 3. *If* **A** *is a nonsingular matrix, then the system of linear equations represented by* **Ax** = **b** *has the* unique *solution* **x** = **A**$^{-1}$**b**.

Section 6.2

Suppose **A** and **B** are nonsingular $n \times n$ matrices. Then

IMP1: **A**$^{-1}$ is nonsingular, and $\left(\mathbf{A}^{-1}\right)^{-1} = \mathbf{A}$.

IMP2: (**AB**) is nonsingular and $(\mathbf{AB})^{-1} = \mathbf{B}^{-1}\mathbf{A}^{-1}$.

IMP3: **A**^{T} is nonsingular and $\left(\mathbf{A}^T\right)^{-1} = \left(\mathbf{A}^{-1}\right)^T$.

IMP4: If c is a nonzero scalar, then $c\mathbf{A}$ is nonsingular and $(c\mathbf{A})^{-1} = \frac{1}{c}\mathbf{A}^{-1}$.

Theorem 4. *If* **D** *is a diagonal* $n \times n$ *matrix having no zero entries on its main diagonal. Then* **D** *is nonsingular, and*

$$
\mathbf{D}^{-1} = \begin{bmatrix} \frac{1}{d_{11}} & 0 & \cdots & 0 \\ 0 & \frac{1}{d_{22}} & \cdots & 0 \\ \vdots & \vdots & \ddots & \vdots \\ 0 & 0 & \cdots & \frac{1}{d_{nn}} \end{bmatrix}.
$$

Theorem 5. *An upper (lower) triangular* $n \times n$ *matrix having no zero entries on its main diagonal is nonsingular and its inverse is also upper (lower) triangular. If an upper (lower) triangular matrix has at least one zero entry on its main diagonal, then it is singular.*

Theorem 6. *If* **S** *is a nonsingular symmetric matrix, then* **S**$^{-1}$ *is also symmetric.*

Theorem 7. *Let* **K** *be an* $n \times n$ *skew-symmetric matrix. If* n *is odd, then* **K** *is singular. If* n *is even and* **K** *is nonsingular, then* **K**$^{-1}$ *is also skew-symmetric.*

If **A** is a nonsingular matrix and k and l are any integers, and c is a nonzero scalar, then

RFE1: $\mathbf{A}^{k+l} = \mathbf{A}^k\mathbf{A}^l$

RFE2: $\left(\mathbf{A}^k\right)^l = \mathbf{A}^{kl}$

RFE3: $\left(\mathbf{A}^T\right)^k = \left(\mathbf{A}^k\right)^T$

RFE4: $(c\mathbf{A})^k = c^k\mathbf{A}^k$.

CPM: If **A** is a nonsingular matrix, and if **B** and **C** are two linear combinations of integer powers of **A**, then **BC** = **CB**.

Section 6.3

Theorem 8. *If* $\mathbf{A}$ *is an orthogonal* $n \times n$ *matrix, then* $\mathbf{A}$ *is nonsingular and* $\mathbf{A}^{-1} = \mathbf{A}^T$.

Theorem 9. *(Size of a Mutually Orthogonal Set of Vectors) (SMOSV) A set of mutually orthogonal n-dimensional nonzero vectors cannot contain more than n vectors.*

Chapter 7: Geometric Transformations with Matrices

Section 7.2

CMT1: If T_1 and T_2 are two matrix transformations represented by the matrices $\mathbf{A}_1$ and $\mathbf{A}_2$, respectively, then the transformation T resulting from the composition of first applying T_1 followed by applying T_2 is represented by the matrix $(\mathbf{A}_2\mathbf{A}_1)$; that is, the product of the two matrices, where the product is in the order that the transformations were applied *going from right to left.*

CMT2: If M is a matrix transformation and T is the translation by the vector $\mathbf{t}$, then, for every vector $\mathbf{v}$, $(M \circ T)(\mathbf{v}) = M(\mathbf{v} + \mathbf{t}) = M(\mathbf{v}) + M(\mathbf{t})$.

CMT3: If M is a matrix transformation and T is the translation by the vector $\mathbf{t}$, then, for every vector $\mathbf{v}$, the result of applying the conjugation of M by T to $\mathbf{v}$ is $(T^{-1} \circ M \circ T)(\mathbf{v}) = M(\mathbf{v}) + M(\mathbf{t}) - \mathbf{t}$.

ConjTrans: A transformation of the plane centered at a point (a, b) can be obtained by conjugating the corresponding transformation *centered at the origin* with the translation by the vector $\mathbf{t} = [-a, -b]$.

Index

For further Sales of concerns and information please contact our
HD representative CPCH supraunraelfraudscom Taylor & Francis
Verlag (mph), Landbergstraße 25, 80374 München, Germany